Understanding War

The War Trilogy

Volume I: *Winning at War*

Volume II: *War Wisdom*

Volume III: *Understanding War*

To the patrons of Seilong

The Borndan

in recognition of its

fine staff!

With all best

wishes,

Chris Porhalu

Understanding War

An Annotated Bibliography

Christian P. Potholm

University Press of America,® Inc.
Lanham • Boulder • New York • Toronto • Plymouth, UK

4501 Forbes Boulevard, Suite 200, Lanham, Maryland 20706
UPA Acquisitions Department (301) 459-3366

Unit A, Whitacre Mews, 26-34 Stannary Street,
London SE11 4AB, United Kingdom

Printed in the United States of America
British Library Cataloguing in Publication Information Available

Library of Congress Control Number: 2016935686
ISBN: 978-0-7618-6773-9 (cloth : alk. paper)—ISBN: 978-0-7618-6774-6 (electronic)

∞™ The paper used in this publication meets the minimum requirements of American National Standard for Information Sciences Permanence of Paper for Printed Library Materials, ANSI/NISO Z39.48-1992.

After 50 years of teaching, this work is dedicated to
all the students of Tufts, Dartmouth, Vassar, and especially Bowdoin
who have enriched my life so deeply over these wonderful years.
You are the purpose, the inspiration and the reward.
Thank You

Contents

Preface

This work completes my long-gestated war trilogy. The first book, *Winning at War*, finds a value-free conceptual framework in order to analyze success in war irrespective of time, place, or culture. The second, *War Wisdom*, looks at a dozen themes that appear in many diverse cultures across both time and space. This volume, *Understanding War*, exposes students to some of the vast array of works dedicated to the study of war and warfare, that most persistent of phenomena.

My list of acknowledgments for this work is long, indeed. There are many who assisted in putting *Understanding War* together: Lynne Atkinson, who runs the Department of Government with such efficiency and goodwill, and her able assistants: Jordan Goldberg, Tina Curtin, Derek Brooks, Chase Taylor, Ryan Holmes, Sophie Springer, Steve Robinson, Charley Allen, Louisa Diaz, Liam Killion, Stevie Lane, Sam Miller, Nick Daniels, Ben Bristol, Aidan Penn, Grégoire Faucher, and Gibson Harley. I appreciate also the ongoing assistance over the years of Liz Weston and Joe Calvo.

No list of gratitude for this project would be complete without a full rendering for the marvelous staff of the Bowdoin Library, under the direction of Marjorie Hassen and including Judy Montgomery, Jaime Jones, Barbara Harvey, Amy Heggie, Leanne Pander, Joan Campbell, Richard Lindeman, Guy Saldanha, Greg Stowe, Patricia Myshrall, Alida Snow, Barbara Levergood, and Carmen Greenlee, who facilitate scholarship for all. In this regard, I am either the Bowdoin Interlibrary Loan Program's best customer or its worst nightmare, but this work would not have been possible without their generous and ongoing assistance, delivered with great patience and restraint despite often intense provocation. Bowdoin is most blessed to have such a group of dedicated professionals serving the research needs of students and faculty.

I also express thanks to Claude Berube and Williamson "Wick" Murray for all their earlier assistance on the companion volume *Winning at War* and especially to Wick for sharing his *Military History: A Selected Bibliography*, which he produced for the Institute for Defense Analyses in 2003. That fine and comprehensive work stimulated me to bring together all my notes on war books, starting with the earliest periods of human history, and to begin to fill in the many gaps in my knowledge of all things war. Appreciation also goes to Patrick Rael, who provided valuable insights into various topics and considerations, as well as to David "Broo" Parmelee, who served his country and his clan so well and who always is there to discuss military matters.

As always, our enduring thanks to the men and women of the armed forces and those who study them for their input as well, particularly Bob Whelan, Frank C. Mahncke, and especially Bowdoin's stalwart US Marine, Army, Navy, and Seal contingents, including Gil Barndollar, Katie Forney Petronio, Jack Dingess, Ian Merry, Dave Donahue, Pack Janes, Chris Pelletier, Eddie Yoo, Sage Santangelo, "Boomer" Repko, Gibson Hartley, Mac Caputi, and Brendan Lawler.

And finally a most special debt of gratitude is due to Jed Lyons, who had such faith in the War Trilogy from its inception and helped guide each volume to its conclusion. His generosity, hard work, and excellent direction resulted in the swift publication of *Understanding War* to coincide with my fiftieth year of teaching. Jed heads a wondrous team, including Jackie Hicks, Julie Kirsch, Beverly Shellem, Niki Guinan, Robert and Cynthia Swanson, and especially Holly Buchanan (whose input on design and supervision was greatly appreciated)—all of whom academic publishing is blessed to have available.

Finally, and as ever, I acknowledge the love and support of my wife, Sandra Quinlan Potholm, who makes all this effort worthwhile.

My debts are many and heartfelt.

I

Analyzing Warfare

This selection of topics and annotated bibliography of self-contained pieces on war wisdom, war novels, and war films is designed to be a companion piece to the first book in this war trilogy, *Winning at War*. It provides a conceptual framework for understanding warfare regardless of time, space, or culture.

War is such a vast and unruly topic that it is necessary to apply some conceptual framework so that one can begin to organize the material. We therefore take as a trope the image of Mars, the god of war. Using this device, we can look at war in a moral vacuum, outside the fog of nationalism and ideology and "my country right or wrong" or "my country always wrong in war." By using the god of war as an analytic tool, we can also circumvent the often confusing and usually obscuring sense of who is "right" and who is "wrong."

It must be said that it is an intellectual tragedy that military and strategic matters are today really studied in-depth only at a few academic locations, such as the US military academies and war colleges, plus Yale, Duke, MIT, Georgetown, Tufts, Johns Hopkins, Ohio, and Norwich Universities. Unfortunately, a very large percentage of American students are trying to come to grips with the realities of global history and life without a firm grounding in one of the most widespread and enduring phenomena in all of human history.

It is therefore regrettably necessary for more and more individuals to study war on their own with some help from a bibliography such as this, one that is designed to be explorative, not definitive or unnecessarily value-centric. This section introduces the interested reader to both the template of Mars and its many applications.

Chapter One

Why Study Military Matters?

> "The formal study of military history remains the orphaned child of the college and the university."—Victor Hanson

> "Man makes war in his own image."—Chester Wilmot

> "Whether one likes it or not, war has played for better or for worse a fundamental part in the whole process of historical change."—Michael Howard

> "War is a plague but a ubiquitous one, worthy of much careful study."—William Langer

> "Anyone who clings to the historically untrue—and thoroughly immoral—doctrine that violence never settles anything, I would advise to conjure up the ghosts of Napoleon Bonaparte and the Duke of Wellington and let them debate it. The ghost of Hitler would referee. . . . Violence, naked force, has settled more issues in history than has any other factor, and the contrary opinion is wishful thinking at its worst. Breeds that forget this basic truth have always paid for it with their lives and freedom."—Robert Heinlein

> "War appears to be as old as mankind, but peace is a modern invention."—Henry Maine, *Early History of Institutions*

Why study wars? It would seem ludicrous to many that such a question even needs to be asked. The ubiquitous nature of war across time and space, throughout so many cultures and so many peoples, would seem to obviate even the asking of it. Hopefully, a mere glance at the works listed here will make it clear how important and even essential it is to study war. Whether one opposes war or accepts it, there is a vital task to be served by understanding what war is all about and how readily and for how many reasons humankind has resorted to the killing of other humans. Studying war is always to be reminded of W. T. Sherman's dictum that combat is the "dearest school on earth." Any hard-won knowledge derived from studying war must surely be worth its weight in true coin.

Caroline Alexander, *The War That Killed Achilles: The True Story of Homer's Iliad and the Trojan War* (New York: Viking Press, 2010). Homer's *Iliad* was written around 750–700 BCE (about a war between the Greeks and the Trojans that probably took place in 1250 BCE). Since then, according to the author, for the next 3,250 years, 94 out of every 100 years in the following centuries have seen large-scale conflicts in one or more parts of the world. Reason enough to carefully examine the events that occupy only the last few months of that ten-year war because the "*Iliad*'s evocation of war's devastation, then is as resonant today—perhaps especially today—as it was in Homer's Dark Age." Note also that she suggests, "The greatest war story ever told commemorates a war that established no boundaries, won no territory, and furthered no cause." It simply captured the essence of humankind's perceived need to make war, the marked inability of people who tired of war to stop it, and the devastating consequences of war. A most useful study.

Susan Wise Bauer, *The History of the Ancient World* (New York: W. W. Norton, 2007). If you ever, even for a moment, disbelieved that war is tightly woven into the DNA of humans, read this volume. It is also a fine, once-over, light introduction to history and shows a firm grasp of the rise and fall of empires, states, and rulers and the centrality of warfare to the human condition.

Chris Bishop and Ian Drury, *Battles of the 20th Century* (London: Hamlyn, 1989). This lavishly illustrated volume (800 photographs) examines 90 different battles. Some are seminal (Midway), some are illustrative (the rape of Nanking), and some are just British nationalism ("Incredible Victory: The Battle of the River Plate"), but reading through the various battles, you will certainly see ample evidence of the need to figure out why humans put themselves into so many situations at the behest of Mars.

Jeremy Black (ed.), *The Seventy Great Battles in History* (London: Thames and Hudson, 2005). I was well into this lavishly illustrated book of important battles when I realized the ability of humans to paint Technicolor portraits of death and dying and how deep-seated the urge to accept war must be in order for thousands and thousands of years of strife and destruction to be accepted as normal. A powerful argument, I believe, for the need to study war and its connection to humankind.

Geoffrey Blainey, *The Causes of War* (New York: Free Press, 1988). A wide-ranging look at war and peace, making a major attempt to categorize different types of war ("long wars," "shorter wars," and "the mystery of wide wars"). He finds peace a bigger mystery than war because it is so often threatened by war. I found book 2, "The Web of War," the most useful portion of the work.

William Boyd, *Brazzaville Beach* (New York: William Morrow, 1990) (F). An ironic novel chronicling an American researcher who insists chimpanzees are totally pacific, while another trained observer clearly witnesses "war" between two bands and the massacre of one (a "little Trojan War"). The tale is nicely set against the backdrop of a contemporary war in a fictional Angola, with three sides and a ritual pointlessness to the human "battles." The chief researcher insists that "war among chimps cannot be" but is finally forced to admit that it "can be" and indeed "is." The whole work suggests strongly that the fault, dear Brutus, does lie in our genes as well as ourselves and our culture. A stimulating read for those interested in the nature of war and of humankind.

Gerard Chaliand, *The Art of War in World History: From Antiquity to the Nuclear Age* (Los Angeles: University of California Press, 1994). From the Battle of Kadesh through the battles of the Greeks, Romans, Chinese, Indians, Byzantines, Arabs, Persians, Ottoman Turks, and Mongols to the cultures of the present, this 1,000-plus-page tome is an ode to the ubiquity and longevity of the culture of war and those who practice or study it or do both.

Edward S. Creasy, *Fifteen Decisive Battles of the World: From Marathon to Waterloo* (New York: Da Capo Press, 1994). An unabridged version of the first 1851 edition, this volume is a milestone in the history of warfare, concentrating as it does on 14 battles as decisive moments in 2,300 years of world history. It is certainly one of the most famous books of that century, and its approach to the impotence of battles as turning points in human history was in vogue for well over a century. Syracuse, Chalons, Tours, Hastings, Blenheim, and Saratoga are included in the battles deemed worthy of our attention. A very important work despite its Eurocentric focus.

Note: The selection of a particular battle as a "hinge of history" or a "decisive event" in shaping future events inherently contains the notion of counterfactual analysis or what would have happened if the battle had turned out differently. See for example, R. Cowley (ed.), *What If? The Foremost Military Historians Imagine What Might Have Been* (New York: Putnam, 1999). These are a fine range of engaging essays, including ones of a Persian victory at Salamis in 480 BCE, a Spanish naval win in 1588, Napoleon prevailing at Waterloo in 1812, and a Japanese victory at Midway in 1942. Makes you think. A good read.

Samuel W. Crompton, *100 Wars That Shaped World History* (San Mateo, CA: Bluewood Books, 1997). Although quite Eurocentric, this slender volume gives quite a look at why military history needs to be studied and digested.

———, *100 Battles That Shaped World History* (San Mateo, CA: Bluewood Books, 1997). More of the same, with some breezy insights. Again, light on non-European battles.

Marcus Cunliffe, *Soldiers and Civilians: The Martial Spirit in America 1775–1865* (Boston: Little, Brown, 1968). At first glance, this might be an odd place for this volume, but in teaching war over many years, I have become convinced that Americans in general, and this generation of students in particular, can benefit from seeing how deep and long-lived is civilian distrust of the military. Looking at "professionals" and "amateurs" in the American war for independence and the early Civil War, the author finds that the regular army was "disliked by many citizens on principle and in practice" as far back as the founding of the republic. Nevertheless, their craft needs to be studied.

Paul Davis, *100 Decisive Battles* (London: Oxford University Press, 1999). One of the better books of this type, as it has a more global perspective and excellent bibliographies on the various battles for interested students. From Megiddo to Badr to Hakata Bay, the global obsession with war is well documented.

Jared Diamond, *The World until Yesterday* (New York: Viking, 2012). Useful is his conclusion in the section "About Many Wars": "Thus it could not be claimed that some societies are inherently or genetically peaceful, while others are warlike. Instead it appears that societies do or don't resort to war, depending on whether it might be profitable for them to initiate war and/or necessary for them to defend themselves against wars initiated by others." Diamond, unfortunately, is hard-pressed to find "peaceful ones."

Mary L. Dudziak, *War-Time: An Idea, Its History, Its Consequences* (London: Oxford University Press, 2012). Wartime is special; time is out of order. We are used to thinking of it as episodic and temporary, but if it is not "temporary," as in the war on terror, are its "temporary" demands thus made permanent? Or, as Cicero put it, "In time of war, law is silent," and in the post-9/11 world, does this pertain to the indefinite future? Has war, in fact, for Americans become normal life?

R. Ernest Dupuy and Trevor N. Dupuy, *The Encyclopedia of Military History from 3500 B.C. to the Present* (New York: Harper and Row, 1977). Anyone who believes that war is an aberration or simply a few punctuation marks in the flow of human events should read this 1,400-page testimony to the human ability to persist at war from time immemorial (and probably beyond). A great reference work, as well.

J. F. C. Fuller, *The Decisive Battles of the Western World* (London: Paladin, 1972) (2 vols., one covering 480 BCE to 1757, and another, the period 1792–1944). An extensive, wide-ranging look (originally published in 1954) at such battles as Trafalgar, Leipzig, Sedan, Kiev, Stalingrad, and Moscow in the modern period and Salamis, Plataea, Zama, and Hastings in the earlier one. A sweeping analysis and description of the influence of past battles and wars on the history of the world, albeit primarily from a Western perspective.

———, *A Military History of the Western World* (New York: Funk and Wagnalls, 1954) (3 vols.). Why study war? Well, Fuller has an answer:

> Whether war is a necessary factor in the evolution of mankind may be disputed, but a fact which cannot be questioned is that, from the earliest records of man to the present age, war has been his dominant preoccupation. There has never been a period in human history altogether free from war, and seldom are there more than a generation which has not witnessed a major conflict: great wars flow and ebb about as regularly as the tides.

Fuller covers most of them with skill and dispatch, albeit from a Eurocentric perspective for the most part.

Robert Gilping, *War and Change in World Politics* (Cambridge: Cambridge University Press, 1981). Looking at war in the context of international state systems and the factors for or against change and stability, the author concludes, "Hegemonic war historically has been the basic mechanism of system's change in world politics." He draws on the 30 Years' War, the wars of the French Revolution and Napoleon, the wars of Louis XIV, and World Wars I and II.

Robert Greene, *The 33 Strategies of War* (New York: Viking, 2006). A big, sprawling tome exploring the "subtle social game of everyday life" as seen benefiting from a study of the military underpinnings taken from history. U. S. Grant, Sun Tzu, Shaka Zulu, Napoleon, and even Salvador Dali all provide military maxims of use in civilian life. Not for the fainthearted.

Michael Handel, *Masters of War: Classical Strategic Thought* (London: Frank Cass, 1992). Looking in considerable depth at both Sun Tzu and Clausewitz and filling in Mao, Corbett, and so on, the author finds considerable overlap!

Victor Hanson, *Carnage and Culture: Landmark Battles in the Rise of Western Power* (New York: Doubleday, 2001). Makes a strong case for studying battles and war, if only because they have proven to influence so much of the history we often take for granted.

———, "Why Study War?" in his *The Father of Us All: War and History, Ancient and Modern* (New York: Bloomsbury Press, 2010), pp. 3–30. Why indeed? Because humankind has a war side that must be understood and channeled if the mistakes of the past are not to be relived over and over into the indefinite future.

Max Hastings (ed.), *The Oxford Book of Military Anecdotes* (London: Oxford University Press, 1985). Looking at the history of war mostly from an Anglo-American perspective, the editor provides humor and insights that he found in the writings ranging from the Bible to Viscount Slim, from Xenophon to Oliver Cromwell, from Siegfried Sassoon to himself, Max Hastings. Suggests that war is inextricably woven into the history of humankind, which is the "story of man's efforts to kill each other."

Ian Hogg, *Battles: A Concise Dictionary* (New York: Harcourt Brace, 1995). Three thousand years of the history of warfare arranged alphabetically. Hard to see much human progress here, although a lot of people claim there is.

Michael Howard, "The Use and Abuse of Military History," in his *The Causes of Wars* (Cambridge, MA: Harvard University Press, 1984), pp. 188–97. Many in academic circles fear "militarism" and find the study of military history its handmaiden, but as long as the study has "width, depth and context," it is vital to our understanding of ourselves as long as it does not shy away from "complicated and disagreeable" findings. He concludes by quoting Jacob Burckhardt: "The true use of history, military or civil, is not to make men clever for next time, it is to make them wise forever." Amen.

———, *War in European History* (London: Oxford University Press, 2009). A magnificent, overarching account of the interrelationship of European history and war. Howard divides the history of European warfare into war of the knights, mercenaries, merchants, professionals, revolutions, nations, and technologists, providing the reader with perspicacious insights and convenient organizing principles. First written in 1975, it has really stood the test of time.

The Illustrated Encyclopedia of Warfare (London: DK Publishing, 2012). Lavish, eclectic, and with a fine directory of battles, this is a fine work to see the sweep of warfare over the ages and get a sense of the multiplicity of empires, kingdoms, and principalities that practiced warfare seemingly endlessly.

John Keegan (ed.), *The Book of War: 25 Centuries of Great War Writing* (New York: Viking Press, 1999). This comprehensive set of essays from Thucydides and Xenophon to Davy Crockett, the Duke of Wellington, Erwin Rommel, and William Lawrence provides a passing parade of war. If it has been going strong for 25 centuries, then it's probably worth studying.

———, *The Face of Battle* (New York: Viking Press, 1976). The book was quite groundbreaking, as it moved the focus of military history from a "generals and great men of war" approach to the battle as battle. It especially tries to see battles from the perspective of the individual soldier within them by focusing on Agincourt (1415), Waterloo (1815), and the Somme (1916). See especially his "The Usefulness of Military History" and "The Deficiencies of Military History," pp. 22–36, and "The History of Military History," pp. 54–62.

David Keen, *Useful Enemies: When Waging Wars Is More Important than Winning Them* (New Haven, CT: Yale University Press, 2012). In a speculative, personal, and provocative effort, the author presents a number of reasons wars run on for too long: Not all wars are fought to be won, there are wars within wars, wars over natural resources are often too lucrative for those intervening, others provide too much prosperity to some of the internal actors, and some militaries simply want to keep fighting. Even NGOs can inadvertently prolong conflicts; if, for example, they are paying the local Taliban *not* to cause trouble, then there is little incentive for the Taliban to make peace permanently. His tour through Sierra Leone, Angola, Afghanistan, Vietnam, Iraq, Columbia, Rwanda, the Congo, and the Sudan may leave readers with war fatigue of their own.

Raymond Kelly, *Warless Societies and the Origins of War* (London: Oxford University Press, 1996). This anthropologist defines *war* as violence when "social substitution" comes into play; that is, when any man, woman, or child in the collectivity can be a target of violence. This would push the origin of war well back to 35,000 to 10,000 years ago, during the Upper Paleolithic.

James Lacey and Williamson Murray, *Moment of Battle: The Twenty Clashes That Changed the World* (New York: Bantam, 2013). A good collection of mostly Eurocentric selections and a bit of a reach for Objective Peach in the drive for Baghdad in 2003, but on the whole, solid on most, such as Saratoga, Vicksburg, the Marne, and Zama.

William Langer, *An Encyclopedia of World History: Ancient, Medieval, and Modern, Chronologically Arranged* (Boston: Houghton Mifflin, 1956). Open this book at random. Pick a page, any page. Wars abound, all across time and space and culture. "War is a plague but a ubiquitous one, worthy of much careful study." Hence there is a crying need to study it.

Wayne E. Lee, "When Did Warfare Begin?" *MHQ: The Quarterly Journal of Military History* (Winter 2015), pp. 64–71. Argues that "[w]ar does not dominate the archeological record but it does suffuse it" and cites archeological evidence of early human violence going back to 45,000 years ago. He also claims, "Conflict necessitated ever more complex forms of cooperation."

Franklin Margiotta (ed.), *Brassey's Encyclopedia of Military History and Biography* (London: Brassey's, 1994). Although the work is not really very encyclopedic, it does provide the reader with an introduction to some of the most famous generals and wars. It is hard to see significant patterns of inclusion or exclusion within it, and it ends up being quite weak on non-European and American examples.

Jan Mieszkowski, *Watching War* (Stanford: Stanford University Press, 2012). Why study war? Because it is so fascinating, says this author. He looks at the interplay between war and watching war and those who do both and, in particular, the "dissolution of the battlefield's boundaries" in the centuries of total war and war among the people.

Michael Neiberg, *Warfare in World History* (London: Routledge, 2001). Makes the case that the "immediate effects of war can be catastrophic, but war has also proved to be the most important instrument of long term change in world history." You may not agree, but it is useful to engage the author on his own terms.

Amy Oakes, *Diversionary War: Domestic Unrest and International Conflict* (Stanford, CA: Stanford University Press, 2012). Do some leaders deliberately provoke international conflict in order to divert their people from domestic unrest? The author answers in the affirmative but suggests that such efforts often fail and usually do not offer an effective response to domestic discontent.

The case studies and principle examples used to illustrate this process seem something of a farrago: Argentina invading the Falkland Islands in 1982, President Polk sending troops against the Mormons in Utah in 1857, Louis XVI in 1788, Peru in 1988, and the Habsburgs in 1849. This last example seems quite different, in that the "pauper" state of the Austrian Empire actually got Russia to come in and repress its recalcitrant Hungarian subjects for little compensation.

Robert L. O'Connell, *Ride of the Second Horseman: The Birth and Death of War* (London: Oxford University Press, 1995). Tracing war from 5500 BCE, when the "marauders" attacked Nilotic farmers, he sees war as a worldwide and timeless phenomenon because "at the root of human warfare are a bundle of behavior predispositions." Yet he is very hopeful that the collapse of the Soviet Union heralds, if not the end of war, then at least a situation where the "pressure grows to apply military power to stop wars." Bit premature on those assumptions?

Brian Orend, "When Is War Justifiable," *Military History* (October 2007), pp. 18–19. A look at the ways in which war is justified and is justifiable. Some interesting looks at a stubborn human phenomenon. Discouraging if one hopes to rid humans of this plague.

Keith Otterbein, *The Anthropology of War* (Long Grove, IL: Waveland Press, 2009). An anthropologist's take on the origins and persistence of war, especially the two paths—those of hunting bands and those of settled agricultural peoples—to it. Somewhat clotted prose.

Elaine Pagels, *Revelations: Visions, Prophecy, and Politics in the Books of Revelation* (New York: Viking, 2012). Her examination of the book of Revelation (written by John of Patmos around 90 CE following the Jewish War with the Romans) in the canonical Bible shows the ongoing tensions within the Christian movement and the clash between various views of Jesus. She sees the fourth-century (at the Council of Nicaea held after Emperor Constantine declared Christianity the state religion in 313 CE) inclusion of the four gospels, the writings of Paul, and Revelation as an artificial compromise, the "shoehorning in" of the polar opposites of Christ, the all-loving, pacific Son of God, and Jesus, the Divine Warrior coming back to Earth at the head of a victorious army as both judge and war leader.

From the vantage point of *Understanding War*, however, the transformation of Christianity (we have come a long, long way from the Sermon on the Mount) with a projection of warfare indeed up into Heaven before it returns to Earth as an affirmation of humankind's need to include war into the very fabric of apocalyptical life. Cosmic life is thus made to reflect earthly life and vice versa, with war in all its powerful centrality as key to the end of time. Powerful stuff, as it echoes man's search for God and, in the form of Mars, a god's search for man. If Mars can highjack the teachings of Jesus of Nazareth, then he is a powerful force indeed.

Peter Paret, *Clausewitz and the State* (New York: Oxford University Press, 1976). This work, which is densely argued and presented, gives a good background on the life and times of Clausewitz, as well has the notion that "all history is revision." Chapter 4, "What Is War," is worth reading, as is the exploration of the notion that "War is only one expression of state power." The author believes that Clausewitz's insights are neither "time-bound" nor limited by geography.

Bruce Porter, *War and the Rise of the State: The Military Foundation of Modern Politics* (New York: Free Press, 1994). A penetrating history of warfare from a mostly Eurocentric point of view but one that underscores the relevance of military matters to that history. Dedicated to Samuel Huntington.

Richard Preston, Alex Roland, and Sydney Wise, *Men in Arms: A History of Warfare and Its Interrelationship with Western Society* (New York: Harcourt Brace Jovanovich, 1991). This volume makes the case that military history can only be truly understood within its political, economic, social, and technical environments. Straightforward and workman-like, covering war in ancient times to the fall of the Berlin Wall.

The Smithsonian History of Warfare also provides one important reason for studying war: History is so shot full of wars, insurrections, and violent human confrontations. See the books in the series for a quick overview of the particular types of war, such as Daniel Moran, *Wars of National Liberation* (New York: HarperCollins, 2001); Douglas Porch, *Wars of Empire* (New York: HarperCollins, 2000); Brian Reid, *The Civil War and the Wars of the 19th Century* (New York: HarperCollins, 1999); and Gunther E. Rothenberg, *The Napoleonic Wars* (New York: HarperCollins, 2006).

Evan Thomas, "Why Men Love War," *Newsweek*, May 10, 2010, pp. 34–37. The author feels guilty about being a part-time war "liker," but he is convinced that war is really here to stay. See also his favorite case study, *The War Lovers* (New York: Little, Brown, 2010), which asserts that Teddy Roosevelt, Henry Cabot Lodge, and William Randolph Hurst started the Spanish-American War and Thomas Brackett Reed tried to stop it but was unable to do so.

Barbara Tuchman, *The March of Folly: From Troy to Vietnam* (New York: Alfred A. Knopf, 1984). What do the Trojans, the Renaissance popes, and the governments of George III and US presidents Kennedy, Johnson, and Nixon all have in common? The "persistence of error is the problem." War, she believes, comes from tyranny, incompetence, excessive ambition, perversity, or folly. To Tuchman, wars are simply cascades of follies.

Harry Holbert Turney-High, *Primitive War: Its Practices and Concepts* (Columbia: University of South Carolina Press, 1971). This anthropologist defines *war* as occurring when certain practices are evident: tactical formations, command and control, planning, and logistical aspects. These are so central to warfare, he argues, that societies lacking them lie below the "military horizon" and practice "primitive war," not "true war." For Turney-High, genuine or true war, then, occurs some 8,000 years ago but especially after state formation after 3000 BCE.

Alfred Vagts, *A History of Militarism* (New York: W. W. Norton, 1937). The author sees militarism as a "vast array of customs, interests, prestige, actions and thought associated with armies and wars and yet transcending true military purposes" and believes that "militarism displays the qualities of cast and cult, authority and belief."

Simon Van Booy (ed.), *Why We Fight* (New York: Harper and Row, 2010). A collection of quotations from Dickens, the Bible, George S. Patton, and Desmond Morris. Quite a wide range of views as to why we fight, but at its base, a powerful argument for studying war.

Matthew White, *The Great Big Book of Horrible Things* (New York: W. W. Norton, 2012). A huge compendium of manmade disasters, with war, mass killings, repression, and forced starvation illustrating man's inhumanity toward man going back to the dawn of history, with some truly eye-popping statistics (World War II, 60 million; Mao and Genghis Khan, each 40 million; famines in British India, 26 million; Stalin and Taiping Rebellion, each 20 million; and so it goes). Makes a good case for studying such calamities or at least not pretending they don't matter to us as a species.

Dwight Jon Zimmerman, *The Book of War* (New York: Tess Press, 2008). Some battles, some weapons, some generals, some military thinkers, and in the blinking of an eye or two, you've gone from the chariot battle of Kadesh to Operation Anaconda in Afghanistan and realize you've just covered nearly all of human history. And even then, you haven't even reached the ends of the warfare universe.

Chapter Two

Eros and Mars: The Nature of Humankind; or, Making Love *and* War

Over the centuries, many famous and even not-so-famous scholars and activists have looked at warfare and drawn numerous (and often contradictory) conclusions about its origins, uses, and degree of ubiquity in terms of our mutual human history. For me, the short story by Robert Sheckley entitled "Pilgrimage to Earth" (from his 1957 collection of the same name) contains the essence of understanding the nature of humankind in this context. An intergalactic traveler in some distant future travels to planet Earth to sample one of two dimensions of life that humans do best: making love and making war. Lost in much discussion of warfare is the extent to which both are essential polarities of human behavior and, beginning with that primordial tale, *The Iliad*, often intertwined. Also, as Serge Yalichev has pointed out in his *Mercenaries of the Ancient World*, the Sumerian Standard of Ur tablet (now in the British Museum), which dates from 3000 BCE, depicts soldiers in uniforms, carrying spears, axes, and shields, and chariots with drivers and javelin throwers, so that, for at least 5,000 years of human history, war has been intrinsic to that history.

The works listed in this chapter introduce the reader to the many dimensions of warfare and its intrinsic nature and give wide-ranging evidence of its importance to the human condition.

Poul Anderson, *Guardians of Time* (New York: Ballantine Books, 1960) (F). Science fiction at its best, as soldiers from the future engage in time travel in order to keep history the way it turned out. The story about turning the Mongols back from what is now California is superb.

Jeremy Black, *Warfare in the Eighteenth Century* (London: Cassel, 1999). A worthwhile standard text on the nature of war in the 18th century. Quite a bit of useful detail and conceptualization. Black makes a strong case for war as being a primary driver of interstate activity during the period under review.

Max Boot, *War Made New* (New York: Gotham Books, 2007). A fascinating introduction to the way war changes with technology and intellectual advancement over and over and over again. Boot uses many fine case studies to get the reader to appreciate the rhythm of point–counterpoint–new point in weapons development and its eventual impact on tactics and strategy.

Paul Bracken, *Fire in the East: The Rise of Asian Military Power and the Second Nuclear Age* (New York: Harper and Row, 1999). In the spirit of Mackinder and Spykman, he believes that there is a "crisis of room" and the world is now more of a zero-sum game. The 6,000-mile arc from Israel to North Korea is enlarging itself though both population *and* missile growth, producing an "arc of terror."

Joseph Campbell, *The Hero with a Thousand Faces* (New York: Pantheon Books, 1949). Campbell looks at human societies and their many manifestations and finds underlying myths and legends that transcend individual cultures. War-making finds ample justification in those myths.

Robert Citino, *Quest for Decisive Victory: From Stalemate to Blitzkrieg in Europe, 1899 to 1940* (Lawrence: University Press of Kansas, 2002). An incisive look at how war changed in the early years of the 20th century and how the Boer War and the Russo-Japanese War were the precursors of much of that change.

"Comparative Ways of War: A Roundtable," *Historically Speaking* 6, no. 5 (November 2010), pp. 20–30. American, German, and Chinese types of warfare studied as a way of looking at "the way of wars."

Thomas Conlan, *State of War: The Violent Order of the Fourteenth Century Japan* (Ann Arbor: University of Michigan Center for Japanese Studies, 2003). Fascinating account of a most turbulent period in Japanese history.

Bernard Cornwell, *Lords of the North* (New York: HarperCollins, 2007) (F). This series covering the struggles among the Saxons, the Vikings, and the Celts and provides useful insights into the striving for success in battle, even if it means one's death.

Leonard Cottreli, *The Tiger of Ch'in: The Dynamic Emergence of China as a Nation* (New York: Holt, 1962). The long, long search of the Chinese for local hegemony and stability is documented.

Hans Delbruck, *The Dawn of Modern Warfare* (Lincoln: University of Nebraska Press, 1990). The fourth in a four-part series on the history of warfare entitled The Art of War. Early volumes include *The Barbarian Invasions*, *Medieval Warfare*, and *Warfare in Antiquity*. Written in the late-19th and early-20th centuries, this is a fine series against which to check other modern writings about war. Michael Howard has called Delbruck "perhaps the greatest of modern military historians."

Mircea Eliade, *The Sacred and the Profane* (New York: Harcourt, Brace, 1959). Shows the ways in which life and death interact on the human psyche. The author provides a most useful overview of the relationship between ordinary life and special life.

Robert Ferguson, *The Vikings: A History* (New York: Viking, 2009). The violence of the Viking age (9th to 11th centuries) is contained in the Viking oath given to a newborn son along with a sword, "I shall not leave you with any property. You have only what you can provide for yourself with this weapon." A worthwhile summary of the Viking way of life and that of many other warrior cultures.

John France, *Perilous Glory: The Rise of Western Military Power* (New Haven, CT: Yale University Press, 2011). Although the title seems an effort to address contemporary issues, this volume is in fact an excellent history of warfare—and warfare as a whole—showing how humans from so many cultures have worshiped Mars for so long in so many places and how much commonality there is despite the various cultures involved that it is astonishing that so many think war should not be studied.

Azar Gat, *War in Human Civilization* (London: Oxford University Press, 2006). The origins of warfare seem long-standing—and familiar to us when examined, even from afar.

William Golding, *The Inheritors* (London: Faber and Faber, 1955). This is said to be the author's favorite book. For our purposes, the work is a strong voice arguing for a heritage of interspecies rivalry and warfare, just as *Lord of the Flies* argues for a heritage of intraspecies rivalry, as the Neanderthals of the *Inheritors* have their adults killed and their young stolen by modern humans (who in turn fear the Neanderthals). Incidentally, the debate about what happened to the Neanderthals continues to the present.

———, *Lord of the Flies* (New York: G. P. Putnam's Sons, 1954) (F). I have always been amazed that Golding professed to be so surprised at the enduring popularity and importance of this work because it touched in such a primordial and powerful way, the ease with which the thin veneer of civilization can easily be ripped away and how warfare can so readily become the default position for humans. The irony of the ending is relevant, too, for in *Lord of the Flies*, the military arrives *in order to* keep humans from making war on each other.

Stephen Graham, "How the News of War Came to a Village on the Chinese Frontier," in John Keegan, *The Book of War* (New York: Viking, 1999). A brief introduction to the notion of a "distant war," strangely akin to the story of a camel drover in the Australian outback who, when he encounters the two young men going off to fight the Germans in World War I half a world away, looks at them in a very puzzled way and says, "I knew a German once." This moment, captured in the film *Gallipoli*, underscores the absurdity of many motivations for war.

Edward Hamley, *The Operations of War* (Edinburgh: William Blackwood and Sons, 1878). This monumental work by the general who was commandant of the British Staff College after 1870 is one of the most comprehensive overviews of the "modern conditions of war" and is one of the important post-Clausewitzian analyses of the period. Particularly good is his "The Influences of Obstacles" and the firm grounding of the work in many specific battles and campaigns, such as Salamanca (1812), Italy (1796, 1800, 1849, and 1859), Champagne (1814), and the Franco-Prussian War (1870). Seeking to impart "special professional knowledge" in a field where much is needed, the author largely succeeds.

Hammond Historical Atlas of the World (Maplewood, NJ: Hammond, 1997). A marvelous collection of maps by era. A must-read. Many specific references to it appear in this bibliography.

Michael Handel, *Masters of War: Classical Strategic Thought* (London: Frank Cass, 1996). War is timeless and deeply embedded in human culture and history, and his examination of Sun Tzu, Clausewitz, Jomini, Machiavelli, and Mao, among others, shows its commonality.

Victor Hanson, *The Western Way of War: Infantry Battle in Classical Greece* (New York: Knopf, 1989). Hanson asserts the primacy of culture and the supremacy of the Western way of war. It is really too bad Hanson never truly ventured outside his classical training and looked into the non-Western warcraft of so many other peoples. His limited perspective hampers his analysis, as numerous loci tend to support and undercut his central thesis.

Richard Holmes (ed.), *Battlefield: Decisive Conflicts in History* (London: Oxford University Press, 2002). A vast compendium of most Western battles. Many battles listed seem anything but "decisive" however.

Robert Jansson, *Kari's Saga: A Novel of Viking Iceland* (New York: Booksurge, 2007) (F). A wordy but informative novel of the Viking age in Iceland as Christianity arrives and the Nordic world continues its interaction with England, Ireland, and the rest of Europe. Useful for understanding why some humans are suited only for war and military leadership and why they wither and decline as settled farmers.

Donald Kagan, *The Peloponnesian War* (New York: Viking, 2003). A wonderful examination of one of the most studied wars of ancient times by a man who knows it best, in part because of his most careful study of Thucydides. A classic.

John Keegan, *The Face of Battle* (New York: Viking, 1976) and *A History of Warfare* (New York: Knopf, 1993). Two works that introduce the reader to war from a specific point of view as the author tries to bring order out of the seeming chaos of war. Both are big, sweeping studies, and *The Face of Battle* revolutionized the view of humans at war by focusing on the rank-and-file soldiers rather than the generals.

John A. Lynn, *Battle: A History of Combat and Culture* (Boulder, CO: Westview, 2003). A fine and enduring look at war and its relationship with culture, although the connection at times can be questioned.

William C. Martel, *Victory in War: Foundations of Modern Military Policy* (Cambridge: Cambridge University Press, 2007). Looks at modern warfare through the existential lens of what can—and should—constitute "victory."

Lauro Martines, *Furies: War in Europe 1450–1700* (New York: Bloomsbury Press, 2013). Anyone seeking to discover the true horrors of war would do well to start with this powerful and penetrating study. The author gives valuable insights into the "total war" that engulfed many parts of Europe ("A War Mosaic") during the period under review and shows us the horror of the ongoing "harrowing" of war. "Warfare during this period was directed overwhelmingly against civilians in town and country." The sacking of cities (i.e., "the notion of putting a city into a sack and going off with it") and the plundering of the countryside are cases in sharp and even disgusting relief.

You will not read this work and be unmoved. The level of depravity and desertions, the incredible down-drag in the "soft core of logistics: food," the size of the armies plus their camp followers (an army of 20,000 combatants was really a traveling mob of 48,000 to 50,000; p. 161), the role of loot for one and all, the fragility of the armies (for famine, disease, and combat cost armies 25 percent of their strength yearly; p. 170), and the insightful notion that states during this period were almost adjuncts to their armies are all cast in sharp relief. The true face of war is devastation and death. Also intriguing is the notion of how much human excrement and waste from cavalry and baggage horses provided a vast theater of disease. A truly imaginative and worthwhile work.

William H. McNeill, *The Age of Gunpowder Empires 1450–1800* (Washington, DC: American Historical Association, 1989). Links the European colonial expansion with the military revolution that began in Europe and spread throughout the world in varying degrees.

Walter Millis, *Arms and Men: A Study in American Military History* (New York: G. P. Putnam's Sons, 1956). Looks at American military history by conceptualizing the various revolutions that it has undergone, such as "The Managerial Revolution," "The Democratization of War," and "The Hypertrophy of War." Still shows many valid principles invaluable 50 years later.

Kenzaburo Oe, *Nip the Buds Shoot the Kids* (London: Marion Boyars, 1995) (F). *A Lord of the Flies* East, as war spawns the suppression of deviance and the repression abounds. A nice variation on Golding. Frightening, as is the persistence of war, and this Japanese author shows us how fundamental it all is.

Neil Oliver, *Vikings* (London: Weidenfeld and Nicolson, 2012). The Viking era is explored from a variety of perspectives and the Norway/Scotland/Islands interactions are highlighted. Precious little on "The Great Heathen" army's fighting dimensions but homage paid to their "violent, dramatic cosmology."

Geoffrey Parker, *The Military Revolution: Military Innovation and the Rise of the West 1500–1800* (Cambridge: Cambridge University Press, 1988) and *Warfare: The Rise of the West* (Cambridge: Cambridge University Press, 1995). Parker is the scholar who has done more than anyone else to get us to focus on the revolution in military affairs that propelled the West across the globe but only after the dynasties of the West practiced on one another for centuries.

———, "The Western Way of War," in Geoffrey Parker (ed.), *The Cambridge Illustrated History of Warfare* (Cambridge: Cambridge University Press, 1995). A short but pithy introduction to his specialty.

Note: Jeremy Black, arguing with both Geoffrey Parker and Michael Roberts (reviewed later), is quite adamant that no such "revolution" took place during these dates; instead he declares that the period 1660–1815 saw the true revolution in warfare. See his *European Warfare: 1660–1815* (London: UCL Press, 1994). Black concludes, "Military strength was central to this rise in Western power, both within and outside Europe, and was to give shape to the nineteenth century world order."

Stephen Pinker, *The Better Angels of Our Nature* (New York: Viking, 2011). Seeing a "remarkable decline in violence," as he translates earlier casualty figures from war into modern equivalent (Genghis Khan's wars took an estimated 40 million lives, equal to 275 million today based on the population percentage). Pinker sees a desire to get rich and the *rise* in state power as the answers (although he does grant a "de-civilizing" role for Nazi Germany). Sees war declining everywhere. A strangely optimistic book.

Clifford J. Rogers (ed.), *The Military Revolution Debate: Readings on the Military Transformation of Early Modern Europe* (Boulder, CO: Westview, 1995). A lot of counterpoints to Parker and his thesis, plus some alternative aspects of support.

Charlie Schweder, *Man of War* (New York: Hudson Street Press, 2012). Reenactors refight battles from World War II, the American Civil War, 16th-century Poland, Vietnam, and the American Revolutionary War, among others. The fascination of war extends far beyond those who fight for real.

Shaybani's Siyar, *The Islamic Law of Nations* (Baltimore: Johns Hopkins Press, 1966). The 18th-century effort to put the "Territory of Islam" and the "Territory of War" into historical as well as religious perspective. Sees in the Islamic law the "need to establish a world public order based on divine legislation and to enforce it by the jihad." Extensive analyses of "Islamic Law and the Law of Nations," "Safe Conduct," and "On Apostasy." There is a very chilling note on apostates: "If a Muslim apostatizes, Islam would be offered to him, he has either to accept it or be killed at once." Provides a number of insights into the persuasive power of religion to incite conflict.

Brian Francis Slattery, *Lost Everything* (New York: Tom Doherty Associates, 2012) (F). In this grimmest of grim postapocalyptical tales, humans, far from being united by the growing menace of climactic disasters, splinter into smaller and smaller armies until they produce a world that would make even Thomas Hobbes shudder—or appear utopian in outlook. "The war was about everything."

John Sowell, *Conquests and Culture: An International History* (New York: Basic Books, 1998). An early work that, to me at least, is very important in showing how "cultures" change over time in their relationship to warfare.

Michael Stephenson, *The Last Full Measure: How Soldiers Die in Battle* (New York: Crown, 2012). A marvelously powerful tour through the history of warfare (rather paralleling Keegan's *A History of Warfare*) but with exhausting, often heartrending, "on the spot" accounts of the death and the dying, interspersed with very useful statistics. A sample of these includes: (1) By 1918, each British division had machine guns capable of firing the equivalent of 38,000 riflemen (p. 196); (2) 300,000 British soldiers (40 percent) killed on the Western Front were never found (p. 194); (3) disease accounted for 66 percent of all fatalities suffered by the Union and the Confederate armies; and (4) 310,000 of 400,000 Soviet tankers, or 77.5 percent lost their lives during World War II (p. 313). Ends up being a most powerful antiwar book in terms of its impact.

Tim Travers, *The Killing Ground: The British Army, the Western Front and Emergence of Modern Warfare 1900–1918* (London: Allen and Unwin, 1987). Brutal and straightforward account of how war became hyperlethal during the First World War.

Peter Watson, *The Great Divide: History and Human Nature in the Old World and the New* (London: Weidenfeld and Nicolson, 2012). Disappointingly thin on the various natures of warfare throughout time and space but does contain some interesting material on Mesoamerica and the role of bloodletting in flower wars, destructive events, and shell-star (i.e., conquest) wars. Mostly derivative.

John Wyndham, *The Day of the Triffids* (Garden City, NJ: Doubleday, 1951) (F). This marvelous postapocalyptical science-fiction novel is mostly about the walking plant Triffids, which try to take over earth, but its relevance here is in the startled bewilderment of the narrator when humans—fighting to stay alive in the face of blindness, Triffids, and the decline of civilization—begin again to suggest war against their fellows as necessary. "Great God almighty! We've lived through all this—and now the man proposes to start a war." Of course. Why not?

Chapter Three

The Search for Fundamental Truths about War

Sun Tzu, Kautilya, Thucydides, Jomini, Clausewitz, and Other Military Thinkers and Students of War

Many other authors and actors have also written about warfare and warriors and about the nature of war itself, not worrying so much about why we wage war but how to do it successfully and who has made significant contributions in helping us understand both the nature of war and what approaches have proven to be enduringly perspicacious in its applications.

This chapter presents a sample of the broad range of human thinking about war and, for others, thinking about that range. Military thinkers, it should be noted, go in and out of fashion, but taken together, they can illuminate and underscore the amount of time and effort that has been devoted to the study of the nature, causes, and results of war. It is fascinating to see both the timeless relevance of such ancient writers as Thucydides, Kautilya, and Sun Tzu and the equally stimulating arguments of more recent students of warfare, such as Clausewitz, Jomini, Douhet, Corbett, Boyd, Mackinder, and Mahan.

As one examines this area of study, one should always make significant allowance for the assertion of Richard Preston and others: "The history of human society has been punctuated by war; but the study of military history has all too often been undertaken as if war always existed in a vacuum" (Preston, Wise, and Werner, 1956).

Note: I have pulled out the subfield of geostrategic works and commented on them following this main section. This is not an attempt to further segregate geography from considerations about war but rather to highlight its importance despite the disuse into which it has fallen in some circles. With the possible exception of military history itself, no topic has become more understudied within the recent academy, as the filters of race, gender, sexual orientation, and ethnicity have seemingly come to dominate analysis across a broad spectrum of causality.

D. H. Ahir, *Asoka the Great* (New Delhi, India: B. R. Publishing, 1995). This great and legendary Indian ruler and warrior has his story told, and it is a very intriguing story, indeed.

Bevin Alexander, *How Wars Are Won: The 13 Rules of War: From Ancient Greece to the War on Terror* (New York: Crown, 2002). Taking a great deal of analysis and concepts from Sun Tzu ("striking at an enemy's weakness" and "feigning retreat"), the author looks at both history and the present situations in Iraq and Afghanistan with uneven results.

Raymond Aron, *Clausewitz: Philosopher of War* (London: Routledge and Kegan Paul, 1976). A careful dissection of Clausewitz ("A gifted soul"), his life and times and work and its applicability to wars since, including wars of national liberation and the Cold War. Aron is a most gifted exegete himself, concluding, "War is a chameleon."

Roger Boesche, *The First Great Realist: Kautilya and His Arthashastra* (Lanham, MD: Lexington Books, 2003). The best book thus far explaining the political philosophy of Kautilya and the context of his times.

John Boyd, *A Discourse on Winning and Losing* (mimeographed, 1987). A wonderful, even marvelous, mind at work. His OODA loop turns out to be very elegant and very useful in analyzing conflict from the individual to the collective. Boyd truly understands the dynamics of battle and war, and anyone studying *A Discourse* will be well rewarded. A pity he never had a sabbatical to write a book expanding on his impressive ideas.

Bernard Brodie, *Strategy in the Missile Age* (Princeton, NJ: Princeton University Press, 1959). Looks at nuclear war, including preventive war, preemptive attack, massive retaliation, deterrence, and limited war, concluding that nuclear war would be "immeasurably worse" than anything known in the past. Useful in pointing out how deterrence must not fail in a variety of circumstances.

———, *War and Politics* (New York: Pearson, 1974). Looks at the case studies of World War I, World War II, Korea, and Vietnam. He believes that nuclear weapons were a strategic revolution in and of themselves and that rational application of their use would work against actually employing them in war, hence they were a new deterrent making war less, not more, likely in international relations. Associated massive retaliation, then, with keeping the peace, although he later argues that a flexible response to their use could mean nuclear weapons could be used in a limited war.

Greg Cashman, *What Causes War?* (Lanham, MD: Lexington Books, 1992). This work looks at everything from empirical theory to international interaction to studies of anarchy and power and concludes that no single theory suffices and that there are multiple causes of war.

Gerard Chaliand (ed.), *The Art of War in World History* (Berkeley: University of California Press, 1994). All the usual Western suspects are rounded up but added are those from Persia, the Ottoman Empire, Mongolia, India, and the Arab World. I was pleased to see Kautilya included. A stimulating and eclectic collection and a broad canvas of history on which is depicted the long, thrilling, as well as disheartening dimensions of warfare.

Karl von Clausewitz, *On War* (Princeton, NJ: Princeton University Press, 1976). Love it or hate it, Clausewitz's work *On War* is one of the most profound analyses of war ever produced, and although it arose out of his experiences during the Napoleonic era, it remains of crucial importance to our understanding of the many dimensions of war. Remember that Clausewitz was a child of the Enlightenment and believed that all things could be studied, war among them. He was right, and this work belongs in the first rank of war studies. Originally written in German and published after he died, therefore lacking a final editing and "smoothing out," it is a difficult must-read.

Note: For a biting critique of Clausewitz and his theories, see John Keegan, "What Is War?" in his *A History of Warfare* (New York: Alfred A. Knopf, 1993), pp. 3–76.

Note: For a more positive appreciation of Clausewitz, see Peter Paret, *Clausewitz and the State* (London: Oxford University Press, 1976). Analyzes in considerable depth the life and times of Clausewitz while still leaving ample space to do a deep dive into his analysis of the phenomenon of war, concluding that Clausewitz provides a "mastery of reality through understanding."

Note: Perhaps the best introduction to Clausewitz from an exegete's point of view is Michael Howard, *Clausewitz: A Very Short Introduction* (London: Oxford University Press, 2002). Quoting Bernard Brodie, perhaps America's greatest strategic thinker, "His is not simply the greatest, but the only great book about war." Howard argues that there is "no systematic study comparable to that of Clausewitz." Clausewitz is always concerned about the link of policy to action. Howard also emphasizes how Clausewitz is misjudged because what he said was that war should be about the "killing of his courage rather than his men." Points out that Clausewitz personally experienced the climactic battles of Jena, Borodino, and Waterloo! Notes also his big influence on Marx, Engels, Lenin, and Trotsky, especially his emphasis on war as an instrument of policy and policy as the product of certain basic social factors.

Note: However, for the best in-depth look at the life of Clausewitz and the times in which he lived, together with superb military analysis of the military campaigns of the Napoleonic wars involving him, see Donald Stoker, *Clausewitz: His Life and Work* (London: Oxford University Press, 2014). Highlighting Clausewitz's statement, "Politics, moreover, is the womb in which war develops," the author gives the political context of the nations at war. Good emphasis on the importance of the battles of Borodino and Leipzig.

Christopher Coker, *Barbarous Philosophers: Reflections on the Nature of War from Heraclitus to Heisenberg* (New York: Columbia University Press, 2010). A wide-ranging look at war through the lens of various philosophers, almost exclusively Western in nature. Coker puts himself very much in the discussion.

———, *Humane Warfare* (London: Routledge, 2001). The author argues that humanizing war or attempting to not only is ironic but also has distinct limits and often pejorative outcomes. The West tries to reinvent war, but war persists. To remove cruelty from war is to remove effectiveness in fighting those who do not.

Bruce Condell and David T. Zabecki (eds.), *On the German Art of War Truppenfuhrung* (Boulder, CO: Lynne Reiner, 2001). German generals write how to achieve success in warfare through operational assertions. One of the key elements in the German *Truppenfuhrung* was *Auftragstaktik*, or "mission-type orders," which means that officers were given tasks but how those tasks were to be carried out depended on the officers' judgment. This type of tactical flexibility was at the heart of Germany's war-making during World War II.

Charles De Gaulle, *The Army of the Future* (New York: J. B. Lippincott, 1941). A reprint of his 1934 work, this volume is an astonishing tribute to the insights of De Gaulle (of whom Churchill would reportedly say, "Of all the crosses I had to bear during World War II, the Cross of Lorraine was the most burdensome"). Writing about the "principles of which the French did not act—while the Germans did," De Gaulle saw the immense vulnerability of Luxembourg and the Ardennes as well as the fact that "[t]here is a grim relationship between the properties of speed, power and concentration which modern weapons confer upon a well-trained military elite." In short, he, too, saw the power of blitzkrieg and foresaw the war of the future. Quite startling to read with our knowledge of what actually happened.

Don Delillo, "Human Moments in World War II," in his *The Angel Esmeralda* (New York: Scribner's, 2011), pp. 25–43. Declaring "All wars refer back" and "The war tells us it is foolish to believe." But humans have and do and will.

Christine de Pizan, *The Book of Deeds of Arms and of Chivalry*, edited by Charity Cannon Willard and translated by Sumner Willard (University Park: Pennsylvania State University Press, 1999). Written in the 15th century by this Italian-born but French-court author, *The Book of Deeds* resurrects many classical writings on war (especially Vegetius) but provides very useful contrasts between medieval European war practices and those from antiquity, including just war, siege warfare, chivalry, trickery, and subtlety. Many examples from contemporary Europe as well as campaigns of Scipio, Hannibal, and Hanno. A truly amazing work given the time, the place, and the sex of the author.

Giulio Douhet, *The Command of the Air* (Washington, DC: Office of Air Force History, 1985). Reprint of the 1942 compilation of Douhet's earlier works, including "A New Form of War" and "War of 19.." which was published in 1930. This Italian commanded his country's army air force and saw the importance of strategic bombing, total war, and the specter of whole cities destroyed. Ironically, he was court-martialed for his statements that the Italian army was incompetent and unprepared for World War II. Obviously, he missed a number of dimensions—such as the importance of tactical air support, which he derided—but Douhet remains a systematic, overarching visionary who saw how death from the air would affect future wars.

Note: Although not a major author, the British marshal of the Royal Air Force, Hugh Trenchard, sometimes called the "Father of the Royal Air Force," belongs with Douhet and the American general Billy Mitchell as early and significant proponents of air power in warfare.

Ardant du Piq, *Battle Studies: Ancient and Modern Battles* (Harrisburg, PA: Military Services, 1946). Views of war written in the 1870s by a French colonel, this work focuses on the "human element"; Napoleon's "moral element," the élan or "moral force"; and the role it plays in battle. One fears that a lot of Frenchmen died in World War I thinking élan was what mattered when one encountered machine guns. Still, du Piq is correct in focusing on the role of morale in battle, and his point that "man does not enter battle to fight, but for victory" rings true. In fairness to him, Foch and the other generals took much of what he said out of context because, at the end of the day, he maintained, "Mass attacks are incomprehensible. Not one out of ten had been carried to completion and none of them could be maintained against counter attacks." Sure sounds like World War I generals disregarded this dimension. I also like his emphasis on small-unit cohesion: "Four brave men who do not know each other will not dare to attack a lion. Four less brave, but knowing each other well, sure of their reliability and consequently of mutual aid, will attack resolutely. There is the science of the organization of armies in a nutshell."

Edward Mead Earle (ed.), *Makers of Modern Strategy: Military Thought from Machiavelli to Hitler* (Princeton, NJ: Princeton University Press, 1948). Although Earle, a professor at the Army War College, rounds up the usual suspects (Jomini, Clausewitz, etc.), he also has essays from a variety of authors on such often overlooked military analysts as Moltke, Schlieffen, Lenin, Trotsky, Stalin, Ludendorff, and Hitler. In terms of the Führer, Earle sees the "sapping of French power" as essential to his worldview and, having reduced France to vassal status, seems to have gone awry.

Antulio Echevarria, *After Clausewitz: German Military Thinkers before the Great War* (Lawrence: University Press of Kansas, 2000). A wide-ranging but densely packed analysis of German military thinkers in the 19th century. The author believes it wasn't that they underestimated the technological changes as much as the German military thinkers underestimated the ability of other European nations to copy them from the earlier Prussian successes against the Austrians and the French. They also failed to see how difficult it was to command large numbers of troops with all their added firepower over time. Echevarria argues that it was junior officers during early World War I who vaunted the "spirit of the offensive," not their senior counterparts who paid more attention to the realities captured by Clausewitz.

Alan English (ed.), *The Changing Face of War* (Montreal: McGill-Queen's University Press, 1998). A set of Canadian views of strategy, some interesting essays, including ones dealing with the importance of attrition (which decides most wars, not maneuver) and Du can Milne's fine chapter on the Soviet strategic thinker Mikhail Tukhachevsky, of "Deep Battle" fame.

John English, *Marching through Chaos* (Boulder, CO: Praeger, 1999). How armies got so big and so unwieldy and why and how NATO was very fortunate that its warfighting assumptions were never tested in battle. Interesting exploration of the US situation during the Gulf War, pointing out that the

Soviet Union denied Iraq space intelligence of the Left Hook and that was, if not decisive, then at least most salutary. Concludes, "In the final analysis, then as long as there are people on earth, there will be a need for armies to march in good order and discipline through chaos."

Ferdinand Foch, *The Principles of War* (New York: H. K. Fly, 1918). Sees the "novel features" of World War I as being overridden by fundamental tactics. War-making is not an exact science, but "[i]n war, everything is related." His principles include freedom of action, economics of power, strategy, surprise, and protection and have many echoes of Clausewitz: "Modern war, in order to reach its purpose: to impose one's will on the enemy, knows but one means, the destruction of the opponent's organized forces" as he focuses on "decisive battles."

Lawrence Freedman, *Strategy: A History* (London: Oxford University Press, 2013). This substantial, comprehensive, hermeneutic work attempts to examine the various dimensions and history of "strategy," which the author defines as the "art of creating power" (p. xii). Starting with the prehistory of primates and traversing the centuries up to 2010, his wide-ranging observations take the reader across a vast landscape of action—military, political, social, and economic.

The author pays welcome attention to such often overlooked strategists as John Boyd, although there are some striking omissions in the long arc of his survey. For example, the preeminent Indian realist Kautilya is nowhere to be found, and there is precious little about Soviet strategists prior to World War II (such as Marshal Tukhachevsky) and during the Cold War. There is also some puzzling overemphasis on certain practitioners. For example, while David Petraeus gets but one quick reference (and his coauthor James Mattis none, along with Giuseppe Mazzini, Roger Trinquier, etc.), there is a veritable deluge of references to Tom Hayden (15), the SDS (13), even Stokely Carmichael (5).

J. F. C. Fuller, *Armament and History* (New York: Charles Scribner's Sons, 1945). A rapid transit though the ages of valor, chivalry, gunpowder, steam, and oil, concluding with Stalin's quote "He who has the petrol has the Empire."

———, *Memoirs of an Unconventional Soldier* (London: Ivor Nicholson and Watson, 1936). The author of dozens of books about war, one who calls himself a "wandering student" and a "Synthetic-Iconoclast," writes engagingly about all manner of war subjects, including tanks and aircraft and the way war changes. A veteran of the Boer War and World War II, his is quite a life and set of views on warfare over the ages.

W. B. Gallie, *Philosophers of Peace and War: Kant, Clausewitz, Marx, Engels and Tolstoy* (Cambridge: Cambridge University Press, 1979). An in-depth look at all five, with a lot of interconnections made. Engels, it now appears, was constantly looking for revolutionary revolts by soldiers and how wars could influence revolution; and revolution, war.

Azar Gat, *The Origins of Military Thought: From the Enlightenment to Clausewitz* (Oxford: Clarendon Press, 1989). From the Enlightenment comes the notion that man can study all things, including war, and understand them. This tour du horizon includes essays on Machiavelli, Montecuccoli, Jomini, Clausewitz, Archduke Charles, Benenhost, Douchet, J. F. C. Fuller, Little-Hart, Mahan, and Corbett. Informative and stimulating.

Colin Gray, *Fighting Talk: Forty Maxims on War, Peace, and Strategy* (Westport, CT: Praeger, 2007). From war and peace through security to history and the future, the author covers a number of topics in stimulating if somewhat didactic fashion. A couple I liked: "War Works—But Always Has Unintended and Unanticipated Consequences" and his riff on Bernard Brodie's "Strategy wears a dollar sign."

———, *Modern Strategy* (London: Oxford University Press, 1999). Focuses on strategy as the use or threat of military power for political purposes and as the bridge between military power and political purpose. Outlines a number of interesting dimensions, including "The Strategist's Tool Kit" and "The Grammar of Strategy," concluding that, while the conduct and character of war and strategy may change, the nature and function of war and strategy remain the same. A solid Clausewitzian, and I believe for good reason.

Note: Some American generals, such as James Mattis, put Gray in the very front rank of modern strategists.

———, *The Strategy Bridge: Theory for Practice* (London: Oxford University Press, 2010). An ambitious and "bold" overview of 2,300 years of strategic thinkers, with a clear roadmap to his favorites. He declares that strategy can go a "bridge too far" by including so many variables as to be useless to practitioners. Humans can study war and form judgments; they can see through the shape-shifting nature of war and discern its true nature. A strategist doesn't have to be the "best," only better than his or her opponent. Strategy is simply a "quest for control," but it cannot avoid the law of physics or the dictates of geography and psychology. General James Mattis calls him one of the great strategic minds of our age. His selections and rankings:

First Rank:

- Carl von Clausewitz, *On War* (1832–34)
- Sun-Tzu, *The Art of War* (ca. 490 BCE)
- Thucydides, *The Peloponnesian War* (400 BCE)

Second Rank:

- Nicolo Machiavelli, *The Art of War* (1521)
- Nicolo Machiavelli, *The Prince* (1522)
- Nicolo Machiavelli, *Discourses on Livy* (1531)
- Antoine Henry de Jomini, *The Art of War* (1839)
- Basil Liddell Hart, *Strategy: An Indirect Approach* (1921)
- J. C. Wylie, *Military Strategy: A General Theory of Power Control* (1967)
- Edward Luttwak, *Strategy: The Logic of War and Peace* (2001)

Third Rank:

- Bernard Brodie, *The Absolute Weapons: Atomic Power and World Order* (1946)
- Bernard Brodie, *Strategy in the Missile Age* (1959)
- Bernard Brodie, *War and Politics* (1973)
- Thomas Schelling, *The Strategy of Conflict* (1960)
- Thomas Schelling, *Arms and Influence* (1966)

Other Contenders:

- Julius Caesar, *Commentaries* (44 BCE)
- Alfred Thayer Mahan, *The Influence of Sea Power upon History 1660–1783* (1890)
- Julian Corbett, *England in the Seven Years War* (1907)
- Julian Corbett, *Some Principles on Maritime Strategy* (1911)
- J. F. C. Fuller, *Armament and History* (1946)
- John Boyd, *A Discourse on Winning and Losing* (1987)
- Martin van Creveld, *The Transformation of War* (1991)
- Martin van Creveld, *The Culture of War* (2008)

Michael Handel, *Masters of War* (London: Frank Cass, 1986). A comprehensive overview of Sun Tzu, Machiavelli, Clausewitz, Mao Tse-tung, Jomini. A useful introduction and comparison.

Victor Davis Hanson (ed.), *Makers of Ancient Strategy* (Princeton, NJ: Princeton University Press, 2010). A good introduction to the ancient world of war and the writings and conclusions of those contemporaries whose work and impact survive the ages. Weak on non-Western thinkers, however, which leads the author to overpraise some of them. Hanson would have *loved* Kautilya if he'd ever read him.

Beatrice Heuser, *The Evolution of Strategy: Thinking War from Antiquity to the Present* (Cambridge: Cambridge University Press, 2010). A tour d'horizon from antiquity to the present. The author sees the persistence of the Napoleonic paradigm of "violent conflict aiming at military victory" through time but argues that ethical approaches to war have always been present and need to be reasserted now, for "we can only win ultimately if the other side benefits as well." Still, she warns us throughout about the "frailty of human logic."

Michael Howard (ed.), *The Causes of Wars* (Cambridge, MA: Harvard University Press, 1984). A collection of essays, including *The Theory and Practice of War*, with such insightful quotations as (1) "The alternation of war and peace has been the very stuff of the past"; (2) "War has been throughout history a normal way of conducting disputes between political groups"; and (3) "The trouble is that if we are to regard war as pathological and abnormal, then all conflicts must be similarly regarded; for war is only a particular kind of conflict between a particular category of social groups, sovereign states."

———, *The Invention of Peace: Reflections on War and International Order* (New Haven, CT: Yale University Press, 2000). Beginning with Sir Henry Maine's insight "War appears to be as old as mankind but peace is a modern invention," Howard sees Emmanuel Kant as one of the first to have a recipe for peace but accepts the notion that in the modern era, "[i]t is no cause for wonder that once war became more popular it should have become more violent."

———, *The Theory and Practice of War* (Bloomington: Indiana University Press, 1965). A collection of essays ranging from ones on Jomini and Clausewitz to others on Soviet military doctrine and the making of the Israeli army, all in honor of Liddel Hart's 70th birthday.

Victoria Tin-bor Hui, "The Dynamics of International Politics in Ancient China," in her *War and State Formation in Ancient China and Early Modern Europe* (Cambridge: Cambridge University Press, 2005), pp. 54–108. Interstate interactions during 626–254 BCE are in the spotlight with lots of detail.

Henri Antoine Jomini, *The Art of War* (El Paso, TX: El Paso Norte Press, 2005). A Swiss contemporary of Clausewitz who needs to be read along with him. Whereas Clausewitz saw war, at least in part, as art, Jomini was more inclined to view it as a constellation of fixed verities or science. "Methods changes, but principles are unchanging." Jomini looks at the Napoleonic wars and how they were fought and seeks to codify them and put them as a guide to military practice in the future. He did agree with Clausewitz, however, in terms of the importance of the "employment of masses upon the decisive points." For Jomini, who was with Napoleon (and then the Russian czar), Napoleon simply was returning to eternal verities of war (not as Clausewitz, who argued that Napoleon and the French Revolution was a revolution in warfare). Also Clausewitz placed more emphasis on chance and friction, whereas Jomini thought adherence to the eternal principles could mitigate them. Jomini was a strong advocate of civilian control over the military where war was concerned, whereas Clausewitz favored more military control over the actual war.

Note: As Mark M. Smith points out in *The Smell of Battle, the Taste of Siege* (London: Oxford University Press, 2014), the order, rationality, and scientific nature of Jomini's observations were widely taught at West Point prior to the outbreak of the Civil War and appealed to many, such as P. G. T. Beauregard (topographer), Erwin McDowell (architect), and Stonewall Jackson (mathematician), who liked to view the military situation with precision.

Gerald Jordan (ed.), *Naval Warfare in the Twentieth Century 1900–1945: Essays in Honour of Arthur Marder* (London: Croom Helm, 1971). A wide-ranging set of essays on naval science, affairs and operations, Naval philosophies, and the naval arms race are highlighted by Lord Mountbatten's praise of Professor Arthur Marder, who he said has produced the "finest history of the Royal Navy." Bar none? Quite a compliment, and the geopolitical connotations are indeed central and vital.

John Keegan, *The Art of War: War and Military Thought* (Washington, DC: Smithsonian Books, 2000). A breezy introduction to military thinking from Sun Tzu to the present. Kautilya and John Boyd are major omissions, and he is far too harsh on Clausewitz than is healthy given the ubiquity of Clausewitzian thought and prescription.

B. H. Liddell Hart, *Strategy: An Indirect Approach* (New York: Praeger, 1956). Using the wide arch of history, Liddell Hart concludes with a look at strategy (how to win wars) and grand strategy (why you fight one in the first place). Believes strongly that, "For the spirit of barbarism can be weakened only during a cessation of hostilities, war strengthens it—pouring fuel on the flame."

Edward Luttwak, *Strategy: The Logic of War and Peace* (Cambridge, MA: Harvard University Press, 1987). The "Logic of Strategy," the "Levels of Strategy," and "Grand Strategy" are all covered with skill and dispatch, as the author concludes, "The study of strategy should be its own reward."

Niccolo Machiavelli, *The Art of War* (New York: Bobbs Merrill, 1965). This sometimes overlooked and underappreciated work is a true gem, even if most exegetes concentrate on *The Prince*. Seeing the civil and military life in positive interaction, Machiavelli provides us with the "first classic of modern military science," with accurate descriptions of warfare under the Romans, the Greeks, and the Swiss. Accurately sees that discipline is the essence of the Roman way of war and the key to the then-modern combination of pikemen and swordsmen. Believed that citizen soldiers were preferable to condottieri, he felt that a "military is always an instrument and a bulwark of Republicanism." Did tend to undervalue the role of artillery and cavalry, but listening to "Fabrizio Colonna" (ironically patterned after the legendary condottieri captain Prospero Colonna) speaking on the art of war is to sit at the feet of a Renaissance master and let his insights gently wash over you by a Florentine fire. Seven chapters covering a variety of subjects, from fighting in winter to fortifying towns, are diverting. Book 3 allows Machiavelli to expound on his basic ideas concerning strategy.

———, *Discourses on Livy*, translated by Harvey Mansfield and Nathan Tarlov (Chicago: University of Chicago Press, 1936). Believed to have been written between *The Prince* and *The Art of War*, this work champions the virtues of a republic and uncorrupted politics and reflects Machiavelli's ruminations on the first 10 books of Livy (out of the 132 he wrote). Calling for a return to Roman *virtue* (virtue), the author looks at the role of *fortuna* (destiny or fate) and the ways it can be managed through audacity and adaptation. Echoes Sun Tzu as well as Levy in stressing "That To Use Fraud in Waging War Is A Glorious Thing." He looks at four types of war: offensive, defensive, limited, and total, favoring the offensive.

———, *The Prince* (New York: Dover, 1992). This 16th-century European should be read side by side with the *Arthashastra* of the 4th-century-BCE Indian war philosopher Kautilya to see Machiavelli as an epigonic realist compared to Kautilya. Machiavelli, however, was spot-on in his focus on Roman discipline in warfare; see his *The Art of War*, reviewed in this chapter.

Harvey C. Mansfield, *Machiavelli's Virtue* (Chicago: University of Chicago Press, 1996). New and old ways to look at what Machiavelli really said as opposed to what many say he said. For Mansfield, Machiavelli is a true realist taking a more balanced view than many others have asserted.

Mao Tse-tung, *The Art of War* (El Paso, TX: El Norte Press, 2005). Somewhat overpraised and understudied, this work shows Mao at both his best (the true revolution caused by his disregard of Marx's central notion concerning the proletariat and its supposed decisive role in war) and worst (much pure propaganda). Many ideas lifted, without attribution, from Sun Tzu and also influenced by the writings of Clausewitz. The Clausewitzian trinity called for the use of violence to stir up the passions; those passions would give the political party the support it needed to generate the military means of success. The military strength gave the political leadership the ultimate triumph.

———, *On Protracted War* (Beijing: Foreign Language Press, 1966). A compilation of a series of speeches from 1938 outlining Mao's views concerning a protracted war as a way that small revolutionary groups could challenge the occupation of the Japanese (and, in time, the existing Nationalist Chinese government). Discusses the three phases of the struggle: (1) strategic defense, (2) strategic stalemate, and (3) strategic counteroffensive. In the first, the counterinsurgents were on the offensive and the insurgents on the defensive; in the second, the insurgents engaged in offensive guerrilla warfare; and in the third, the guerrillas go over to conventional warfare and the counterinsurgents are on the defensive. Mao asserts that the success of the revolutionaries was inevitable: "Therefore it should be repeated that the war will be arduous and protracted. But we are confident that, through the efforts of our Party and the whole people, the resistance already started will sweep aside all obstacles and continue to advance and develop."

Rafael Martinez, *Rome's Enemies (4): Spanish Armies* (Oxford: Osprey Books, 1986). Looks at the Iberian Peninsula (however defined) and the Celtic tribes that inhabited it during the times of Roman and Carthaginian expansion. Good analysis of various tribes and lifestyles and Roman efforts (largely successful over time) to subdue them. Not as much emphasis on the "Spanish sword" as seems desirable.

Brian McGing, *Polybius' Histories* (London: Oxford University Press, 2010). This marvelous study in the Oxford Approaches to Classical Literature series provides the reader with a very good overview of Polybius's 40-volume study of the rise of Rome. Like Xenophon—but unlike Thucydides—he has no problem inserting himself into the events that altered and illuminated the 2nd-century-BCE period. Of considerable importance for our purposes is book 18, in which Polybius gives the reasons for Roman success over the Greeks and, in the process (the battle of Cynoscephalae against the Macedonians had been fought and won in 167 BCE), he highlights the advantages of the Roman infantry sword formation over the Greek phalanx: The phalanx is only invincible in very specific circumstances where flexibility on the part of opponents is not possible, requiring as it does a "ground that is plain and naked and free from obstacles of every kind." He goes on,

> The Roman order, on the contrary is never attended even on such occasions with any disadvantage. Among the Romans every single solder, when he is once armed and ready for service, is alike fitted to engage in any time or place, or on any appearance of the enemy, and preserves always the same power and the same capacity in action whether in separate companies or man to man. As the parts, therefore, in the Roman order of battle are so much better contrived for use than those in the other, so success in action must also be greater in the one than the other.

Duncan Milne, "An Example of Force Development: Tukhachevsky and the Soviet Art of Deep Battle," in *The Changing Face of War: Learning from History*, edited by Allan D. English (Montreal: McGill-Queen's University Press, 1998), pp. 67–72. This remarkable man, a czarist soldier who was captured by the Germans, escaped, joined the Reds, and became an important general, defeating Admiral Aleksandr Kolchak in Siberia and Anton Denikin in the Ukraine, before running out of steam in Poland in 1920. He then played a key role in defeating the Kronstadt uprising. An organizer of the deep battle concept, he influenced Zhukov, the Soviet battle plans in Manchuria in 1945, and eventually the American air/land battle in Iraq in 1991, among others. Executed by Stalin in 1937.

George Modelski, "Kautilya: Foreign Policy and the International System in the Ancient Hindu World," *American Political Science Review*, no. 58 (1964), pp. 549–60. An early article "discovering" Kautilya and still worth reading today.

Richard E. Neustadt and Ernest May, *Thinking in Time* (New York: Free Press, 1986). Some interesting looks at decision making about war and the role of historical analogy in Korea, Vietnam, the Cuban Missile Crisis, the Bay of Pigs, and so on. Highlights E. G. Marshall's notion of making decisions with an eye to the future by seeing "time as a stream" with its "clear sense of the long past from which those futures will come."

Peter Paret (ed.), *Makers of Modern Strategy from Machiavelli to the Nuclear Age* (Princeton, NJ: Princeton University Press, 1986). A very wide-ranging but useful book introducing many who have thought long and hard about war.

———, *Understanding War: Essays on Clausewitz and the History of Military Power* (Princeton, NJ: Princeton University Press, 1992). A series of essays around the concepts of war as seen from Clausewitz and the concepts he raised.

Christian Potholm, *War Wisdom: A Cross-Cultural Sampling* (Lanham, MD: University Press of America, 2015). Twelve themes about war found in many cultures across time and space.

Richard Preston, Sydney Wise, and Herman Werner, *Men in Arms: A History of Warfare and Its Interrelationships with Western Society* (New York: Praeger, 1956). Goodness how these Western writers get carried away with the "uniqueness" of Western warfare. This Eurocentric study is quite weak on numerous dimensions of war, including the importance of the protection of capital from people and rulers and the assumption that there will always be another war. A once-over of classical, feudal, 18th- and 19th-century, and total warfare.

L. N. Rangarajan (ed.), *Kautilya: The Arthashastra* (Baltimore, MD: Penguin Books, 1992). The godfather of all realists can be studied with amazement and awe, lo these many centuries later. Kautilya really needs to come to the attention of more of those who truly study strategy and tactics over the ages.

Kamal Sadiq, "What's Morality Got to Do with It? Benevolent Hegemony in the International System of South Asia," *Comparative Studies of South Asia, Africa and the Middle East* 29, no. 2 (2009), pp. 6–21. How warfare can be detached from morality and how it was by the philosopher who made Machiavelli look like Mr. Rogers.

Maurice de Saxe, *Reveries on the Art of War* (Harrisburg, PA: Dover Books, 2007). One of the most delightful books on war ever written, not the least because of the back story on the author. Maurice was the "eldest of 354 acknowledged illegitimate children of Frederick Augustus, Elector of Saxony and King of Poland." Described as the "worst speller ever known," he entered military service at the age of 12 and became one of the great marshals of France under Louis XV. His many insights in this work ring true over the ages, emphasizing, as he does, "legs over arms," flexibility in battle, discipline (the "soul of armies"), the opening up of the battlefield, and combined arms with cavalry integrated with infantry. De Saxe also said, "Raising troops by fraud is an odious practice" and "Raising troops by force is still worse" in an age when most armies were recruited in such fashions.

———, *Reveries on the Art of War* (Westport, CT: Greenwood Press, 1971). Reprint of the 1757 work. In this original translation, Saxe writes in a difficult-to-read fashion with 18th-century words, phrases, and expressions, so it is a tough slog. But a worthwhile experience if you want to glimpse the 18th-century warfare from a fine practitioner thereof in his own words. Very specific on fortifications, attacks, war in the mountains, wars in forests, and so on. Reminds one of the detail provided by Kautilya in his *Arthashastra*.

See also, Joseph R. Riling, *Baron von Steuben and His Regulations* (Philadelphia: Ray Riling Arms Books, 1966). The complete facsimile of the original "Blue Book" regulations of 1779, which played such a major role in turning the Continental Army into professional soldiers after von Steuben joined Washington's command. The *Regulations* strongly accents "constant drilling," discipline, and small group cohesion. Excellent illustrations of all drills.

Thomas Schelling, *Arms and Influence* (New Haven, CT: Yale University Press, 1966). This is my favorite of his books. He examines the bargaining power of threat and the use of military force. Chapters include "The Principle of the Diplomacy of Violence," "The Idiom of Military Action," and such other notions as "the weapons speak."

———, *The Strategy of Conflict* (Cambridge, MA: Harvard University Press, 1964). The classic study of theories of strategy and conflict, including limited war, deterrence, and game theory contrasted with examples from everyday life. Seems oddly dated today but still introduces important concepts.

Eitan Shamir, *Transforming Command: The Pursuit of Mission Command in the U.S., British, and Israeli Armies* (Stanford, CA: Stanford University Press, 2011). An engaging account of the history of the Prussian-German *Auftragstaktik*, with its emphasis on mission control—the setting of the goals of the missions but allowing the field personnel to decide how to best achieve them. The author finds the process incompletely absorbed by all three militaries, although he suggests that the effort has stimulated reform.

Jon Sumida, *Decoding Clausewitz* (Lawrence: University of Kansas Press, 2008). A thick and convoluted but ultimately useful reappraisal of Clausewitz indicating that the master did not believe that victory in battle did not determine the outcome of all wars and argues that Clausewitz is very sound on guerilla warfare, for it is the attacker (i.e., the occupier) who is eventually at a disadvantage from the defender (the population and the guerillas leaders), with *will* as the key. Quotes Herman Hesse at the end concerning the study of history, "It is a very serious task . . . and probably a tragic one."

Sun Tzu and Sun Pin, *The Complete Art of War* (New York: HarperCollins, 1996). An all-time classic study of war, still so obviously relevant that it has become the bible for the US Marines of today. Needs to be read in conjunction with Clausewitz, however, in order to get the true "go" of warfare.

Romila Thapar, *Asoka and the Decline of the Mauryas* (Oxford: Oxford University Press, 1961). Some evidence that a would-be conqueror should not turn to Buddhism if he wants to stay in that ballgame. But a very useful introduction to the life and times of Asoka and the amazing empire he expanded.

Thucydides, *The Peloponnesian War* (New York: Modern Library, 1982). The father of history analyzes war in a seemingly most modern way. A must-read for anyone interested in learning about the nature of war. It is a truly magical work. See especially the sections covering the war in Sicily (books 5–7) and the Melian dialogues (in book 5, chapter 17).

Note: Strongly recommended as a companion piece to this work is the *Landmark Thucydides*, edited by Robert Strassler (New York: Free Press, 1996). The concise margin notes and especially the myriad of excellent maps really enrich one's understanding of the scope and intricacy of the war.

Richard Simpkin, *Deep Battle: The Brainchild of Marshal Tukhachevskii* (London: Brassey's, 1987). This extraordinary man, confidant to Trotsky, Lenin, and Stalin (who had him shot in 1937 for supposedly plotting a coup), developed the Soviet notion of deep battle, an offensive doctrine that used a holding force and a turning force, with a hinge between to move around the enemy in ever-increasing deep penetrations. Called for "ever strong tank armies in high-tempo maneuver at greater and greater depth." Strategic blitzkrieg and forerunner of the US air/land battle. An innovative genius of envelopment. According to the author, Tukhachevsky was "streets ahead" of the German blitzkrieg proponents, as well as Western postwar military thinkers.

Martin van Creveld, *The Culture of War* (New York: Ballantine Books, 2008). Shows in many, many ways how deep war is embedded in the human psyche and why the movie *The Hurt Locker* turns out to be one of the most perceptive of all war movies in this regard. There are some men and women (and we really don't know how many) who feel themselves only truly alive when they are in harm's way and doing battle.

———, *The Transformation of War: The Most Radical Reinterpretation of Armed Conflict since Clausewitz* (New York: Free Press, 1991). A fine counter to Clausewitz, although he goes too far in claiming Clausewitz's "irrelevancy" because the essence of war—the contest of wills—has not changed at all. Also, van Creveld drastically underestimates the military revolution of unmanned aerial vehicles and other high-tech weapons.

———, *Wargames: From Gladiators to Gigabytes* (Cambridge: Cambridge University Press, 2013). A comprehensive overview of the ways humans have attempted to abstract the elements of warfare and use them as a substitute for war and a means of preparing for it, as well as for the enjoyment of ritualized conflict. Seeks—and largely succeeds—in embedding war even more deeply in the human collective history.

Barry Watts, *Clausewitzian Friction and the Future of War* (Washington, DC: Institute for National Strategic Studies, 1996). Argues persuasively that the technological revolutions sweeping military possibilities will not eliminate Clausewitzian friction due to human limits, process limits, decision loop limits, and the increasing presence of stress.

Harry Yarger, *Strategic Theory or the 21st Century: The Little Book on Big Strategy* (Carlisle, PA: Strategic Studies Institute, 2006). A short (76-page) monograph arguing, "Strategy is fundamentally a choice." He says it is proactive, subordinate to policy, and subordinate to the nature of the environment and requires a holistic perspective, among other things. In fact, the perspective is so holistic that it is almost impossible to imagine any policy maker being able to take it all into account with acute paralysis.

Chapter Four

Geopolitics and the Long Shadow of Geography

"The rule of the sea is indeed a great matter."—Pericles, *The Peloponnesian War*

"Give me six hours control of the Strait of Dover and I will gain mastery of the world."—Napoleon, *Dreadnought*

"Control of the sea means security. Control of the sea means peace. Control of the sea can mean victory."—J. F. Kennedy, *Sea Power*

"The stopping power of water."—John Mearsheimer, *The Tragedy of Great Power Politics*

George Baer, *One Hundred Years of Sea Power: The US Navy 1890–1990* (Stanford, CA: Stanford University Press, 1994). Another study showing the long shadow of Alfred Thayer Mahan and his concept of sea power and geopolitics, as well as the US Navy's efforts to keep relevant at various stages of history. The demise of the Soviet Union seemed to undercut the need for many of its tenets, but the espousal of Mahan's principles by the Chinese military in recent years has given the US Navy a huge new lease on strategic life.

Bernard Brodie, *A Guide to Naval Strategy* (Princeton, NJ: Princeton University Press, 1944). Looks at sea power in the modern world. Essentially World War II–centric but argues for the continuing verities of the importance of sea power, seaborne communications, maritime supply, and command of the seas, concluding, "We must remember always that the basic element of strength in any Nation is not its machines but in its manhood."

———, *Sea Power in the Machine Age* (Princeton, NJ: Princeton University Press, 1941). Written before Pearl Harbor, this volume looks at steam warships, iron-hulled warships, new armor and new ordinance, torpedoes, mines, submarines, and aircraft, giving the reader a sense that an era has passed and another one is ongoing, but Jomini's words are still telling: "Methods change, but principles are unchanging."

Julian Corbett, *Some Principles of Maritime Strategy* (London: Longmans Green, 1911). The author of a number of books on naval strategy, including *Drake and the Tudor Navy* and *The Campaign of Trafalgar*, Corbett provides a look at grand strategy, applying the principles of Clausewitz and Jomini to the naval warfare and exploring the "theory that the command of the sea must always be the objective"; the "theory of means," or the "constitution of the fleets"; and the "theory of method," or the concentration, dispersal, and destruction of forces in the tradition of the "weapons speak." A classic. The highlights:

1. Be flexible in naval strategy (defensive and offensive).
2. Accent combined arms (British navy enables Wellington to succeed in Portugal and Spain).
3. Use local superiority to gain operational advantage; don't rely on a single climactic battle.
4. Think tactically, operationally, and strategically in a seamless web.

Note: Corbett believes, "Theories of war can only increase the effective power of conduct," but we must guard "against over-valuation," and "naval strategy" (which determines where and how the fleet should be deployed) is different from "maritime strategy," which involves a "war in which the sea is a substantial factor."

Note: US naval strategy in the Pacific during World War II had elements of Mahan (Midway) but more of Corbett (Guadalcanal).

Seth Cropsey, *Mayday: The Decline of American Naval Supremacy* (New York: Overlook Duckworth, 2013). An interesting combination of naval history, naval theories, the rise and fall of naval powers, and a clear and even poignant call for revitalization of the US Navy (declining rapidly to fewer than 300 ships, with no end in sight despite rising international obligations and threats). Some points to ponder: "No state that has allowed its sea power to decline has succeeded at recovering it"; Britain's 40-ship navy is the same size today as it was when Henry VIII was king in 1547; stabilizing the American shipbuilding industry is the key to the future; and a fresh look at alternatives in the area of force projection, such as much smaller, faster carriers and a slew of attack submarines to threaten Chinese global commerce. Interesting potpourri of observations are in this volume.

Colin S. Gray, *The Leverage of Sea Power: The Strategic Advantage of Navies in War* (New York: Free Press, 1992). Following Mahan, the author argues that "sea power has been the midwife of victory" and that "maritime command or control is thus critically important." He combs history, from the failure of Athens at Syracuse through the Carthaginians, the Romans, Napoleon, Hitler, and the Soviets, to show how "secure seas for friendly use and deny their use to the enemy" is crucial. Even uses the example of Byzantium, which was only taken three times from 1200 to 1453: the first two times during the Fourth Crusade and finally by the Turks in 1453. Cites Herbert Richmond on World War II, "Sea power did not win the war itself; it enabled the war to be won," as a fitting overview. Considerable relevance for Kaplan and others looking at the United States vis-à-vis China and India in the coming decades.

Colin S. Gray and Geoffrey Sloan (eds.), *Geopolitics, Geography and Strategy* (London: Frank Cass, 1999). Thirteen essays on a number of topics ranging from Mackinder and Mahan to the Gulf War and astropolitical analysis. Gray's essay "Inescapable Geography" lays down the intellectual markers: "All politics is geopolitics, all strategy is geostrategy." Williamson Murray also highlights the fact that "[b]oth history and geography will have their revenge on U.S. military forces, strategy and policy in the next century." Prophetic.

Russell Grenfell, *Sea Power* (Garden City, NY: Doubleday, 1941). "Sea Power in Defense," "Sea Power in Offence," and sea power all around, but literally as this was written, the "air menace" was raising its ugly head. But the author predicted that Great Britain would not be conquered by Germany unless German air superiority overcame British naval superiority and maintained the "sovereignty of the seas." Made an interesting point that, for the cost of maintaining the 400,000 British soldiers on the continent, the country could have had 6,000 aircraft, and in the summer of 1941, this seemed like an excellent trade. Agreed, although the French might not.

Robert Kaplan, *Monsoon: The Indian Ocean and the Future of American Power* (New York: Random House, 2010). Pointing out that 90 percent of all goods flow by water and that such key choke points as the straits of Hormuz and Malacca are only open because of hard power and force projection, this work is an admirable example of the continuing impact of geography-centric analysis of global warfare. Many excellent maps showing Chinese efforts to create de facto Chinese enclave ports. In 2012, they ran the most efficient portion of Piraeus in Greece!

———, *The Revenge of Geography: What the Map Tells Us about Coming Conflicts and the Battle against Fate* (New York: Random House, 2012). Contains an excellent overview of the 19th- and 20th-century literature on geopolitics, as well as his reading of the current tealeaves from a variety of viewpoints: Chinese, Indian, Iranian, and US Mackinder, Mahan, Spykman, and others are introduced to the reader in manageable hunks.

Gerry Kearns, *Geopolitics and Empire: The Legacy of Halford Mackinder* (London: Oxford University Press, 2009). The World Island of Europe/Asia became Mackinder's central space-time continuum and led to the formulation that "Who rules east Europe commands the Heartland; Who rules the Heartland commands the World Island (Europe, Asia and Africa). Who rules the World Island commands the World." Mackinder believed in a kind of geopolitical Darwinism. In a closed-space world, it is a zero-sum game for territorial acquisitions, democracy versus authoritarianism, sea power versus land power. The long shadow of his thought continues—from imperialism and Nazi expansionism (also Friedrich Ratzel's "living space" Lebensraum), the Cold War, and now the US–China–India struggle for raw materials.

John Keegan, *The Price of Admiralty: The Evolution of Naval Warfare* (New York: Penguin Books, 1988). A most enlightening overview of sea power from an Anglo-American perspective, the interested reader will find much to admire and enjoy. Unfortunately the case studies—Trafalgar, Jutland, Midway, and the Battle of the Atlantic—are confined to the "modern" period of naval strategy.

Paul M. Kennedy, *The Rise and Fall of British Naval Mastery* (New York: Charles Scribner's Sons, 1976). A classic of this genre. Looks at the elements of sea power and traces the rise of Britain though Pax Britannica and the end of its maritime supremacy due to strategic overreach (Liddell Hart's "strategic overextension") *and* the coming true of the prophecies of Halford Mackinder (as opposed to Alfred Mahan), who said that the future lay with continental powers (United States, Russia, and now China) rather than peripheral, sea-trading nations.

Halford Mackinder, *Democratic Ideals and Reality* (New York: Henry Holt, 1942). Reprint of Mackinder's 1919 work, reprising his notions of "The World Island" and "The Heartland" after World War I. Mackinder divided the world into the Heartland, the Inner or Marginal Crescent (which would later become Nicholas J. Spykman's Rimland), and the Offshore Islands and

Continents. He ends a lengthy discourse on empire and individuals with a plea for the League of Nations and a "global neighborliness." This part is less helpful for our purposes. Mackinder is also most famous for his dictum:

> Who controls eastern Europe rules the Heartland;
> Who controls the Heartland rules the World Island;
> Who rules the World Island rules the World.

A. T. Mahan, *The Influence of Sea Power upon History 1660–1783* (New York: Dover, 1987). A reprint of the original 1890 work, this volume captures the essence of Mahan's importance as a point of reference for sea power and its role in history. Reflecting Mahan's ongoing interest in grand strategy (the goal of national naval policy and strategy and their international context), as well as the science and art of command (the nature of operational decision making in war), the work argues that "travel and traffic by water have always been easier and cheaper than by land." In addition, throughout history, naval superiority, even supremacy, has been essential to the protection of national interests. A country's ability to create and maintain sea power is affected by six principal conditions: (1) extent of territory, (2) population size, (3) national culture, (4) political structure, (5) geographical position, and (6) such physical configurations as natural resources and climate. German Kaiser Wilhelm II put a copy of this work on every German warship!

Note: For a useful analysis of Mahan, see Jon Sumida, *Inventing Grand Strategy and Teaching Command: The Classic Works of Alfred Thayer Mahan Reconsidered* (Baltimore, MD: Johns Hopkins University Press; Washington, DC: Woodrow Wilson Center Press, 1997).

Note: Mahan looked for the decisive surface naval battle à la Clausewitz on land and focused on control of choke points and imperial holdings to prepare for war. His thinking greatly stimulated Germany prior to World War I and Japan prior to World War II, although in both wars, the decisive battles involved far more than naval supremacy in terms of battleships (i.e., submarines and aircraft).

Arthur Marder, *From the Dreadnought to Scapa Flow*, 5 vols. (Oxford: Oxford University Press, 1961–). Together with his three-volume set *Fear God and the Dreadnought*, this author wrote the definitive account of the Royal Navy, for which Lord Mountbatten called him the greatest of naval historians. Incredibly detailed and densely written but with a key eye for insights and irony. Quite a piece of work but now tremendously understudied.

Herbert Richmond, *Statesmen and Sea Power* (London: Oxford University Press, 1946). The author, a British admiral, takes the reader through the history (from Elizabethan times to "The German Wars") of the interaction among the sea, geography, and the statesmen charged with charting a course for British strategic thinking. Quite a good read.

Nicholas J. Spykman, *America's Strategy in World Politics: The United States and the Balance of Power* (New York: Harcourt Brace, 1942). Follower and critic of Mackinder, he states, "Geography is everything," and thus sees the United States as most favored because of direct access to Caribbean, Atlantic, and Pacific. Stresses the importance of the 20–60 north latitude degree belt, the temperate zone. Thinks globally: "Global war, as well as global peace means that all fronts and all areas are interrelated" and stresses the unity of world politics and the world sea. Believed that, after the war, balancing Germany and Russia was the most important dimension and offered the greatest danger.

———, *The Geography of the Peace* (1944). Believes that "Who controls the Rimland rules Eurasia; Who rules Eurasia controls the destiny of the World." Interestingly enough, he saw China (rightly) as not strong enough to control or even to attempt to control the Asian rimlands but would certainly have picked up his ears at hearing today's China talking about the geography of expansion! Because "who control the rimlands of Europe and Asia is the single most important element in the U.S. security situation." His focus on Latin America seems timely with the situation in Mexico deteriorating yearly and Chinese inroads for natural resources growing.

Hew Strachan, *The Direction of War: Contemporary Strategy in Historical Perspective* (Cambridge: Cambridge University Press, 2012). Excellent scope and excellent perspective. Highlights the importance of history for understanding that strategy is about doing things, applying means to ends, not just an end unto itself, as is often the case in "strategic studies." Lots of good Clausewitz and Thucydides "asymmetrical warfare is as old as warfare itself." Strategy is always relative to time and place, but its verities are not. A very useful analysis, one of the better ones in this section.

Robert Strausz-Hupe, *Geopolitics: The Struggle for Space and Power* (New York: George Putnam's Sons, 1942). Saw protracted conflict in the rise of Germany as a threat to both England and the Soviet Union as countries vie—as they have always vied—for the Heartland. A strong global realist, "Geopolitics and the competition for space is enduring—a never ending battle."

Chapter Five

The Template of Mars

After studying war for more than 40 years, I became convinced of the need to develop a conceptual framework, one that could enable scholars to examine the essence of warfare across time and space and various cultures. In the process, I developed the Template of Mars, using the trope of Mars, the god of war. If there were a god of war, I thought, which actions and activities would he or she reward and which would he or she punish?

It was not until I was writing the second volume in this trilogy on war that it dawned on me how significant the notion of Mars is just in terms of the impact of war on religion. While some religions, such as Islam and Hinduism, have war-making lardered into their cosmological insights, even extremely pacific religious aspects of Christianity and Buddhism were all, over time, overcome by the demands of Mars. In terms of war-making, all the major religions have eventually and at least periodically been overcome by the human propensity for war—and for drawing one's gods into one's combats.

In terms of the process of exploring the nature of success in war, I examined seven key variables by which one would examine war regardless of culture or age. These variables include (1) superior weapons and technological entrepreneurship, (2) superior discipline, (3) sustained but controlled ruthlessness, (4) receptivity to military and integrative innovation, (5) the ability and willingness to protect capital from people and rulers, (6) the centrality of superior will, and (7) the belief that there will always be another war.

A word of caution: Many commentators on warfare often stress one of these elements or the other, but in truth, over time, Mars has demanded adherence to these various but interrelated dimensions, and high performance in one area cannot compensate for a lack of attention to the other dimensions.

These are best studied as interlocking and mutually reinforcing: See Christian P. Potholm, "Introduction: The Template of War," in *Winning at War: Seven Keys to Military Victory throughout History* (New York: Rowman and Littlefield, 2009), pp. 1–30. Some additional readings on the subject as well as the sources of many of this chapter's footnotes (see pages 225–26 of *Winning at War*) include:

Michael Howard, *War and the Liberal Conscience* (New Brunswick, NJ: Rutgers University Press, 1978). The liberal worldview that reason and action can change the world for the better is challenged by the evidence of World War I, Fascism, and Communism.

Robert Kaplan, *Imperial Grunts: The American Military on the Ground* (New York: Random House, 2005): "Injun Country," pp. 3–16, "Yemen," pp. 17–38, and "Mongolia," pp. 91–130. Kaplan provides the reader with an ongoing and perspicacious set of insights into the American Empire in the field. A must-read.

SUPERIOR WEAPONS AND TECHNOLOGICAL ENTREPRENEURSHIP

An important—but often exaggerated—element of war-making that captures the attention of students of war is superior weapons technology. Which side has the best weapons and which side makes the best use of innovation in weapons and weapons systems are often the focus of scholarly attention. Yet, it is very important to remember that the mere possession of superior weapons is less important than the extent to which those weapons are integrated into the polity's armed forces and its warriors are trained effectively in their uses. It is also axiomatic that most militaries are both conservative and risk adverse when it comes to the acquisition of new weapons, and the integration of newer technologies are often impeded by the very militaries that could benefit from them the most. Superior weapons technology, therefore, needs to be examined in terms of both its relevance for the military in question and its local and regional opposition. For an introduction to this topic, see chapter 2, "Superior Weapons and Technology Entrepreneurship," in *Winning at War*, pp. 31–50. Additional readings on the subject, as well as the sources of many of this chapter's footnotes (see pp. 226–31), include:

David A. Armstrong, *Bullets and Bureaucrats: The Machine Gun and the United States Army, 1861–1916* (Westport, CT: Greenwood Press, 1982). Fascinating study of how the US Army failed to see the potential of the

Gatling, Maxim, and machine gun until World War I thrust its killing power upon the battlefield.

Max Boot, "Triumph of Prussian Technology and Tactics," *MHQ: The Quarterly Journal of Military History* 19, no. 1 (Autumn 2006), pp. 56–57. The Prussian "needle gun" helps win the Austro-Prussian War and revolutionizes infantry warfare.

Brian M. Carney, "Air Combat by Remote Control," *Wall Street Journal*, May 12, 2008, A-13. UAVs are beginning to be appreciated by the mainstream press.

George Crile, *Charlie Wilson's War: The Extraordinary Story of the Largest Covert Operation in History* (New York: Atlantic Monthly Press, 2003). How the Stinger missile helped to turn the tide against the Soviet forces in Afghanistan.

Robert Drews, *The End of the Bronze Age: Change in Warfare and the Catastrophe, ca. 1200 B.C.* (Princeton, NJ: Princeton University Press, 1993). An amazing recreation of revolution in military affairs brought on by the iron sword wielded by infantrymen in an age when chariot nobility dominated.

John Ellis, *The Social History of the Machine Gun* (New York: Random House, 1975). A startlingly incisive look at the history of the gun, which wreaked such havoc on infantry formations.

Felipe Fernandez-Armesto, *Pathfinders: A Global History of Exploration* (New York: Norton, 2007). How the explorers in a variety of societies changed the history of humankind, not always for the better.

Julia Keller, *Mr. Gatling's Terrible Marvel* (New York: Viking, 2008). A truly game-changing weapon that led to a revolution almost as great as that created by the bolt action rifle itself.

Ben Kiernan, *Blood and Soil: A World History of Genocide and Extermination from Sparta to Darfur* (New Haven, CT: Yale University Press, 2007). Not a pretty picture; in fact, one that will depress you to no end. Humans can truly be horrid and horrible and rather easily led to do very, very bad things—so Kiernan argues.

Ian Knight, *The Zulus* (Oxford, UK: Osprey, 1989). Many fascinating pictures and drawings featuring the rise of the Zulu nation and its immense outreach known as the Mfecane, "the Crushing." Chaka really was quite a military genus.

John Laband, *The Rise and Fall of the Zulu Nation* (New York: Arms and Armour, 1997). An amazing story that should be taught to all those who would understand warfare and the relativity of power.

Angus McBride, *The Zulu War* (Oxford, UK: Osprey,1976). A short but well-illustrated account, especially useful for showing troop movements and the British square at Ulundi.

Donald R. Morris, *The Washing of Spears* (New York: Simon and Schuster, 1965). The best and liveliest account of the Zulu nation and its warfare. Written at a time when Eurocentric analysis of warfare dominated the academy, this was an important advance in terms of both depth and scope.

Peter Padfield, *Maritime Supremacy and the Opening of the Western Mind* (Woodstock, NY: Overlook Press, 2000). A groundbreaking look at how the West opened its mind and went out to conquer the world. Not that the world asked to be conquered, you understand.

Constance Pleshakov, *Tsar's Last Armada: The Epic Voyage to the Battle of Tsushima* (New York: Basic Books, 2005). A truly monumental failure and a firm indication that the dictates of Mars could be adopted by cultures other than those in the West.

Darrell H. Semitis, "Japanese Naval Transformation in the Battle of Tsushima," *Military Review* (November–December 2004), pp. 261–319. How the Japanese worshiped at the shrine of Mars, and it paid off hugely in this crucial battle. Belies the notion that worshiping Mars is or was confined to Europeans.

P. W. Singer, *Wired for War: The Robotics Revolution and Conflict in the 21st Century* (New York: Penguin, 2009). Some interesting insights into the newest revolution in military affairs.

John M. Taylor, "The 'Japanese Nelson' Crushes the Russians," *MHQ: The Quarterly Journal of Military History* 21, no. 2 (Winter 2009), pp. 8–19. Tsushima was the battle, and Admiral Togo was the leader who revolutionized the Japanese military and directed this savage defeat of the Russians, who had sailed halfway around the world to lose their entire fleet.

Leonard Thompson, "The Zulu Kingdom and the Mfecane," in *A History of South Africa* (New Haven: Yale University Press, 1996), pp. 80–87. The rise of the Zulu nation, their revolution in military affairs, and their startling impact on southern Africa.

Peggy Warner, *The Tide at Sunrise: A History of the Russo-Japanese War, 1904–5* (London: Frank Cass, 1974). A good overview of the entire war, which presaged so much of the future course of armed conflict.

SUPERIOR DISCIPLINE

> "Collective defense can be so compelling—so addictive, in fact—that eventually it becomes the rationale for why the group exists in the first place."—Sebastian Junger, *War*

> "Love, or bond of comradeship, 'Here is Bill; I will go or stay as he does.'"—Joshua Chamberlain

> "Sir, I'm getting a hell of a lot more sleep than I did in Germany on training exercises."—2nd Armored Cavalry Regiment tanker during the VII Corps battles smashing the Republican Guards in 1991

> "The coward's fear of death stems in large part from his incapacity to love anything but his own body. The inability to participate in others' lives stands in the way of his developing any inner resources sufficient to overcome the terror of death."—J. Glenn Gary, *The Warriors*

As obvious as it may seem, superior discipline in warfare is often overlooked or underestimated when theoreticians and practitioners look at the ingredients for success in warfare. For me, discipline remains one of the most important, if not the most important, ingredient in the Template. It is not simply the following of orders but also the following of which orders would have been given in the absence of a higher military authority.

Throughout time and space, superior discipline has proven itself over and against superior technology or superior will, and the essence of superior discipline remains central to the study of warfare. As the autodidact Chinese general Zeng Guofan put it while he was defeating the Taiping Rebellion, "The way to achieve victory is to be found in men, not arms."

For an introduction to this topic, see chapter 3, "Superior Discipline, *Winning at War*, pp. 51–68. An interesting variation on the role of hunting in promoting discipline is found in Tim Blanning, *The Pursuit of Glory* (reviewed later), especially "Court and Country," pp. 393–422. Additional readings on the subject, as well as the sources of many of this chapter's footnotes (see pp. 234–37), include:

Donovan Campbell, *Joker One: A Marine Platoon's Story of Courage, Leadership and Brotherhood* (New York: Random House, 2009). Takes the reader through the melding of a unit and its transformation in Iraq during the battle for Ramadi, when his unit, Joker One, goes from being regarded as an *awat* (Arabic term for soft, sugary cake that crumbles easily to the touch) to the strongest tribe.

Bruce Condell and David Zabecki (eds.), *On the German Art of War: Truppenfuhrung* (Boulder, CO: Lynne Reiner, 2001). Why the Germans, whether on the defensive or the offensive, against this foe and that foe, inflicted 20 to 50 percent more casualties on their opponents during World War II. A valuable look at the hows and whys of success in warfare and how it became institutionalized in one armed force.

Dora L. Cost and Matthew E. Kahn, *Heroes and Cowards* (Princeton, NJ: Princeton University Press, 2008). Which stresses, strains, and reinforcements help to make individuals into one or the other in warfare.

Martin van Creveld, *The Art of War: War and Military Thought* (London: Cassell, 2000). A tour of some of the major military thinkers from a variety of cultures, albeit still heavy on European views of the phenomenon.

———, "Educating War" and "Games of War," in his *The Culture of War* (New York: Ballantine Books, 2008), pp. 6–84. How, and why, war is so tenaciously intertwined with culture.

———, *Fighting Power: German and U.S. Army Performance, 1939–1945* (Westport, CT: Greenwood Press, 1982). This Israeli soldier professor comes to the conclusion that the Germans, however hateful their racial views, were more effective at fighting than their opponents in virtually all types of situations.

Jared Diamond, *Guns, Germs, and Steel: The Fates of Human Societies* (New York: Norton, 1999). Food production, guns, germs, and steel help explain the differing rise and fall of complex human societies, with some gaining at the expense of others, usually to their considerable detriment.

Nathaniel Fick, *One Bullet Away: The Making of a Marine Officer* (New York: Houghton Mifflin, 2005). Dartmouth graduate serves a tour in Afghanistan and one in Iraq and says that his training in the California desert was more difficult.

Frederick the Great, *The Art of War* (New York: Da Capo Press, 1999). Rediscovering the Roman virtue of discipline.

Felix Gilbert, "Machiavelli: The Renaissance of the Art of War," in *Makers of Modern Strategy from Machiavelli to the Nuclear Age*, edited by Peter Paret (Princeton, NJ: Princeton University Press, 1986), pp. 1–31. Good and succinct analysis of Machiavelli and his ideas about warfare.

Christopher Hamner, *Enduring Battle: American Soldiers in Three Wars, 1776–1945* (Lawrence: University of Kansas Press, 2011). Faced with "gruesome violence" and the "unavoidable circumstance of combat," how soldiers cope. Looking at the Battle of Cowpens (1781), Shiloh (1852), and the Huertgen Forest (1944), the author traces the changes in tactics—from linear to group dispersal—and finds that training and rote muscle memory, leadership, comradeship, and being well armed all help soldiers cope, but he remains amazed "at how many endured combat for so long." Me too.

Flavius Josephus, *The Jewish War*, translated by G. A. Williamson (New York: Penguin Books, 1959). An insightful look at warfare in the Roman world by an eyewitness. With only a little imagination, the reader can see the legacy of sustained ruthlessness that endures in the region and beyond.

Ian Knight, *The Zulus* (Oxford, UK: Osprey, 1989). Many fascinating pictures and drawings featuring the rise of the Zulu nation and its immense outreach known as the Mfecane, "the Crushing." What a military genius Chaka was.

John Laband, *The Rise and Fall of the Zulu Nation* (New York: Arms and Armour, 1997). An amazing story that should be taught to all those who would understand warfare and relativity.

Angus McBride, *The Zulu War* (Oxford, UK: Osprey, 1976). A short but well-illustrated account, especially useful for showing troop movements and the square at Ulundi.

J. E. Lendon, "Roman Siege of Jerusalem," *MHQ: The Quarterly Journal of Military History* 17, no. 4 (Summer 2005), p. 7. How the siege was actually accomplished.

John A. Lynn, "Soul of the Sepoy," *MHQ: The Quarterly Journal of Military History* 17, no. 2 (Winter 2005), pp. 46–55. Exporting the virtues of discipline and drill and their successful adaptation in India.

Alan Marshall, *Oliver Cromwell—Soldier: The Military Life of a Revolutionary at War* (London: Brassey's, 2004). Quite possibly the Royalists would have won the English Civil War had not Cromwell created the New Army with a conscious emphasis on discipline.

Evan Mawdsley, *The Russian Civil War* (New York: Pegasus Books, 2008). See especially pages 3–44. The Bolsheviks are losing the war until they take discipline and officer talent from the playbook of the Old Regime.

Lord Moran, *The Anatomy of Courage* (London: Constable, 1945). Based on his service as a British medical officer in World War I at the front, the author seeks to find out "how courage is born and how it is sustained in a modern army of free people." He concludes that courage is not common and has to be taught—by selection, discipline, pride in arms and leadership—concluding, "But in this life the individual shrinks to nothing, he has no longer the right to an opinion, only loyalty to the regiment matters."

Williamson Murray, MacGregor Knox, and Alvin Bernstein (eds.), *The Making of Strategy: Rulers, States, and War* (Cambridge: Cambridge University Press, 1994). A wide-ranging and insightful series of essays ranging from the Peloponnesian War, Chinese 14th-century strategy, England, France, the United States, and Germany. Strong Eurocentric focus.

Michael Reynolds, *Men of Steel* (New York: Sarpedon, 1999). The SS are justly reviled for their murders and concentration camp genocide, but the author believes Waffen-SS field divisions fought so well in Normandy that their training and élan need to be studied and understood by viewing their performance in action.

Kaushik Roy, "Military Synthesis in South Asia: Armies, Warfare, and Indian Society, c. 1740–1849," *Journal of Military History* 69 (July 2005), pp. 651–90. A much-needed look at how local Indian military traditions and those of the imported British and French were blended by various groups in the 18th and 19th centuries.

C. P. Stacey, *Quebec, 1759: The Siege and the Battle*, edited by Donald Graves (Toronto: Robin Brass Studio, 1959). Generals Wolfe and Montcalm battle for North America; a detailed window into the art of war in Canada, mid-18th century. There was more to the winning than just the battle on the Plains of Abraham as the siege and bombardment brought Montcalm to come out for a pitched battle. The author feels that "disciple and graining had their true effect. This was the fundamental factor." In other words, Wolfe's troops were more disciplined and professional.

US Marine Corps, *Warfighting* (Washington, DC: Department of the Navy, 1997). The Marine Corps has discovered Sun Tzu, and it shows. Quite a change in strategy from "hi diddle diddle, straight up the middle," which prevailed as late as 1991.

Bing West and Ray L. Smith, *The March Up: Taking Baghdad with the 1st Marine Division* (New York: Bantam Books, 2003). A good, fast read about the fastest and longest blitzkrieg in history, with superior discipline at the heart of it.

Admiral Sandy Woodward, *One Hundred Days: The Memoirs of the Falklands Battle Group Commander* (Annapolis, MD: Naval Institute Press, 1992). Give the high technology of the Argentines, their interior—and near-by—lines of communications, and other advantages, the commander believes it was a "dammed close run thing." States that if the Argentines had sunk or severely damaged the British aircraft carrier *HMS Hermes*, then the invasion would have had to be called off. Discipline and training made the difference in the Falkland (Malvinas) War.

Evan Wright, *Generation Kill* (New York: G. P. Putnam's Sons, 2004). The *Rolling Stone* reporter nailed this one, an outstanding portrait of men at war. The author originally was disdainful of the "killers" he was riding with, but on the lunge to Baghdad in 2003, he felt safest with the lad who fired the quickest and the most rounds in the unit, although though the hair-trigger lad may have killed a few camels and even some Iraqi civilians in the process.

SUSTAINED BUT CONTROLLED RUTHLESSNESS

Chapter 4, "Sustained but Controlled Ruthlessness" in *Winning at War*, pp. 69–102, is the longest chapter in the book and one of the most difficult for many to accept, as it deals with the mass killing that accompanies all war. Many find it difficult to see war in its true amoral light, preferring to believe that war can be waged without civilian casualties and without cruelty. But as W. T. Sherman so famously put it, "War is cruelty. There is no use trying to reform it. The crueler it is, the sooner it will be over."

Yet, indiscriminate brutality may have a reverse impact, delaying the end of a war by provoking a response that leads to more and more killing on both sides. Therefore, it is important to add *controlled* to the lexicon of Mars and ending up, as this chapter does, by focusing on the willingness to do whatever is necessary to prevail, even if that means cruelty to one's enemy and even one's own people.

This vital and often misunderstood dimension of warfare must be carefully explored and examined. Some additional readings on the subject, as well as the sources of many of this chapter's footnotes in *Winning at War* (see pp. 237–43), include:

Paul Allen, "Subduing the Seminoles," *MHQ: The Quarterly Journal of Military History* 12, no. 3 (Spring 2000), pp. 54–63. These were some stubborn Native Americans and could have served as role models for others.

Thomas B. Allen and Norman Polmar, "Gassing Japan," *MHQ: The Quarterly Journal of Military History* 10, no. 1 (Autumn 1997), pp. 38–45. The US military had other weapons ready to go if the atomic bombs didn't work. And always remember that General Curtis LeMay considered the atomic bomb an interruption to his in-process plans to saturate the country from one end to the other with liquid fire. LeMay's plans, if carried out, would have killed far more Japanese than Hiroshima and Nagasaki did.

Fred Anderson, *Crucible of War: The Seven Years War and the Fate of Empire in British North America 1754–1763* (New York: Alfred Knopf, 2000). One of the first "world wars" is fought out in Europe, Asia, the Caribbean, and North America.

Nicolson Baker, *Human Smoke: The Beginning of World War II, the End of Civilization* (New York: Simon and Schuster, 2008). Author believes all strategic bombing of civilians is bad and immoral and doesn't work. Baker has never liked strategic bombing, and it shows. He is justifiably outraged over the use of area bombing but not very persuasive about its alternatives. Baker seems bound and determined at all points to avoid the realities of the dictum "War serves war."

David Bergamini, *Japan's Imperial Conspiracy* (New York: William Morrow, 1971). A huge, wide-ranging, even sprawling indictment of Emperor Hirohito and his complicity in the Japanese atrocities during World War II. A convincing case is made and placed in historical context with the imperial heritage on display from 50 CE onward. With hindsight, it seems clear that the emperor was very fortunate to have General Douglas MacArthur as his protector and to have the Soviet Union beginning the Cold War when it did.

Laurence Bergreen, *Columbus: The Four Voyages* (New York: Viking, 2011). An in-depth look at the life and times of Columbus, providing the first extensive biography of the last 50 years. Estimates that, as a result of the arrival of the Spanish in the Caribbean, 50,000 Native Americans took their own lives. This book is not for the faint of heart.

Dale Blair, *No Quarter: Unlawful Killing and Surrender in the Australian War Experience, 1915–1918* (Charnwood, Australia: Ginninderra Press, 2005). A forthright account of Australian killings of surrendering Germans, surrendered Germans, wounded Germans, and Germans in general during the trench warfare of World War I, justified as "men caught up in the madness of war."

Martin Caidin, *The Night Hamburg Died* (New York: Ballantine Books, 1960). A somewhat lurid yet sobering account of the 10 days of bombing during July 1943 that reduced Hamburg to rubble in a fiery inferno, killing more than 75,000 people. He compares the 100 acres of Coventry utterly destroyed with the 6,000 acres of Hamburg. The author points out that "Operation Gomorrah" produced an effect that "[n]ot even Hiroshima and Nagasaki . . . could match the utter hell of Hamburg" because it went on for so long. Indeed, Albert Speer personally went to Hitler and said that a half-dozen cities treated in this fashion would have ended the war.

Richard J. Chacon and Ruben G. Mendoza (eds.), *North American Indigenous Warfare and Ritual Violence* (Tucson: University of Arizona Press, 2007). Ritual violence saved Native American lives, until they came up against Europeans in war, and then it cost them many, many lives.

Thomas M. Coffey, *Iron Eagle: The Turbulent Life of General Curtis LeMay* (New York: Thomas Crown, 1986). A life devoted to proving Nicolson Baker wrong. LeMay, a legendary figure in American military history, gets a sympathetic hearing from this biographer.

Haruko Taya Cook, "Nagano 1945: Hirohito's Secret Hideout," *MHQ: The Quarterly Journal of Military History* 10, no. 3 (1998), pp. 44–47. Preparations were made for Japan fighting on even after an American invasion of the home islands.

Bernal Diaz, *The Conquest of New Spain* (Baltimore, MD: Penguin Books, 1963). A striking eyewitness account by a man who accompanied Cortez on his successful campaigns to overthrow the Aztec Empire. No quarter given in that effort.

T. R. Fehrenbach, *Comanches: The Destruction of a People* (New York: Knopf, 1979). But only after a long period of prosperity, expansion, slave trading, and success. The story of the Comanches is a substantial amendment to the axiom that Native American peoples improperly worshiped the god of war.

Michael Fellman (ed.), *The Memoirs of W. T. Sherman* (New York: Penguin Books, 2000). A marvelous insight into one of the North's best and most consistent generals, who brought the winning whirlwind strategy to inflict on the South.

John E. Freling, *A Wilderness of Miseries: War and Warriors in Early America* (Westport, CT: Greenwood Press). Patterns of warcraft among Native Americans are carefully delineated.

Mark Grimsley, *The Hard Hand of War: Union Military Policy toward Southern Civilians, 1861–1865* (Cambridge: Cambridge University Press, 1995). Bringing the war home to the slave-owning South and punishing the civilian population along with the Southern military. This strategy saved Lincoln's presidency, got him reelected, and finally defeated the rebellion of the Southern states.

S. C. Gwynne, *Empire of the Summer Moon: Quanah Parker and the Rise and Fall of the Comanches, the Most Powerful Indian Tribe in American History* (New York: Scribner, 2010). Faithfully captures the long sweep of Comanche history and destiny for their history; offers a substantial amendment to the notion that the Native Americans were easily or quickly defeated and destroyed. The population of Texas, for example, had to go from 15,000 in 1836 to more than 600,000 in 1860 before anyone could challenge their control over the vast territory of Comancheria. The author skillfully tracks the rise of the Comanches from 1625 to 1750, calling it "one of the great social and military transformations in history," by which the Comanches became supreme, defeating in turn many other Indian tribes (including the Apaches), the Spanish, the Mexicans, and American settlers.

David Hackett Fisher, *Champlain's Dream* (New York: Simon and Schuster, 2008). An excellent look at North American colonization by Champlain and the world of the Native Americans he encountered. Well written and illuminating.

Pekka Hamalainen, *The Comanche Empire* (New Haven, CT: Yale University Press, 2008). How one tribe actually expanded and for a long time and prospered when faced with various European invaders—Spanish, Mexican, Texan, and American. Good lessons, albeit retroactive ones for other Native Americans.

Victor Hanson, "The Wages of Suicide: Okinawa April 1–July 2, 1945," in *Ripples of Battle* (New York: Doubleday, 2003), p. 12ff. Hanson makes a powerful point that the Battle of Okinawa set in motion military decision making that would, by its very nature, insist on the utilization of the atomic bomb.

Max Hastings, *Armageddon: The Battle for Germany, 1944–1945* (New York: Knopf, 2004). The Russians take revenge on Nazi Germany with true sustained ruthlessness.

———, *Retribution: The Battle for Japan, 1944–1945* (New York: Knopf, 2007). Harsh account of the destruction of Japan and how it was well deserved. He sees Hiroshima and Nagasaki on a continuum with the earlier firebombing of Hamburg and Tokyo.

Isabel Hull, *Absolute Destruction: Military Culture and the Practice of War in Imperial Germany* (Ithaca, NY: Cornell University Press, 2005). How a belief system began, which led, during World War II, to "horrible extremity."

Lance Janda, "Shutting the Gates of Mercy: The American Origins of Total War, 1860–1880," *Journal of Military History* 59, no. 1 (January 1995), pp. 7–26. The American Civil War produced hardened fighters who took the lessons of that war west and applied the sustained ruthlessness of Sherman and Sheridan to the Plains Indians.

Francis Jennings, *The Invasion of America: Indians, Colonials and the Cant of Conquest* (Chapel Hill: University of North Carolina, 1975). It was an invasion, and it was successful, in part because the Native Americans failed to put up a united front, and the various tribes, with few exceptions, could not or would not match the demands of Mars.

Peter Kiss, "Incredible War Machine in the East," *Military History* (December 1986), p. 38ff. The Mongols set the gold standard for success in war while practicing sustained ruthlessness.

Steven LeBlanc and Katherine E. Register, *Constant Battles: The Myth of the Peaceful, Noble Savage* (New York: St. Martin's Press, 2003). Casts in sharp relief the notion that Native Americans lived in harmony with nature and each other until the arrival of the Europeans.

Wayne E. Lee, "Peace Chiefs and Blood Revenge: Patterns of Restraint in Native American Warfare, 1500–1800," *Journal of Military History* 71 (July 2007), pp. 701–41. Why Native Americans had a hard time appreciating the true war aims of the Europeans.

J. E. London, "Roman Siege of Jerusalem," *MHQ: The Quarterly Journal of Military History* 17, no. 4 (Summer 2005), pp. 6–15. The destruction of the Jewish revolt is studied and highlighted.

John K. Mahon, *History of the Second Seminole War 1835–1842* (Gainesville: University Presses of Florida, 1991). The Seminoles were one of the few Native American groups who waged so successfully for so long, causing 14 percent of casualties among regular army troops, a higher percentage than US losses in the Civil War. Speaking of the Civil War, this "grueling, gloryless war" produced, among others, Generals William T. Sherman, George Thomas, Braxton Bragg, George Meade, Joseph Johnson, and John Pemberton.

Note: The Seminole War cost more American military casualties than all of the post–Civil War fighting against the Plains Indians.

Patrick M. Malone, *The Skulking Way of War: Technology and Tactics among the New England Indians* (Lanham, MD: Madison Books, 2000). Shows why the Native Americans were truly unprepared to match the Europeans in warfare, and this had less to do with their level of technology than their centuries-old style of war.

Daniel Mandell, *King Philip's War: Colonial Expansion, Native Resistance and the End of Indian Sovereignty* (Baltimore, MD: Johns Hopkins University Press, 2010). The period 1620–1670 in New England shows the astonishing failure of the Native Americans to unite against the European invaders. Indian uprisings, especially in King Philip's War, were seen as God's retribution for Puritan sins, and at the end, the Europeans punished loyal and even Christian loyal Indians, as well as "enemy" Indians. Ironically, the author points out how the Puritan colonies in Massachusetts also lost sovereignty vis-à-vis the English crown. Native Americans recoiled at the European style of warfare as being "too furious, and slays too many men."

Charles Mann, *1491* (New York: Alfred Knopf, 2005). A useful overview of what North and South America looked like while Columbus was still trying to get royal backing for his expedition. Makes the case that there were many more Native Americans and their civilizations were more advanced and their economies more vibrant than most of us thought before reading this book.

———, *1493* (New York: Alfred Knopf, 2011). His follow-up work deals with the grand theme of the "Columbian Exchange," with its far-reaching implications for Asia, Europe, Africa, and the Americas. Excellent material on the constant warfare between the Africans and Indians and Europeans in the interior of Brazil, Peru, Central America, and Mexico, as well as the northern coast of South America. See especially his "Forest of Fugitives."

John McCain, *Faith of My Fathers* (New York: Random House, 1999). Everything that McCain has or has not done, been or become, since his Vietnam experiences seems tinged by them. A devastatingly harsh indictment of the American political leadership and how they influenced the war.

Alfred C. Mierzejewski, *The Collapse of the German War Economy 1944–1945* (Chapel Hill: University of North Carolina Press, 1988). What really did in the German war production was a combination of factors, including heavy strategic bombing.

Williamson Murray and Allen R. Millet, *A War to be Won: Fighting the Second World War* (Cambridge, MA: Belknap Harvard, 2000). A straightforward and highly readable account based on very significant scholarship. Wide ranging and informative.

Robert Charles Padden, "Cultural Change and Military Resistance in Araucanian Chile, 1550–1730," *South Western Journal of Anthropology* 13, no. 1 (Spring 1957), pp. 103–21. A fascinating account of a people who refused to submit to the Spanish invaders for almost 200 years and who adapted ritual cannibalism to turn the Christian sacrament into an ironic inversion, with one of their chiefs telling the Spanish general, in effect, "I ate your predecessor and I will eat you."

Mary Palevsky, *Atomic Fragments: A Daughter's Questions* (Los Angeles: University of California Press, 2000). A very incisive and far-reaching look at the people who created the atomic bomb and then saw it used in action and the moral soul-searching of some who helped make it possible. A very useful addition to the literature of "sustained ruthlessness" and the various ways of looking at its application. Note how prominent the Mongol parallels are utilized.

Edouard Perroy, *The Hundred Years War* (London: Oxford University Press, 1917). A noted French historian traces the long and devastating relationship between Edward Plantagenet III and Philip VI of Valois and their descendants who managed to make war for more than 100 years, from 1328 onward. The Hundred Years' War, he feels, ended feudal war and shows sustained ruthlessness par excellence. (Also cited in the section on European medieval warfare).

Nathaniel Philbrick, *The Last Stand* (New York: Viking, 2013). A very insightful account of Custer, Sitting Bull, and the Battle of the Little Bighorn as potent symbols of the underlying theme of sustained ruthlessness. General George Cook, command of the Wyoming Column, talking about his use of Indian scouts from the tribes he was decimating: "It is not merely a question of catching them better with Indians, but of a broader and more enduring aim—their disintegration."

Andrew J. Rotter, *Hiroshima: The World's Bomb* (New York: Oxford University Press, 2008). Horrified Rotter is and horrified he remains as he suggests the far-reaching impact of the bomb.

R. J. Rummel, *Death by Government* (London: Transaction, 1995). A truly horrendous account of all the humans killed by their own—and other—governments.

Howard Russell, *Indian New England before the Mayflower* (Hanover, NH: University Press of New England, 1980). Many good insights, although perhaps a bit overly positive about intra-Indian warfare.

Dudley Saward, *"Bomber" Harris* (London: Buchan and Enright, 1984). A former staffer for Air Marshal Sir Arthur Harris, the architect of the mass bombing of German cities, writes a praiseful but still incisive account of the British air campaign during World War II. From the mass bombing of Cologne in May 1942 to those of Hamburg in July 1943 and Dresden in February 1945, Saward makes the strong case that the bombing was very effective, citing as proof the statements of Goebbels, Speer, and Milch, which would seem to carry some weight. Also makes the telling point seldom seen elsewhere that the German antiaircraft operations required upward of 900,000 men and women, forces that could not be used elsewhere. This constituted a "second front" long before the Normandy landings in June 1944.

Armstrong Starkey, *European and Native American Warfare, 1675–1815* (Norman: University of Oklahoma Press, 1998). Quite a telling contrast here, and Starkey didn't even employ the term *sustained ruthlessness*.

Ian K. Steele, *Warpaths: Invasions of North America* (New York: Oxford University Press, 1994). Good accounts of Indian ways of war and how there were common styles of warfare spread across two continents and among many disparate peoples.

Daniel Swift, *Bomber Country: The Poetry of a Lost Pilot's War* (New York: Farrar, Straus, and Giroux, 2010). The heartfelt story of a man who follows the life and death of his paternal grandfather (a pilot who flew Lancaster bombers with the 83rd Squadron of the Royal Air Force), interspersed with poetry about the war, bombing, and criticisms of those who bombed. I found two dimensions to be of considerable import. One is the assertion from the US Strategic Bombing Survey of "The Effects of Strategic Bombing on German Morale," which, not surprisingly, concluded that "bombing seriously depressed the morale of German civilians" but then goes on to assert that there wasn't much more depression in German cities, which were heavily bombed compared to those that were only lightly bombed, concluding, "The maximum morale effect of dropping a given tonnage of bombs on Germany would have been attained by lighter raids as widely distributed as possible, rather than by concentrated heavy bombings in limited areas." The second is a quote from Josef Goebbels's diary from 1943: "One can only repeat about air warfare: we are in a position of almost helpless inferiority and must grin and bear it as we take the blows from the English and Americans."

J. Adam Toose, *The Wages of Destruction: The Making and Breaking of the Nazi Economy* (New York: Viking, 2007). Albert Speer once said that the Allied bombing cut German wartime production in 1943 by 10 percent and in 1944 by 20 percent. Toose gives many statistics that argue the impact was even greater.

C. Dale Walton, *The Myth of Inevitable U.S. Defeat in Vietnam* (London: Frank Cass, 2002). A strongly revisionist view of the US effort in Vietnam. Well worth a read.

David Weber, *Barbados: Spaniards and Their Savages in the Age of Enlightenment* (New Haven, CT: Yale University Press, 2005). The Enlightenment did not extend to Protestant Spaniards, suspect Catholic Spaniards or Jewish Spaniards, so why would it possibly extend to Native Americans?

Joseph Wheelan, *Terrible Swift Sword: The Life of General Philip H. Sheridan* (New York: Da Capo Press, 2012). Ever since at age 12 I read a landmark book about him, "Little Phil" has been one of my heroes. This biography, although light on his personal life, really captures the essence of the man. His fierce warriorhood (he always led from the front); his espousal of all-out war (but then taking the side of the Indians against the Indian agents); and his insistence that Yellowstone National Park be protected from speculators, hunters, and railroad men show a man for all seasons.

James H. Willbanks, *Abandoning Vietnam: How America Left and South Vietnam Lost Its War* (Lawrence: University Press of Kansas, 2008). Willbanks believes this and makes a good case for his belief. A wide-ranging and very interesting set of assertions.

James Wilson, *The Earth Shall Weep: A History of Native America* (New York: Atlantic Monthly Press, 1998). Broad, sweeping, sympathetic work detailing the decline in Native Americans from 7 to 10 million to 250,000 in 400 years. The whole sordid tale but, regrettably, no mention of Mars.

RECEPTIVITY TO MILITARY AND INTEGRATIVE INNOVATION

> "There is nothing more difficult to take in hand, more perilous to conduct, or more uncertain in its success, than to take the lead in the introduction of a new order of things."—Niccolo Machiavelli

The role of innovation is the central theme of chapter 5, "Receptivity to Military and Integrative Innovation," in *Winning at War*, pp. 103–30. That chapter explores the ways in which new weapons *and* new ideas come into practice and usage by the militaries of the world. One might assume that militaries throughout history have always wanted to have the latest weapons or use the latest techniques. Yet the reverse has far more often been the case. Most militaries have been very conservative in their willingness to adopt new weapons and new tactics, often waiting until long after their utility has been demonstrated.

This has been true for military hardware. New weapons and weapons systems have often been resisted, sometimes on social grounds, sometimes on assumptions about loss of power. For example, the French nobility steadfastly resisted the introduction of the longbow into their armed forces, and the Japanese samurai during their feudal period outlawed firearms once they saw how effective they were—and how easily peasants with no social or military status could acquire the skill to utilize the new weapon.

This has also been true for military tactics and strategy. With new weapon capabilities—such as the increased killing range of rifles over smooth-bore muskets—comes a need for new tactics. The horrendous slaughter of the American Civil War was due in part to the failure of both armies to adopt to the realities of the newer weapons introduced earlier and during the war.

Therefore, the way in which polities, societies, and militaries accept innovation and integrate new weapons and new tactics and new strategies into their operations is one of the most important keys to understanding the course of war over time. Some additional readings on the subject, as well as the sources of many of this chapter's footnotes (see pp. 243–48), include:

Stephen Alford, *The Watchers: A Secret History of the Reign of Elizabeth I* (New York: Bloomsbury Press, 2012). An engrossing account of the information flow that shaped English, Spanish, papal, and French strategy in the age of the Protestant Reformation and the rise of the heretic English monarchy, which many wanted to see destroyed.

Roy Appleman, *Disaster in Korea: The Chinese Confront MacArthur* (College Station: Texas A&M, 1989). Straight and steady is this account, and it pulls few punches in ascribing blame for the fiasco, although it is less fulsome in its praise of the genius behind Chinese success General Peng than others might argue.

Juliet Barker, *Agincourt: Henry V and the Battle That Made England* (New York: Little, Brown, 2005). A splendid account of one of the most widely studied battles indicating how important it was to the sense of being English.

Shelford Bidwell and Dominick Graham, *Fire-Power: British Army Weapons and Theories of War 1904–1945* (London: Pen and Sword Classics, 2005). Gives good insights into why, for example, the British, who developed the first effective tanks, failed to see their strategic value until shown the way by the Germans in France during 1940.

Antoine Bousquet, *The Scientific Way of Warfare: Order and Chaos on the Battlefields of Modernity* (New York: Columbia University Press, 2009). An attempt by humankind to bring science to the battlefield in order to impose order on its chaos. Not always successful.

Niccolo Capponi, *Victory of the West: The Great Christian-Muslim Clash at the Battle of Lepanto* (New York: Da Capo Press, 2006). Capponi sees 1572 as a turning point in world history and gives many insights into why it was and why the battle turned out the way that it did.

Richard Clarke, *Sting of the Drone* (New York: St. Martin's Press, 2014) (F). In addition to being great fiction, this book looks at the ebb and flow and counteractions to the introduction of a new weapon system, in this case the armed unmanned aerial vehicle (UAV), or drone, and also explores the morality of using a particular weapon. Always remember the social and religious efforts to get the crossbow banned from European wars, as at the Second Lateran Council of 1139, which supposedly prohibited its use against Christians while allowing its use against Muslims and others.

Thomas M. Coffey, *Iron Eagle: The Turbulent Life of General Curtis LeMay* (New York: Crown, 1986). Curtis LeMay gets his due and then some in this sprightly written biography. LeMay's dramatic change in tactics and strategy would turn out to be very significant in the final defeat of Japan. The author also suggests that LeMay regarded the atomic bomb as a diversion, already planning, as he did, for the burning of most of Japan to the ground without its use.

Robert Coram, *Boyd: The Fighter Pilot Who Changed the Art of War* (New York: Harper Books, 2002). A must-read for all those who seek to understand the down drag of bureaucratic inertia when it comes to warfare.

Roger Crowley, *Empires of the Sea: The Siege of Malta, the Battle of Lepanto, and the Contest for the Center of the World* (New York: Random House, 2008). Muslims and Christians clash in an ongoing struggle for supremacy in the Mediterranean basin. Perhaps this work has considerable relevance for the future.

John Francis Guilmartin Jr., *Gunpowder and Galleys: Changing Technology and Mediterranean Warfare at Sea in the Sixteenth Century* (Cambridge: Cambridge University Press, 1974). See review in chapter 15.

David Halberstam, *The Coldest Winter: America and the Korean War* (New York: Hyperion, 2007). Doesn't really break much new ground but is a well-written and absorbing account of the war that still baffles us and whose refusal to truly end remains ever present.

Grant T. Hammond, *The Mind of War: John Boyd and American Security* (Washington, DC: Smithsonian Books, 2001). A brilliant, transformative genius whose OODA loop remains of enormous conceptual importance.

Victor Hanson, "The Market—Or Capitalism Kills," in his *Carnage and Culture* (New York: Doubleday, 2001). Hanson makes a strong case that capitalism is as good at war as it is at making material things.

Robert Hardy, *Longbow: A Social and Military History* (London: Sutton, 2006). Everything you ever wanted to know about this transformative (for the Welsh and the English, at least) weapon and especially why it never caught on many other places.

Charles Hilbert, "Samurai Slaughtered at Nagashino," *Military History* 60, no. 4 (October 1996), pp. 63–68. The battle that turned Japanese samurai against the "fiery weapons."

Lord Kinross, *Ataturk* (New York: William Morrow, 1965). A sympathetic but defensible account of the decline of the Ottoman Empire, the rise of the

Young Turks, the war of independence, and the rise of the Turkish Republic, all tied up with the life of Kamal Ataturk.

Matthew Morton, *Men on Iron Ponies: The Death and Rebirth of the Modern U.S. Cavalry* (DeKalb: Northern Illinois University Press, 2009). How the US CIA and Special Forces worked with the Northern Alliance to defeat the Taliban.

Williamson Murray and MacGregor Knox (eds.), *The Dynamics of Military Revolution 1300–2050* (Cambridge: Cambridge University Press, 2001). A variety of essays showing how the various militaries adapted or did not adapt to the military revolutions provides the reader with much to think about.

For an interesting historical novel about the struggle for the Mediterranean basin, see William Napier, *Clash of Empires: The Red Sea* (London: Orion Books, 2012) (F).

Anthony Pagden, "Turning the Ottoman Tide," *MHQ: The Quarterly Journal of Military History* 20, no. 4 (Summer 2008), pp. 8–17. Suleiman the Magnificent fails to capture Vienna after taking Hungary and threatening the Holy Roman Empire. The whys are as interesting as the hows.

Noah Perrin, *Giving Up the Gun: Japan's Reversion to the Sword 1543–1879* (Boston: David R. Godine, 1979). Fascinating account of how a very successful innovation, the Portuguese-imported harquebus, enabled one Japanese clan to defeat another, but the very powerful impact it would have wreaked on the Samurai culture (and supremacy) turned the military culture against it. Shows the difficulty of introducing new weapons to military cultures, which are set in their ways, often requiring huge shocks to adapt.

Stephen Rosen, *Winning the Next War: Innovation and the Modern Military* (Ithaca, NY: Cornell University Press, 1999). The march of military innovation has not ceased, nor, it seems, will it.

Thomas Sowell, *Conquests and Cultures: An International History* (New York: Basic Books, 1998). Much-needed sanity interjected into the relationship between topical and timely hegemony and continuing hegemony.

John Stoye, *The Siege of Vienna* (London: Collins, 1964). How Suleiman the Magnificent was thwarted at the gates of Vienna and, if truth be told, within the city.

Barry Watts and Williamson Murray, "Military Innovation in Peacetime," in *Military Innovation in the Interwar Period*, edited by Williamson Murray and Allan R. Millett (Cambridge: Cambridge University Press, 1996). Great Britain, France, Germany, Italy, and others react to changes in the challenges of military innovation.

THE ABILITY AND WILLINGNESS TO PROTECT CAPITAL FROM PEOPLE AND RULERS

> "Wealth is usually needed to underpin military power, and military power is usually needed to acquire and protect wealth."—Paul Kennedy

> "Only in the dreams of Elizabethan officials can war be conducted as a business."—George M. Thomson

Chapter 6 of *Winning at War* (pp. 131–46) explores the role played by wealth and its allocation in determining success in war. It outlines a number of different dimensions, including the most obvious, the size of a particular entity's gross domestic product (or, in ages past, tribute from colonies and subordinate peoples or states).

But in addition, it examines the ways in which Mars smiles on entities that allocate the proper amount to military preparedness; the relationship between civilian and military expenditures for this especially can tell us a lot about potential, even likely, success in war. It is important in this regard, however, to look carefully at the way in which military expenditures are spent. Regardless of the size of the defense appropriations, how is that amount allocated? Often money allocated for war-making is frittered away by a top heavy, kleptocratic military, which results in a great imbalance between money spent on troops, training, and military hardware and that spent on officers' salaries and corruption at all levels.

Thus, from the point of view of Mars, logistics in warfare is vital, as are available funds for training, technology, and innovation. However, a close examination is required to see the extent to which any polity is getting its money's worth for future defensive or offensive warfare. Some additional readings on the subject, as well as the sources of many of this chapter's footnotes (see pp. 248–51), include:

Jurgen Brauer and Hurbert Van Tuyll, *Castles, Battles, and Bombs: How Economics Explains Military History* (Chicago: University of Chicago Press, 2008). Surprisingly dull on balance. But it is important to realize that castles, while expensive, were also less expensive than not having one.

Roger Crowley, *City of Fortune: How Venice Ruled the Seas* (New York: Random House, 2012). A marvelous—and cautionary—tale of how Venice founded, maintained, and eventually began to lose its Stato da Mar, the commercial and military empire that included Zara, Scutari, Corfu, Crete, Negroponte, Salonika, and Kaffir, stretching along the Adriatic, the Mediterranean, and the Black Sea, on to trading with Mamluks in Palestine and Egypt. Through wars with Genoa, trade with the Mongols, and the relentless pressures of the Ottoman Turks, Venice held its own for centuries. Crowley has written a marvelous work showing the need for protecting the capital from people and the penalties of failing to do so. Everything the city bought or sold came by sea during the three periods under review: 1000–1204, "The Merchant Crusaders"; 1204–1500, "The Princes of the Sea"; and 1400–1503, "The Rising Moon" of the Ottoman Turks. And the sea required, at all times, the projection of force to protect commercial interests.

———, *Empires of the Sea: The Siege of Malta, the Battle of Lepanto, and the Contest for the Center of the World* (New York: Random House, 2008). See previous review.

Niall Ferguson, *The Cash Nexus: Money and Power in the Modern World, 1700–2000* (New York: Basic Books, 2001). Money not only makes the world go around, but it also makes war go around.

Holger H. Herwig, "The Battlefleet Revolution, 1885–1914," in *The Dynamics of Military Revolution 1300–2050*, edited by Williamson Murray and MacGregor Knox (Cambridge: Cambridge University Press, 2001), pp. 114–31. Great Britain refused to be pushed into second place, and Germany unnecessarily provoked Great Britain by an ill-conceived and strategically dead-end naval arms race.

Robert Kaplan, *Hog Pilots and Blue Water Grunts* (New York: Random House 2006). The American armed forces in global battle space, part 2. More outstanding insights. Another must-read by anyone trying to understand America's global situation today.

———, *Imperial Grunts* (New York: Random House, 2004). The American armed forces in global battle space, part 1. Outstanding insights. A must-read.

Kevin Shillington, *History of Africa* (New York: St. Martin's Press, 1989). Gives the reader a good sense of the major aspects of African history, although somewhat weak on the strategy, tactics, and weapons of its various militaries throughout time and space.

THE CENTRALITY OF SUPERIOR WILL

"Violence is the heart and secret soul of the sacred." Rene Girard

Any study of war should have at its heart an examination of the sources of will, which make humans willing to fight and kill others of their species. Girard truly captures the intrinsic dynamism of the power of "the sacred" that drives humans to kill their fellows in large numbers, be that "sacred" religion, ideology, or nationalism—whatever is held to be "sacred" to that society or polity at that time and place. Chapter 7 in *Winning at War* (pp. 147–66) provides an overview of the major wellsprings of superior will in warfare.

Ranging from such well-known dimensions as love and greed to religion, ideology, and a love of battle, this chapter looks at over a dozen in-depth, providing a broad spectrum of the many reasons warfare has flourished as a human activity over the millennia. Its conclusions are bleak: There are so many causes for war and so many people over the eons willing to employ them that history looks like a series of wars punctuated by brief periods of peace rather than, as is often alleged, the other way around.

This is a challenging subject, requiring a broad sweep of sources to tap its inherent fertility. Some additional readings on the subject, as well as the sources of many of this chapter's footnotes (see pp. 251–54), include:

David Bachrach, *Religion and the Conduct of War c. 300–1215* (Woodbridge, UK: Boydell Press, 2003). War buttressed by religion; religion bolstered by war. A nasty mix, one still looming large on the current world scene, as it has been since organized religion was invented and implemented.

David Bell, *The First Total War: Napoleon's Europe and the Birth of Warfare as We Know It* (Boston: Houghton Mifflin, 2007). A superb analysis of a very important phenomenon, the nationalization and militarization of the people, making for modern war.

David Bergamini, *Japan's Imperial Conspiracy* (New York: William Morrow, 1971). How and why Japan started World War II in China and subsequently engulfed the entire region in a mad lunge for empire and domination.

David Carrasco, *The City of Sacrifice: The Aztec Empire and the Role of Violence in Civilization* (Boston: Beacon Press, 1999). Sacred cities demand sacred violence, and in looking at the Aztec capital of Tenochtitlan (from 1325 to 1521), the author finds "prodigious acts of violence," including "monumental human sacrifice," and argues that these killings were simply ongoing (and partial) debt payments to the gods. Montezuma's eating of the flesh of the victims of these payments, he adds, were “mythic morsels.” Barry Lopez may need to modify his paean to the “performance space” of Tenochtitlan.

Iris Chang, *The Rape of Nanking: The Forgotten Holocaust of World War II* (New York: Penguin, 1998). The horrors of killing upward of 300,000 people in China somehow lacks cogency for many today. The author tries to rectify that omission.

Fred Donner, *The Early Islamic Conquests* (Princeton, NJ: Princeton University Press, 2007). When Islam was, indeed, spread by the sword, starting with Muhammad’s conquest of Medina and Mecca. Shows the centrality of war and violence to the spread of Islam. Significant parallels with Judaism and Christianity after the 4th century.

John Dower, *War without Mercy: Race and Power in the Pacific War* (New York: Pantheon, 1987). How the war in the Pacific became a war without quarter, and Hiroshima, far from being an outrider event, belongs in the center of the conflict.

Nigel Eltrimingham, *Accounting for Horror: Post-Genocide Debates in Rwanda* (London: Pluto Press, 2004). Somehow it seems easier to blow someone up with a bomb rather than hack them to death with a machete, but Rwanda proved otherwise.

Drew Gilpin Faust, *This Republic of Suffering: Death and the American Civil War* (New York: Alfred Knopf, 2008). The space-time continuum that enabled soldiers who were about to die believe that they would see their loved ones again in the future.

Mark Felton, *Slaughter at Sea: War Crimes of the Imperial Japanese Navy* (Annapolis, MD: US Naval Institute, 2008). More about Japan’s Imperial dictums and where they led. Not a nice story.

Ronald Fraser, *Napoleon’s Cursed War: Popular Resistance in the Spanish Peninsular War* (London: Vescoe, 2008). How the Catholic Church, the British army, and ordinary Spanish men and women drove the French from Spain in a process Napoleon cursed until the day he died.

Daniel Fromkin, *A Peace to End All Peace* (New York: Henry Holt, 1989). A must-read for anyone wishing to understand today’s conflicts in the Middle East. The author documents how the many decisions made during World War I and in 1922 set in motion the long-term bitterness, conflicts, and animosities writ large. Also, a good source for the Template element “There

Will Always Be Another War." The title comes from Archibald Wavell's comment on the leaders of Britain, France, and the United States: "After 'The war to end wars' they seem to have been pretty successful in Paris at making a 'Peace to end Peace.'"

Pani Grahi, *India's Partition: The Story of Imperialism in Retreat* (London: Routledge, 2004). And the cruel imperatives of ethnicity and religions writ large. A truly monumental failure of statesmanship and diplomacy and a vicious indictment of human's inhumanity toward other humans.

Nisid Hajari, *Midnight's Furies: The Deadly Legacy of India's Partition* (New York: Houghton Mifflin Harcourt, 2015). A good blow-by-blow account of how failed leadership—Jinnah, Gandhi, Nehru, Patel, and Mountbatten, among others—exacerbated primordial attachments and failed to stop the imperatives of religion and ethnicity, the negative of which remain on the Indian subcontinent to this day.

Chris Hedges, *War Is a Force That Gives Us Meaning* (New York: Public Affairs Publishers, 2002). A strongly worded and closely argued piece. Not for the faint at heart.

Douglas Hill, "Kill Them All . . . God Will Recognize His Own," *MHQ: The Quarterly Journal of Military History* 9, no. 2 (1997), pp. 98–109. Religious fervor and military cynicism at its best as Christians slaughter Christians for the greater glory of God.

Norman Housley, *Fighting for the Cross: Crusading in the Holy Land* (New Haven, CT: Yale University Press, 2008). A very good account of the Crusades and the conflicting motivations of those who fought in them.

———, *Religious Warfare in Europe 1400–1536* (London: Oxford University Press, 2002). Despite his disclaimer to the contrary, the author makes a good case for the notion of the "Age of Religious War," stretching from the council of Clermont in 1095 to the Peace of Westphalia in 1648. Certainly during the period under his closer review, religious motivation provides many, many instances of will sustainment in warfare.

John Keegan, *Victory at Any Cost: The Genius of Vietnam's General Vo Nguyen Giap* (London: Recon Press, 1976). Keegan can't help but admire Giap's dedication to victory with sustained ruthlessness and anything else that helped him accomplish his goals.

Yasmin Khan, *The Great Partition* (New Haven, CT: Yale University Press, 2007). British imperial power leaves, and the fiery eruptions of ethnicity and religion well up—with disastrous results and precious little statesmanship shown by those who should have anticipated the pogroms.

MacGregor Knox, "Mass Politics and Nationalism as Military Revolution: The French Revolution and After," *The Dynamics of Military Revolution 1300–2050*, edited by Williamson Murray and MacGregor Knox (Cambridge: Cambridge University Press, 2001), pp. 57–73. An insightful look at how the French Revolution changed the nature of warfare by making it possible to arm and mobilize the masses in military formation, all in the name of nationalism.

Alan J. Kupermann, *The Limits of Human Intervention: Genocide in Rwanda* (Washington, DC: Brookings Institution Press, 2001). The horror of genocide in Rwanda is intensified by the ordinariness of the means and repetitiveness of its occurrence.

Dominique Lapierre and Larry Collins, *Freedom at Midnight* (New York: Viking, 2001). The end of British India in 1947 seen from many different points of view. Very engagingly written, so much so that it is possible to lose some of the horror of the partition.

John Micklethwait and Adrian Wooldridge, *God Is Back: How the Global Revival of Faith Is Changing the World* (New York: Penguin Press, 2009). This book is groundbreaking in its scope and message. Quite a surprise to many and a phenomenon that helps to explain the continuing power and ubiquity of sectarian violence.

T. P. Schwartz-Barcott, *War, Terror, and Peace in the Qur'an and Islam: Insights for Military and Government Leaders* (Carlisle, PA: Army War College Foundation Press, 2004). Muhammad, after all, was a wartime commander as well as religious prophet. How could the legacy of Islam be otherwise?

Peter Silver, *Our Savage Neighbors: How the Indian Wars Transformed Early America* (New York: W. W. Norton, 2008). The author argues that it was the demonization of the Native Americans that ended up uniting the immigrants from the various European countries and made them proto-Americans.

David Livingston Smith, *The Most Dangerous Animal, Human Nature, and the Origins of War* (New York: St. Martin's Press, 2007). The author claims and illustrates that war is too widespread in too many cultures throughout time to just be an aberration.

John G. Stoessinger, *Why Nations Go to War* (New York: Wadsworth, 2005). A liberal perspective on the re-reoccurrence of war. The author means to hold individuals and political and economic systems responsible.

Yuri Tanaka, *Hidden Horrors: Japanese War Crimes in World War II* (Boulder, CO: Westview Press, 1987). There were many, and the vast majority went unpunished. Highlights the amazing lack of collective guilt inflicted on the Japanese when compared to that on the Germans.

Thucydides, *The Peloponnesian War* (New York: Modern Library, 1951). The historian who started it all and saw war with stunning clarity, a clarity that shines through the ages until the very present. A truly awe-inspiring work. Cannot really be summarized here, but even a partial reading will be very rewarding for the interested student.

Martin van Creveld, *Command in War* (Cambridge, MA: Harvard University Press, 2003). What makes for solid generalship and how command and control plays such a vital role in warfare.

Geza Vermes, *Christian Beginnings: From Nazareth to Nicaea AD 30–325* (New York: Penguin, 2012). Shows the evolution of Christian thought that enabled this religion to transform itself from one of the most peaceful to one of the most warlike, becoming the state religion of the Roman Empire under Constantine in 313. The portions on the Apostolic Fathers (Clement, Ignatius, Polycarp, Hermas, and Diognetus) and the "Three Pillars of Wisdom" (Tertullian, Clement of Alexandria, and Origen) are worth the price of admission. A stunning tour de force based on considerable linguistic prowess.

W. Montgomery Watt, *Muhammad at Medina* (Oxford, UK: Clarendon Press, 1956). How Muhammad ruled and why he was successful in his efforts to create a new society inspired by Allah. It is interesting that so many who make pronouncements about Islam overlook the dimensions of the polity and society (they were one in the same) that existed when Muhammad ruled.

THE BELIEF THAT THERE WILL ALWAYS BE ANOTHER WAR

"Peace itself is war in masquerade."—John Dryden, *Absalom and Achitophel*

Perhaps the most important portion of *Winning at War* is chapter 8, "There Will Always Be Another War" (pp. 131–46), which argues that Mars has favored those nations and people who do not assume that war is a one-time aberration but an intrinsic part of the human condition. Its conclusions run counter to the positions held by many. But on balance, the notion that nations who prepare for war are less likely to suffer defeat—or pay a higher price for victory—than those who do not seems axiomatic, and the history of the United States would seem to underscore this proposition. Some additional readings on the subject, as well as the sources of many of this chapter's footnotes (see pp. 255–57), include:

Ken Alibek, *Biohazard: The Chilling True Story of the Largest Converted Biological Weapons Program in the World—Told from Inside by the Man Who Ran It* (New York: Random House, 1999). An amazing tale of duplicity and treachery, as well as a lack of control over weapons systems in a supposedly totalitarian state.

Philip Bobbitt, *The Shield of Achilles* (New York: Knopf, 2002). Argues that, over the last six centuries, states have been the optimal institution for waging war and creating peace, but now they are giving way to "market-states." Interesting take on the "Long War" of 1914–1990 between the forces of Communism and those of the market states.

Roger Boesche, *The First Great Political Realist: Kautilya and His Arthashastra* (Lanham, MD: Lexington Books, 2002). A very important thinker, one who produced one of the great political books from the ancient world. Gives power to realism and is a fine counter to the writings of Sun Tzu and Sun Pin. Kautilya should be studied far more than he is.

Mark Bowden, *Black Hawk Down: A Story of Modern War* (New York: Atlantic Monthly Press, 1999). The incident—because, in terms of scale, it was only a small event—in Mogadishu when 2 American Black Hawk helicopters were shot down and 18 American soldiers killed in 1993 turned out to be momentous in terms of its long-term impact. The Clinton administration terminated the Somali mission soon thereafter, and Osama bin Laden concluded that the United States could be driven from the Middle East. Too often, the sheer drama of the book and the movie it spawned overshadows the true strategic implications of the events depicted.

John Broder, "Climate Change Seen as Threat to U.S. Security: Pentagon Studies Issue: Crises May Destabilize Regions and Require Military Action," *New York Times*, August 9, 2009, pp. 1ff. The military sees threats, opportunities, and budgetary payoffs in the hot new topic of climate change.

Nigel Cliff, *Holy War: How Vasco Da Gama's Epic Voyages Turned the Tide in the Centuries-Old Clash of Civilizations* (New York: Harper, 2011). It is 1498, and Europe is still on the defensive. Against all odds, it is the Portuguese who break through to the spice lands, bringing with them European "Christian" values (including the Inquisition) as they search for spices and Christians. The author claims it was the crusading impulse as well as the desire for wealth that drove them. With no tradition of naval warfare, the Indian Ocean was made for European maritime expansion (although the competition would end up *among* Europeans, not between them and the Muslim seafarers). The Europeans would have a permanent naval presence in these waters from 1593 until the present.

George Crile, *Charlie Wilson's War* (New York: Atlantic Monthly Press, 2001). An excellent study of the Soviet defeat in Afghanistan, although relying heavy on the American perspective and exaggerating the connection between the Stinger missile and the Soviet decision to withdraw. It certainly seems to have speeded up that process, but Mikhail Gorbachev had already decided to somehow extricate the crumbling Soviet Union from this foreign adventure by the time the Stingers were introduced.

John A. English, *Marching through Chaos: The Descent of Armies in Theory and Practice* (Westport, CT: Praeger, 1996). Interesting volume, which concludes by quoting Michael Howard: "Nothing has occurred since 1945 to indicate that war, or the threat of it, could not still be an effective instrument of state policy. Against people who were not prepared to defend themselves it might be very effective indeed."

Thomas Friedman, *The Earth Is Flat* (New York: Farrar, Straus, and Giroux, 2005). Very global and very hopeful in outlook, this work argues that so much of humankind is now connected that conflict is likely to lessen over time. He may turn out to be correct, but for now, he amazingly discounts the role of the global military in keeping the global world order. Quite Pollyanna-ish concerning the coming struggle for natural resources, including but not limited to fossil fuels.

Francis Fukuyama, *Another Bloody Century: Future Warfare* (London: Weidenfeld and Nicolson, 2005). "Barring a magical political or ethical elixir . . . [w]ar and warfare will always be with us: war is a permanent feature of the human condition."

———, "The End of History?" *National Interest*, no. 16 (Summer 1989), pp. 3–18. One fears the author got a little ahead of himself in his exuberance over the breakup of the Soviet Union, but his article continues to reverberate through various discussion cycles, so it needs to be examined. History did not end in 1089 or indeed 1991.

———, "Reflections on the End of History, Five Years Later," *History and Theory* 34, no. 2 (May 1995), pp. 27–43. A wildly optimistic worldview largely overridden by events but one that is often present in academia.

Samuel Huntington, "The Clash of Civilizations?" *Foreign Affairs* 72, no. 3 (Summer 1993), pp. 22–49. Realism proven correct in large part, although Huntington perhaps simplifies too much. Often paired with Fukuyama's "The End of History" in a point-counterpoint array.

Donald Kagan, *On the Origins of War* (New York: Doubleday, 1995). A fascinating study of the outbreaks of such wars as the Punic and World War I. Well written and researched and offering fresh insights to both the process of going to war and the miscalculations that affect decision makers in the run-up to war.

Robert Kagan, *The Return of History and the End of Dreams* (New York: Alfred Knopf, 2008). The realist school is alive and well when looking at the world today. This author gives a good overview of the realist worldview as seen flourishing in the new century.

Parag Khanna, *The Second World: Empires and Influence in the New Global Order* (New York: Random House, 2008). Sees a major role for the European Union, Japan, and regional groupings, all of which he believes will undermine American hegemony going forward. Seems to drastically underestimate the importance of "hard power" going forward.

Andrew F. Krepinevich, *7 Deadly Scenarios: A Military Futurist Explores War in the 21st Century* (New York: Bantam Dell, 2009). An intriguing set of alternative futures are examined. Not for the faint of heart.

Colonel Qiao Liang and Colonel Wang Ziangsui, *Unrestricted Warfare: China's Master Plan to Destroy America* (Panama City: Pan American, 2002). Despite its lurid title, this is a must-read, as these very perspicacious writers saw clearly that it was not the "end of history" and that warfare had simply taken on a multitude of forms that would aid the weak against the strong. Amazingly prescient.

Joseph Maiolo, *Cry Havoc: How the Arms Race Drove the World to War 1931–1941* (New York: Basic Books, 2010). How even Stalin was shocked at "red militarism" and how it spread. The Italians knew they were "[m]oving among giants." Germany rearmed to get ready for the next war, while the French and British tried to have guns and butter. Ultimately, time got speeded up by events, and everybody rushed pell-mell to war. Although the author decries the role of the arms race, he ends up making a strong case for its efficacy and enduring demands, concluding, "[T]he rules can change, and so can the stakes, but the race goes on."

Anthony Pagden, *Worlds at War: The 2,500 Year Struggle between East and West* (New York: Random House, 2008). For those who think the Christian and the Islamic worlds are destined for peace and mutual understanding, this book offers an overview that is far more pessimistic and that asserts that there had been an ebb and flow of one religion versus the other since the 8th century.

Harry R. Yarger, *Strategy and the National Security Professional: Strategic Thinking and Strategy Formulation in the 21st Century* (New York: Praeger, 2008). An in-depth look at the mindset of those seeking to make sense out of national security in the present age by looking carefully at national security issues in the past.

Chapter Six

The Decision to Drop the Atomic Bomb; or, Hiroshima as Seen through the Lens of Mars

The subject of the use of atomic bombs during World War II by the United States has been among the most controversial in the history of weapons utilization. This is largely the case because of the perceived need to inject morality into what was a strategic decision. The controversy remains vibrant even today, with vigorous proponents and opponents quite convinced of the rightness of their arguments. There is thus considerable advantage in taking a new look at the decision itself to understand the contexts within which it was taken and the strategic alternatives available at the time.

There is also the sense that the dropping of the atomic bombs changed warfare. Whether this is actually true is a matter of some debate. What is much more definitive is the notion of Michael Howard that the dropping of the atomic bombs on Hiroshima and Nagasaki involved two extra-European powers and thus witnessed the end of an era of European expansion and dominance that had lasted since Christopher Columbus and Vasco da Gama first projected European power into the New World and Asia at the end of the 15th century.

Some important readings on the subject include:

Thomas B. Allen and Norman Polmar, "Gassing Japan," *MHQ: The Quarterly Journal of Military History* 10, no. 1 (1997), pp. 38–45. The US military had other weapons ready to go if the atomic bombs didn't work. And always remember that General Curtis LeMay considered the atomic bomb an interruption to his in-process plans to saturate the country from one end to the other with liquid fire. LeMay's plans, if carried out, would have killed far more Japanese than Hiroshima and Nagasaki did.

Iris Chang, *The Rape of Nanking: The Forgotten Holocaust of World War II* (New York: Basic Books, 1977). In December 1937, the Japanese killed upward of 300,000 Chinese civilians, or more than were killed at Hiroshima, Nagasaki, and the firebombing of Tokyo combined. The author is angry with them and the rest of the world for never making the Japanese pay or even admit guilt.

The Committee for the Compilation of Materials on Damage Caused by the Atomic Bombs in Hiroshima and Nagasaki, *Hiroshima and Nagasaki: The Physical, Medical and Social Effects of the Atomic Bombings* (New York: Basic Books, 1986). A truly wrenching volume showing the horrendous damage done to human beings by the blast, fire, and radiation. The pictures alone will searingly make you understand why war is bad for human beings and nuclear war is particularly bad for those affected.

Haruko Taya Cook, "Nagano 1945: Hirohito's Secret Hideout," *MHQ: The Quarterly Journal of Military History* 10, no. 3 (1998), pp. 44–47. How the Japanese military planned to keep fighting even after US forces invaded the island.

Hoito Edoin, *The Night Tokyo Burned* (New York: St. Martin's Press, 1987). Helps put Hiroshima and Nagasaki in perspective. The firestorms, death, and destruction were much greater in Tokyo than in either of the atomic-bombed cities. Sixteen square miles of Tokyo was leveled, as 250 tons of bombs per square inch were dropped, doing more damage than the 1923 earthquake and killing upward of 100,000 people. Makes the point that Osaka, Nagoya, Yokohama, and many other Japanese cities were burned to the ground as well. After Tokyo was set afire, Japanese prime minister Kinaki Kiosk pledged to win the war.

Richard B. Frank, *Downfall: The End of the Imperial Japanese Empire* (New York: Random House, 1999). From the Japanese perspective, the true and decisive battle for World War II was supposed to start when the United States invaded the Japanese home islands, not before. A massive strategic misassumption but one that American planners had to deal with at the end of the war.

Victor Hanson, "The Wages of Suicide: Okinawa April 1–July 2, 1945," in *Ripples of Battle* (New York: Doubleday, 2003), p. 12ff. Hanson points out that more people were killed in the battle for Okinawa than in the atomic bombings of Hiroshima and Nagasaki. American decision makers were subsequently traumatized by the notion of "10, 20 Okinawas" if the Japanese home islands were invaded.

Max Hastings, *Retribution: The Battle for Japan, 1944–1945* (New York: Knopf, 2007). An outstanding summary of the final battles that sealed the fate of the Japanese Empire and left more than a million of its solders still in the field in Manchuria, the rest of China, Indonesia, and various islands in the Pacific. Well written and powerful.

John Hersey, *Hiroshima* (New York: Random House, 1946, 1986). Most Americans, if they have read any book about Hiroshima, have read this one, and there's a good reason for it. It is carefully, even tenderly, told and contains the reactions of the Japanese who were bombed. The descriptions of death, destruction, flesh falling off, vomiting, and being buried alive are all very poignant, and most remarkable is the survivors' dedication to peace to make sure this fate is never anyone else's.

Wayne E. Lee, *Barbarians and Brothers: Anglo-American Warfare, 1500–1865* (London: Oxford University Press, 2011). Looking at "restraint" versus "atrocity" and "terror" versus "mercy," the author examines the English pacification efforts in Ireland, the English of both sides during their civil war, and in action against Indians and American colonists. He finds the English generally treated barbarians worse than brothers. Not surprising when you think of it.

Wilson D. Miscamble, *The Most Controversial Decision: Truman, the Atomic Bombs and the Defeat of Japan* (Cambridge: Cambridge University Press, 2011). "Was it necessary?" and "Was it right?" These two major questions are asked and answered. There were, the author argues, eight Japanese warlords who were making the decisions about the war. There is no sign they would have listened to anything else. He also makes the good point that Truman does indeed have blood on his hands, but he stopped the more holistic bloodletting that was projected at 250,000 killed by the Japanese per month from July 1945 on, when one includes Burmese, Javanese, Chinese, Filipinos, and Koreans, as well as Americans and Japanese. He believes that the Japanese were responsible for between 17 and 24 million dead (p. 115). The dropping of the atomic bombs ended that incredible slaughter, although this element is seldom included in the calculations when one argues against the use of the atomic bomb.

Takashi Nagai, *The Bells of Nagasaki* (Tokyo: Kodansha International, 1949). A vivid and very moving account of a Japanese doctor who survives the initial blast and is an eyewitness to the devastation and destruction, even as he helps some of the survivors in the "world of the dead."

Andrew J. Rotter, *Hiroshima: The World's Bomb* (London: Oxford University Press, 2009). Makes the case that nuclear weapons, or at least their use, has become part of the human condition, and everybody has a stake in their past, present, and future.

William Styron, "Rat Beach," *New Yorker*, July 20, 2009, p. 62ff. For the US forces, anything done to end the war before they had to invade Japan was a gift from God. The atomic bombs qualified.

Steven Walker, *Count Down to Hiroshima* (New York: Harper, 2005). Covers, almost hour by hour, the three weeks leading up to the dropping of the atomic bomb and the moment when the "world changed," providing a "peep into hell," in the words of one observer. Ten thousand times hotter than the surface of the sun, the blast destroyed 10,000 buildings and 75,000 people (including 25,000 Koreans, many of them forced laborers, and 10 American POWs, several of whom were stoned to death by irate Japanese after the bomb struck). Ominously, after describing the full damage, the author points out that the 1952 H-bomb was "thousands" of times more powerful than "Little Boy" and "Fat Boy," which fell on Hiroshima and Nagasaki, respectively.

Peter Wyden, *Day One: Before Hiroshima and After* (New York: Simon and Schuster, 1984). Quite a good account of the men and women who made the bomb, deployed it, and used it in action, as well as how it felt to those upon whom it was dropped. The author concludes, "The nuclear dawn brought false promise, but the day is not done."

Chapter Seven

Mars Is a Jealous God, Part 1

The Lessons of Iraq War I; or, Clausewitz Vindicated—For the Last Time?

Positioned here, this chapter may seem out of place, unless one chooses to see the Vietnam War and the First Iraq (or Gulf) War as a part of a continuum starting in 1965 and running to 1991, which, I believe, is the proper way to appreciate it. It is almost as if, in a true history of warfare, both wars were mirror images of each other. The armed forces of the United States and their civilian leaders made many, many mistakes in the former, and made many, many excellent military decisions in the latter. Along the way, the American military completely remade itself in terms of discipline, technology, and strategy. In terms of the god of war, the two wars show a strong contrast in each, whereby the winner followed, almost to the letter, the tenants of the Template and the loser failed to do so.

This chapter looks at how the ends of wars contain the tactical and strategic seeds for the next war. It also explains how a new utilization of the concepts of Clausewitz (including centers of gravity, the fog of war, the friction of war, culminating points, and *Schwerpunkts*) by US military planners led to a strategy in which "hi diddle diddle, straight up the middle" was replaced by one more in keeping with his findings. The result was one of the most astonishing outcomes in military history, for both the success of the battlefield action and the failure of the political results. This chapter argues that there is a considerable need to see different wars within single geographical and historical matrixes, such as Iraq I (1991), Iraq II (2003), Iraq III (2003–2012), Afghanistan I (2001), and Afghanistan II (2002–present).

Rick Atkinson, *Crusade* (Boston: Houghton Mifflin, 1993). A carefully written and most useful introduction to the US-led expedition to liberate Kuwait. Well written and attention holding.

Sherman Baldwin, *Iron Claw* (New York: William Morrow, 1996). A naval pilot describes action in the Persian Gulf during the First Iraq War. Some useful insights into below-decks action, as well as combat.

John R. Ballard, *From Storm to Freedom: America's Long War with Iraq* (Annapolis, MD: Naval Institute Press, 2010). A very useful volume that correctly sees the First and Second Iraq Wars as part of a space-time continuum that needs to be examined as a whole. He also perceptively sees the Second Iraq War as having two distinct segments: the set-piece—the blitzkrieg—and the initially smoldering and then savage outbreak of the insurgency that followed. While many would argue with his conclusion, ("the conflict ended with the major objectives set for it largely achieved and with a very different, and much improved future possible for the people of Iraq"), all can benefit from his carefully constructed time line of events.

Sir Peter de la Billiere, *Storm Command* (London: HarperCollins, 1992). The top British commander in the Iraq I War provides his perspective, and it is a most useful one. Straightforward and free from vain glory. A refreshing account.

Tom Carhart, *Iron Soldiers* (New York: Pocket Books, 1994). For armored warfare buffs, this is a must. It is the story of the 1st Armored Division ("Iron Soldiers") and their incredible record-setting blitzkrieg through the Tawakalna, Medina, and Adnan Republican Guard Divisions (plus elements of 11 other Iraqi divisions). At a cost of 4 M-1A1 Abrams tanks, 2 Bradleys, and 4 Humvees, they destroyed 440 tanks, 485 armored personnel carriers, and 1,383 trucks. The battle group suffered but 4 deaths in a battle where American planners feared 20,000 casualties. It is an amazing story and one that is known to few.

Dick Cheney with Liz Cheney, "Desert Shield" and "Desert Storm," in his *In My Time: A Personal Memoir* (New York: Threshold Editions, 2011), pp. 181–239. As secretary of defense, Cheney is instrumental in the fight to liberate Kuwait, and his insistence on changing the proposed strategy of "hi diddle diddle, straight up the middle" saves lives and is proven correct. Perhaps his finest hour.

Tom Clancy, *Every Man a Tiger* (New York: G. P. Putnam's Sons, 1999). General "Chuck" Horner provides invaluable insights into the bombing campaign that helped make Coalition ground action such a success. After Vietnam, says Horner, "We had to learn how to be an air force all over again."

———, *Into the Storm* (New York: G. P. Putnam's Sons, 1997). Clancy gets inside the mind and heart of General Fred Franks Jr., who lost a leg in Vietnam and fought his way back against many odds to lead the "Left Hook" of the Coalition offensive to liberate Kuwait and did a good job of it, despite the interference and strident prodding of "Stormin' Norman." Very good maps of the ground war.

Stephen Duncan, *Citizen Warriors: America's National Guard and Reserve Forces and the Politics of National Security* (Novato, CA: Presidio, 1997). Useful look at the role of the "new" citizen warriors of the United States in Desert Shield and Desert Storm and an overview of the All-Volunteer Total Force Policy. On balance, a vast improvement over the previous American way of war, at least according to Mars.

Editors of Military History, *Desert Storm* (Leesburg, VA: Empire Press, 1991). Many good insights and overarching themes about the war in 1991 that projected the United States into the region as never before and where it still is today.

David Finkel, *The Good Soldiers* (New York: Sarah Crichton Books, 2009). A sad, depressing account of the soldiers who fought against the insurgency in Iraq before the tide turned. Their efforts were eventually vindicated, but in the "instant" history of today, who in the wider world knows or cares? A pity.

General Tommy Franks, *American Soldier* (New York: Regan Books, 2004). The top American general in Iraq II describes the war before it became Iraq III, by which time he had wisely retired. Not a great general but one with a superb sense of timing.

Lawrence Freedman and Efraim Karsh, *The Gulf Conflict 1990–1991* (Princeton, NJ: Princeton University Press, 1992). Good, comprehensive, overall assessment.

Richard Hallion, *Storm over Iraq: Air Power and the Gulf War* (Washington, DC: Smithsonian Institution Press, 1992). A bit over the top in terms of hyping the ascendency of air power in warfare, calling it a "revolution," but the author puts the First Gulf War campaign in a good historical context.

John Keegan, *The Iraq War* (New York: Alfred Knopf, 2004). Once-over lightly from one of the top military historians. Not much new here. One wishes Keegan would write fewer books but ones with more of his personal insights and knowledge.

David Kilcullen, *Out of the Mountains: The Coming Age of the Urban Guerilla* (London: Oxford University Press, 2013). The author sees a "megatrend" of insurgencies moving into the cities. These huge, ungovernable (and unprotectable) urban agglomerations (8 of the top 10) are within 50 miles of the sea. Cites the examples of Mumbai (2008); Mogadishu (1992); Kingston (2010); and Yefress, Libya (2011). He concludes, "It's time to disengage ourselves—body and mind—out of the mountains."

Robert Leonhard, *The Art of Maneuver: Maneuver-Warfare Theory and the AirLand Battle* (New York: Ballantine, 1991). Focuses on the notion of defeating the enemy *through* maneuver and looks at the Gulf War as a triumph of that concept and a vindication of the AirLand Battle but honestly concludes that it has difficulties in a war set in a multipolar environment as well as low-intensity conflicts. Written at a time when others were talking about the "end of history," this work seems most prescient.

Bernard Lewis, *The Crisis of Islam* (New York: Random House, 2004). A realistic and in-depth analysis of Islam in historical perspective. Much disparaged by liberal critics.

Douglas Macgregor, *Warrior's Rage: The Great Tank Battle of 73 Easting* (Annapolis, MD: Naval Institute Press, 2009). A gripping, searing account of the 2nd Armored Cavalry Regiment, whose 4,000 soldiers and 350 armored vehicles were the spearhead of the VII Corps smashing of the Republican Guards during the first Iraq War. Bemoaning the fact that it was the last one and blaming the "unhealthy patterns of caution, indecision and inaction," the author seems to ignore the commitment President George Bush made to the UN and the Coalition partners *not* to drive on Baghdad. The author is correct, however, in asserting the Coalition *could* have done so. Franks's VII Corps alone had 1,100 tanks and 2,000 other armored vehicles at their disposal, to say nothing of complete air supremacy.

Andy McNab, *Bravo Two Zero* (New York: Dell, 1993). A rough-and-tumble first-person account of the First Iraq War by a member of British special forces before being captured. Accurately describes this dimension as "[d]irty, deadly and rough outside the rules."

Judith Miller, *Saddam Hussein and the Crisis in the Gulf* (New York: Random House, 1990). How Hussein planned, executed, and eventually disengaged himself from the invasion of Kuwait and who aided and abetted him. Quite a triumph out of quite a mistake.

Molly Moore, *A Woman at War* (New York: Charles Scribner's Sons, 1993). Back when it was unusual for American women to be at war and writing about it. Clearly shows women in harm's way and worthy of equal opportunity if they so choose.

David Morris, *Storm on the Horizon: Khafji—The Battle That Changed the Course of the Gulf War* (New York: Free Press, 2004). A battle that should be much studied because it foretold the rest the war and its aftermath. Saddam Hussein believed that the monumental thrashing his forces suffered in the battle was actually a victory.

Williamson Murray and Kevin Woods, *The Iran-Iraq War* (Cambridge: Cambridge University Press, 2014). Anyone seeking to understand Iraq I and II needs to examine this amazing insight into the mind of Saddam Hussein. Based on captured documents, it describes and analyzes the eight-year long Iran-Iraq War, showing Saddam's mindset for starting it, his misperceptions, and those of Ayatollah Khomeini. Seeking to take advantage of upheaval in Iran and the decimation of the Iranian armed forces by Khomeini, Saddam fancifully launched a war that quickly bogged down. Human waves of believers from Iran then counterattacked over and over and over until both sides were exhausted. The authors conclude Saddam simply was able to purchase 10 to 12 billion in arms yearly, while Khomeini used human waves.

General William G. Pagonis, *Moving Mountains* (Boston: Harvard Business School Press, 1992). The author once again proves that logistics win wars. Ninety percent of US supplies and equipment came to the Gulf War by sea, so it was just as well Saddam didn't attack during the sixth months it took to get all the gear to the Gulf.

Kenneth Pollack, *The Threatening Storm* (New York: Random House, 2002). Another expert who believed there were weapons of mass destruction in the hands of Saddam Hussein and helped to provide information to buttress the case for war. The author continues to offer op-ed pieces in the *New York Times* and *Wall Street Journal*, almost as if he had gotten the advice correct. Quite an irritating pundit, really.

Tim Pritchard, *Ambush Alley: The Most Extraordinary Battle of the Iraq War* (New York: Ballantine, 2005). Based on eyewitness accounts, this work covers the Marine "Tip of the Spear" Task Force Tarawa and its attack to capture the Euphrates River bridges at Nasiriyah during March 23, 2003, when new and untried troops met heavy resistance, tried new tactics, and suffered horrendously from A-10 "Blue on Blue" fratricide. Illustrates and vividly captures the fog, friction, and chaos of war, leading the author to conclude, "War is a confusing, bloody mess," but the Marines carried their objectives.

Keith Rosenkranz, *Vipers in the Storm* (New York: McGraw Hill, 1999). A close look at the ground-support missions flown by the F-16, the workhorse of the First Gulf War. Many specifics about flying at night in close support situations where fractions of seconds make the difference between life and death for those you are trying to help and those you are trying to hurt.

Norman Schwarzkopf, *It Doesn't Take a Hero* (New York: Random House, 1992). Full of bombast and glory-seeking, this volume does give an insight into the general who took credit for the "Hail Mary" pass offensive that won the war, even though he initially pushed hard for a "hi diddle diddle, straight up the middle" strategy, a strategy that was wisely rejected by messieurs Powell and Cheney.

Keith L. Shimko, *The Iraq Wars and America's Military Revolution* (Cambridge: Cambridge University Press, 2010). A lucid overview of the space-time continuum between the Vietnam War and the present war in Afghanistan using the concept of the revolution in military affairs (RMA) as a unifying device.

William L. Smallwood, *Strike Eagle: Flying the F-15E in the Gulf War* (Washington, DC: Brassey's, 1994). The fighter jocks and how they scared off those Iraqi pilots who thought they were in the big leagues. But it was a Saudi pilot alone who scored a double kill in the first Gulf War.

———, *Warthog: Flying the A-10 in the Gulf War* (Washington, DC: Brassey's, 1993). Amazing story of one of the best close-support aircraft of all time, including the German JU-87, and how the Air Force wanted to take it out of service after it proved its value. As of 2015, it is still flying and doing a most important job in the American fight against ISIS.

Jay Stout, *Hornets over Kuwait* (Annapolis, MD: Naval Institute Press, 1997). US naval aviation in action in the First Gulf War as seen through the eyes of a carrier pilot. A lot of yanking and banking by the guys who know how to do that in the air and on land.

US Marine Corps, *Warfighting* (Washington, DC: Government Printing Office, 1997). Drawing on hundreds of years of warcraft as well as some insights from Sun Tzu, the US Marines get ready to keep doing what they have always done: fight hard and let history draw its own conclusions.

US News and World Report, *Triumph without Victory* (New York: Random House, 1992). Points out that liberating Kuwait did nothing about overthrowing Saddam Hussein and yet is weak in describing how the United States was to have effected such change without physically occupying Baghdad or disregarding the UN resolution by which the Coalition was bound.

Alex Vernon, with Neal Creighton Jr., Greg Downey, Rob Holmes, and Dave Trybula, *The Eyes of Orion: Five Tank Lieutenants in the Persian Gulf War* (Kent, OH: Kent State University Press, 1999). Absorbing account of five tank lieutenants in the 24th Infantry Division (see its exploits in World War II and Korea) and their experiences in the Persian Gulf War. The accounts of tank warfare in the middle of the night, in the middle of a thunderstorm, slashing and driving forward, and shooting on the run are truly illuminating. The descriptions of "fighting pure" at Jalibah airfield alone is worth the price of the book. Not as professionally written as Carhart's *Iron Soldiers* but a worthwhile read.

Steven Vincent, *In the Red Zone: A Journey into the Soul of Iraq* (Dallas: Spence, 2004). Writing in 2004, the author may have had a hard time finding a main-line publisher because his views that the United States was right to intervene in Iraq and the war would not necessarily end with the collapse of the country were definitely in the minority. But there is great profundity in his statement, "On 9-11, the despair and self-loathing at the center of the Muslim world was unleashed on America. On that day, tribal Islam became *our* problem too."

Bruce Watson (ed.), *Military Lessons of the Gulf War* (Novato, CA: Presidio Press, 1991). There were many of them, including "don't stop the war too soon when you are winning" and "don't let the enemy win at the negotiating table what he or she cannot win on the battlefield" and "never, never be in such a rush to make peace that you lose the prospect to make it lasting."

Bob Woodward, *The Commanders* (New York: Random House, 1991). The leaders who led the US and Allied coalition that liberated Kuwait but failed to remove Saddam Hussein from power and who, in the rush to end the war quickly, sowed the seeds for future and ongoing strife.

———, *Plan of Attack* (New York: Simon and Schuster, 2004). Bush the hero leads the United States to victory in the first Afghan campaign, despite a reluctant military and because of an eager CIA. Within a few years, the roles would be reversed with disastrous results.

Chapter Eight

Is the Template Obsolete in the Face of Insurgencies?

> "Wars consisteth not in battle only, or in the act of fighting; But in a tract of time, wherein the will to contend by battle is sufficiently known; and therefore the notion of time, is to be considered in the nature of time."—Thomas Hobbes, *Leviathan*

> "Rebellions can be made by two percent active in the striking force and 98 percent passively sympathetic."—T. E. Lawrence, *Seven Pillars of Wisdom*

> "You have the watches but we have the time."—Pathan saying

> "Pinpointing the end of an insurgency is like pinpointing the end of a marriage."—Owen West, *The Snake Eaters*

> "The paramount need . . . to gain the hearts and subdue the minds of Americans"—General Henry Clinton, quoted in *Invisible Armies* (note: 1776!)

Some argue that "wars among the people" represent a new form of warfare, but the initial set of readings in this chapter strongly suggest that it is simply a new version of an old, old paradigm of warfare. This assumption is vigorously challenged in chapter 10 of *Winning at War* (pp. 195–224). Fortunately for the reader, there are a plethora of sources listed here that argue both points of view. In addition to the works listed in this chapter, interested readers should also consult the additional readings on the subject cited in chapter 10's footnotes (see pp. 258–59).

Terrorists, partisans, guerrillas, insurgents, irregulars, revolutionaries, special warfare practitioners, special operations adherents. Because of the plethora of works on this subject, I have divided this section into three major parts: "Theoreticians and Practitioners," "Overviews of Insurgencies," and "Some Illustrative Case Studies: Showing the Experience of Insurgency and Counterinsurgency across Cross-Cultural Lines." All are suggestive rather than exhaustive in content.

THEORETICIANS AND PRACTITIONERS

Paul Aussaresses

Paul Aussaresses, *The Battle of the Casbah* (New York: Enigma Books, 2002). An unvarnished account of the French counterinsurgency efforts during the Battle of Algiers. The author worked with Roger Trinquier in suppressing the FLN in the capital and argues for the inclusion of summary execution and torture, concluding, "I don't think I ever tortured or executed people who were innocent." Tactical efficiency but strategic inefficiency as it turned out. But it does answer in the affirmative whether torture "works" in the tactical sense. The FLN Algiers network was rolled up big time. Aussaresses lectured at Ft. Bragg in the 1960s and influenced those who would form the US Phoenix Program in Vietnam.

Michael Bakunin

Michael Bakunin, *Statism and Anarchy* (Cambridge: Cambridge University Press, 1996). The foremost theoretician of Russian anarchists took inspiration from Hegel as he became anti-Austrian, anti-German, anti-Semite, and anti-Marxist through a long career of unpaid loans, prison, banishment to Siberia, and escape, as well as some years in Italy and Switzerland. Enamored with "popular spontaneity," he was also intrigued by "revolutionary dictatorship." Ironically, the anarchists of Russia would be among the first to criticize the Bolsheviks and the first to be liquidated by them. Marx thought Bakunin a deviant for placing "will" ahead of "economic determinism" and had him expelled from the International.

John W. Beebe

John W. Beebe, "Beating the Guerrilla: Korea 1950," in *Military Review* 25, no. 2 (December 1955), pp. 66–72. An amazingly prescient and unappreciated treatise on winning a counterinsurgency campaign based on Operation Ratkiller (December 1951–March 1952) through Operation Trample (December 1953–June 1954). Too bad the American military never paid attention to this article and its conclusion:

> It would appear desirable to give more instruction in antiguerrilla operations in service schools, and more training in this subject to units. Antiguerrilla plans should be prepared by every command faced with the possibility of a guerrilla threat. Recommendations for the adoption of political, psychological and economy polices designed to prevent the rise of a guerrilla force or to minimize its effect should be pressed vigorously on the government concerned.

C. E. Callwell

C. E. Callwell, *Small Wars: Their Principles and Practice* (Lincoln: University of Nebraska Press, 1996). One might think that this 1896 study of the spread of imperialism "on the ground" would not be very fruitful in studying modern-day insurgencies. But in fact, it is chock full of useful insights: "Seize what the enemy prizes most," "strike hard," and "pursue relentlessly" are some of his advice. He places great emphasis on the human element, not technologies in dealing with insurgents *before* they get up a head of steam. Should be required reading for those operating in Somalia, Yemen, Mali, and the rest of the Sahel.

Fidel Castro

Fidel Castro and Ignacio Ramonet, *My Life* (New York: Scribner, 2006). A stream-of-consciousness account of his life and times and political philosophy. Gives some "Lessons for a Guerrilla War" (pp. 205–14) that show how much ad hoc adjustment he and his tiny band made. He claims never to have used terror and assassination and focuses on paying for supplies and never putting innocents in harm's way. While declaring himself a Marxist-Leninist, he stresses his ideological movement from the utopian socialism (and the liberation inspiration of Marti and Sandino) to radical socialism. And he denies being influenced in any serious way by Ho, Giap, or Mao. In fact, he exhibits most enthusiasm for Ernest Hemingway and his *For Whom the Bell Tolls*, declaring, "And we always went back to it, consulted it, to find inspiration, even when we were already guerrillas." I think he is being very candid when he says, "We revolutionaries learned the art of war up there (in the Sierra) by fighting." This "learn as you go," ad hoc exercise sounds like it could have been crushed with even a minimal, competent counterinsurgency effort, as Castro was captured a number of times and let go!

Robert Quirk, *Fidel Castro* (New York: W. W. Norton, 1993). A straightforward but long narrative of his life showing his most improbable rise to revolutionary status. "The Rebel Victory" shows astonishing ineptness on the part of the Cuban government, before and during his Sierra Maestra days. Makes very little of his revolutionary doctrine, which appears to be highly derivative.

Carl von Clausewitz

Carl von Clausewitz, "The People in Arms," in his *On War* (New York: Alfred A. Knopf, 1993), pp. 578–84. This short, often overlooked chapter on insurgency is very pithy, asserting that "we consider a general insurrection as simply another means of war—in its relationship, therefore to the enemy." He stresses both the importance of the country's conditions ("must be rough and inaccessible") and the nature of the people ("The national character must be suited to that type of war"). From the point of view of the insurgent, a "general uprising, as we see it, should be nebulous and elusive, its resistance should never materialize as a concrete body."

Christiaan De Wet

Christiaan Rudolf De Wet, *Three Years' War* (New York: Charles Scribner's Sons, 1903). Inside a Boer commando with gusto, this is a fascinating account. Especially good are his chapters "The English Swarm over Our Country" (pp. 82–91) and "Was Ours a Guerrilla War?" (pp. 225–29). De Wet claims they were not guerrillas because Great Britain simply annexed their country.

Frederick Engels

Frederick Engels, *The Peasant War in Germany* (New York: International, 1926). Engels looks at the Reformation and the German Peasant Revolt of 1525 through the eyes of class warfare and sees the "feudal masters," the princes, and the princes emerging as victorious. Good lessons for would-be counterinsurgents, one imagines.

David Galula

A. A. Cohen, *Galula: The Life and Writings of the French Officer Who Defined the Art of Counterinsurgency* (Santa Barbara, CA: Praeger, 2012). Few authors can seamlessly combine military biography and exegete excellence concerning military doctrine. This volume shows that Cohen excels at those tasks. *Galula* is a superb combination of a delving into Galula's life together with a careful, intellectually rigorous examination of his doctrine and a location of its place in the literature of insurgency. Simply put, the author believes Galula is a "superstar," in a class with Clausewitz and Mao. Moreover, Cohen argues persuasively that any army that is expected to operate overseas today and tomorrow will require firm grounding in the knowledge of insurgency and counterinsurgency. Stresses the need to provide security and control (thus a seamless blend of population-centric and enemy-centric focus) and the need to prove to the bulk of that population that the counterinsurgents have the "will, the means and the ability to win."

David Galula, *Counter-Insurgency Warfare: Theory and Practice* (New York: Frederick Praeger, 1964). Quite the period piece this, written at the height of "Mao mania," but captures the proportion of the military to the political (20 percent to 80 percent) in successful revolution or counterrevolution. Galula, with his experience in Algeria, has become much studied and wildly praised in counterinsurgency circles. Especially well worth reading are the parts on liquidating guerrilla "remnants." For the "Galula Doctrine," some of the major points, according to Cohen, reviewed previously, are: (1) the primacy of protection and its concomitant control aspects; (2) the blending of the population and enemy-centric strategies; (3) protection as a means for a comprehensive strategy of "clear, hold, build"; (4) counterinsurgency personnel always having the holistic as well as the tactical goals in mind; and (5) the clear sine qua non that irregular warfare requires both determination and understanding by those in command, plus more patience than normal in waging war.

Note: Thomas Rid, "The Nineteenth Century Origins of Counterinsurgency Doctrine," *Journal of Strategic Studies* 33, no. 5, pp. 727–58, is especially useful in pointing to the developments of French pacification in the 19th century, most notably those of Marshal Joseph-Simon Gallieni in Tonkin and Madagascar (against the Hovas, among others) and Marshal Thomas Bugeaud in North Africa against various Arab and Berber tribes. The *razzia* (violent, lightning-fast raids against enemy population centers) and the *bureau arabe* (political officers with a "rather unmilitary skill-set") coordinate military action and civic development as part of the overall colonial structure brining locals into the administration. Gallieni's policy of clear and hold led to his "oil slick" analogy, later appropriated by others as the "oil spot" approach. The French system in these cases thus became very population centric. The population was to be convinced that their interests were better served by working with the French than by taking a chance with the insurgents.

———, *Pacification in Algeria: 1956–1958* (Santa Monica, CA: RAND, 1963). If only American military planners had given this work the attention it deserves—first during Vietnam, then in Iraq, and finally in Afghanistan! Having observed the revolutions and attempted revolutions in China, Malaya, the Philippines, and Indochina, Galula ended up leading counterinsurgency operations in the Kabylia area, east of Algiers, and came up with four laws for successful counterinsurgent operations: (1) The object is the population, not territory ("People are the terrain of the war"); (2) support of the population must be organized (i.e., it does not grow organically); (3) the "minority supporting the government will be followed by the majority, only if the counter-insurgency is seen as the ultimate victory" (shades of John Vance in Vietnam!); and (4) "[s]eldom is the material superiority of the counterinsurgency so great that the entire country can be covered" (clear and hold!). He concludes, so rightly, "A counter insurgency is never lost *a priori* because of a supposedly unpopular regime." This is powerful stuff. He died in 1967. Read this book to understand true counterinsurgency.

Ann Marlowe, *David Galula: His Life and Intellectual Context* (Washington, DC: SSI Monograph, 2010). Calling Galula a "theoretical genius" for his accent on population-centric warfare ("The population had to be protected, controlled, won over, and isolated from the rebels.") and his emphasis on strong, honest, efficient administration, one that balances the use of the "carrot with the stick." Born into a Jewish family in Tunisia (who had been there long before colonization by the French), he moved to Morocco; was educated in France at St. Cyr; served the Vichy French and then the Free French; posted to China (an early student of Mao's theories and practices); married an American diplomat; served in Greece and later Algeria (as a captain, then major); retired from the army in 1962; and then served as consultant to NATO, the US military and RAND Corporation, and others. Reborn in *U.S. Army Field Manuel 3–24* (COIN) of 2006. Wrote *The Tiger's Whiskers* under the pen name Jean Caran in 1965 and was the basis for the character "Jean Leone" in Seymour Topping's 1999 novel *The Peking Letter.*

Giuseppe Garibaldi

Ron Field, *Garibaldi: Leadership, Strategy, Conflict* (Oxford, UK: Osprey, 2011). A snappy, breezy introduction to the red-shirted Garibaldi and the many who led them through thick and thin. Garibaldi fought in Brazil, Uruguay, much of Italy (1849–1870), and later France during the Franco-Prussian War at the end of an International Brigade. He also offered his services to the North during the Civil War. Garibaldi was for universal suffrage, emancipation of women, and the abolition of ecclesiastical property, in addition to his perspicacious espousal of insurgency tactics. Marx was quite dismissive, calling him a "taxi driver" for turning the Kingdom of the Two Sicilies over to Victor Emmanuel.

Charles Forbes, *The Campaigns of Garibaldi in the Two Sicilies: A Personal Narrative* (London: William Blackwood and Sons, 2010). A Kessinger Publishing Legacy reprint of the 1862 edition, this breezy book gives us a sense of the Garibaldi "crusade" as he and his Red Shirts fought "Papal and Bourbonic Oppression." He was a doer and an inspirer: "To arms, then! Let us by one blow put an end to our chronic misfortunes. Let us show the world that this is truly the land once trodden by the Great Roman race." He and they did!

Giuseppe Garibaldi, *The Autobiography of Giuseppe Garibaldi*, 3 vols. (New York: Harold Fertig, 1971). This reprint of Garibaldi's 1889 autobiography is a vast, sprawling work, with his comments on Mazzini, Cavour, the papal states, war in Uruguay, Sicily, Italy, and France. An amazing set of insights coupled with considerable personal trials and tribulations.

Giuseppe Garibaldi, *Rules for the Conduct of Guerrilla Bands.* "Insurrection—by means of guerrilla bands—is the true method of warfare for all nations desirous of emancipating themselves from a foreign yoke. . . . It is invincible, indestructible." Guerrillas need "to constantly attack, do mischief, and retire." Garibaldi was to inspire revolutionaries across Europe and the Americas with simple but timeless prescriptions and his heroic style of leadership.

Vo Nguyen Giap

Cecil B. Curren, *Victory at Any Cost: The Life and Times of Vo Nguyen Giap* (London: Brassey's, 1997). Giap wrote the book on sustained ruthlessness, and it paid off; still he used lives with truly reckless abandon. Curren, in the midst of great admiration, does suggest he was "not a nice person." He excelled in using time and space in order to defeat a technologically superior foe.

Vo Nguyen Giap, *How We Won the War* (London: Recon Press, 1976). Giap's later press interviews, however, give a more nuanced and even contradictory view from his vantage point during the war. He was a remarkable blend of theoretician and practitioner, and interestingly enough, his major mistakes seem to have come more from adherence to the former than decision making as the latter.

———, *People's War, People's Army* (New York: Bantam, 1968). The classic account of the Vietnam War by its chief strategist. Some sugarcoating of his disasters, but some of his later accounts were more honest and forthright. Shows how much he owes Clausewitz as well as Mao.

Nathanael Greene

"We fight, get beat, rise and fight again" was his motto, and Greene also said—and this proved to be extremely valid, for not just the success of the American insurgency, but the Russian, Chinese, and Vietnamese insurgencies as well—"You may strike a hundred strokes, and reap little benefit from them, unless you have a good army to take advantage of your success." This combination of elements, and the strategy that accompanies them, gives the insurgents the best chance of eventual ultimate success. Lord Cornwallis paid him the ultimate tribute when he called him "as dangerous as Washington."

Note: The University Press of North Carolina has printed an extensive set of Greene's middle-of-the-war letters. See Richard Showman (ed.), *The Papers of General Nathanael Greene*, 3 vols. (Chapel Hill: University Press of North Carolina, 1983), covering the period 1777–1779.

Terry Golway, *Washington's General: Nathanael Greene and the Triumph of the American Revolution* (New York: Henry Holt, 2006). An amazing story of an American autodidact who was initially refused entrance into the Rhode Island militia because of his limp but subsequently was appointed general by the legislature of Rhode Island and served his apprenticeship under Washington at Boston, Long Island, New York, Valley Forge, Philadelphia, Trenton, and Princeton and during the Foraging War, where guerrilla and partisan tactics surrounded the more famous set-piece battles. But it was only after General Horatio Gates was crushed by the British in August 1780 at Camden, South Carolina, that he became commander of the southern wing of the Continental Army. He improvised and adopted guerrilla warfare, becoming the "shadow of an army." He drove Cornwallis to distraction with forced marches, dividing his soldiers when required; lightning moves; and a focus on living to fight another day as he marauded all over North Carolina, South Carolina, and parts of Virginia. Greene bled Cornwallis in a series of narrow losses or draws (Guilford Courthouse and Eutaw Springs) and ultimately drove him to seek solace in Yorktown, marching there from Wilmington, North Carolina. Later, the British would withdraw from Charleston, holding only Savannah in the south. Greene was given several plantations by a grateful South Carolina, and this nonpracticing Quaker used slaves to try to make a go of it before his untimely death. His efforts to capture the "hearts and minds" of those undecided during the insurgency and his sense of fairness toward the Tories should be studied, as well as his military accomplishments.

Lt. Colonel Paul Montanus, *A Failed Counterinsurgency Strategy: The British Southern Campaign—1780–1781: Are There Lessons for Today?* (Carlisle, PA: US Army War College, 2005). The British had a flawed assumption, namely that there was a large Loyalist population in the Carolinas, but instead, the British occupation "inflamed the population and pushed them towards the rebels rather than pacify them." The author offers hope that the United States will avoid their mistakes in 2005 Iraq: a brutal civil war, a lack of patience, a limited number of British regulars available in theater, a lack of British restraint, and a "failure to Americanize" the counterinsurgency.

James Swisher, *The Revolutionary War in the Southern Back Country* (Gretna, GA: Pelican, 2008). An interesting account of how the American rebels went from set-piece battles to insurgency and then, as their local successes grew, from partisan bands into a coherent coordinated pattern with an eventual field army as guerrilla warfare (led by Daniel Morgan, Francis Marion, and Nathanael Greene) morphed into mobile warfare, eventually winning such major battles as King's Mountain, Cowpens, and Guilford Courthouse and driving Charles Lord Cornwallis to hunker down in Yorktown, Virginia—with disastrous results for the British cause. See especially "The Rise of Partisan Warfare" (pp. 173–202). Earlier, the British, seeking to rebound from their defeat at Saratoga (1777) and the entrance of the French (1778), decided to take the war to the south, capturing Savannah and Charleston and moving inland to take advantage of the large number of Loyalists.

Note especially the campaigns of Francis "The Swamp Fox" Marion, as well as Daniel Morgan's brilliant victory at Cowpens (1781).

Ernesto "Che" Guevara

Che Guevara, *Che: The Diaries of Ernesto Che Guevara* (New York: Ocean Press, 2009). Sadly for our lad, his lover was working for the KGB, and the CIA was monitoring the KGB's transmissions. Not a good model for one's security protocol if one is a guerrilla trying to elude capture and death.

———, *Guerrilla Warfare*, 3rd ed. (Wilmington, NC: SR Books, 1997). I have always found this 1960 work to be strangely counterintuitive and contradictory. One the one hand, Guevara's totally bungled final effort in Bolivia (his mistress was a KGB agent, and her transmissions to that august body were monitored by the CIA, and he was trying to start a revolution among a people whose language he did not speak, etc.) is somewhat off-putting as an example of how to do things. On the other, his conclusions are clear, concise, and very useful: "1. Popular forces can win a war against the army. 2. It is not necessary to wait for all conditions for making revolution; the insurrection can create them. 3. In underdeveloped America the countryside is the basic arena for armed fighting."

Aylmer L. Haldane

General Aylmer L. Haldane, *The Insurrection in Mesopotamia* (London: W. Blackwood and Sons, 1922). An in-depth look at the 1920 insurrection/insurgency and how it was defeated by a rare combination of stick, carrot, and persistence, plus good leadership. This campaign was the first time air power was used against insurgents, according to Jonathan Roth.

Ernest Hemingway

Ernest Hemingway, *For Whom the Bell Tolls* (New York: Charles Scribner's Sons, 1946). For many, his best novel and one that captures the human dynamics in small partisan units as well as the general feelings of being at war. But it is also important as a guerrilla "how-to" manual as declared in Fidel Castro's *My Life* (reviewed previously in this chapter). While Castro undoubtedly took his ideology from Lenin, he claims that he and his compatriots referred to this work early and often, especially on how guerrillas should behave toward the population they were trying to liberate. Especially illuminating is the assumption that, during insurgency, you can practice and sustain humanity in action.

Note: Then, in reading Seymour Topping's *The Peking Letter* (New York: BBS, 1999), one discovers yet another indication of the breadth of Hemingway's impact on revolutionaries. On page 245, for example, when the Chinese heroine Lillian notices he has been reading *For Whom the Bell Tolls*, the dedicated Communist Han announces, "It's a good piece of revolutionary literature." Probably too bad for tens of millions of Chinese that Mao didn't keep the more humanistic version of revolution depicted there.

Albrecht Kesselring

Albrecht Kesselring, "The Partisan War in Italy," in his *The Memories of Field Marshal Kesselring* (London: William Kimber, 1953), "pp. 224–33. An excellent overview of the difficulties and dimensions of guerrilla warfare for a many known as "Smiling Albrecht," Kesselring directed a fair amount of counterinsurgency work in Yugoslavia and Italy during World War II.

Note: For a broader perspective of the German (and Italian) counterinsurgency efforts in the Balkans more generally, see the US Army, *German Anti-Guerrilla Operations in the Balkans (1941–1944)* (Washington, DC: Center of Military History, 1989). Lays out a thinly veiled appreciation for the very complex nature of the insurgency, counterinsurgency and interinsurgency struggles across a large chessboard, concluding that one in seven of those in German uniforms were casualties. Concludes that the Germans were always undermanned in their operations and that some of their substitutions—Italians, Croats, Serbian "quislings," and Bulgarians—actually made the situation worse. "Smiling Albrecht" had the right formula perhaps but could never overcome the demands of Operation Barbarossa in implementing it.

Charles Krulak

Charles Krulak, "Strategic Corporal: Leadership in the Three Block War," *Marines* 28, no. 1 (January 1999), pp. 26–33. How and why the surge worked and how the tip of the spear is often ahead of the general in charge of spear-throwers.

Edward G. Lansdale

Edward G. Lansdale, *In the Midst of Wars: An American's Mission to Southeast Asia* (New York: Fordham University Press, 1991). This is quite an amazing account of a man at the center of counterinsurgency in the Philippines and in the early (and later) stages of American involvement in the Vietnam War. Modest and self-effacing (although exuding an occasional whiff of "post hoc, ergo propter hoc") despite a lengthy career in the OSS, the US Air Force, and the CIA, Lansdale believed that the secret to counterinsurgency was adherence to, and promotion of, the American ideals of the Bill of Rights and the Declaration of Independence and by stressing true democratic values, ideas, and processes. His emphasis on avoiding venality in government, civic action, clean water, freedom of political expression, local defense, and soldiers as "brothers" to the villagers all ring true. I also like his saying that the best weapon against a guerrilla is a knife, and the worst, an airplane. Lansdale clearly distrusted the later American emphasis on raw military power as an end in and of itself. Lansdale finally served as a special assistant to the American ambassador Henry Cabot Lodge from 1965 to 1968.

It is important to realize how much even attempting Lansdale's prescriptions for successful counterinsurgency depend on the quality of leadership available to those implementing reforms. For example, compare the man of the people, "on the stump" style of Ramon Magsaysay in the Philippines with that of the "Emperor," Boa Dia, in Vietnam, who attempted to run that country from the French Riviera, eventually appointing Ngo Dinh Diem as prime minister in 1954. Fictional accounts of Lansdale (portrayed as quite naïve) and his exploits appear in Graham Greene's *The Quiet American* and William Lederer and Eugene Burdick's *The Ugly American.*

———, *Lessons Learned—The Philippines, 1946–1953* (Washington, DC: Department of Defense, 1962). Short and anecdotal but spot-on for at least half of successful counterinsurgency, with the accent on the political-psychological approach: "It was not the number of dead Huks. It wasn't the money or hardware. It was the ever increasing number of Filipinos dedicated to making their democracy live." Lansdale was the much maligned and mocked "hero" of *The Quiet American* and *The Ugly American* books and films, but his emphasis on having every solider be a civic action agent and on the tried and true appeal of the US Constitution and Bill of Rights remain extremely relevant for winning counterinsurgency struggles in the 21st century.

Jean Larteguy

Jean Larteguy, *The Centurions* (New York: Dutton, 1961), *The Praetorians* (New York: Dutton, 1963) and *The Hounds of Hell* (New York: Dutton, 1966) (F). All provide startling and useful insights into the revolution and counterrevolution in Vietnam, Algeria, and the Congo. Written in medias res, they contain very important looks at the process of insurgency and counterinsurgency, with the combination of romantic comradeship and Gallic cynicism, which only the French seem capable of simultaneously sustaining for long periods of time

T. E. Lawrence

T. E. Lawrence, "Science of Guerrilla Warfare," in "Guerrilla," in *Encyclopedia Britannica*, 11th ed., edited by B. H. Liddell Hart (New York: Sears Roebuck, 1929**[AQ1: Please verify addition of publisher and location.]**). This succinct but important piece sketches out what worked against the Turks when employed by the Arabs led by Lawrence. Among the stipulations: eccentric, fluid movements are best; range should be favored over force (there is "fire in movement"); keeping the enemy tied down is as effective as killing them; insurgencies should be diffuse, kinetic, amorphous; and you need to get inside the minds of the enemy, your own troops, and your home front. A useful overview, but at the same time, there appear to be some significant dimensions seldom present in various insurgency situations, such as a lack of discipline, the "on the job training" of guerrilla leaders, and the low levels of integration among the Arab tribes (no "new men" come out of the rebellion).

———, *The Seven Pillars of Wisdom: A Triumph* (Garden City, NY: Doubleday, 1926). This previously privately printed memoir contains detailed descriptions of the campaign Lawrence and the Arabs waged against the Turks during World War I. Lots of material on specific tactics as well as strategy employed in the "naked desert, under the indifferent heaven." Truly a marvelous read by a brilliant, adaptive, occasionally narcissistic, romantic English schoolboy-turned-desert-fighter who used guerrilla tactics and actions to aid the Arabs in the struggle against the Turkish Empire and help the British during World War I. Sage words for all counterinsurgents and insurgents, using the Arabs as a case study: "Do not try to do too much with your hands. Better the Arabs do it tolerably than that you do it perfectly. It is their war and you are to help them, not to win it for them." Randall Woods has called this work a "long paean to just war and self-sacrifice to the point of self-immolation."

———, "Twenty Seven Articles," in *The Arab Bulletin*, August 20, 1917. Short, pithy, and actually quite humorous: "A slave brought up in the Hejaz is the best servant, but there are rules against British subjects owning them so they have to be lent to you." Much good advice for embedded political or military officers, including "[d]o not try to do too much with your own hands. Better the Arabs do it tolerably than that you do it perfectly. It is their war, and you are to help them, not to win it for them." This quote is seldom coupled with the next sentence, which provides an enduring caution for those who would seek to educate indigenous counterinsurgents: "Actually also, under the very odd conditions of Arabia, your practical work will not be as good as, perhaps, you think it is." Good advice, that, for many lands and peoples!

James Schneider, *T. E. Lawrence and the Arab Revolt* (New York: Bantam, 2011). According to Thomas Riggs, the author deserves great credit for suggesting that T. E. Lawrence "put guerrilla warfare within the context of modern industrialized warfare" by taking military thought away from the "dominant Western paradigm of annihilation." However, he was to sacrifice the freedom of the Arabs on the altar of British and French imperialism.

Jeremy Wilson, *Lawrence of Arabia: The Authorized Biography of T. E. Lawrence* (New York: Atheneum, 1990). The author and editor of eight previous works on T. E. Lawrence writes a huge, sprawling, inclusive biography. For our purposes, the section "The Years of Conflict 1914–1922" shows insurgency and counterinsurgency in Syria, Iraq, and Saudi Arabia as the Arabs revolt against the Turks.

Lenin

Vladimir Lenin, *What Is to Be Done?* (Moscow: Marxist Internet Archive, 1902). Lenin argues for a "party of a new type," one able to revolutionize the working classes, one that will be the "vanguard of the proletariat" to be centralized and dedicated to revolution, not reform.

Paul von Lettow-Vorbeck

Paul von Lettow-Vorbeck, *My Reminisces of East Africa* (Nashville, TN: Battery Classics, 1990). This 1920 memoir provides fascinating detail of the

World War I campaign in German East Africa, Portuguese East Africa and Rhodesia as the author led his Schutztruppe of 3,000 Germans and 11,000 African askaris from 1913 until *after* the November 1918 armistice. See especially chapter 7, "Guerrilla Warfare and Further Preparations." Lettow-Vorbeck actually had extensive counterinsurgency experience when, from 1904 to 1906, he was involved in the suppression of the Herero and Hottentots in what is now Namibia. He emphasizes the importance of moving rapidly and continually, using what was available, and always avoiding battle with "overwhelmingly hostile pressure." The loyalty of his African troops was—and is—legendary.

Bert "Yank" Levy

Max Hastings, "Setting Europe Ablaze," in his *Winston's War: Churchill 1940–1945* (New York: Alfred Knopf, 2010), pp. 364–82. A stunningly honest account of how futile guerrilla and partisan warfare was during World War II, concluding with Mark Mazower, "Only in the USSR did German counter-terror fail." But nowhere in Europe, even in Yugoslavia, did resistance operations avert the need for regular forces to defeat those of the Nazis.

Bert "Yank" Levy, *Guerrilla Warfare* (Washington, DC: Government Printer, 1942). Readers wishing to get an introduction to guerrilla warfare, one served up in a snappy, upbeat, even joyful way, could do far worse than sample the wit and wisdom of Bert "Yank" Levy. Levy, a Canadian veteran of many regular and irregular missions in the Middle East, Dunkirk, Nicaragua, and Spain, writes a primer for the British Home Guards in 1942, making suggestions to his charges as to how to defeat the upcoming Nazi invasion of the British Isles. In the process, the author catches the spirit and verve of small-scale guerrilla operations ("wasp warfare") throughout time and space.

Note: Reading this book as a teenager, I thought it was pretty much a "one-off" personal account with no real historical significance. But recently, I discovered that Churchill had indeed created a "stand-behind" force of 3,500 men called "The British Resistance Organization" (which lasted from July 1940 until November 1944 and was based at Coleshill Estate, Oxfordshire) to conduct guerrilla operations in the event of a Nazi invasion. See "Churchill's Secret Army Revealed," *BBC News* (April 2011), p. 14.

Note: See also the stunningly detailed and illuminating work Arthur Ward, *Churchill's Secret Defense Army: Resisting the Nazi Invaders* (South Yorkshire, UK: Pen and Sword, 2013). In the fear-haunted summer of 1940, with the Battle of Britain raging overhead, only 30 miles separated Hitler's army from England. Many, if not most, thought in terms of when, not if, an invasion was coming. In this context, the Home Guard and other defense units were formed to deal with the crisis. The author lovingly (and with excellent illustrations) documents these times by focusing on the "secret British underground army," known as the Auxunits, who operated in secret from their hidden pillboxes and hundreds of "Operational Bases" (OBs) or "hides." In the event of a German invasion, they were to "stay behind" in the OBs for several weeks, providing intelligence reports to the regular army and, after the Germans military had passed, emerging to do sabotage. Britain's "guerrillas in waiting" were organized and directed by Military Intelligence Research (MI-R). In the overarching need to "do something," some of these efforts seem somewhat slap-dash and amateurish, although they were taken most seriously by participants. The Auxunits were finally disbanded in 1944, although public recognition did not come until 1996, when a newspaper article proclaimed "Honour at Last for the Elite Force Whose Hour Never Came." The former secret commandos were secret no more. This is a marvelous, even powerfully nostalgic, read and a fine testament to the men and women who did their bit when their country faced maximum danger.

Mao Tse-Tung

Jeremy Ingalls, *Dragon in Ambush: The Art of War in the Poems of Mao Zedong* (Lanham, MD: Lexington, 2010). Twenty of Mao's poems are presented as being rich in symbolism and hidden meaning. They show Mao's personal goal of becoming "yi ren" ("one man, de facto emperor of China") and clever manipulator of Marxist doctrine, Chinese nationalism, and Stalinist methods and provide his "rationale for ruthlessness."

Mao Tse-Tung, *The Art of War* (El Paso, TX: El Norte Press, 2005). Somewhat overpraised and understudied, this work shows Mao at both his best (the true revolution caused by his disregard of Marx's central notion concerning the proletariat and its supposed decisive role in war) and worst (much pure propaganda). Many ideas lifted, without attribution, from Sun Tzu and also influenced by the writings of Clausewitz. The Clausewitzian trinity called for the use of violence to stir up the passions; those passions would give the political party the support it needed to generate the military means of success. The military strength gave the political leadership the ultimate triumph.

———, *On Guerrilla Warfare*, translated by Samuel B. Griffith (Urbana: University of Illinois Press, 2000). The essence of warfare according to Mao, who showed his historical genius by turning Marx and Lenin on their heads after the failed Communist effort in Shanghai in 1927, taking his forces on the "Long March." Interestingly enough and often overlooked are his strong statements: "The outcome of the war depends mainly on regular warfare" and "History knows many peasant wars of the 'roving rebel' type, but none of them ever succeeded." By that he meant that the stage of guerrilla warfare required (at the proper time) the movement from guerrilla warfare to main-unit, regular warfare. This principle was closely adhered to (on a number of occasions) by the North Vietnamese and Viet Minh, who moved back and forth among the various modes of warfare.

———, *On Protracted War* (Beijing: Foreign Language Press, 1966). A compilation of a series of speeches from 1938, outlining Mao's views concerning a protracted war as a way that small revolutionary groups could challenge the occupation of the Japanese (and, in time, the existing Nationalist Chinese government). Talked about the three stages or phases of the struggle: (1) strategic defense, (2) strategic stalemate, and (3) strategic counteroffensive. In the first, the counterinsurgents were on the offensive and the insurgents on the defense; in the second, the insurgents engaged in offensive guerrilla warfare; and in the third, the guerrillas go over to conventional warfare and the counterinsurgents are on the defensive. Mao asserted that the success of the revolutionaries was inevitable, "[t]herefore it should be repeated that the war will be arduous and protracted. But we are confident that, through the efforts of our Party and the whole people, the resistance already started will sweep aside all obstacles and continue to advance and develop."

———, *Selected Writings of Mao Tse-Tung* (Beijing: Foreign Language Press, 1967). More thoughts by the man who led the takeover of the world's most populous country by first guerrilla and then conventional warfare. His emphasis on treating the civilians properly echoes earlier assertions by Mazzini and others.

Francis Marion

Robert Bass, *Swamp Fox: The Life and Campaigns of General Francis Marion* (New York: Henry Holt, 1959). From August 1780 to September 1781, "he alone" kept eastern South Carolina from being pacified while Marion served as a corps commander under Nathanael Greene. A "ragged little guerrilla" who proved "bold, elusive and an armed will-o'the-wisp," Marion was regarded by some as "second only to Washington" as a Revolutionary War hero. His tactics and strategy were mostly self-taught, and his operations serve as a model for insurgents everywhere.

Hugh F. Rankin, *Francis Marion: The Swamp Fox* (New York: Thomas Y. Crowell, 1973). Although this account is devoid of any broader analysis of guerrilla warfare and fails to put Francis Marion in any historical context of partisan military strategy, its detailed descriptive account gives the reader a skirmish-by-skirmish appreciation for the "American Cincinnatus," who only showed brightly from 1780 to 1782, but that was enough to influence the outcome of the American Revolution. Marion had a keen appreciation that patriotism during a revolution was a "fluid emotion." His hit-and-run "hornet-style" attacks had the "stamp of ruthlessness," but he also accepted those Loyalists who had changed sides. Marion's impact is best measured by the British themselves: "Our army will be destroyed by damned driblets. . . . America is an ugly job . . . a damned affair indeed."

Mason Weems and Peter Horry, *Life of General Francis Marion* (New York: Lovell's Library, 1882). Horry, who fought side by side with Marion, wrote a simple account in 1802 about his life and times, and Weems turned it into an incredibly whimsical, laudatory praise poem declaring that "two of the noblest thunderbolts of her wars, George Washington and Francis Marion," deserve our undisguised praise. Also highlights Marion's inclusive, forgiving approaches to the Loyalists who recanted and was against the wholesale deportation of them.

Giuseppe Mazzini

Giuseppe Mazzini, "Rules for the Conduct of Guerrilla Bands," in *Life and Writings of Joseph Mazzini* (London: Smith, Elder, 1891), pp. 361–78. An amazingly valuable text giving extremely good advice concerning rigorous discipline, respect for the people, tight control over all aspects of guerrilla activities by a political "Centre of Action," and the universal rights of man make him among the most prescient of guerrilla theoreticians. Some selections that have endured 150 years of testing include:

1. "Guerrilla warfare may be considered as the first stage of a national war."
2. "The political mission of the bands is to constitute the armed apostolate of the insurrection."
3. "Every band should be a living programme of the morality of the party."
4. "Guerrilla bands are the precursors of the nation, and endeavor to raise the nation to insurrection. They have no right to substitute themselves for the nation." No matter who populates them, guerrilla bands are always supposed to show tolerance and respect and be inclusive as they seek the sympathy of the people.

Ho Chi Minh

Ho Chi Minh, *Down with Colonialism!* (London: Verso, 2007). Admitting that "theoretical innovation was not his forte," Ho followed Lenin, and especially Mao, to the letter, although he got much more mileage out of the notion of nationalism than either of them. He was quite disappointed at the lack of proletariat solidarity on his decolonization efforts, and his September 2, 1945, declaration of independence shows indebtedness to the American version. Most telling is his essay "Instructions Given at a Conference on Guerrilla Warfare." Here, "Uncle Ho" states plainly and unequivocally, "The aim of guerrilla warfare is not to wage large-scale battles and win big victories but to nibble at the enemy, harass him in such a way that he can neither eat nor sleep in peace, to give him no respite, to wear him out physically and mentally and finally to annihilate him." Well worth reading. Too bad more US strategists didn't take the time to digest his thoughts and background.

David Petraeus and James Mattis

Victor Davis Hanson, "Iraq Is 'Lost'" in his *The Savior Generals: How Five Great Commanders Saved Wars That Were Lost, From Ancient Greece to Iraq* (New York: Bloomsbury Press, 2013), pp. 190–237. Hanson compares Petraeus to the brilliant but unrewarded Byzantine general Belisarius and gives him credit for reversing military doctrine and fighting truly enormous political headwinds both outside and within the American military. Acid portraits of those who wished to make self-fulfilling prophecies out of military challenges.

David H. Petraeus and James N. Mattis, *FM 3–24: Counterinsurgency* (Washington, DC: Department of the Army, 2006). The men who led the successful surge in Iraq and who now face the challenges of Afghanistan lay out the most comprehensive and yet specific dimensions of counterinsurgency.

Note: Lieutenant-Colonel John Nagl, who worked closely with Petraeus on FM 3-24 said of the many books that were important to its writing, "[P]erhaps none was as important as David Galula's *Counterinsurgency Warfare: Theory and Practice*" (Cohen, Galula, p. 243).

Red Cloud

Bob Drury and Tom Clavin, *The Heart of Everything That Is: The Untold Story of Read Cloud, an American Legend* (New York: Simon and Schuster, 2013). Red Cloud, the Lakota Sioux war leader, conducted guerrilla warfare (Red Cloud's War, 1866–1868) against American encroachment of Native American lands in what is now Montana, South Dakota, and Wyoming.

Jerry Lazar, *Red Cloud: Sioux War Chief* (New York: Chelsea House, 1995). One of the few Native American leaders who not only saw the menace but also figured out how to defeat it, but he would be undone by the failure of other tribes to apply sustained ruthlessness in battle and beyond.

R. Eli Paul (ed.), *Autobiography of Red Cloud: War Leader of the Oglalas* (Helena: Montana Historical Society, 1997). Much information about his early years and his views of how and why the whites must be stopped.

Beneys Reitz

Beneys Reitz, *Commando: A Boer Journal of the Boer War* (Pretoria: Cru-Guru, 2008). Another insurgency crushed by standard methods that have worked for millennia. Reitz, however, gives the reader a good sense of the tactics and strategy of guerrilla warfare taken on the fly and how the counter-insurgency was forced to make major changes in strategy and tactics in order to defeat the Boers. Reitz was there and shows considerable insights into the process.

Robert Rogers

Robert Rogers, *Journal of Major Robert Rogers* (Worcester, MA: American Antiquarian Society, 1918). Available on microfilm in the Bowdoin Library, interested students can get a real feel for the strategic and tactical dimensions of insurgency and counterinsurgency from this quite amazing source.

John F. Ross, *War on the Run: The Epic Story of Robert Rogers and the Conquest of America's First Frontier* (New York: Bantam, 2009), especially "Epilogue," pp.451–66 which traces Rogers's heritage for the Special Forces and Rangers of today. The lessons of what the French called "la petite guerre" remain. Who knew—or cares to know—that the United States was an incubator of insurgent strategies and tactics.

Mark Wulff, *Robert Rogers' Rules for Ranging Service: An Analysis* (Westminster, MD: Heritage Books, 2006). Rogers's 28 rules are decorticated and deconstructed (and, in many cases, acted out in the field). Taken directly from Rogers's journals, they show the tactical brilliance of his efforts during the French and Indian War, from avoiding ambushes to avoiding fords; flexibility; and an ongoing accent on maneuver and speed. There is a definite lack of political or ideological baggage (or glue) in these useful tactical admonishments.

Note: Rogers's impact runs along an astonishingly long arc through time and space. For example, Randall Woods, in his *Shadow Warrior: William Colby and the CIA* (New York: Basic Books, 2013), believes it is central to the development of William Colby and his long and distinguished (if controversial) career in the CIA. The author also analyzes Roberts's impact on the novels of Kenneth Roberts, such as *Northwest Passage* (Garden City, NY: Doubleday, Doran, 1937).

Rupert Smith

General Rupert Smith, *The Utility of Force: The Art of War in the Modern World* (New York: Alfred A. Knopf, 2007). This widely read—and perhaps overpraised—book comes out of the author's experience in NATO, Northern Ireland, Kosovo, and Bosnia. He believes that "war among the people" truly changes the nature of warfare and that there has been a "paradigm shift" from armies to strategic confrontations among a broad range of combatants. Irregular warfare requires a shift into this new mode, where civilians are involved as combatants, victims, bystanders, and objects to be protected. Warfare is now overtly and overwhelmingly "population centric." Protecting the civilian population is now the first order of business, putting counterinsurgency into this mode of operation.

Sun Tzu

The works of Sun Tzu are explored in chapter 3.

Robert Thompson

Robert Thompson, *Defeating Communist Insurgency: Experiences from Malaya and Vietnam* (New York: Praeger, 1966). The Malay counterinsurgency experience following World War II might have been a most useful model for Vietnam as well as numerous other comparable situations, but that turned out not to be. The author, who was an important player in both situations, outlines some of the might-have-beens had the lessons of Malaya been truly learned. This work is well worth studying in order to get a sense for the *sine qua nons* that were required in the Malay situation and that need to be duplicated for successful counterinsurgencies elsewhere. Some of Thompson's most important principles for successful counterinsurgency include (1) the government must have a clear political aim to establish and maintain a free, independent, and united country; (2) it must function according to law and not apply that law capriciously; (3) it must have an overall plan, including political, social, economic, administrative, and policy elements, not simply security and military operations; (4) the government must give priority to defeating the range of political subversion, not just the guerrilla fighters of that subversion; and (5) in the guerrilla phase of counterinsurgency, it must secure (and hold) its base areas first. The secret, he believes, is for the government to take the initiative away from the guerrillas and force them to go on the defensive. Thompson then focuses on a central military truth (not espoused by those who feature "search and destroy" in massive sweeps): "It is not often realized that in guerrilla warfare guerrilla forces are just as vulnerable to ambushes as the government forces."

Note: Thompson's primary contribution to the literature and practice on insurgency has to do with his highly successful "strategic hamlet" program that worked very well in Malaya and would eventually work in Vietnam, when the overarching strategic shift to "clear and hold" took place from 1968 to 1972 and beyond. In Vietnam during the earlier period, however, while Thompson and President Ngo Dinh Diem favored putting security first and economic development second, the CIA chief William Colby and Prime Minister Ngo Dinh Nhu favored putting economic development first. Obviously considerable confusion and cross-purpose implementation resulted.

Robert Thompson, *Make for the Hills: Memories of Far Eastern Wars* (London: Leo Cooper, 1989). His long career in insurgency (China and Burma during World War II) and counterinsurgency (Malaya, Vietnam, and Thailand) make for an unusual skill set. Credits the Chindits for helping to save Imphal and the resettlement and arming of 500,000 Chinese in rural areas as central to success during "The Emergency." His contributions to counterinsurgency more generally are quite noteworthy, including (1) his theme of counterpenetration vital to disrupt penetration; (2) his accent on the notion that prevention is better than curing when it comes to government-inspired ills; (3) his belief that various successful techniques learned in war are likely to be forgotten (or resisted by others faced with similar situations); and finally (4) his statement "In strategic terms if an insurgent movement, at a cost which is infinitely acceptable, can impose costs on the government which are not indefinitely acceptable, then although it may be losing every battle, it is winning the war" (p. 94). Sage advice dispensed over a long period of time. Some listened. Some did not.

———, *Revolutionary War in World Strategy: 1945–1969* (New York: Taplinger, 1970). A global tour de horizon focusing on the role of the Soviet Union and China in fomenting local and world revolution. In calling for "No More Vietnams," he is not talking about avoiding counterinsurgency but in doing it smarter and more effectively. Two major points well worth remembering: (1) Without truly massive Soviet aid, it is difficult to imagine North Vietnam overrunning South Vietnam; and (2) "There are some who will require standards of perfection for theoretical government which they do not expect from their own. There can be more arrogance in liberalism than there is in power." Thompson also blames Europe for failing to help the United States adequately in the defense of freedom.

Josip Broz Tito

Milovan Djilas, *Tito: The Story from Inside* (London: Weidenfeld and Nicolson, 1981). This is a most engaging work that provides an insider's guide to this metal worker turned revolutionary turned party comrade in Moscow turned World War II insurgent turned ruler who liked palaces.

———, *Wartime* (New York: Harcourt Brace Jovanovich, 1977). This work focuses carefully and minutely on World War II in Yugoslavia. See especially "The People's Uprising" and "The Civil War within a War." There is also a finely grained, odd description of a prisoner exchange with the Germans in 1943 to rescue Tito's wife, as well as engaging descriptions of his various meetings with Stalin. Djilas concludes, "Revolutions begin new epochs, whose directions no one can foresee, let alone determine."

Fitzroy Maclean, *Eastern Approaches* (New York: Time Life Books, 1949). This splendid read by one of those intrepid British explorers and men of action contains an entire section (part 3, "The Balkan War") about Tito, who was very impressive. Josip Broz's nickname was prophetic in the extreme: "Tito" in Serbo-Croatian means "You do that. You do this." Gives an excellent insight into Tito's insurgency philosophy—flexibility with firm belief in Communist ideology as a driver of successful revolution with true acceptance of a union of all classes to defeat the Germans. Maclean's description of "Operation Ratweek," the harassment of the retreating Germans in 1944, is one I had not seen elsewhere. Very well written by a most engaging personality about another most engaging personality.

Note: Maclean also wrote a highly acclaimed biography of Tito, one that indicates his independent streak. See his *Heretic: The Life and Times of Josip Broz-Tito* (New York: Harper, 1957).

Josip Broz Tito, *Josip Broz Tito: Military Thought and Works* (Belgrade: Vojnoizdavacki Zavod, 1982). A splendid look at a theoretical and practical genius at war in his own words. Some of his assertions are a must-read for the study of insurgency and counterinsurgency: The entire country is a battle space; there must be a union of all classes; people, not historical forces, must lead the revolution; rural orientation is essential, and the insurgents must avoid set-piece battles but also urban fighting; and offense is the key, as is command and control. Marvelous maps show progress in this Hobbesian world of war during World War II in Yugoslavia. Tito's partisans had to defeat the Germans as well as their "collaborationist forces," such as the Croats, Slovenes, Albanian, and Serbian "quislings," among others. This book is a treasure trove for anyone wanting to understand the theory and the practice of insurgency.

Note: The accomplishments of Tito during the insurgency phase are matched by his holding "Yugoslavia" together while he lived. For what he was up against, see Dimitrije Djordjevie and Stephen Fischer-Galati, *The Balkan Revolutionary Tradition* (New York: Columbia University Press, 1981). The "powder keg of Europe" had a vast, excruciatingly long, and harrowing tradition of revolt, revolution, and struggle against any and all hierarchical powers, be they Ottoman; Austrian; or, later, Soviet or, indeed, Serbian.

Roger Trinquier

Roger Trinquier, *Modern Warfare: A French View of Counterinsurgency* (New York: Frederick Praeger, 1964). Written in 1961 by the French military figure who fought in both Vietnam and then Algeria, masterminding the highly successful counterinsurgency effort in Algiers, this is a marvelously insightful work, and in 2012 as in 1961, its conclusions still ring true. "Modern warfare" for Trinquier is subversive or revolutionary warfare, making a changed situation in which "[w]arfare is now an interlocking system of actions—political, economy, psychological, military—that aims at the *overthrow of the established authority in a country and its replacement by another regime*." In "modern warfare" the sine qua non of victory is the "unconditional support of the population," while the enemy is not the insurgents' military forces but an "armed clandestine organization whose essential role is to impose its will upon the population." That clandestine organization must be destroyed despite the difficulties of doing so by a democratic, lawful society. Page 62, for example, contains a good comparison between the traditional army and the guerrillas, as well as detailed instruction of how to penetrate the clandestine organization with carrots; sticks; and, if necessary, torture. Information is the key to success, as is a "net of fine mesh" thrown over the population as the clandestine organization is identified and destroyed. In conclusion, the author strikes a positive note, arguing, "[O]ur repeated failures result solely from poor employment of our resources." General Petraeus and others seem to have read and absorbed the principles of modern warfare with a deep understanding of their ongoing verities—albeit with a sustained overlooking of the ingredient of torture.

U.S. Army and Marine Corps

Andrew Birtle, *U.S. Army Counterinsurgency and Contingency Operations Doctrine 1860–1941* (Washington, DC: Center of Military History, US Army, 2002). A long and sometimes loving look at all the American efforts against insurgents, both domestic and foreign over the long sweep of history. A surprisingly large number of cases were successful for the counterinsurgents. A very important, even encyclopedic, work but one that would not be well received by those who believe that insurgency is inevitable.

———, *U.S. Army Counterinsurgency and Contingency Operations Doctrine 1942–1976* (Washington, DC: Center of Military History, US Army, 2006). "Stability operations" and constabulary efforts are viewed, including efforts against the German "Werewolves" in Germany after World War II (for most of the war, the United States worked *with* the insurgents in many countries). Many ad hoc efforts lead the author to conclude that "half force, half administration" is a good mix. Quite interesting on the Chinese and Korean "civil wars," among others. Although he never followed his own advice, for example, Chang called for "7 parts politics, 3 parts military" in any effort to defeat guerrillas.

Richard Shultz, *The Marines Take Anbar: The Four-Year Fight against Al Qaeda* (Annapolis, MD: Naval Institute Press, 2013). Anyone wishing to understand insurgency and especially counterinsurgency should read this book. A must. Consider some of the lessons learned: Population-centric warfare is different; understanding the cultural context is essential; armed groups are complex and diverse and have exploitable cleavages; counterinsurgency must be adapted to time and place, but it is still war; counterinsurgency requires skilled advisers; indigenous police are a key force multipliers; counterinsurgency is an intelligence-led fight; and nonkinetic lines of operations are the way forward. Note: The reading is slow going because it captures the ebb and flow of a four-year struggle.

US Marine Corps, *Small Wars Manual* (New York: Skyhorse, 2009). A reprint of the 1940 version of this extraordinary work, first published in 1935. Distilling the lessons of US Marine experiences in the Philippines, Haiti, Nicaragua, and elsewhere, this highly readable account is a classic. Its 15 chapters cover a broad range of subjects, from organization and training to the disarmament of the population to supervising elections. Stressing the importance of knowing the population with which one is working, it states what should always be the golden rule of counterinsurgency: "The motive in small wars is not material destruction It is usually a project dealing with the social, economic and political development of the people." It also urges morphing from military functions into police ones as soon as the "lawless elements" are reduced in strength. Too bad this work was not read—and appreciated—by more American policy makers over the years since it was published.

Orde Wingate

Note: In reading Simon Anglim's "Orde Wingate in the Sudan, 1928–1933: Formative Experiences of the Chindit Commander," *RUSI Journal* (June 2003), pp. 97–100, one learns that Wingate was involved in counterinsurgency in the Sudan and Palestine but also insurgency against the Italians in Ethiopia and the Japanese in Burma, giving him ample opportunity to view the phenomenon from both sides.

Leonard Mosley, *Gideon Goes to War* (New York: Charles Scribner's Sons, 1955). A marvelously intimate, essentially positive book by a reporter who interviewed Wingate many times in the middle of the war. Gives good insights into his fanaticism, outsider status, and violent attachments (to Zionism for example) and casts him as a military genius of sorts (irregular warfare, long-range penetrations, and highly mobile guerrilla warfare) but one who was tough on his troops, the Chindits suffering huge losses in Burma (and leaving many behind to either commit suicide or be captured by the Japanese). Also a tad rough on his animals—all 25,000 camels he used in the drive to liberate Addis Ababa perished. This is one reason the Americans encouraged him to use aircraft for the second Chindit assault. Unfortunately, Wingate was killed in a plane crash shortly thereafter.

OVERVIEWS OF INSURGENCY

John Arquilla, *Insurgents, Raiders, and Bandits: How Masters of Irregular Warfare Have Shaped Our World* (Chicago: Ivan R. Dee, 2011). Irregular warfare abounds as practiced for and against the Plains Indians, the Chechens in the Caucasus Mountains, and such practitioners as Americans Robert Rogers and Nathanael Greene, Spanish Francisco Espoz y Mina, Russian Denis Davydov, South African Christiaan de Wet, German Paul von Lettow-Vorbeck, and the British counterinsurgent operator Frank Kitson.

Ian Beckett, *Modern Insurgencies and Counter-Insurgencies: Guerrillas and Their Opponents since 1750* (London: Routledge, 2001). Wide-ranging study shows the extent and length of successful counterinsurgency *and* insurgency efforts.

Phillip Bobbit, *Terror and Consent: The Wars for the Twenty-First Century* (New York: Alfred Knopf, 2008). A close look at how the century begins in terms of war. Some useful material on the role of insurgency and counterinsurgency.

Max Boot, *Invisible Armies: An Epic History of Guerrilla Warfare from Ancient Times to the Present* (New York: Norton, 2013). In the vast sweep of history, only one in five insurgencies is successful, and the author asserts, "Guerrilla warfare is as old as mankind." From the Aetolians versus Athens in 426 to the 450-year struggle of the Scots against the English. Lots of good material here, but because of the scope, much is necessarily superficially presented. Boot conflates terrorists, partisans, guerrillas, irregulars, special operations, and others like them to provide an overarching and impressive montage of the history of guerrilla warfare. His data base running back to 1775 and extensive bibliography are of considerable value. This is one of the best overviews of the historical background of a great variety of insurgencies.

James Corum, *Bad Strategies: How Major Powers Fail in Counterinsurgency* (Minneapolis: Zenith Press, 2008). Fails to understand successful counterinsurgencies and the lessons they teach but good at showing problems faced by those who would conduct them.

Robert Jackson, "Sovereignty and Its Presuppositions: Before 9/11 and After," *Political Studies* 55 (2007), pp. 271–73; *Sovereignty: The Evolution of an Idea* (Cambridge: Polity, 2007). The nation-state is under siege from many quarters, including micro-militancy. Substate nationalism is a constant incubator of insurgency throughout the centuries.

Anthony James Joes, *Victorious Insurgencies: Four Rebellions That Shaped Our World* (Louisville: University of Kentucky Press, 2010). Looking at the successful insurgencies in China, Vietnam, Cuba, and Afghanistan, he underscores the huge role played by the Japanese in aiding the Communists in both China and Vietnam and the key role played by the United States in abandoning Batista and conceding Cuba to Castro long before the eventual fait accompli. Ends up asserting—correctly, in my view—that a greatly weakened counterinsurgency capability of the target state turns out to be the independent variable in successful or unsuccessful insurgencies. A stimulating and often convincingly revisionist read.

Robert Kaplan, *Imperial Grunts: The American Military of the Ground* (New York: Random House, 2005), especially "Afghanistan," pp. 184–256; "Horn of Africa" pp. 273–307; and "Iraq," pp. 307–70. Kaplan spends a lot of time with US forces in the field and provides many useful insights.

David Kilcullen, *The Accidental Guerrilla* (London: Oxford University Press, 2009). Not all insurgents set out to become revolutionaries, or why armies need anthropologists, according to the author.

John Mackinlay, *The Insurgent Archipelago: From Mao to bin Laden* (New York: Columbia University Press, 2009). Makes a number of assumptions about what the insurgents have in common but overlooks some the differences among their motivations *and* their world. Seeking to examine the new forms of insurgency and their connections around the world, Mackinlay bifurcates the history of insurgency and counterinsurgency into pre-Mao and post-Mao periods, seeing Mao's highly political "war among the people" as a true watershed phenomenon. Currently, he sees the new aspects of insurgency in the transformative patterns introduced by the Internet, economic deregulation, the rise of feral militias, and the ongoing mutations of global insurgency prototypes, and he points out that the latter is "not a uniquely Islamist or Muslim phenomenon." He also decries "one size fits all" efforts at counterinsurgency and the "overly kinetic" and heavy military approach, spending considerable effort explaining the counterterrorism (but not counterinsurgency) of the UK Home Office's CONTEST effort and contrasting it with the US efforts under the Operation Enduring Freedom rubric. Only by focusing on the new and changing nature of an "archipelago of individuals, cells and communities," he believes, can the dynamics of insurgency be understood and adequate strategies for counterinsurgency developed.

Daniel Marston and Carter Malkasian (eds.), *Counterinsurgency in Modern Warfare* (New York: Osprey, 2008). Some good essays here on a complicated and multidimensional subject. A broad array of authors and typologies.

Mark Moyar, *Triumph Forsaken: The Vietnam War 1954–1965* (London: Cambridge University Press, 1988). Revisionist and very powerful. A must-read. One of the best works from this perspective. Shows a strong argument that just because one insurgency succeeded doesn't mean it *had* to succeed.

John Nagl, *Learning to Eat Soup with a Knife: Counterinsurgency Lessons from Malaya and Vietnam* (Chicago: University of Chicago Press, 2007). Counterinsurgency is messy and slow and requires a great deal of constant learning and relearning in order to be effective.

Robert Pape, *Dying to Win: The Strategic Logic of Suicide Terrorism* (New York: Random House, 2005). Why suicide bombers have some huge advantages in tactical situations but are less powerful in strategic ones.

Peter Paret, *French Revolutionary Warfare from Indochina to Algeria* (New York: Praeger, 1964). Explains the French doctrine of counterinsurgency and provides an interesting contrast where the military solution for Algeria was actually quite successful when compared to the military applications in Vietnam, although the outcome was the same. An interesting conclusion of the French: "It is a mistake to think that revolutions can be won by negotiations." Adamant that the internal rebels "[m]ust be cut off from foreign assistance." Again, we end up with the importance of the will of the counterinsurgency country, government, and civilian population.

John Robbs, *Brave New War: The Next Stage of Terrorism and the End of Globalization* (New York: John Wiley and Sons, 2007). Certainly this is something "The World Is Flat" Friedman should read.

Richard H. Shultz and Andrea J. Dew, *Insurgents, Terrorists, and Militias: The Warriors of Contemporary Combat* (New York: Columbia University Press, 2006). They are widespread, hard to eradicate, and a threat to the traditional nation-state. Disorder and more disorder seem in order for the world.

Robert Taber, *The War of the Flea: A Study of Guerrilla Warfare Theory and Practice* (New York: Citadel Press, 1970). Wild claims about near inevitability for guerrilla warfare: "What is to stop it?" Forty years later, we can say, "Quite a bit actually."

SOME ILLUSTRATIVE CASE STUDIES: SHOWING THE EXPERIENCE OF INSURGENCY AND COUNTERINSURGENCY ACROSS CROSS-CULTURAL LINES

Algeria

Alf Heggoy, *Insurgency and Counter Insurgency in Algeria* (Bloomington: Indiana University Press, 1972). Ideologically charged sources and proximity to events make lasting analysis difficult, but the author is quite adroit at showing how reform came too late, was too little and halfhearted, and was met with too much resistance from the colons to be effective. Still, "[b]y the end of 1957 and certainly by the middle of 1958, the nationalists had all but lost the military conflict within Algeria." Sanctuaries in Morocco and Tunisia saved the day for the revolutionaries as the FLN broke the will of the French in the metropole. General Massu's 10th Paratroop Division won the Battle of Algiers militarily but turned Algerians against France in the process and the French against the Algerian war. As late as June 1950, 171,000 Algerians were still fighting alongside the French.

Paul Henissart, *Wolves in the City: The Death of French Algeria* (New York: Simon and Schuster, 1970). An engaging account of the three-cornered struggle to determine the future of Algeria during the final year of the war with the French government, the Algerian National Liberation Front (FLN), and the *pied noire* settler *Organization Armee Secrete* (OAS). The failed OAS putsch in 1961 enabled de Gaulle to skillfully marginalize the *pied noires* when the air force failed to join the revolt. Generals Jeane-Jacques Susini, Maurice Challe, Raoul Salan, and Jacques Soustelle are all prominently featured. When 850,000 *pied noires* fled to France in 1962, "French Algeria died badly," claims the author.

Angola and Mozambique

It is interesting to note that the Portuguese held onto their colonial possessions longer than Holland, Britain, and France, who, while they occasionally won counterinsurgency struggles (i.e., Malaya, Syria, Madagascar, and Oman), lost many more (Indochina, Indonesia, Cyprus, Palestine, Algeria, etc.).

The Angolan situation is unusual, too, for the many-sided nature of its conflict: Portugal against insurgents (UNITA, MPLA, FLNA) and then the insurgents as counterinsurgents (MPLA and Cuba vs. UNITA and SA).

In Mozambique, the forces of the *Frente de Libertacao de Mocambique* (FRELIMO) was successful as insurgents and as counterinsurgents defeating *Resistencia Nacion al Mocambicana* (RENAMO) during the long civil war (1976 to 1990) that followed independence in 1974.

The interested reader will want to compare with the Irish rebellion and civil war for complexity. Both Angola and Mozambique showed the same pattern as Ireland; that is, insurgents win round 1, become counterinsurgents, and defeat the next group of insurgents.

James Ciment, *Angola and Mozambique: Post-Colonial Wars in Southern Africa* (New York: Facts on File, 1997). Rebel insurgencies in Angola and Mozambique are crushed, not always by conventional counterinsurgency techniques, as UNITA and RENAMO lost both international aid and standing.

William Finnegan, *A Complicated War: The Harrowing of Mozambique* (Los Angeles: University of California Press, 1992). A long, discouraging look at the terrible disasters that struck so many Mozambiquians during decolonization and the civil war that followed, leaving no fewer than 10 million *afetodos* (affected ones) and 3 million *deslocados* (refugees), as RENAMO punished the country from one end of it to the other, with devastating results. In fact, the author claims, "The cruelty of RENAMO mesmerized everyone in Mozambique." Apt title. The country was so bad off that India sent 150 aid workers. *A luta continua* (the struggle continues).

Edward George, *Cuban Intervention in Angola* (London: Frank Cass, 2005). An outstanding example of a misunderstood principle. Yes, the Cuban "revolutionaries" were supporting the MPLA "revolutionaries," but the Cuban intervention was a classic example of a successful *counterrevolution* strategy, defeating UNITA and their South African allies in set-piece battles.

Thomas Henriksen, *Revolution and Counterrevolution: Mozambique's War of Independence* (London: Greenwood Press, 1983). How FRELIMO won the war and Portugal lost the war, but with a close eye to the might-have-beens rather than assumptions of what had to happen. Good examination of the insurgency strategy.

John Marcum, *The Angolan Revolution* (Cambridge: MIT Press, 1978). The classic study of the histories of the MPLA, FLNA, and UNITA and an analysis of how all insurgencies are not created equal and how not all are deemed worthy of ultimate success. Very useful to understand the fighting that followed independence and how guerrilla strategies had to be modified to fit different times and places.

Eduard Mondlane, *The Struggle for Mozambique* (London: Zed Books, 1983). With emphasis on ideology and fitting Mondlane and FRELIMO into the "global struggle" against capitalism, this book is not as useful as it might have been had it included a more in-depth analysis of the military strategies of FRELIMO.

Marius Scheepers, *Striking inside Angola with 32 Battalion* (West Midlands, UK: Helion, 2012). Badly written and confusingly presented in tiny type, this work nevertheless gives focus to an insurgency attempt within the context of a left "revolutionary" government. The MPLA, backed by Cuba and the Soviet Union, and the forces of UNITA, backed by South Africa and the United States (with the South African war against SWAPO in the then South West Africa also in full sway), from 1966 to 1988 culminated in a deep, conventional strike into Angola, one that was eventually driven back by the forces of the government. The chapter on "Guerrilla Warfare," pp.132–36, contains quite timely advice for counterinsurgents elsewhere.

Brazil

Euclides Da Cunha, *Rebellion in the Backlands* (Chicago: University of Chicago Press, 1944). "Brazil's finest book" depicts the crushing of the Canudos rebellion, 1896–1897, with "guerrilla warfare in its pristine form." A cry from the heart on the courage of the insurgents but a depiction of another insurgency crushed, brutally.

Burma

Field-Marshal Viscount William Slim, *Defeat into Victory: Battling Japan in Burma and India, 1942–1945* (New York: Cooper Square Press, 2000 reprint). Main force *and* guerrilla action defeat the Japanese on the ground as British forces learned the elements of counterinsurgency by practicing insurgency during World War II. Orde Wingate's irregular "Chindits" are given less credit for their efforts.

China

Jay Taylor, *The Generalissimo: Chiang Kai-shek and the Struggle for Modern China* (Cambridge: Harvard University Press, 2009). Gives many reasons Mao was very fortunate to have Chiang as an adversary and why revolutionaries have to have considerable help from their counterinsurgency foes if they are to prevail. Mao was truly blessed.

Dan Topping, *The Peking Letter: A Novel of the Chinese Civil War* (New York: Public Affairs, 1999) (F). On one level this is a fairly predictable, even formulaic, love story with a brave, idealistic, and naïve American military man doing daring things to gain the love of his beautiful Chinese girlfriend, who, at the end of the day, does not go with him when he is deported but she hopes to "meet again." But on another level, it is a quite useful portrait of China during the collapse of the Nationalist government in 1949 and the coming to power of the Chinese Communists. Chiang (a "Confucian more than a Christian or a modern man") is about to hive off to Formosa with enough military and gold to survive. The Communists, led by Lin Bao and Mao, are poised to take over the country, having been successful in Manchuria and in the titanic battle of the Huai-Hai (with hundreds of thousands of Nationalists killed or captured), and as they take over Peking, they are beginning to take retribution on landowners, speculators, and running dogs of imperialism, including missionaries. There is the whiff of panic in the air, massive corruption of governmental officials, lost military actions, and the seemingly inevitable flow of revolutionary change. And through it all, the United States generally acts in a bumbling and indecisive manner ("[b]lundering about . . . full of goodwill, but not quite understanding us"), caught up in a massive misjudgment of the relationship between Stalin and Mao as well as overblown domestic fears of Communist infiltration. I found most interesting of all the acidly etched portrait of Jean Leone, the French military attaché in Peking—cynical, worldly, perspicacious, and militarily astute. Surely Leone is a thinly disguised version of David Galula, who would go on to become a leading theoretician (and practitioner) of counterinsurgency during and after his experiences in Indochina, China, and Algeria and whose wisdom would be recycled yet again in David Petraeus and James N. Mattis, *Counterinsurgency*, FM 3-24 (2006), a counterinsurgency masterpiece. All in all, a worthwhile read by a former *New York Times* reporter who witnessed this history as it was being made.

Odd Westad, *Decisive Encounter: The Chinese Civil War 1946–1950* (Stanford, CA: Stanford University Press, 2003). A judicious, balanced, and quite readable account of the Chinese civil war, including a formidable summary of the rise and ultimate success of the CPP. Some very useful insights into the Communists' failure to take Jinmen (Quemoy) and their success in taking Hainan.

England

Note: Frederick Engels pointed to the 1381 "Great Rising" as a precursor to the momentous European events of 1848.

Maurice Ashley, *General Monck* (Totowa, NJ: Rowman and Littlefield, 1977). A worthwhile account of the able general George Monk, who fought for the crown, parliament, the lord protector, and Oliver Cromwell, as well as Charles II, who rewarded him with a dukedom. Fighting against the Dutch as "General of the Sea," and in France, Ireland, Scotland, and England, he was a professional soldier par excellence, and his life and time illustrate the need to be politically nimble during troubled times. His accent on morale, discipline, and rewarding loyal soldiers remains valid today.

R. B. Dobson, *The Peasants' Revolt of 1381* (New York: Macmillan, 1970). A most useful collection of primary documents on all phases of the 1381 and 1382 events (the wave of repression). Helpful in getting the reader to see matters through contemporary eyes and to make sure we realize (1) there were only 10,000 involved, not the 50,000 to 60,000 claimed elsewhere, (2) the leaders' "courage was better than their causes," (3) it was not simply a peasant revolt but was supported by many of the "large ecclesiastical proletariat of late medieval England," (4) there was an "inherent fragility of the movement," and (5) being a "member" of the *Magna Societes* (Great Society) was enough to get one hung by the numerous juries that tried so many after the revolt collapsed.

Alistair Dunn, *The Great Rising of 1381: The Peasant's Revolt and England's Failed Revolution* (Stroud, UK: Tempus, 2002). A sprightly written account of the Peasant Revolt of 1381, also known as "The Great Rising." The author points to the antecedents of famine; the black death; and especially the poll taxes of 1377, 1379, and 1380, as well as the youth of King Richard II (age 14). Breaking out in eastern and southern England (starting in Kent and Essex) and moving to London, the revolt was essentially "among the poorest elements of rural and urban society." John Battle, Wat Tyler, and Jack Straw emerged as leaders in a "plurality of its patterns of violence." Eventually the revolt failed, the author believes, because it lacked a military apparatus with which to blunt that of the aroused noble and royal men at arms because the peasants believed the king when he signed the freedom charters dissolving feudalism and because of the draconian measure of pacification by the judicial commissions and the military.

Dan Jones, *Summer of Blood: The Peasants' Revolt of 1381* (New York: Harper, 2009). The clearest and most well-written accounts of this "flash rising of the humblest" I have read. Jones gives the reader a good picture of June 1381 and its momentous events. Wat Tyler come through clearly as an inspiring leader but one in over his head and one who, after the uprising has made great inroads, seems to lose his way during the confrontation with Richard II, pushing for the end of not just serfdom but lordship as well. Initially, the totally paralyzed nature of the government is in evidence as the mobs rampaged through London, capturing the tower and sacking the Savoy, the largest palace complex in England at the time, killing many, including close counselors of the king. From the point of view of the political center, the hero is William of Walworth, the mayor of London, who stabs and wounds Wat, goes to London to get reserves, and then comes back to behead him (for which he is knighted by Richard II). More than 1,500 were killed in the counterterror of the government, whose retribution by armed knights was considerable. The author makes it obvious that even highly motivated peasants and artisans could not stand against the heavy-horse style of warfare once that got ignited.

Philip Lindsay and Reg Groves, *The Peasants' Revolt 1381* (London: Hutchinson, 1950). "One of the greatest popular uprisings in Medieval History" is examined in considerable detail. The portrait of medieval life for 90 percent of the population is devastating. The authors weave in the elements of the black death, John Wycliffe, John Ball, Wat Tyler, and the spontaneous rising and desire for "The Great Fellowship" or "The Great Society." Amazingly, the boy king (previously known as the "Black Prince") had no standing army or police force to counter, so he signed various documents, reneging as soon as he acquired the forces to crush the rebellion. He did so with a vengeance, concluding, "You were and you are serfs and shall remain in bondage, not that of old but in one infinitely worse." Thousands were slain to make his point, and it was small comfort that feudalism would eventually come to an end. The authors rightly praise the heroism of the human spirit in the attempt they so carefully record. A very interesting example of insurgency in the Middle Ages.

Ethiopia

W. E. D. Allen, *Guerrilla War in Abyssinia* (Baltimore, MD: Penguin, 1943). Written in the middle of World War II by a British captain with Force Gideon, this work gives a nice flavor for the total effort (it was one of the first successes of the war for the British), its small size, its logistical difficulties, and the way in which guerrilla warfare progresses from success to success. Illustrates one definite pattern for liberation with changing local authorities who swing to the winning side. Also shines light on the major guerrilla activity around Lake Turkana and the territory to its north.

Anthony Mockler, *Haile Selassie's War: The Italian-Ethiopian Campaign 1935–1941* (New York: Random House, 1984). A comprehensive overview of not just the Force Gideon but also the earlier war and the other dimensions of the liberation of Ethiopia. Very good on Ethiopian players and intrigues, as well as the interplay among the various British participants. Sees Orde Wingate as a "Lawrence found."

Leonard Mosley, *Gideon Goes to War* (New York: Charles Scribner's Sons, 1955). See the earlier review in this chapter for the life and times of Orde Wingate, who led Force Gideon in its successful guerrilla campaign against the Italians during World War II.

C. P. Potholm, *Liberation and Exploitation: The Struggle for Ethiopia* (Washington, DC: University Press of America, 1986), especially "The Battle for Ethiopia," pp. 56–72. Force Gideon is placed in the larger context of the liberation of the whole Horn of Africa and the role of the emperor of Ethiopia as the Italians give way to guerrillas and conventional forces of combined arms. This monograph arose out of an MALD thesis at the Fletcher School of Law and Diplomacy under the direction of Professor John Spencer (cited in chapter 4).

John H. Spencer, *Ethiopia at Bay: A Personal Account of the Haile Sellassie Years* (Algonac, MI: Reference Publications, 1984). A wide-ranging mixture of strategic and historical overviews by an American legal advisor with unparalleled access to the emperor of Ethiopia for more than 40 years, this work gives extensive background information on the challenges to the government of Haile Sellassie from within and without the country mounted by friends and allies as much as by enemies. A fascinating set of insights into the great power politics that were at play, giving overarching drama to local events on the ground. As one of John Spencer's students at the Fletcher School of Law and Diplomacy, I can now appreciate even more fully his perspicacious insights and ever-clear logic based as they were on such real-world, real-time experiences. He was truly an eyewitness to, and significant player in, the tumult of the war and postwar years, when Ethiopia had to fend off not just the Italian occupation but also the postwar designs of Great Britain and even (briefly) South Africa.

Greece

> "I was sent to Greece, where I expected to see another communist insurgency snowballing. I saw instead its defeat, a revealing experience."—David Galula, *Pacification*

Andre Gerolymatos, *Red Acropolis, Black Terror: The Greek Civil War and the Origins of Soviet-American Rivalry, 1943–1949* (New York: Perseus Books, 2004). Good introduction to the Greek civil war of this century (remember the Peloponnesian War), which contains in it the effective methods of counterinsurgency that prevailed in this contest. Because Stalin and Churchill rather cynically assigned Greece a 90-percent-to-10-percent "sphere of influence" to Great Britain, his title is a little off-putting. In any case, the Soviet Union did not provide major support for the *Ellinikos Laikos Apelefherotikos Straos* (ELAS), or National People's Liberation Front, during the civil war. The Soviet Union was too busy solidifying its hold on the other countries it occupied in Eastern Europe during and after the war.

Mark Mazower, *Inside Hitler's Greece: The Experience of Occupation 1941–44* (New Haven, CT: Yale University Press, 1993). An in-depth look at the dynamics of "The Chaos of the New Order" as well as "Free Greece" and "A Cemetery Awash in Blood," as the author sorts out the many-sided struggles during World War II that led inexorably to the outbreak (and indeed the outcome) of the Greek civil war. A very worthwhile read.

India

William Dalrymple, *The Last Mughal: The Fall of a Dynasty: Delhi, 1857* (New York: Knopf, 2007). Early jihadist activity with Muslim, Hindu, and Christian dimensions and mutual ferocity! Seeing the Indian mutiny of 1857 in this light is both illuminating and thought-provoking. The author presents the astonishing figure that only 7,796 of its 139,000 sepoys stayed loyal to Britain (p. 244), mostly Gurkhas and Sikhs. It would appear that the insurgency was really defeated by the lack of leadership provided by the last Mogul Zafar, who was 80, and by sectarian divides within the Hindu and Muslim fighters and leaders. Astonishing levels of violence in the counterinsurgency, as British troops often slaughtered everyone in their path as they recaptured Delhi. Indeed, one of their fiercest commanders, John Nicholson ("This Great Imperial Psychopath"), led a hell column of particular intensity, although he was finally killed in battle.

Note: For a stark portrait, seen through the eyes of a true British romantic, of the Indian mutiny of 1857 and the activities of the East India Company leading up to it, lardered through and through with the multidimensional racism of the day, see the novel by John Masters, *Nightrunners of Bengal* (New York: Carroll and Graf, 1951).

Michael Edwards, *Red Year: The Indian Rebellion of 1857* (London: Hamish Hamilton, 1973). Makes the point that the rebellion was not totally quashed until 1859 and that the educated classes, English and Indian, by and large supported the British. Afterward, the number of white officers per sepoy unit was increased, and there was white control over artillery. Interestingly enough, at the height of the rebellion, universities were established at Kolkata, Mumbai, and Chennai, universities that would produce a lot of the leaders of Indian independence in the 20th century.

Ireland (Including the Irish Civil War)

The Irish rebellion and subsequent revolution and civil war could be places in the sections on "insurrections," "revolutions," or "civil war" (or even putting the works on Michael Collins under the section on strategic insights and approaches, especially his use of urban terror), but its oft-cited "specialness" and its religious and colonial aspects seem to call out for inclusion in the broader category of revolutionary upheaval and the intertwined natures of insurgency and counterinsurgency.

Note: The Irish rebels claimed they adopted their motorized "flying columns" from their studies of the Boer War.

Note: The Irish experience is an excellent opportunity to examine the usefulness and effectiveness of sustained ruthlessness by both sides (not just in the rebellion but also in the civil war that followed) because both utilized torture, assassinations, and urban and rural terror to try to win their causes. Interesting contrasts between the British and the Irish revolutionaries during the insurgency or the Irish War of Independence and then between the Irish Free State and the IRA during the civil war.

Max Boot, "Shiners and Peelers: The Irish War of Independence 1919–1921," in his *Invisible Armies: An Epic History of Guerrilla Warfare from Ancient Times to the Present* (New York: Norton, 2013) (reviewed earlier in this chapter), pp. 246–65. Rightly puts the Irish risings within the context of insurgency and counterinsurgency and is astute in making the comparison with the elements with which the insurgents became, in the guise of the Irish Free State, harsh counter-insurgents in the tradition of the Vendee after the French Revolution, when the revolutionaries became the harshest of counterrevolutionaries. The British seemed to think they were fighting a conventional war (from the time of the Easter Rising onward), but later, the Irish Free State Provisionals knew they were fighting an insurgency (against the IRA Republicans) and acted accordingly.

Tim Coogan, *Michael Collins: A Biography* (London: Hutchinson, 1990). The "Big Fellah" gets a lot of good press here, portrayed as hot-headed but shrewd and given a lot of help by the English. For example, after the Easter Rising of 1916, he is captured, but instead of being put with those to be kept in jail for life, he is allowed to wander across the room to join those who will soon be amnestied! Or when his female cousin is put in charge of the crown's secret coded messages, leaving the Big Fellah to exclaim, "In the name of Jesus, how did these people ever get an empire?" One does wonder. Collins's astute use of urban terror and manipulation of British public opinion gets an agreement but that in turn leads to civil war, as Irish factions face off. Collins says, after "Wading through Blood" and before he enters "The Mouth of Flowers," "Ah whatever happens my own fellow countrymen won't kill me." Wrong on that, he was, but the author believes that the "Little Fellah," Eamon de Valera, did not order him killed. Interesting. The ruthlessness of Big Fellah was stark, however.

James Dingley, *The IRA* (New York: Praeger, 2012). Accenting the Irish Republican "theology" as well as history and philosophy, the author asserts that the core of Irish nationalism was "Gaelic Catholicism which was anti-liberal democratic, anti-science, and anti-modernity." Interesting that the Provisional IRA killed nearly 1,800 people from 1970 to 1997 in a failed attempt to attach the Ulster counties to the Irish Republic before George Mitchell was able to arrange a peace accord. Sees a resurgent IRA challenging the status quo in Northern Ireland in the future.

Stephen Duffy, "An Unrecognizable State of War: Confronting the Irish Insurgency, 1918–1921," *Journal for the Liberal Arts and Sciences* (Fall 2009), pp. 66–83. The author calls it a "prototype of the low-intensity conflict" as "indigenous nationals" defeat a much stronger opponent, an opponent who fails to judge the political as well as the military situation. Once again, political forces and lack of will in the opponent's homeland pressured a negotiated settlement.

T. Ryle Dwyer, *Big Fellow, Long Fellow: A Joint Biography of Collins and de Valera* (London: Gill and Macmillan, 1998). A largely descriptive account of "one unwanted" and the "other cherished" odd couple who, although destructive enemies at the end, nevertheless propelled the Irish into rebellion, revolution, and ultimately civil war. It is easy to imagine how the terrorism, assassination, torture, and armed strife of the rebellion turned upon itself in the civil war (June 1922–May 1923) as the pro-treaty Irish National Army fought (and defeated) the Irish Republican Army. The civil war may have cost more Irish lives than the rebellion and revolution against the British.

———, *The Squad* (Cork, Ireland: Mercier Press, 2005). An utterly fascinating tale of Michael Collins's assassination and terror group known simply as "The Squad." The author asserts that Collins was the "originator of modern urban terrorism" as he and The Squad assassinated informers and British police as well as soldiers with the avowed aim of getting the British to overreact (which they did) in order to gain the sympathy of the broader Irish population, the bulk of which was quite passive when the rebellion began in 1916. Pitiless, remorseless, and systematic was the terror and counterterror that it provoked. Amazing how freely the Irish rebels moved around in plain sight, often on bicycles.

J. B. E. Hittle, *Michael Collins and the Anglo-Irish War: Britain's Counter-Insurgency Failure* (Washington, DC: Potomac Books, 2011). Looking at the 1918–1922 struggle through the lens of intelligence and counterinsurgency, the author suggests that Great Britain did little right and much wrong (there were 20 attempts on the life of the British viceroy, Sir John French). I wish he had continued to cover the Irish civil war in more depth, analyzing the successful counterinsurgency of the Irish Republic against the IRA. Also, why was torture so effective on one side and not on the other?

Joseph McKenna, *Guerrilla Warfare in the Irish War of Impendence, 1919–1921* (London: McFarland, 2003). Stating that the Anglo-Irish War was a "typical guerrilla war," the author does a county-by-county and even town-by-town account, especially of the flying columns (used by both sides). Believes that the IRA made it impossible for Britain to hold the territory and claims the "overthrow of British rule in Ireland became the model for the overthrow of British imperialisms in Africa, Asia and elsewhere." Also claims Meacham Begin, Mao, Giap, Castro, and Che Guevara all studied the Irish guerrillas.

Charles Townsend, *The British Campaign in Ireland 1919–1921* (London: Oxford University Press, 1975). The British had false hopes after the Easter

Rising and failed the political challenge of Sinn Fein and that of the hard men of the IRA. Drifting and losing control of the battle space, their counterbrutality turned out to be counterproductive. But at base, he blames the failure to develop an effective intelligence operation.

Kenya

Peter Baxter, *Mau Mau* (Solihull, UK: Helion, 2012). Points to the land crisis among the Kikuyu and the gang killings on both sides ("gangs and pseudo-gangs"). Accents the very important role played by the Kikuyu Home Guard and other tribes, such as the Wakamba and Masai. "Mau Mau was a disease which had been eradicated and would never be revived again."

Daniel Branch, *Defeating Mau Mau, Creating Kenya: Counterinsurgency, Civil War, and Decolonization* (Cambridge: Cambridge University Press, 2009). An excellent and very useful study of the uses of terror and counter-terror in insurgency and counterinsurgency, using the 1951–1960 Mau Mau uprising as a case study. Rightly emphasizes the crucial role played by the "loyalist" Kikuyu (many more of them fought with the colonial government as against it). While most accounts downplay the role of the loyalists, this author accents their centrality, and his chapter 3, "Defeat of the Insurgency," is well documented and very illuminating. Also points out how the Lari massacres of 1953 were central to the entire process.

Caroline Elkins, *Imperial Reckoning: The Untold Story of Britain's Gulag in Kenya* (New York: Henry Holt, 2005). This wildly inflated (she claims "hundreds of thousands" killed in an "incipient genocide" compared to the actual figure of 25,000) and mission-driven account underplays the role of the loyalists to the point of missing the point that the divisions within the Kikuyu ethnic unit rather than its solidarity directly affected the outcome. Misses entirely the civil war dimensions as well as the failure of the Mau Mau to encompass an overarching true nationalist or ideological justification. Still, she somehow won a Pulitzer Prize for this work.

Elspeth Huxley, *The Flame Trees of Thika: Memories of an African Childhood* (London: Chatto and Windus, 1882). This charming volume of reminiscences captures the dual nature of colonialism in Kenya and in general. "Surely that's the whole point of our being here," Tilly remarked. "We may have a sticky passage ourselves, but when we've knocked a bit of civilization into them, all this dirt and disease and superstition will go and they'll live like decent people for the first time in their history." Tilly looked flushed and excited when she said this, as if it were something dear to her heart. "That's is not the whole point of my being here," Alex Wilson put it, during a pause that followed. "I didn't come to civilize anyone. I came to escape from the slavery one has at home if one doesn't inherit anything. I mean to make a fortune if I can. Then I shall go home and spend it. If that helps to civilize anyone I shall be delighted but surprised."

Jomo Kenyatta, *Facing Mount Kenya: The Tribal Life of the Gikuyu* (New York: Vintage Books, 1962). Anthropologist and future president of independent Kenya, Jomo Kenyatta writes about the cultures of the Kikuyu. Useful background to understand the decolonization struggle (with its many facets) and postindependence political problems.

Wunyabari Maloba, *Mau Mau and Kenya: An Analysis of a Peasant Revolt* (Bloomington: Indiana University Press, 1993). The classic, and in many ways the most useful, book on the Mau Mau uprising. Good chapters on the British and Mau Mau military strategies, and the work also does a good job of separating the legitimate African grievances and the complicated struggle for the hearts and minds of the Kikuyu (especially the roles of the Home Guard and Loyalists). The author concludes that the "guerrillas failed to link their military exploits to political mobilization" but that many of their basic grievances and goals actually became actualized at independence, ironically by such leaders as Tom Mboya and others in the Kenyan African Union, who did no fighting.

Robert Rurak, *Something of Value* (Garden City, NY: Doubleday, 1955) (F). Even more hyperbolic and lurid than the Elkins account but one written from the white settler point of view, this novel does capture the central role of oathing and counteroathing in the struggle and the horrific and dehumanizing violence on both sides. Also good on the formation of the Home Guard, the anti-insurgent roles of the Masai, Wakamba, and Turkana and the use of counterterror in an insurgency situation. While the book is fundamentally racist in tone, it nevertheless captures the true feelings of many during the times and ironically exposes the torture, murder, and executions of the Europeans, thus foreshadowing Elkins by many decades (although that was surely not Rurak's intent). Incidentally, the author was banned from independent Kenya.

Malaya

Noel Barber, *The War of the Running Dogs: The Malayan Emergency 1948–1960* (New York: Weybright and Talley, 1971). A 12-year struggle showing once again that security (the British cycled in 100,000 troops over the period) *and* winning hearts and minds were both required.

John Coates, *Suppressing Insurgency: An Analysis of the Malayan Emergency 1948–1954* (Boulder, CO: Westview Press, 1992). A true classic and must-read. Amazing detail and careful chronicling of the process of winkling out dozens and dozens of insurgents with "patience, patience, patience." Great charts and maps capture the true nuts and bolts of the campaign, including "Average of Battalion Elimination by Racial Group 1953–54" and "Sorties by (type of) Aircraft."

Donald Mackay, *The Malayan Emergency 1948–60, the Domino That Stood* (London: Brassey's, 1997). A "state of emergency" was, in fact, a successful counterinsurgency war. The author gives considerable credit to Sir Henry Gurney and the poor leadership of the Communist Chin Peng rather than Sir Gerald Templer, calling Peng a "pedantic" follower of Mao's teachings. Mackay is careful not to overpromise the lessons of the Malayan situation.

Robert Stubbs, *Hearts and Minds in Guerrilla Warfare: The Malayan Emergency 1948–1960* (London: Oxford University Press, 1989). The term *hearts and minds* first popularized during this war as the Chinese Communist guerrillas tried to move into the postwar world. One problem they faced was the fact that only 40 percent of the population of Malaya was Chinese and there were no dependable sources of outside aid, but the proper emphasis on non-military—as well as military—activities does seem transferable.

Oman

John Akehurst, *We Won a War: The Campaign in Oman 1965–1975* (Southampton, UK: Wilton, Salisbury, Wiltshire, 1982). Wonderful case study of counterinsurgency and how it was won with Iranian troops under British direction. This is a very much understudied example of how to defeat a guerrilla force. See the other works on Oman listed here for additional information.

Calvin Allen, *Oman: The Modernization of the Sultanate* (Boulder, CO: Westview Press, 1987). The winning of an important but almost never studied successful counterinsurgency situation.

Calvin Allen and W. Lynne Rigsbee, *Oman under Qaboos: From Coup to Constitution 1970–1996* (London: Frank Cass, 200). Local leadership proved up to the task of counterinsurgency. A lesson for many futures.

John Peterson, *Oman's Insurgencies: The Sultanate's Struggle for Supremacy* (London: SAQI, 2007). A well-researched, carefully presented case study that should be a must-read for those who would understand counterinsurgency. This is a very important amendment to many insurgency paeans.

John Townsend, *Oman: The Making of a Modern State* (New York: St. Martin's Press, 1977). How insurgency was defeated in a small state by an able ruler with some help from his friends, including some troops provided by the shah of Iran.

The Philippines

James Arnold, *The Moro War: How America Battled a Muslim Insurgency in the Philippine Jungle, 1902–1913* (New York: Bloomsbury Press, 2011). The United States had done insurgency before and prevailed, even against Muslims. The author shows how the masters of the Civil War and the conflicts on the Plains ultimately prevailed, although it took a long time, and it was not an uplifting endeavor. "Disarming Moros was a sad and messy business." Unfortunately, the lessons of the conflict were lost before 1954 and 2003.

Peter G. Gowing, *Mosque and Moro: A Study of Muslims in the Philippines* (Manila: Philippine Federation of Christian Churches, 1964). This short study concludes, rather pessimistically, "Warfare, alternately hot and cold, has raged between Muslims and Christians in the Philippines for well-nigh 400 years, and peace has not yet been made."

Miguel J. Hernandez, "Kris vs. Krag," *Military History* (June 2006), pp. 58–65. Over a period of 15 years (1902–1917), US forces crush the Islamic Moro insurgency in the Philippines and, judging from the pictures accompanying this article, very harshly, including massacres. Note: The Kris was the curved knife favored by the Moros; Krag, the US army rifle of the day.

Anthony Jones, "Counterinsurgency in the Philippines 1898–1954," in Daniel Marston and Carter Malkasian (eds.), *Counterinsurgency in Modern Warfare* (Oxford, UK: Osprey Books, 2008). A must-read, for it shows the prices that have to be paid if an insurgency is to be defeated. The subtext, of course, is the question as to whether those prices are worth it.

Marc Leepson, "The Philippine War: Successful Counterinsurgency or a Grab for Territory," *Military History* (November 2007), p. 60. It was both! Article presents both sides of the issue before taking sides in this worthwhile piece.

Adrian Linn, *The Philippine War 1899–1902* (Lawrence: University Press of Kansas, 2000). Linn marvels at how 25,000 American soldiers "cowed and terrorized" 7 million people. He finds that the "insurrectos" were defeated by their bad leadership, sustained ruthlessness, civic action, and army-led social reform. Well worth reading.

———, *The U.S. Army and Counterinsurgency in the Philippine War 1899–1902* (Chapel Hill: University of North Carolina Press, 2005). A "nasty little war" is examined through the actions of the US Army in four districts of Luzon. Shows how the US Army made ongoing adjustments to changing tactics and strategy of the guerrillas.

David Silbey, *A War of Frontier and Empire: The Philippine-American War, 1899–1902* (New York: Hill and Wang, 2007). An illuminating work about when American was a real empire and defeated several insurgencies in one location of that empire with lessons for today from both aspects.

Russia (19th Century)/Soviet Union (20th Century)

Lesley Branch, *The Sabres of Paradise* (New York: Carroll and Graf, 1960). Engaging story of the Russian conquest of Shamyl, imam of Dagestan, and his fight to preserve the independence of the Caucasus region. Useful tale of counterinsurgency as well. Russia continues to have difficulty with the region even today.

Moshe Gammer, *The Lone Wolf and the Bear: Three Centuries of Chechen Defiance of Russian Rule* (Pittsburgh: University of Pittsburgh Press, 2006). A useful introduction to Russian conquest, czarist rule, Soviet domination, and the continuing resistance to Russian rule. Saves much scorn for those who collaborate with the imperialists: "Tremble in your cages when we go out hunting! Because, more than any bear, we wolves hate dogs." Useful sections on Chechnya's declaration of independence in 1991, the first Russo-Chechen war (1994–1996) and the second Russo-Chechen War (1999–2005). A long struggle not yet concluded, the author notes.

Kenneth Slepyan, *Stalin's Guerrillas: Soviet Partisans in World War II* (Lawrence: University Press of Kansas, 2006). Insurgents and partisans in USSR-occupied territory, some under the control of Moscow, others now. Insights into the day-to-day struggle against the Germans and their allies and the tensions within and among partisan banks. Excellent chapter, "The Crisis of Partisan Identity, 1943," with superb maps on pages 189 and 192. Partisans went from 100,000 to 181,000 during that year. Good section on women partisans in that chapter.

Leo Tolstoy, *Hadji Murad* (New York: Dodd Mead, 1912) (F). A novel set in the Caucasus in 1851 tells the story of a devote Muslim Chechen warrior, Hadji Murad, who comes to the Russians when the head of the resistance fighters, Shamil, quarrels with him and takes his family hostage. Hadji Murad offers to help the Russians pacify the area if they will exchange prisoners with Shamil. But due to bureaucratic failure, jealousy in the capital, a Great Russian feeling of cultural superiority, and general inertia, his offer is not accepted. Hadji Murad then tries to flee and is cut down by Cossacks, his bloodied head brought back to the Russian garrison. Vivid accounts of search-and-destroy missions, the failure to show generosity in counterinsurgency operations, and a lack of appreciation of the nature of medieval warfare all make this an outstanding set of insights into irregular warfare and the thin line between banditry and guerrilla campaigns.

Rhodesia/Zimbabwe

Ian Pringle, *Dingo Firestorm: The Greatest Battle of the Rhodesian War* (Cape Town: Zebra Press, 2012). Illustrative of a growing number of memoirs about war in southern Africa, this work captures the small-scale nature of the war and its many international dimensions. For example, the author does not see Great Britain's animus toward the Rhodesian Unilateral Declaration of Independence (UDI) as ensuring the success of the insurgents nor the split between ZAPU and ZANU with the Soviet and Chinese support, respectively, but rather the decision of South Africa to make political common cause with Zambia after the Rhodesians had prevailed militarily. The Dingo operation involved 184 troops and 61 aircraft and destroyed (at least temporarily) ZANLA bases at Tembue and Chimoio during November 1977, killing more than 1,000 ZANU personnel in the process.

South Africa

Martin Dugard, "Farmers at Arms: Second Boer War 1899–1902," *Military History* 26, no. 1 (2010), pp. 26–33. Shows how counterinsurgency took three years to break the back of Boer resistance. Sobering figures: 26,521 women and children and 1,421 men died in British concentration camps compared with total British war casualties of 21,942 and between 7,000 and 9,000 Boer combatants.

Byron Farwell, *The Great Anglo-Boer War* (New York: W. W. Norton, 1990). Why people who study counterinsurgency never look at the Anglo-Boer War is quite beyond me. It's all here: the difficulties, the opportunities, the sine qua nons.

Denis Judd and Keith Surridge, *The Boer War* (New York: Palgrave Macmillan, 2003). The war, its insurgency, and counterinsurgencies and its foretelling of World War I are displayed in easy-to-read fashion. Especially useful is their chapter, "The Methods of Barbarism" (pp. 187–96), looking at the methods used by the British to choke off support for the Boer commandos. "When is war not a war? When it is carried on by methods of barbarism in South Africa," claimed an English critic, but the British military was having none of that sort of criticism, as they simply wanted to get the job done.

Thomas Pakenham, *The Boer War* (London: Weidenfeld and Nicolson, 1979). The Boer was a classic example of a popular insurgency and an effective counterinsurgency. Contains many important elements of the latter, including the fact that the "loyalists" consisted of one-fifth of the Boer population who fought with the British as well as an effective mix of clear-and-hold *and* search-and-destroy as the Afrikaner commandos were hunted down at the same time some of the population was put in "concentration camps" to separate the people from the guerrillas.

Christopher Wilkinson-Latham, *The Boer War* (Oxford, UK: Osprey, 1977). A useful set of photographs captures the precursors to the heavy casualties of trench warfare, which would be such an important dimension of World War I, and the development of mobile guerrilla warfare and how to suppress it (and at what cost).

Sri Lanka

Gordon Weiss, *The Cage: The Fight for Sri Lanka and the Last Days of the Tamil Tigers* (London: Bodley Head, 2011). A very long-running insurgency is eventually crushed by violent and cruel counterinsurgency tactics. Not regarded as a model for antiguerrilla tactics by some.

United States (a Potpourri from the 18th–21st Centuries)

It is always quite amusing to hear pundits talk about the United States and its lack of experience with guerrilla warfare until recently. In fact, Americans have had an extremely long and varied experience with insurgency *and* counterinsurgency, beginning with the French and Indian Wars, against the British during the Revolution, and many additional conflicts up to the present. It is important to remember that not remembering the lessons of insurgency and counterinsurgency is not the same as not learning them in the first place. This section of readings teases the reader into looking at just a sample of the many examples within the American experiential frame.

Note: For example, much insurgency experience lay in the American Revolution, which was won by a combination of guerrilla *and* main-force battles. That long and bloody experience would eventually cost 1 percent of the existing population, making it the second-most-costly war in US history, the first being the American Civil War, with 1.6 percent of the population losing their lives (Boot, *Invisible Armies*, p. 77).

Note: For other, more traditional, works on the American Revolution and its background, see chapter 18.

T. H. Breen, *American Insurgents, American Patriots* (New York: Hill and Wang, 2010). A much-needed revisionist account of the American Revolution, underscoring the huge role played by those proto-Americans who, as insurgents, set the stage for the successful revolution. Breen's contention that the real revolution took place between late 1774 and April 1775, with the breaking of imperial bonds by local Committees of Safety and culminating at Lexington and Concord, where the "killing of militiamen by an army of occupation permanently transformed the American political landscape," even before the Declaration of Independence. This account rings true and does much to put the American Revolution rightly at the heart of the spectrum of the world's most important revolutions.

Henry Clinton, *The American Rebellion: Sir Henry Clinton's Narrative of His Campaigns, 1775–1782*, edited by William Wilcox (New Haven, CT: Yale University Press, 1954). Edited by William Wilcox and with an indispensable foreword by him, these remembrances are vital to our true understanding of how the Americans succeeded. British commander Clinton served longer than any other major figure (save Washington), from Bunker Hill to Yorktown, with campaigns in New York, Rhode Island, Philadelphia, and the South. Clinton engineered the most complete British victory of the war at Charleston but failed utterly to match the strategic moves of the Americans. Was often timid and vacillating but nevertheless grasped the essential facts that the British army had to operate out of major ports and was ill-suited for defeating partisan warfare.

Charles Cornwallis, *Correspondence of Charles, First Marquis Cornwallis*, edited by Charles Ross (Cambridge: Cambridge University Press, 2013). His letters (originally published in 1858) as collected here provide a useful counterpoint to all that we as Americans have ingested concerning our revolution and give valuable insights into Cornwallis as a military leader. After the revolution, he became governor general of India and lord lieutenant of Ireland.

Christopher Dickey, *Securing the City: Inside America's Best Counterterror Force—the NYPD* (New York: Simon and Schuster, 2009). A straightforward account with some good insights. I learned that, when there is an emergency or a crisis, the NYPD can "surge" 176 patrol cars into the trouble spot, one car from each precinct. Also, it would appear that the NYPD had a much better liaison with the CIA and foreign intelligence networks than with the FBI! Commissioner Ray Kelly gets very high marks from the author.

Ronald Hoffman, Thad Tate, and Peter Albert (eds.), *An Uncivil War: The Southern Backcountry during the American Revolution* (Charlottesville: University Press of Virginia, 1985). A wide-ranging in-depth set of essays looking at how British counterinsurgency efforts ultimately backfired despite major military victories and how the "Liberty Men" eventually frustrated their efforts to enlist Loyalist support and crush the opposition in rural Virginia, Georgia, and South and North Carolina.

Richard Ketchum, *The Winter Soldiers* (Garden City, NY: Doubleday, 1973). Focusing on 1776 and early 1777, the author asserts that, on Christmas Eve 1776, the goal of independence never seemed more distant or problematical and only Washington's vision and victories at Trenton and Princeton prevented the end of the Continental Army. Breezy, readable, and with good illustrations, this work is a nice introduction to this first phase of the war. Also notes how General Howe went into winter quarters prematurely in December 1776 and might have been more successful if he'd gone after Washington wherever he went. These were truly the "times that try men's souls," and indeed, the "snow was tinged with blood."

Mark Kwasny, *Washington's Partisan War 1775–1783* (Kent, OH: Kent State University Press, 1996). A very useful and illuminating account of Washington's skillful use of state militias in consort with the Continental Army during his eight-and-a-half-year struggle. Makes the case that his approach was complex and nuanced and he came to highly value partisan guerrilla warfare, not only in the southern theater, but also in the area from New Jersey to Connecticut. The militia partisan activity drove the British and their Hessian mercenaries to distraction; "[l]ittle war was continuous war," as they sought the decisive battle that would end the revolution. Early search-and-destroy strategy failed.

Peter Mansoor, *Baghdad at Sunrise: A Brigade Commander's War in Iraq* (New Haven, CT: Yale University Press, 2008). Very insightful by a general who was there when the war turned. Now a professor at Ohio State University, one of the few places in American academia where military history is still studied and prized.

James McCarjar, *Lost Victory: A Firsthand Account of America's Sixteen Year Involvement in Vietnam* (Chicago: Contemporary Books, 1989). Strongly revisionist and well worth reading. One doesn't have to accept the author's basic premise to see why abandoning analysis after 1968 leads many observers and scholars to miss the realities of counterinsurgency, even in the eventual failure of that effort.

Andre O'Shaughnessy, *The Men Who Lost America* (New Haven, CT: Yale University Press, 2013). A highly readable and enjoyable account of the British military and political leaders and what they did or didn't do to defeat the American insurgents. A good example of how successful revolutions depend on the mistakes of the counterinsurgents as well as the good judgment calls of the insurgents. The biggest mistake, says the author (and a number of the participants), was overestimating the number and commitment of the Loyalists. The military strategies of John Burgoyne, William Howe, Lord Cornwallis, and Henry Clinton get higher marks than usual, while the political strategists in England, especially King George III and secretary of state for colonies Lord John Germain, get more criticism. Interestingly enough, the political leaders seemed driven by a great fear of a domino effect, namely that losing the American colonies would mean losing the sugar islands of the Caribbean, which were, at the time, much more valuable to England. Often forgotten, too, were the large and important contemporary victories elsewhere—Gibraltar, India, British Caribbean.

Ted Morgan, *Valley of Death: The Tragedy at Dien Bien Phu That Led America into the Vietnam War* (New York: Random House, 2010). Well-written and very powerful account of the 1953–1954 set of events that reverberate down to the present day. The odd thing was a lot was going right for them before the French decided on such a disastrous search-and-destroy strategy.

Neil Sheehan, *A Bright Shining Lie: John Paul Vann and America in Vietnam* (New York: Vintage Books, 1989). Sheehan's book can be read two ways: (1) why the United States was doomed in Vietnam and (2) how the United States could have won if Vann's strategy had been implemented much, much earlier. Vann was certainly right about one thing when he said, "I don't know if security is 10% or 90% of the solution but I know it's the first 10% or the first 90%." Spot-on he was and is.

Lewis Sorley, *A Better War: The Unexamined Victories and Final Tragedy of America's Last Years in Vietnam* (New York: Harcourt Brace, 1999). Makes a strong case that the United States lost the war in Washington, not in

South Vietnam, and indicates that the 1968–1972 period was one of great progress in counterinsurgency.

———, *Thunderbolt: General Creighton Abrams and the Army of His Times* (New York: Simon and Schuster, 1992). The life and times of General Abrams, whose counterinsurgency strategy in Vietnam turned out to be the correct one and, although forgotten for a long time, reemerged with the surge in Iraq.

John J. Tierney, *Chasing Ghosts: Unconventional Warfare in American History* (Washington, DC: Potomac Books, 2007). Contrary to widespread belief, the United States has had a great deal of experience—and success—in waging counterinsurgency throughout its history. In the Philippines, for example, after World War II, a judicious combination of democratization and force defeated the Communist Hukbalahap rebellion. Incidentally, Colonel Edward G. Lansdale, who was senior adviser to the US Military Advisory Group (USMAG) and to the Philippine defense minister (and later president) Ramon Magsaysay, became the much-ridiculed central figure in the novels *The Ugly American* and *The Quiet American* (see chapter 40) for his efforts in Vietnam despite his earlier success in the Philippines.

Bui Tin, "How North Vietnam Won the War," *Wall Street Journal*, August 3, 1995, A8. Penetrating and insightful and revisionist enough to raise eyebrows.

Note: There are many other books dealing with insurgency reviewed under a variety of topics. See chapters 18, 22, 31, and 32.

Chapter Nine

The American "Empire" and the Dangers That Lie Ahead

How does one assess the American "Empire"? Does such a thing exist? Is there an actual "empire" in the sense the Great Britain and France had empires in the 19th century? If there is a real American Empire, is or would that be a good thing; that is, would the world be better off without it? Is it a bad thing; that is, would the world be better off with it? What would the world of tomorrow look like if it either didn't exist or remained only in a truncated, far less powerful state?

Any true understanding of the problems facing America in the 21st century can be seen by the very need to put its challenges in separate categories for analysis. In terms of today's American foreign and military policy, it is like playing chess on 30 different interlocking game boards simultaneously. Note, too, that "imperial overreach" or not, the United States now has 24 of the world's 34 aircraft carriers and, in terms of force projection, is the only great power capable of declaring—and meaning—the whole globe is a battle space. This has both positive and negative aspects, but it is a realistic factor of enormous import for at least the new term.

There are many opinions about the American Empire and its options in the future. We have broken the many challenges faced by the United States going forward (such as "The Nuclear Dimension," "China," "Failed States," etc.).

By way of introduction, here are a few preliminary works that provide widely different points of view on the subject:

THE EMPIRE AS EMPIRE (OR HEGEMON)

If there is an "American Empire," then it certainly doesn't seem to fit the classic pattern of a true empire, with overseas colonies administered directly by the mother country. Nor does it seem to fit the very useful definition of *imperialism* by William Harris: "The behavior by which a state or people takes and retains supreme power over other states or peoples or lands." Nevertheless, the United States does have worldwide interests that require the stationing of troops overseas, often at the insistence of the host country—South Korea, Japan, Germany, and Kuwait come to mind.

The works in this section explore what this phenomenon should be called, make the case for and against calling the present international order an "American Empire," debate whether it is a good or a bad situation, and often suggest a way forward for the evolution of the current American force projection posture. Personally, I favor the employment of the term *hegemon* instead of *empire*, as in "America as hegemon." This avoids wrangling over terminology and the appropriateness of the term *empire*.

There are a wide range of opinions on these and other aspects of world power configurations and the role of the United States in those configurations listed here, including:

Thomas K. Adams, *The Army after Next: The First Postindustrial Army* (Stanford, CA: Stanford University Press, 2008). Good on technology, somewhat weaker on strategic concerns and visions.

Edward Atkeson, *The New Legions* (Lanham, MD: Rowman and Littlefield). This retired US major general argues that the United States needs to recruit and train its own foreign legion and gives a global tour de horizon as to where this legion will work. He points to the more than 30,000 foreigners who have already joined the US Army in order to gain citizenship. Foreign recruits now make up only 5 percent of all recruits.

Thomas Barnett, *The Pentagon's New Map: War and Peace in the Twenty-First Century* (New York: G. P. Putnam's Son, 2004). Seeking a vision for the future, the author divides the world into "The Functioning Core" (United States, Europe, Japan, China, etc.) and "The Non-Integrating Gap" (Africa, Middle East, failed states wherever they are found). There are 2 billion people disconnected in "The Gap." Barnett argues, "The 'good war' is the won war, not a war waged forever." May need to consult the Salafists on that. Interesting holistic approach, "war within the context of everything else." A worthwhile take on the American "Empire" and its challenges to think ahead of the next day's events. Just skip the bizarre and outlandish conclusion of the book, where the author says Kim Jong-il is going to be removed, the

Iranian reformers are going to win, and the Chinese military represents no strategic threat. More than a decade later, movement toward these outcomes seems limited at best.

Peter Beinart, *The Icarus Syndrome: A History of American Hubris* (New York: HarperCollins, 2010). America, he believes, suffers from the "hubris of dominance," the "hubris of toughness," and the "hubris of reason," albeit not always at the same time!

William Blum, *Rogue State: A Guide to the World's Only Superpower* (Monroe, ME: Common Courage Press, 2004). Very fierce indictment of the United States on all fronts. The author is sure the United States is the problem, not the solution, for many of the world's ills.

Max Boot, "The New American Way of War," *Foreign Affairs* (July/August 2003), pp. 41–58. Nimble and light but shows that the whole world is a battle space.

Thomas C. Bruneau, *Patriots for Profit: Contractors and the Military in the U.S. National Security* (Stanford, CA: Stanford University Press, 2011). A short introduction to the current situation, in which the ratio of private contractors to military personnel has risen from 1:5 to 1:6 in the Revolutionary, Korean, and Gulf Wars to 1:1 in the Balkans peacekeeping and the wars in Iraq and Afghanistan. The author decries the "confusion, ineffectiveness and inefficiency of private security contracting as it is today."

James Carroll, *House of War: The Pentagon and the Disastrous Rise of American Power* (New York: Houghton Mifflin, 2006). Harsh indictment of the "House of War" and how it is always in search of enemies. A trenchant analysis, albeit one a bit light on an exploration of the alternatives.

Thomas J. Christensen, *Worse Than a Monolith: Alliance Politics and Problems of Coercive Diplomacy in Asia* (Princeton, NJ: Princeton University Press, 2011). A closely argued and densely packed assertion that, contrary to common assumptions, intra-alliance splits and divergence are more difficult to deal with than more monolithic alliances.

Niall Ferguson, *Colossus: The Price of America's Empire* (New York: Penguin, 2004). Author sees the United States as morally, and soon to be financially, bankrupt in trying to keep its "empire" going.

Thomas Friedman, *The World Is Flat* (New York: Farrar, Straus, and Giroux, 2005). Argues that, in the present and coming world, the integration of nations and peoples will trump the aspects that divide us.

Adrian Goldsworthy, "Conclusion: A Simple Answer" and "An Even Simpler Moral," in his *How Rome Fell: The Death of a Superpower* (New Haven, CT: Yale University Press, 2010), pp. 405–24. The United States is not Rome.

Eric Hobsbawm, *Empire: America, War, and Global Supremacy* (New York: Pantheon, 2008). Is it really an "empire"? By traditional definitions, it is not, argues the author.

David Hunt, *They Just Don't Get It* (New York: Random House, 2005). An angry colonel spouts off on the "War on Terror," sometimes with great clarity, always with considerable vehemence.

Robert D. Kaplan, *The Coming Anarchy* (New York: Random House, 2000). A realist's realist, the author's essays on the end of the Cold War ring true and perceptive, especially his concerns over American efforts to project power while at the same time encouraging the growth of democracy, pointing out that Hitler and Mussolini emerged from democratic political systems. Having traveled the back spaces and seen power vacuums up close and personal, he sees a steady decline for the nation-state.

———, "Columbia" and "The Philippines," in his *Imperial Grunts: The American Military on the Ground* (New York: Random House, 2005), pp. 39–90 and pp.130–84. With the grunts on the ground in two separate but seemingly strangely linked venues, one gets a sense for the far-flung responsibilities of the United States, responsibilities that seldom show up in the daily "box score."

———, *Monsoon: The Indian Ocean and the Future of American Power* (New York: Random House, 2010). Kaplan suggests a radical reorganization of the way one looks at the world and finds that the Indian Ocean is becoming more and more central to the future of the world, as China expands "vertically" and India expands "horizontally" and the United States and Great Britain remain on Diego Garcia.

Edward Luttwak, *The Grand Strategy of the Byzantine Empire* (Cambridge, MA: Harvard University Press, 2010). Much good advice here if US decision makers were to follow the "operational code" of the Byzantines, whose eastern Roman Empire lasted more than 800 years longer than the Roman Empire in the west. In a nutshell, Luttwak advocates ditching the analogy with Rome and adopting the one with the 1,000-year-long Byzantine Empire.

Kishore Mahbubani, *The New Asian Hemisphere: The Irresistible Shift of Global Power to the East* (New York: BBS, 2008). With considerable glee, the author sees and blames US blunders and disasters for the massive power realignment to China and India. Israel and United States taken to task, Iran and the Soviet Union given every benefit of doubt. Still, the conclusion, albeit trite, makes sense: "Pragmatism is the best guiding spirit we can have as we venture into the new century."

Michael Mandelbaun, *The Case for Goliath* (New York: Public Affairs, 2006). See especially his chapter on "The Future," in which he says about other countries' views of American hegemony, "They will not pay for it, they will continue to criticize it, and they will miss it when it is gone."

Mark Mazzetti, *The Way of the Knife: The CIA, a Secret Army, and a War at the Ends of the Earth* (New York: Penguin, 2013). This *New York Times* reporter is bewildered by all the changes with the Pentagon getting into the spying business and the CIA into the war-making business, with its world-wide drone operations, and both of them subcontracting out violence across the globe. Where are the allegiances of the subcontractors? "Do they lie with the flag? Or do they lie with the bottom line?" At the end of the day, the author outlines the holistic requirements of a hegemon determined not to be phased by the whole world being a battle space and the irritating turf battles that accompany its multiple missions.

Walter Russell Mead, *Power, Terror, Peace and War* (New York: US Council on Foreign Affairs, 2004). Overview of the forces that are shaping the 21st century and the limits of US power to thwart all of them combined, albeit with the need to thwart many of them. The future requires more US presence in the world, not less, if the global order is to stay intact.

Michael Morell, *The Great War of Our Time* (New York: Twelve, 2015). The author, a former CIA deputy director, does an excellent job in showing the complexity and intricacies of the dangers confronting Goliath and makes a strong case for the many things he believes the CIA does correctly. Controversial but substantial, especially for those unaccustomed to examining the international espionage subculture. Well worth reading.

Ian Morris, *Why the West Rules—for Now* (New York: Farrar, Straus, and Giroux, 2010). Traces the upward trajectory of the West: In 1500, its ascendency was not likely; in 1650, it was not clear; in 1800, it was likely. Today, its situation has the ambiguities of the fusion of America and China ("Chimerica") and projects great arcs of uncertainty from Africa through the Middle East to non-Han China and ends with the hope that "[h]istorians needed to explain the differences that divide humanity and how we can prevent them from destroying us." Ambitious tasking, I would say.

Andrew Mumford, *Proxy Warfare* (Cambridge, UK: Polity Press, 2013). This short, sprightly written work focuses on proxy warfare, the indirect engagement of a third party in a domestic or international conflict. The author asserts that this is "warfare on the cheap" and a "missing link in contemporary war and security studies." Using interesting case studies, including the Spanish civil war, the activities of Iran in Iraq since 2003, the Angolan civil war, and various East African wars, the author illustrates both the dynamics and the intents of those who would sponsor proxy warfare. Citing the rise of China and its doctrine of "unrestricted warfare" and the growing US use of unmanned aerial vehicles (UAVs) across the globe as well as other factors, he concludes that proxy warfare has a great and growing future due to its perceived economic (low cost) and political (no "direct" involvement) advantages, as nation-states and even nonstate actors consider it useful for ideological, religious, strategic, and national reasons.

Sean Naylor, *Relentless Strike: The Secret History of the Joint Special Operations Command* (New York: St. Martin's Press, 2015). Wide-ranging and mostly based on open sources, this volume looks at the American Empire in action all across the globe: Grenada, Lebanon, Iraq, Somalia, Syria, Afghanistan, Niger, Iran, and Yemen, among others. It shows a tremendous commitment to global reach and action beyond the conventional battlefield.

William Plaff, *The Irony of Manifest Destiny: The Tragedy of American Foreign Policy* (New York: Walker, 2010). Argues strongly that America's foreign policy had been blinded by anti-Communism and now radical jihadism and has taken it upon itself to cure the ills of the world in a "subordination of ethical values to an ideology of national triumphalism." In fact, he calls it a "colossally militarized but morally nugatory global mission." He may be right, he may be wrong, but he sure lost me when he claimed that George W. Bush started speaking to President Chirac of France in tongues about "Gog and Magog." The author may have been in Paris far too long.

C. P. Potholm, "Democracy Armed," in *The Bowdoin Forum: Journal of International Affairs* 5, no. 1 (Fall 2001), pp. 49–64. Argues for the need for a hegemon and the greater need for America to be that hegemon.

Gideon Rose, *How Wars End: Why We Always Fight the Last Battle* (New York: Simon and Schuster, 2010). Why does the aftermath of war get "short shrift" when planning a war? Is it because "[y]ou shoot the wolf closest to the sled"? In any case, the author bemoans the fact that "wars are difficult to close out even when they are started well" and hopes that, in the future, America will clarify its goals and plans for what happens after the next war ends.

David Sanger, *Confront and Conceal: Obama's Secret War and the Surprising Use of American Power* (New York: Crown, 2012). Surprising indeed. I have never read a book that so blatantly exposes American covert activities as *they are in progress*. Of course, a cynic might assume that a *New York Times* reporter would want to make President Obama look good in an election year and the White House would welcome a portrait of a tough, serious, can-do, destroy-our-enemies president, but this "investigation" into such secret programs as "Olympic Games," a cyberwar attack on Iran's nuclear effort, is almost beyond belief as insider after insider spills his or her guts on the record. There is comic relief, as well as a most serious point, when Secretary of Defense Robert Gates suggests a new strategic approach to Tom Donilon, the most garrulous national security advisor in American history. Gates says simply, "Shut the f—— up." Sound advice, that, but by then, the horses were out of the barn and into the next country. Amazing. It is obviously most difficult to maintain even a truncated version of the American "Empire" with this sort of operational process in motion.

Jeremy Scahill, *Dirty Wars: The World Is a Battlefield* (New York: Nation Books, 2013). Reviewed later this chapter under "Failed and Fragile States."

Mark Steyn, *America Alone: The End of the World as We Know It* (Washington, DC: Regency, 2006). A bleak and somewhat overwrought view of the future. One hopes.

Douglas C. Waller, *Big Red: The Three-Month Voyage of a Trident Nuclear Submarine* ((New York: HarperTorch, 2002). An authentic and deeply stimulating study of a three-month cruise of a Trident submarine, the *USS Nebraska*, on patrol. Longer than the Washington Monument is tall and costing $2 billion, the *Nebraska*—and every other Trident submarine—automatically becomes the fifth-largest nuclear power every time it goes to sea, and its crew members get medals for their cruises before they leave because, if a nuclear war broke out, there would be no home port to which to return. Sobering.

Micah Zenko, *Between Threats and War: U.S. Discrete Military Operations in the Post-Cold War World* (Stanford, CA: Stanford University Press, 2010). A most illuminating study of 36 discreet military operations (DMOs), ranging from Bosnia in 1994 and the Sudan in 1998 to Yemen in 2002, and enforcing a no-fly zone in Iraq (1991–2003). The author points out that, although only half of these efforts were successful, they continue to be favored more by US civilian authorities than their military counterparts.

Tony Zini, *The Battle for Peace* (New York: Macmillan, 2006). A thoughtful American general looks at the new century and sees hope, challenge, and the need for new ways for the United States to achieve its goals in a much changed environment.

THE NUCLEAR DIMENSION

The nuclear age began with the dropping of the bombs on Hiroshima and Nagasaki, and nuclear weapons soon became the "gold standard" of great power status, as Great Britain, France, the Soviet Union, and China acquired those weapons in the 1940s and early 1950s. The Cold War saw the dangers of proliferation—and the strategies and assumptions to prevent it—many of which are heaped with irony. In this section, I look at the ways in which the United States seeks to cope with that history of nuclear weapons and strategy and the present realities and future threats that can come from the nuclear dimension. As there is significant overlap with the sources cited previously, a mere sampling of the many works on the nuclear and related dimensions are listed here:

James Carroll, *House of War: The Pentagon and the Disastrous Rise of American Power* (New York: Houghton Mifflin, 2006). The author makes a strong case, if a somewhat unbalanced one, that the Pentagon engages in self-fulfilling prophecies about global threats and indeed is constantly "in search of enemies."

Noah Feldman, "Islam, Terror and the Second Nuclear Age," *New York Times Magazine*, October 29, 2006, pp. 50ff. Presents a truly horrifying future, where those who want to preserve their present international system will have to get it right over and over and over, whereas the terrorists only have to get it right once.

William Langewiesche, *The Atomic Bazaar: The Rise of the Nuclear Poor* (New York: Farrar, Straus, and Giroux, 2007). Truly disturbing look at the proliferation of nuclear weapons, nuclear waste, and nuclear desires.

Joseph Nye, *The Future of Power* (New York: Public Affairs, 2011). There is "hard" power and "soft" power and now "smart" power, which Nye believes should be used more often. He asserts that the United States is not in terminal decline and can use its smart power—persuasion, clever use of both soft and hard power, and more modest goals—to play its future role (or roles) with more aplomb and with less self-loathing than previously.

Kevin Phillips, *American Dynasty: Aristocracy, Fortune and the Politics of Deceit in the House of Bush* (New York: Viking, 2003). A very harsh and telling indictment of both Bush presidencies and their global goals and tactics.

Ron Rosenbaum, *The Road to a Nuclear World War III* (New York: Simon and Schuster, 2011). The author highlights the possible role Israel and Israel's enemies may play in launching the bombs or missiles that could well set off a limited nuclear war. Stipulates that the Israelis, like the Soviets before them, may have a nuclear doctrine in place preparing for both a second-strike and a first-strike capability. Ominous.

Rear Admiral Bill Sullivan, "Fighting the Long War: Military Strategy for the War on Terrorism," lecture delivered at Mississippi State University, December 2005. A very powerful, chilling, and stern assessment of the future. Only time will tell if Sullivan is an alarmist or a prophet. As of 2015 and the rise of ISIS and the Islamic state, one could realistically lean toward prophet.

CHINA

> "We are the masters now."—Chinese economist, 2010

> "The South China Sea is the Mittleuropa of the twenty-first century."—Robert Kaplan, *Asia's Cauldron*

China's shadow looms large over the United States today as well as over the world and seems likely to project a future of an even larger shadow, whether it is in terms of the "gathering in" of Taiwan, the suppression of separatist movements in Tibet and Sinking, or the global competition for scarce resources. Yet much of Chinese history since the 16th century has been nonexpansive in nature. Is that historical posture changing? What are the implications of such possible changes, and how are they viewed and likely to be viewed by the Philippines, Malaysia, Vietnam, South Korea, and Japan? What are the implications of all of this for the United States and the rest of the world? The works cited here presents the reader with both answers and additional questions.

Note: In 2011, China reportedly spent at least $111 billion on its internal security apparatus and at least $106 billion on it military.

Dennis Blasko, *The Chinese Army Today* (London: Routledge, 2006). China's People's Liberation Army and its new "Long March" to military modernization is summarized in the following statement: China has force and is prepared to use it. It wants other states to know that its potential use of force is a credible threat.

Paul Bracken, *Fire in the East: The Rise of Asian Military Power and the Second Nuclear Age* (New York: HarperCollins, 1999). Argues that the great civilizations of the past—China, India, and Persia—are now reasserting themselves with great strategic import. He cites the 70,000 nuclear weapons amassed during the First Nuclear Age (p. 131) and wonders about the future in the "rim of terror." A useful and early wake-up call to the Western strategists who still believe the West can control its own destiny.

Richard Bush and Michael O'Hanlon, *A War Like No Other: The Truth about China's Challenge to America* (New York: John Wiley and Sons, 2007). From "Thinking about the Unthinkable" to "Why China Could Not Seize Taiwan," the authors provide a great deal to think about, including the notion that just because a war could break out over Taiwan doesn't mean that it will involve the United States directly. They believe that Taiwan might be able to repulse an invasion without any US help. At the same time, they suggest that, at the operational level, China does not have a policy against the first use of nuclear weapons.

Jung Chang, *Wild Swans: Three Daughters of China* (New York: Doubleday, 1991). Ten million copies sold—or more by now. A poignant memoir of life under the Manchus, the Japanese, the Nationalists, and the Communists and the survival of a family as turbulence, war, and revolution overwhelm the society again and again. Powerful.

Tai Ming Cheung, *China's Entrepreneurial Army* (London: Oxford University Press, 2001). The birth, rise, and destruction of the military-business complex, as 15 years of People's Liberation Army excesses led to corruption, nepotism, and smuggling.

———, *Fortifying China: The Struggle to Build a Modern Defense Economy* (Ithaca, NY: Cornell University Press, 2009). How, after fits and starts and ideological impediments, China eventually came to leverage its civilian and military technologies into arms and weapons by applying entrepreneurial methods. The questions for all: Is the "Catch Up" offensive or defensive in nature, and will China's global space industry (including its antisatellite test in 2007) and its dual-use economy provide a healthy or unhealthy challenge to the United States? For his part, the author concludes, "The next two decades may well mark China's coming of age as a top-tier military technological power."

William S. Cohen, *Dragon Fire* (New York: Forge, 2006) (F). The Chinese blind a satellite, as of January 2007. Quite an exciting read, with lots of heroics and gunplay plus some geopolitical lessons to be learned. Much of what he envisioned has played out.

John Cooper, *Playing with Fire: The Looming War with China over Taiwan* (Westport, CT: Praeger, 2006). The author is convinced that, despite the fact that 90 percent of people in China and the United States would oppose such a war, it is likely to occur. Many decision makers in China, he argues, do not believe that the United States has the will to continue to be a world hegemon, so unless Taiwan goes quietly, even enthusiastically, US law would require military action to prevent a Chinese takeover. Sobering.

Anthony H. Cordesman and Martin Kleiber, *China's Military Modernization* (Washington, DC: Center for Strategic and International Studies, 2007). Many useful charts and tables. Long on data, short on projections or predictions, except to say that Chinese motivations are "far from clear."

Yong Deng and Fei-ling Wang (eds.), *China Rising: Power and Motivation in Chinese Foreign Policy* (Lanham, MD: Rowman and Littlefield, 2005). Eleven essays on such diverse topics as nationalism, image building, and multilateral diplomacy, concluding that the Taiwan "overhang" remains troubling and potentially explosive. Also asserts that a "militarily potent Chinese keeps U.S. policy makers sober and cognizant of the need to secure China's cooperation rather than attempting to coerce it."

Frank Dikotter, *Mao's Great Famine: The History of China's Most Devastating Catastrophe 1958–1962* (New York: Walker, 2011). Dikotter is one of the first scholars ever to get a look at Chinese actions that tally up to the "worst man-made catastrophe—ever." Forty-five million people died, he claims, making Mao more deadly than Hitler or Stalin, especially with his "own" people. This is a searing and overwhelmingly powerful indictment of a man who was such a hero to so many for so long.

Randall Doyle, *America and China: Asia-Pacific Rim Hegemony in the Twenty-First Century* (Lanham, MD: Lexington, 2007). Highly praised by Howard Zin and Noam Chomsky, the author is at pains to develop a "pivot area," a new global heartland, lying between Russian Siberia and northern Australia and between the "stans" and Taiwan. Herein lies 30 percent of the world's population and 20 percent of its GDP. Fortunately, he does not see China–US hegemonic wars as "inevitable."

Roger Faligot and Remi Kauffer, *The Chinese Secret Service* (New York: William Morrow, 1989). These two French authors paint a most frightening portrait of the Chinese secret police and spies, concluding that the tradition goes back 2,500 years to the time of Sun Tzu. Today, they claim, the security and intelligence community plays an enhanced political role in the decision making of the People's Republic of China.

Rob Gifford, *China Road: A Journey into the Future of a Rising Power* (New York: Random House, 2007). Something of a disappointment as both a travelogue and an analytic work. The author did move 3,000 miles from Shanghai to Kazakhstan across what he calls the "Last Great Empire," encountering many ordinary people and finding out about their lives. This account has precious little about the military modernization of China, except to say that it has a long way to go. I did like his dichotomy of China watchers divided into the "Panda Huggers" and the "Dragon Slayers." Depressing portrait of Chinese degradation of the environment, including air, water, and soil pollution on a gargantuan scale.

Yefim Gordon and Dmitriy Komissarov, *Chinese Aircraft: The Chinese Aviation Industry since 1951* (Manchester, UK: Hikoki, 2008). Lavishly illustrated and containing excruciating details about China's fighters, bombers, helicopters, missile strikes, and UAVs, this work is weak on overall analysis and a useful conclusion, but it does illustrate China's increasing role in global arms sales, such as those to Bangladesh, Myanmar, Pakistan, Iran, and Namibia.

Mel Gurtov and Peter Van Ness (eds.), *Confronting the Bush Doctrine: Critical Views from the Asia-Pacific* (London: Routledge Curzon, 2005). And they are overwhelmingly critical! Blaming Bush for just about everything and anything, including "bungling the end of the Cold War on the Korean Peninsula." Despite such foolishness, this volume makes some telling points against the Bush doctrine of preemption and the consequences of unilateralism.

John Hershey, *White Lotus* (New York: Buccaneer Books, 1994) (F). Written in 1965, this work can be read on many levels: an alternative history, a contemporary parable, a generalized work on exploitation of one group by another, and a strong indictment of the international slave trade, but it is reviewed here as a possible way to understand deep-seated Chinese resentment at being exploited by the West for so long.

Peter Howarth, *China's Rising Sea Power: The PLA Navy's Submarine Challenge* (London: Routledge, 2006). China now has the largest submarine fleet in the world, and its Kilo and Song class boats are "fast, quiet and well-armed." Their navy gives Chinese leaders a sense of confidence that they could defeat the United States in the Taiwan theater. More ominous perhaps, "Beijing is continuing to develop the means to affect a military solution to the Taiwan issue."

Martin Jacques, *When China Rules the World: The End of the Western World and the Birth of a New Global Order* (New York: Penguin, 2009). In perhaps something of an "overreach" itself, the author sees China coming to global hegemony in keeping with its "natural" status over time. Claiming that China is a civilization as much as a state and a civilization-state rather than a nation-state, he projects continued economic, military, and ideological growth for the foreseeable future, becoming irresistible by 2050. Good historical background that seems convincing concerning his overall thesis of the rise of China as an international player of note, if not hegemony. Most useful is his list of the major periods in Chinese imperial history:

Imperial

- Eastern Zhou, 771–256 BCE
- Warring states, 403–221 BCE
- Quin, 221–206 BCE
- Han, 206 BCE–220 CE
- Tang, 618–907 CE
- Northern Song, 960–1125 CE
- Southern Song, 1127–1279 CE
- Yuan (Mongols), 1279–1368 CE
- Ming, 1368–1644 CE
- Wing (Manchus), 1644–1912 CE

Postimperial

- Nationalist, 1913–1949 CE
- CCP, 1949–present

Yang Jisheng, *Tombstone: The Great Chinese Famine 1958–1962* (New York: Farrar, Straus, and Giroux, 2012). The author claims that the Great Leap Forward cost 36 million Chinese their lives. Thirty-six million is a figure that puts the Nazis to shame (and is down from the figure that others have cited, 45 million). He blames Mao directly, as well as the CCP, and says this is the "worst famine in human history." Particularly poignant is Yang's description of the death of his own father by starvation. For some strange reason, this book is currently banned in China.

Robert Kaplan, "China's Two-Ocean Strategy?" in his *Monsoon: The Indian Ocean and the Future of American Power* (New York: Random House, 2010), pp. 277–93. Kaplan rightly sees the Indian Ocean and South China Sea (indeed the entire western Pacific), as an area of growing importance, not just for China (85 percent of her oil and natural gas will come through the Straits of Malacca into the indefinite future), but also for a rising India and a somewhat diminished America. A most useful and stimulating analysis.

Ronald C. Keith (ed.), *China as a Rising World Power and Its Response to Globalization* (London: Routledge, 2005). Seven authors write interesting essays about China in transition, but only one focuses on military matters.

Henry Kissinger, *On China* (New York: Penguin Press, 2011). The old master reviews Chinese history and tells a fascinating tale of Mao—and Deng's—chilling realism and hypernationalism, as they look at geopolitics through the lens of an atavistic and historical fear of the Soviet Union. Amazing stuff, much of it inside baseball and well worth reading.

David Lampton, *The Three Faces of Chinese Power: Might, Money and Minds* (Los Angeles: University of California Press, 2008). "China explicitly aims to be a great power of comprehensive strength," as it is in a race to maintain internal stability in the face of rising demand and the belief that America has strategic "overreach" to worry about and thus is a target of some opportunity.

Colonel Qiao Liang and Colonel Wang Xiangsui, *Unrestricted Warfare* (New York: Pan American, 2004). Brilliantly incisive look at the new forms of warfare at a time when many were writing about the end of history. Truly forward looking about the nature of war in the 21st century.

Edward Luttwak, *The Rise of China vs. the Logic of Strategy* (Cambridge, MA: Harvard University Press, 2012). A high-strategy overview of the People's Republic of China and its recent swing toward international militancy and territorial assertion. The author believes that this is a sign of "great-state autism," which seems to dovetail with his assertions about the United States and Soviet Union. Chinese great-state autism has led to a number of countries reassessing both their relations with China and the need to find a strategic counterweight to Chinese assertions of hegemony. Vietnam, Australia, Indonesia, the Philippines, Singapore, South Korea, and Japan, as well as the United States, have all reacted accordingly.

James Mulvernon, *Soldiers of Fortune: The Rise and Fall of the Chinese Military-Business Complex, 1978–1998* (London: M. E. Sharpe, 2001). Forced divestiture brings the People's Liberation Army full circle.

David Shambaugh, *Modernizing China's Military: Progress, Problems and Prospects* (Los Angeles: University of California Press, 2002). Good chapter on "Threat Perceptions." He sees "pockets of excellence" within an overall Chinese force structure, with a long way to go before it can challenge US hegemony worldwide, but not so far in terms of "area denial." Author is quite clear, however, that the "Taiwan Strait is not the English Channel and this would not be a D-Day" if the People's Liberation Army decides to invade Taiwan.

Jonathan Spence, *The Search for Modern China* (New York: W. W. Norton, 1990). See especially part 4, "War and Revolution," pp. 435–618, for an easy-to-understand description of World War II in China, as well as its "baffling" civil war. Excellent maps and good material on the little-studied Japanese Ichigo offensive in 1944, which hurt the Nationalist cause then and later. Makes the point that there were still 2.1 million Japanese soldiers in China at the time the war ended.

Robert G. Sutter, *Chinese Foreign Relations: Power and Policy since the Cold War* (Lanham, MD: Rowman and Littlefield, 2008). Sees the rise of China both challenging the United States and also providing a hemisphere-wide need for Asian countries to hedge their bets. Sees the United States as continuing to be the "security guarantor and vital economic partner" as the power of China grows.

Michael D. Swaine, Andrew Yang, and Evan Medeiros, with Oriana Mastro (eds.), *Assessing the Threat: The Chinese Military and Taiwan's Security* (Washington, DC: Carnegie Endowment for International Peace, 2007). Concludes that "U.S. force posture in the Asia-Pacific region may soon be inadequate to deter a Chinese use of force against Taiwan" and that challenge will only grow, perhaps being irresistible by 2015.

US House of Representatives Select Committee, *U.S. National Security and Military Commercial Concerns with the People's Republic of China* (the Cox Report) (Washington, DC: Regency, 1999). An amazingly and extensively documented account of Chinese espionage penetration of US strategic weapons and commercial activities, including thermonuclear devices, multiple reentry vehicles (MRVs), satellites, high-performance computers, and cyberwarfare dimensions, among others. Makes a strong case that China is not a US strategic partner but a formidable rising power with an agenda hostile to the United States.

George Walder, *China: A Wolf in the World?* (London: Gibbon Square, 2008). The author makes fun of the myopic view that "Beyond Hyde Park all is a desert." He believes that the West has an "illusion of options," and he remains fearful of China's directions, believing that the country has not yet faced the true madness of the Mao era, and this "historical falsity" needs to be resolved forthwith.

David Wise, *Tiger Trap: America's Secret Spy War with China* (Boston: Houghton Mifflin, 2011). Chinese espionage and US counterespionage over the past 40 years. Chinese emphasis on the penetration of military technology, the author argues, constitutes a "substantial treat" because "[i]t's pervasive, ubiquitous and constant."

Toshi Yoshimara and James R. Holmes, *Red Star over the Pacific: China's Rise and Challenge to U.S. Maritime Strategy* (Annapolis, MD: Naval Institute Press, 2010). A seminal work on Chinese maritime strategic doctrine, melding Mao and Mahan, with considerable implications for a strategy of following a "shoot the archer, not the arrow" set of assumptions and focusing on the dependence of carrier battle groups on the Aegis system. Probing the vulnerability of the Aegis with anti-ship ballistic missiles (ASBMs) based *on land* and electromagnetic pulses to degrade (two nonnuclear options) its capability are explored. A must-read for anyone interested in US–Chinese military interaction, conceptual or real.

Qiu Ziaolong, *Years of Red Dust: Stories of Shanghai* (New York: St. Martin's Press, 2010) (F). Although on one level this is simply a delightful series of short stories chronicling the life and times of one small city lane in Shanghai, on another, it is a devastatingly disturbing look at the ease with which the population of China has been made to adopt truly dramatic ideological and worldview changes at the behest of the central government.

FAILED AND FRAGILE STATES

"I am a Somali. In Somalia, the gun is our government."—teenage Somali pirate

There are often harsh choices for the international community when states fail. Why are the world's powers unable to stop pirates from Somalia? How does the international community try to put back together a state that has disintegrated? The near-term future, with the economic and military rise of China and India, requires a lot of recalibration on the part of many military planners, but the continuing and growing problem of states that do not have physical control over the territory and populations within their borders seems likely to grow and grow as a problem for defense planners. Yemen, for example, as Robert Kaplan notes, has a population of 22 million, and that population, in turn, has 80 million guns.

Some writers cited here, such as Norman Davies, argue that the term *failed states* is actually a misnomer because they still continue to have international standing, but others take a more benign view of the term. In any case, I have titled this section "Failed and Fragile States" in order to cover more intellectual and historical ground with it.

Jon Lee Anderson, "Letter from Mogadishu: The Most Failed State," *New Yorker*, December 14, 2009, pp. 64ff. A truly awe-inspiring mess of a former country reduced to local savagery and the rule of the AK-47. Tragic and perhaps the wave of the future for many other failed states.

Tim Butcher, *Blood River: A Journey to Africa's Broken Heart* (New York: Grove Press, 2008). A country, the Congo, the size of western Europe, and the author crosses it with no sense of ever being in an organized state. In fact, it is as lawless and dangerous as it was in 1874, when H. M. Stanley shot his way across the continent. Millions of people have been killed since 1960, and the writ of the state does not yet extend to the borders defined on the map.

———, *Chasing the Devil: The Search for Africa's Fighting Spirit* (London: Chatto and Windus, 2010). Retracing Graham and Barbara Greene's trek in 1935, Butcher moves overland between Freetown, Sierra Leone, through Liberia and Guinea and Liberia again to reach Monrovia. In the process, he experiences and describes a world far from statehood, let alone nationhood. It is amazing how little life has changed in some of those areas during the last 70 years in terms of the creation of a viable nation-state. For examples, see Barbara Greene, *Too Late to Turn Back* (London: Settle and Bendall, 1938), and Graham Greene, *Journey without Maps* (New York: Penguin Books, 2007), for parallel views of earlier visits.

Norman Davies, *Vanished Kingdoms: The Rise and Fall of States and Nations* (New York: Viking, 2011). Arguing that successful statehood, in fact, is rare and "state demise" far more widespread, he presents a number of European examples, from Tolosa of the Visigoths (418–507), to Byzantium (330–1453), to the CCCP of the Soviet Union (1924–1991). Today, he points to Somalia, Chad, Moldova, and Azerbaijan as entities where power is or has ebbed from the political center.

C. Christine Fair and Sarah Watson, *Pakistan's Enduring Challenges* (Philadelphia: University of Pennsylvania Press, 2015). A stimulating set of essays on topics from drones to democracy, from challenging economics to a self-defeating worldview, from an expanding nuclear arsenal to backfiring sub-state actors, as well as the changing interests of Saudi Arabia and China. The Pakistani military leadership may not be up to the various tasks. The reset for the United States and Pakistan provided by 9/11 was blundered away by the Pakistani military, their dismal record matched only by that of US policy makers during the same period. Sobering.

Nuruddin Farah, *Crossbones* (New York: Riverhead Books, 2011) (F). Somali author Farah captures the essence of a failed state in this novel about the controlled chaos that is the Horn of Africa, with Somalia divided into Somaliland, Punt, and various portions of southern Somalia beset by factions, warlords, Ethiopians, Americans, and Shabaab Salafists, where anyone at any time can be killed *kutiri-kuteen*, or "on hearsay, no evidence." He laments, with no little horror, the status of his beloved land and people.

Howard W. French, *A Continent for the Taking: The Tragedy and Hope of Africa* (New York: Random House, 2005). A *New York Times* reporter covers the disintegration of his ancestral homeland with fortitude.

John Gimlette, *Wild Coast: Travels on South America's Untamed Edge* (New York: Alfred A. Knopf, 2011). Guyana, Suriname, and French Guiana are presented as disintegrated or disintegrating once one leaves the capital or chief port. It appears as if most of these "countries" are effectively beyond the control of the central political authority and the bush is the true king.

Imtiaz Gul, *The Most Dangerous Place* (New York: Viking, 2010). This is an excellent source from someone who has covered the region for many years. "The Most Dangerous Place" is the Federally Administered Tribal Areas (FATA) on Pakistan's border with Afghanistan. The author describes in convincing detail all the dangers, problems, and difficulties posed by FATA and concludes, "It is not the few thousand armed militants and criminals who make FATA the most dangerous place, but those silent millions who look up to the militants as daring followers of God and Islam, out to challenge the wayward and corrupt Western world," but he does believe that the "state of Pakistan, its military and civil bureaucracy is stronger than most people recognize." One hopes.

Roger Kaplan, "Unity and Anarchy," in his *Monsoon* (New York: Random House, 2010). Kaplan most usefully introduces the notion of Somalia being really three Somalias: a true state of Somaliland in the northwest; the proto-pirate state of Punt in the northeast; and "Somalia" in the south, on which most of the newspaper accounts are centered. For an earlier, most positive view of what might have been, see C. P. Potholm, "The Somali Political System," in his *Four African Political Systems* (Englewood Cliffs, NJ: Prentice Hall, 1970), pp. 181–229.

Ryszard Kapuscinski, *The Shadow of the Sun* (New York: Alfred A. Knopf, 2011). While a testament to the soaring nature of the human spirit—a gigantic pothole in Nigeria becomes a village-sustaining profit center—this work is very depressing from the perspective of the failure of so many of Africa's governments to provide any of the basics of statehood for the people for whom it has responsibility. And those do not include others, such as Liberia, Nigeria, Sierra Leone, and Rwanda, who have suffered "bloodcurdling disintegration."

Martin Meredith, *The Fate of Africa: From the Hopes of Freedom to the Heart of Despair* (New York: Public Affairs, 2005). The author seems rightly despondent about Africa's future although not always willing to identify some key variables among the myriad of complex causes for that failure. However, even without direct guidance from him, if American, you will be grateful for the likes of Washington, Adams, and Jefferson, none of whom ever tried to make themselves "emperor for life" or even "president for life."

Dan Morrison, *The Black Nile* (New York: Viking, 2010). Despite all the turmoil, violence, and decay experienced by the author, this account has a flat, epigonic quality to it. Still, it does show the vast stretches of Africa not under anybody's central government and the ongoing wars that have already lasted four or five decades. Depressing.

Oliver North, *American Special Heroes in Special Operations* (Nashville, TN: Fidelis Books, 2010). Short accounts of American Special Operations personnel in Iraq, Afghanistan, and elsewhere, who are changing the nature of warfare and even the definition of the term *war*. Given the rising numbers of failed states and areas within existing states where the writ of that state does not run, it would appear that this portion of the military calculus can only grow in importance.

Mark Owen, *No Easy Day: The Autobiography of a Navy Seal* (New York: Dutton, 2012). The strangely epigonic story of the man from Seal Team 6, which eventually cornered Osama bin Laden in his compound in Abbottabad and killed him. Suggests that, when terrorism gets franchised and metastasizes, it is harder to stop. Still, there is a sense of closure as they remove the body and cart it to Afghanistan for forensic investigation and verification.

William Reno, *Warfare in Independent Africa* (Cambridge: Cambridge University Press, 2001). A useful tour de horizon with a number of observations of use in determining how failed states become failed states, almost regardless of what type of rebel begins the progress and no matter how many types join the effort.

Bruce Riedel, *Avoiding Armageddon* (Washington, DC: Brookings Institution Press, 2013). A very impressive and important look at America's relationships with India and Pakistan since their independence. Gives the United States credit for defusing possible nuclear war on several occasions but concludes, "The major thesis of this book and its review of American diplomacy in South Asia for the last seven decades is that the United States has been unable to achieve most of its goals in the region." Sobering. Very.

Jeremy Scahill, *Dirty Wars: The World Is a Battlefield* (New York: Nation Books, 2013). The notion that the whole world is now a battle space is clearly on display in this catalogue of US antiterrorist actions across the globe. A wide-ranging and detailed account provides insights into US and other countries' antiterrorist efforts in Yemen, Pakistan, Afghanistan, Iraq, and Somalia across two decades of the war on terror. The work persuasively argues that this type of "dirty" warfare, which was begun under President Bush, has actually been legitimized and expanded during the two terms of President Obama. Now, "[u]sing drones, cruise missiles and Special Ops, the United States has embarked on a mission to kill its way to victory" (p. 516). By choosing security over civil rights, the author concludes, the United States has changed the nature of the conflict so that the "war on terror has become a self-fulfilling prophecy," and he ends by asking, "How does a war like this ever end?" (p. 521). Although obviously concerned by the lack of widespread public opposition to drone strikes and other more numerous extralegal operations, the author unfortunately provides no actionable alternatives applicable to that war (and the opposition tactics that go with it).

Jason K. Stearns, *Dancing in the Glory of Monsters: The Collapse of the Congo and the Great War of Africa* (New York: Public Affairs, 2011). Five million dead since 1966, hundreds of thousands of innocent women raped, the Great War of the Congo belongs, the author demands, with World Wars I and II, the Cultural Revolution, and Rwandan genocide for true examples of horror. Stearns compares the situation to the Hobbesian (Hobbes was writing in the wake of the Thirty Years' War) "all against all" worldview, for there are 40-some different wars all mixed up in the Congo's last 50 years. Somehow this is summed up in President Mobutu's words to his army, "You have guns, you don't need a salary" (p. 116), and the author's cry of despair, "We are returned to Joseph Conrad's notion that the Congo takes you to the heart of darkness, an inscrutable and unprovable mess" (p. 328).

Paul Theroux, *Last Train to Zona Verde* (Boston: Houghton Mifflin, 2013). Traveling from South African shanty towns to the isolation of Namibia north of "the Fence," Theroux eventually finds an intriguing and complex mix of a gigantic oil-fueled kleptocracy in Cabinda and Luanda and a totally failed state in rural Angola. Chilling and sad.

———, *The Lower River* (Boston: Houghton Mifflin, 2012) (F). Only someone who had deep and true awareness of places like those found on the Lower River—and for whom they represent an overwhelming sense of the hell—could have written such a work of Hobbesian and Golding impact. The Lower River could be where "Malawi" meets "Mozambique," but because neither is truly real, it could be many other places. Truly horrifying claustrophobia and sheer helplessness once the state, even a corrupt state, has vanished or regressed into the bush.

ISLAMIC RADICAL SALAFISTS AND LOCAL *AND* GLOBAL INSURGENCIES

Here, one gets an opportunity to review books that look at Islam and the various religious, spiritual, and military aspects of jihad as a function of intra-Muslim conflict as well as inter-Muslim conflict.

Note: Other authors suggest that cooperation existed in the past and may well occur in the present and future. See, for example, Ian Almond, *Two Faiths, One Banner: When Muslims Marched with Christians across Europe's Battlefields* (Cambridge, MA: Harvard University Press, 2009). The author shows the fault lines in both Christianity and Islam over the ages and indicates that cooperation, not conflict, has had its moments.

A. J. Arberry, *The Koran Interpreted* (New York: Macmillan, 1955). It is very important that nonbelievers of Islam have a sense of what is and what is not in this most sacred of books for believers. This book of "divine rhetoric" serves as the basis for the faith of well over a billion followers, so it is crucial that it be read by all. Arberry is a good place to start.

Karen Armstrong, *Muhammad: A Prophet for Our Time* (New York: HarperCollins, 1992). A decade or two ago, it would have been hard to make the case for Armstrong's title, but time and events have overridden that notion. The author casts Muhammad in the positive light, and her work serves as a good introduction. A bit light on his military efforts, however.

Reza Aslan, *No God but God: The Origins, Evolution, and Future of Islam* (New York: Random House, 2005). Puts Islam in its historical context and shows the great variety within its heritage.

Daniel Benjamin and Steven Simon, *The Next Attack: The Failure of the War on Terror and a Strategy for Getting It Right* (New York: Times Books, 2005). A most gloomy assessment.

Peter Bergen, "From the Shadows: An Oral History of Osama bin Laden," *Vanity Fair*, January 2006, pp. 112ff. A surprisingly illuminating look at bin Laden and the background of his alienation.

Gregory Bolich, "Terrorism in the Ancient Roman World," *Quarterly Journal of Military History* (Spring 2006), pp. 52–59. While there was general peace throughout much of the Roman Empire, there was also a surprisingly large number of rebellions and insurgencies.

Andrew Bostom (ed.), *The Legacy of Islamic Anti-Semitism: From Sacred Texts to Solemn History* (New York: Prometheus Books, 2008). Not a pretty picture but a well-documented one.

Christopher Caldwell, *Reflections on the Revolution in Europe: Immigration, Islam and the West* (New York: Doubleday, 2009). A truly amazing book highlighting the demographic and psychographic trends in Europe, which project a decidedly threatened future.

Natana J. Delong-Bas, *Wahhabi Islam* (Oxford: Oxford University Press, 2004). The basis for the Salafist revival. Not going away soon, claims the author, and he seems spot-on in this regard.

Stephen Duncan, *A War of a Different Kind: Military Force and America's Search for Homeland Security* (Annapolis, MD: Naval Institute Press, 2004). Pointing out that the war against terror is only the second truly global war in US history, the author suggests that protecting the homeland is of crucial, overriding importance and will require new attitudes, new laws, new force structures, and new operations to deal with the great complexity of the new world disorder.

Franz Fanon, *The Wretched of the Earth* (New York: Grove Press, 1963). In the 1960s, this book of essays about decolonization and the psychology thereof was one of the most widely read works on college campuses. Largely overridden by events (as well as their culminations), it did cause me while rereading it in 2013 to discover strong resonances in the psychology of present-day radial Salafism, in particular the relevance in Jean-Paul Sartre's introduction of the role of "suppressed, mad fury," "bitterness and spleen," and the "ever present desire to kill us." Maybe this Algerian writer, once overpraised and then largely forgotten, is due for a dust-off.

Fawaz Gerges, *The Far Enemy: Why Jihad Went Global* (Cambridge: Cambridge University Press, 2006). Insightful and worth looking at in terms of jihadist motivation.

Vartan Gregorian, *Islam* (Washington, DC: Brookings Institution, 2003). Islam in perspective. The author traces 1,400 years of Islam with skill and dexterity. The most important takeaway is the diversity within the religion and the extent to which internal divisions do much to both shape and moderate its threat to the West, Christianity, *and* modernity. Islam is a mosaic, not a monolith.

Lee Harris, *The Suicide of Reason: Radical Islam's Threat to the West* (New York: Basic Books, 2007). A harsh indictment of a religion that seems to have avoided the humanistic impulses of the Enlightenment and Reformation.

Karen Elliott House, *On Saudi Arabia: Its People, Past, Religion, Fault Lines—and Future* (New York: Alfred A. Knopf, 2012). A useful, if somewhat repetitious, discussion of Islam in action in the Kingdom, which she depicts more as an extended family ruled by the Al Sauds than a real country, showing the extent to which the rulers use Islam as both carrot and stick, but the author feels the long-term future is bleak.

Daniel Klaidman, *Kill or Capture: The War on Terror and the Soul of the Obama Presidency* (Boston: Houghton Mifflin Harcourt, 2012). Obama sticks—or is forced by political realities to stick—with many of the Bush-era antiterrorist actions and actually greatly expands the drone program, "awed by the precision and lethality of strikes," authorizing hundreds a year from 2010 on, compared with Bush's nine Pakistani strikes between 2004 and 2007. One interesting unintended consequences of the decision to close Guantanamo and the difficulties of trying any of its high-value prisoners has been the new rule under the Obama war on terror: "killing was a lot easier than capturing."

Bernard Lewis, *The Crisis of Islam: Holy War and Unholy Terror* (New York: Random House, 2004). Sees Islam as facing a historic hinge, especially the militancy of a minority affecting the more pacific views of the majority.

———, *What Went Wrong?* (Oxford: Oxford University Press, 2002). A realistic, if unpopular, look at Islam and the failure of the Muslim world to modernize or experience its own Enlightenment.

Roel Meijet (ed.), *Global Salafism: Islam's New Religious Movement* (New York: Columbia University Press, 2010). An unusually fine collection of essays on Salafism, its history, its present in many forms and locations, and its range of futures. Makes an important distinction between "jihadi Salafism" and other strains and projects deep and terrible struggles between Sunni Salafism and Shi'is. Al-Qaeda's Creed and Path are well represented.

George Nafziger and Mark Walton, *Islam at War: A History* (Westport, CT: Praeger, 2003). There have been a lot of times in the past where followers of Islam were at war with the followers of Christianity, and yet, there were times of cooperation or more fragmented patterns, such as some Christians fighting some Muslims with help from other Muslims and vice versa.

Vali Nasr, *The Shia Revival: How Conflicts within Islam Will Shape the Future* (New York: Norton, 2006). Shia and Sunni fault lines appear to be as strong and as dangerous as "Muslim–Christian" fault lines.

Eric Schmitt and Thom Shanker, *Counterstrike: The United States of America's Secret Campaign against al Qaeda* (New York: Henry Holt, 2011). Traces the evolution of US strategy against international terrorism. Discovers that al-Qaeda is "as anal as the Nazi's in keeping records" and believes in "following the money" because supporting a terrorist is cheap but supporting a terrorist network is not. Schmitt and Shanker see that terrorists are likely to eventually succeed in another attack on the US homeland and concludes somewhat realistically, if pessimistically, "It will be impossible to end terrorism."

Richard H. Shultz Jr. and Andrea J. Dew, *Insurgents, Terrorists, and Militias: The Warriors of Contemporary Combat* (New York: Columbia University Press, 2006). Another overview of the various elements currently at play in the unconventional field that we now call "war among the people."

Ilija Trojanow, *Mumbai to Mecca* (London: Haus, 2007). The hajj seen up close and personal, with spiritual grace and physical discomfort.

US Marine Corps, *Warfighting* (Washington, DC: Government Printing Office, 1997). Heavily influenced by Sun Tzu and its own history, the US Marines zero in on effective counterinsurgency.

Joby Warrick, *The Triple Agent: The al-Qaeda Mole Who Infiltrated the CIA* (Garden City, NY: Doubleday, 2011). A cautionary tale but not one with enough power to say "cancel the mission." In intelligence gathering, as in life, it is said, "shit happens." This work does underscore, though, that the drones are priceless.

AFGHANISTAN: WATCH ON THE DANUBE REDUX

Some articles and books about Afghanistan insist that Afghanistan "has never been conquered." Please pay no attention to such a simplistic and antihistorical assertion. For much of human history, Afghanistan has been conquered—by the Persians, Alexander the Great (whose successors ruled it for 200 years), Genghis Khan, Tamerlane, and of course the British, who did lose one war (1842) but came back to defeat the Afghans (1878–1889) in another.

It is only recently, in 1984, when the United States and Pakistan intervened to assist the mujahideen Afghans with Stinger missiles and other logistical support did the Afghans defeat a modern army, but in 2001–2002, that flow was reversed when the alliance of the United States and Northern Alliance swept the Taliban from the board. Of course, none of this matters to the process of the current Afghan insurgency, unless one believes the pseudohistory of the territory never having been conquered, in which case it may matter a great deal. Of course, all of this begs the question as to why anyone would want to try to conquer, or even succeed at conquering, Afghanistan, given its seemingly ungovernable nature. But of course, stranger things have happened with great frequency throughout history.

Jamil Ahmad, *The Wandering Falcon* (New York: Riverhead Books, 2011) (F). A somewhat disjointed but poetic view of life in the highlands of Afghanistan and Pakistan and the codes of honor that make it difficult for modern society, including modern military society, to intrude with success.

Rodric Braithwaite, *Afgantsy: The Russians in Afghanistan 1979–89* (London: Profile Books, 2011). A seemingly objective look at the Russians who fought in Afghanistan. The Soviets expected to be there only six months, had "tactics without strategy" thereafter, and fought "messy, piecemeal and confused battles" only to pull out. A searing look at the Soviet 40th Army formed to fight that war, with some praise for the special units but catching the dry rot of the Soviet system for conscripts. With 2.4 percent of the total Soviet force killed in the war, 11 percent were wounded and an astounding 69 percent suffered from such severe illness as typhoid, cholera, and hepatitis. Debunks the myth that the US Stinger missile "won" the war, claiming, "Gorbachev took the decision to withdraw from Afghanistan a full year before the first Stinger was fired."

George W. Bush, "Afghanistan," in his *Decision Points* (New York: Crown, 2010), pp. 180–221. The former president traces his overview of the situations in Afghanistan and what he did and didn't do about them. "Afghanistan is one of those places you have to see to understand."

George Crile, *Charlie Wilson's War: The Extraordinary Story of the Largest Covert Operation in History* (New York: Atlantic Monthly Press, 2003). How the United States and Pakistan helped the mujahideen defeat the Soviets and then walked away from Afghanistan in the aftermath. Triumph and tragedy in one volume.

William Dalrymple, *Return of a King: The Battle for Afghanistan, 1839–42* (New York: Alfred A. Knopf, 2013). A lavish, deeply researched account of the British effort to play at "The Great Game" and their ill-fated efforts to control Afghanistan by reestablishing Shah Shuja ul-Mulk on the throne. Sold out by the British once they had established him in Kabul, he was incisive when he was less angry about their duplicity than "simply baffled by the amazing stupidity of his allies." The British disaster in 1841, in which virtually the entire force was wiped out, is often overhyped as a lesson against foreign efforts in Afghanistan. It should be remembered that the British were, in the Second Afghan War, to come back and successfully occupy Kabul, although after destroying it, they, too, withdrew. Page 346 gives lie to the painting *The Survivor* because thousands actually did. The book ends with an Afghani warlord wondering if the Chinese will be next.

Note: The first British force did not travel lightly, according to the author on page 137, bringing with them 1,000 European soldiers; 14,000 East Indian Company sepoys; and 38,000 camp followers. The baggage alone required 30,000 camels (including 300 just to carry the officers' wine). This is a masterful work, relying heavily on Afghani, Indian, and Persian sources, as well as those of the Europeans.

Christine Fair, *Fighting to the End: The Pakistan Army's Way of War* (London: Oxford University Press, 2014). If only American and other nations' policy makers would read this work and take it to heart. She portrays a Pakistan that is insecure, myth making about an Indian destruction of their country, and threatened by a "strong" or even coherent Afghanistan as well as by its own created Frankenstein proxies, but it is an "army with a country." A most perfidious ally (to anyone, just ask the Saudis in 2015), Pakistan, she says, is a "purely greedy state," one that never wants turmoil in the region to cease. Most ominous of all, it promotes "jihad under a nuclear umbrella" and "can never be placated." Grim reading.

Charlotta Gall, *The Wrong Enemy: American in Afghanistan 2001–2014* (Boston: Houghton Mifflin, 2014). The Pulitzer Prize–winning author immerses herself in the country and concludes that it is too important to abandon but its problems cannot be solved without getting the government of Pakistan to stop supporting the Taliban. Very positive toward General Stanley McChrystal and his policies while he was there.

Robert Grenier, *88 Days to Kandahar: A CIA Diary* (New York: Simon and Schuster, 2014). The CIA station chief in Pakistan writes cogently, if somewhat verbosely, about US involvement on the ground in Afghanistan, concluding, "We think we are finished with Afghanistan. But Afghanistan may not be finished with us." Insightful, on the ground for the advent of Hamid Karzai, and a must-read for any of the conservative and liberal talking heads who make such a mockery out of foreign policy in this decade. A most cautionary tale.

Michael Hastings, *The Operators: The Wild and Terrifying Inside Story of America's War in Afghanistan* (New York: Penguin, 2012). By the *Rolling Stone* reporter who "got General Stanley McChrystal fired." Of course, General McChrystal got himself fired—for gross stupidity, if nothing else—but Hastings seems like someone you would not really trust under most circumstances. Some great local color: The American troops refer to the ISAF (International Security Assistance Force in Afghanistan) as "I Suck at Fighting." An equal-opportunity savager!

Robert Johnson, *The Afghan War of War: How and Why They Fight* (London: Oxford University Press, 2011). Provides good background on and insights into the First Anglo-Afghan War (1839–1842); the Second Anglo-Afghan War (1878–1881); the Pashtun Rising (1897–1898); the Third Afghan War (1919); the Soviet War (1978–1989); and the Civil War, the Taliban, and the Insurgency (1990–2011), pointing out that "Afghans are not culturally determined in their actions but are reactive and adaptive." Good point to ponder for those arguing for immutability of people, places, and things.

Sebastian Junger, *War* (New York: Twelve, 2010). Junger spent a lot of time on the very fringes of the war (which turned out to be where most of the combat was) in Afghanistan. He writes lovingly of the camaraderie in war, in which Chosen Company suffered 80 percent casualties over time. In the end, his take on war is that soldiers fight for their buddies. Shows in sharp relief why "clear and hold" is the proper strategy for counterinsurgency.

Hugh Pope, *Sons of the Conquerors: The Rise of the Turkic World* (New York: Overlook Duckworth, 2005). War is in the very DNA of the steppe people and their remarkable success over time, as well as their possible reemergence.

Diane Preston, *The Dark Defile: Britain's Catastrophic Invasion of Afghanistan, 1838–1842* (New York: Walker, 2012). The long, deadly, 90-mile retreat from Kabul in the dead of winter, with a complete breakdown in discipline and logistics (2,000 camels struggling through the snow until captured or killed by Afghans). After a successful drive to the capital and the installation of Shah Shuja, a greatly reduced force of one Queen's regiment, seven regiments of Bengal native infantry, and thousands of camp followers was massacred on the way home, a single survivor sent on to detail the disaster. Hardly a model on which to base future campaigns or historical assessments.

Doug Stanton, *Horse Soldiers: The Extraordinary Story of a Band of U.S. Soldiers Who Rode to Victory in Afghanistan* (New York: Scribner's, 2009). This is a must-read for those trying to understand the euphoric victory in Afghanistan that followed the dark night of 9/11.

Robert Wilkinson-Latham, *North-West Frontier 1837–1947* (Oxford, UK: Osprey Books, 1977). A long history of fighting, feuding, conquest, and rebirth, covering the three Afghan wars (two of which the British won), ending with the British pullout in 1947, leaving the local inhabitants "puzzled" that the British had not been defeated, so why would they give up the land? Reading this account will help you understand the ongoing situation in

Afghanistan and the truly Sisyphean task of bringing "democracy," even pseudodemocracy at the national level, to the country.

Bob Woodward, *Obama's Wars* (New York: Simon and Schuster, 2010). A truly scary book. Obama seems professorlike and is surrounded by a "team" that appears dysfunctional at best. For example, Messieurs Gibbs and Axelrod are described by one and all as irritating "water bugs" flitting from subject to subject with no depth of field. But much more important is the portrayal of Pakistan—focused on India, fragmented, with parts of its intelligence services helping the United States and others directing terror groups (such as Lashkar-e-Taiba, the "Army of the Pure," and the Haqqani Network)—with the United States paying the bill for both. Incidentally, when the United States is worried about civilian deaths from drone air strikes, the president of Pakistan replies with some vehemence, "Kill the seniors. Collateral damages worries you Americans. It does not worry me." Interesting statistic: The cost of one US soldier in Afghanistan per year is $250,000; the cost of one Afghan soldier, $12,000. Of course, you are reasonably sure at whom the US soldier will be pointing his or her gun. Most frightening are the detailed plans for COOPEX 2010, a scripted war game in which terrorists detonate a small, crude nuclear device in a US city and the United States then hits 120 already preset targets in Pakistan.

Note: The ongoing conflict in Afghanistan is more extensively covered in chapter 32.

II

Topics in War

These are some important topics that time did not permit us to explore fully during the course of a semester study of war or in a single volume, such as *Winning at War*, but that are worthy of further attention. The works in this part are listed under each topic and are meant to be illustrative, not exhaustive, so please check in the library and on the web for additional volumes and references. These works provide a useful springboard into their respective subjects.

Chapter Ten

Early Warfare

Much literature on war in the West begins with the Greeks and the Romans, but warfare obviously predates those historical upsurges. Some authors point to 3000 BCE as the beginning of organized warfare. Others archeologists push it back to 10,000 or even 35,000 BCE, depending on the definition of *warfare* chosen. Robin Yates (reviewed in this chapter), for example, pushes the Chinese experience with warfare to 28,000 BCE.

Certainly, by the time of the Sumerian civilization (2600 BCE), "true" warfare (by almost any recognized definition) was clearly identifiable in the archeological record. Garrett G. Fagan, for example, points out that tablets and fragments, such as the sculpture fragment from Girsu in Iraq (2600 BCE), the Royal Standard of Ur (2500 BCE), and inlay figures from Mari (2500–2400 BCE) clearly show armed conflict of an organized nature, and the written evidence of the Sumerian King Lists or the Epic of Gilgamesh and Akka (2100 BCE) indicate that fighting between cities was by then by not remarkable. He also suggests that the Assyrians during the early Iron Age (900 to 612 BCE) created and used for conquest one of the most formidable and multifaceted war machines, one ranking with the Romans, Macedonians, and Mongols in terms of complexity, organization, and military effectiveness.

Thus it is very clear that irrespective of when recognizable forms of warfare began, war has been an intrinsic part of the human condition since recorded history. In this section, there are quite good descriptions of the Sumerians, Assyrians, Hittites, Egyptians, Sea Peoples, and others, as well as anthropological studies that concern themselves with the origins of war and how war differs from other forms of intraspecies violence. See also the works reviewed in chapter 1. There are also studies dealing with the big upheavals that transcended regions and peoples, such as the chariot revolution in the second millennium BCE, when the Indo-Europeans on chariots burst into what is now Greece, Europe, Iran, India, and parts of China, or when the chariot nobilities in many Mediterranean politics thus established were overcome by armed infantry with iron weapons.

Note also an excellent Internet source for ancient warfare: http://www.fordham.edu/HALSALL/ancient/asbook.asp.

David Abulafia, *The Great Sea: A Human History of the Mediterranean* (London: Oxford University Press, 2011). Originally purchased for the illumination it could cast on the early worlds of the Sea Peoples and others, it turned out to be a massive compendium of the Mediterranean through all ages, concluding, "The Mediterranean thus became probably the most vigorous place of interaction between different societies on the face of this planet, and it has played a role in the history of human civilization that has far surpassed any other expanse of sea."

Simon Anglim, Phyllis G. Jestice, Rob S. Rice, Scott M. Rusch, and John Serrati, *Fighting Techniques of the Ancient World, 3000 BC–AD 500: Equipment, Combat Skills and Tactics* (London: Amber Books, 2002). Conveniently broken up into sections on infantry, mounted warfare, command and control, and siege warfare (often overlooked in other books on this period), this text provides a useful introduction to ancient warfare. Although overall it is a bit light on the pre-Greek dimensions, it nevertheless has some useful insights into the siege of Lachish in 701 BCE by the Assyrians.

Susan Wise Bauer, *The History of the Ancient World* (New York: W. W. Norton, 2007). A very serviceable introduction to war during the dawn of civilization (Sumer, Egypt, Indus Valley, China). While her main focus is on the rise of kingship by force, there is considerable emphasis on battles qua battles. Well worth reading also to put a particular battle (or rise of a dynasty or people) in perspective.

Robert Bigelow, *The Dawn Warriors: Man's Evolution toward Peace* (Boston: Little, Brown, 1969). Sees the link between cooperation and conflict ("cooperation for conflict"). "We are without doubt the most cooperative and the most ferocious animals that have ever inhabited the earth." All civilizations, he feels, grew "by a clash of arms." Now the question becomes, Can we adapt for cooperation without conflict? The long arc of conflict is deeply embedded in our early-human amalgamations.

Note: For an updated view of this argument, see Chris Stringer, *Lone Survivors: How We Came to Be the Only Humans on Earth* (New York: Henry Holt, 2012). Stringer gives a much-needed overview to all the new scientific knowledge about our prehistoric ancestry, including the role of both inter- and intraspecies conflict and cooperation.

Alfred Bradford, *With Arrow, Sword, and Spear: A History of Warfare in the Ancient World* (Westport, CT: Praeger, 2001). A fine, and indeed illuminating, study of war in Egypt, Assyria, Persia, China, India, Greece, and Rome (both in the West and East). Useful maps and a good narrative flow. Good introduction to the subject.

Cypian Broodbank, *The Making of the Middle Sea* (London: Oxford University Press, 2013). Fascinating overview of the Mediterranean world up to the classical Greek period. Microregions and megatrends are traced with historical and archeological input. Concludes, albeit grudgingly, "The link between change modes of fighting and the circulation of metals is inescapable." Puts the "Sea People" hypothesis in a broader cultural and historical context. A fine work.

O'Brien Browne, "Modern Insights into an Ancient Uprising," *MHQ: The Quarterly Journal of Military History* 15, no. 2 (Winter 2003), p.78–85. Browne argues that the Akkadian Empire in 2260 BCE was the first world's first empire (including the city-states of Nineveh, Ur, Kish, etc.), and its creation and maintenance depended on armed might. He asserts, "Civilization and war go hand in hand," and points out that the composite bow, rudimentary "war carts," and phalanx formations came out of that Tigris and Euphrates confluence.

Trevor Bryce, *Hittite Warrior* (Oxford, UK: Osprey, 2007). Quite an extensive and well-documented look at the Late Bronze Age Great Kingdom of Hatti and the Hittite way of war from the 14th to the 13th centuries BCE. Good insights into their three-man chariots and the importance of the bow and spear to their war-making. This book makes very good use of archeological findings.

———, *The Kingdom of the Hittites* (London: Oxford University Press, 1998). A skillful delineation of the Hittite Kingdom, which for 500 years was the dominant power in Anatolia (from 1700–1200 BCE). Centered east of Ankara, Hatti spread into all directions and seems to have come a cropper in the Late Bronze Age due to the arrival of the Sea Peoples (although the author does not totally accept the Drews military revolution except as a contributing factor). Weak on the military of the Hittites, however.

Note: In terms of linguistics, Joel Arthur is quite persuasive in marshalling support for E. H. Sturtevant's earlier thesis that Hittite is more a "sister" of the Proto-Indo-European languages than a "daughter." See his PhD dissertation "The Indo-Hittite Hypothesis" (University of California, 1971).

R. Brzezinski and M. Mielczarek, *The Sarmatians 600 BC–AD 450* (Oxford, UK: Osprey, 2002). An interesting people (or peoples, starting with the Alans) originally from north of the Black Sea, whose 12- or 14-foot spears (known as the *contus*) gave their cavalry a fearsome reputation. They fought the Romans, who initially tried to keep them beyond the Danube, but then hired them as effective mercenary auxiliaries. The pictures in this volume, especially from the later period, suggest their heavily armed lancers as prototypes for the eventual European heavy-horse form. Their light cavalry used bows and swords.

Brian Carey, Joshua Allfree, and John Cairns, *Warfare in the Ancient World* (London: Pen and Sword, 2007). Covers warfare from the late Stone Age to the Iron Age. Some interesting materials and excellent maps, not just on Egyptians, Hittites, Assyrians, and Persians, but follows the story through the Greeks and Romans. Good, short sketches of various battles, such as Megiddo, Quadest, Marathon, Plataea, Issus, Cannae, Zama, Cynoscephalae, and Teutoburg Forest.

Lionel Casson, *Ships and Seamanship in the Ancient World* (Princeton, NJ: Princeton University Press, 1971). A very detailed overview of naval activities in Egypt, Mesopotamia, and the eastern Mediterranean (from shipbuilding to sailing and rowing), concluding, "The ancient mariners of the Mediterranean can claim credit for most of the major discoveries in ships and sailing that the Western World was to know until the age of steam."

D. V. Cernenko, *The Scythians 700–300 BC* (Oxford, UK: Osprey, 2008). From its heartland in the Crimea and the plains of what is now the Ukraine, these horse people, armed with small leather-and-metal overlapping armor plates, spears, bows, swords, and daggers, expanded and contracted over more than 400 years, defeating at various times at least elements of the Assyrians, Medes, Egyptians, Persians, and Greeks (after the death of Alexander, who died in 323 BCE). Eventually they were hard-pressed by another Asiatic horse people, the Sarmatians.

Eric Cline, *1177 B.C.: The Year Civilization Collapsed* (Princeton, NJ: Princeton University Press, 2014). An attempt to put the passage of the late Bronze Age into the early Iron Age in global perspective. Argues that the Sea Peoples were only one part of the disintegration of the international trading system, what he calls the "first truly global age." The Egyptians were able to survive, but the Hittites were overcome, as were many other peoples. Believes there were waves of Sea Peoples, beginning in 1207 BCE and culminating in 1177 BCE with their defeat by the Egyptians. Good map of sites destroyed on pages 110 and 111, from Thebes to Crete to Cyprus to Megiddo to Aleppo (and one of the Troys).

———, *The Battle of Armageddon: Megiddo and the Jezreel Valley from the Bronze Age to the Nuclear Age* (Ann Arbor: University of Michigan Press, 2000). Armageddon, which is a corruption of the Hebrew *Har Megiddo* (Mount of Megiddo), and the city of Megiddo lie astride the Via Maris, one of the most important trade and travel routes between Egypt and the Euphrates. From 2350 BCE until the present, the author counts 34 major battles. Especially useful are the descriptions of the battles in Syria-Palestine during the Bronze and Early Iron Ages and the sense of how geography plays such an important role in warfare.

Arthur Cotterell, *Chariot: From Chariot to Tank: The Outstanding Rise and Fall of the World's First War Machine* (Woodstock, NY: Overlook Press, 2005). Amazing how machines of war do not change their basics, except very slowly.

Barry Cunliffe, *Europe between the Oceans 9000 BC–AD 1000* (New Haven, CT: Yale University Press, 2010). This is a rare and very useful book covering the whole sweep of early warfare. Although the author is an archeologist and not specifically interested in warfare as such, this work provides a broad overview of Europe during the time of palace kingdoms; Mycenae, Greece; the rise of the Sea Peoples; and the mass movements of people to, through, and within Europe.

Raffaele D'Amato and Andrea Salimbeti, *Sea Peoples of the Bronze Age Mediterranean c 1400 BC–1000 BC* (Oxford, UK: Osprey, 2015). Makes a strong case that at least some of the Sea Peoples came from Achaean groups (as shown by their warcraft). Relies heavily on Egyptian sources for the text. Good use of illustrations showing women as well as male warriors.

Doyne Dawson, *The Origins of Western Warfare: Militarism and Morality in the Ancient World* (Boulder, CO: Westview Press, 1996). Sees a distinction between "primitive warfare," ritual practice, and "civilized warfare" and its adaption to serve as a political instrument but ends up declaring that "war is an act of human and divine justice, as the avenging of wrongs."

Hans Delbruck, *Warfare in Antiquity*, vol. 1 of his *The Art of War* (Lincoln: University of Nebraska Press, 1990). The first edition of this volume was published in 1900 and was long a standard in the field. Very useful but often heavy going for those coming to warfare for the first time. Marvelously detailed.

Philip de Souza (ed.), *The Ancient World at War: A Global History* (London: Thames and Hudson, 2008). A wide-ranging collection of essays that round up the usual suspects (Greece, Macedonia, and Rome) but then provide glimpses of war in the Andes, Mesoamerica, South Asia, and among the Celts in Iberia. A most useful addition to the literature of early warfare.

Robert Drews, *Early Riders: The Beginnings of Mounted Warfare in Asia and Europe* (London: Routledge, 2004). How it took a very long time for cavalry to supplant chariots, but once they did, they did so permanently, almost irrespective of the culture and the location of that culture.

———, *The End of the Bronze Age: Change in Warfare and Catastrophe, ca 1200 BC* (Princeton, NJ: Princeton University Press, 1993). Makes a powerful case that infantry, armed with new Iron Age swords, small shields, and light body armor, overthrew the previously dominant chariot war culture and chariot nobles.

Garrett Fagan, *Great Battles of the Ancient World*, 5 DVDs (The Great Courses, 2005). An excellent series of introductory lectures on the sweep of ancient warfare. Very insightful on Sumerian, Assyrian, Egyptian, Greek, and Roman war. Absolutely devastating toward Victor Hanson and many of his theories of the origins of Western warfare.

Kaveh Farrokh, *Shadows in the Desert: Ancient Persia at War* (Oxford, UK: Osprey, 2007). A gorgeous, lavish, and colorfully illustrated work devoted to war and the rise and fall of the various Persian and related empires, from the Babylonian, Median, Achaemenid, Parthian, and Sassanian periods. Lots of battles and wars with growth and decline clearly and concisely delineated with considerable cultural background. Quite good on the military for a basic introductory text. Anyone wishing to understand contemporary Iran should read this book. Superior maps of all of the above.

Arther Ferrill, *The Origins of War: From the Stone Age to Alexander the Great* (London: Thames and Hudson, 1985). Many chapters are intriguing, especially "Prehistoric Warfare," "Ancient Near Eastern Warfare: The Copper-Bronze Age," and "Assyria and Persia: The Age of Iron."

Nic Fields, *Soldier of the Pharaoh: Middle Kingdom Egypt 2055–1650 BC* (Oxford, UK: Osprey, 2007). This work makes a number of interesting points: (1) With deserts to the west and east and the cataracts to the south, in this time frame, Egypt was relatively insulated from invasion; (2) without horses or chariots during this period, force projection into the Levant or south was limited; (3) Egyptian armies did a lot of public work projects instead of just practicing martial arts and actual warfare; and (4) the Egyptian navy was very important to moving troops up and down the Nile for quick strikes rather than prolonged invasions. Weapons were ordinary bows, battle axes, and spears.

———, *Troy c.1700–1250* (Oxford, UK: Osprey, 2004). A somewhat disjointed account of the numerous "Troys" (at least nine) that flourished near the Turkish village of Truva at Hisarlik on the eastern side of the Dardanelles. Obviously, this strategic location was raided and sacked many times from the early Bronze Age (2900 BCE) to the fall of Homer's and Virgil's Tory in 1260 or 1250 BCE and beyond. This work has very useful depictions of the city itself (quite small), the attack of the Trojans on the Mycenaean fleet at Besilea Bay, and an alternative model for the Trojan Horse, which I found illuminating.

Richard Gabriel, *The Culture of War: Invention and Early Development* (New York: Greenwood Press, 1990). See especially "The Middle Passage: War in the Bronze Age," "The Iron Army of Assyria," and "Persia: The Emergence of Logistics," pp. 35–82.

———, *The Great Armies of Antiquity* (Westport, CT: Praeger, 2002). For many, the first half of the book, the portion dealing with war from 2500 BCE to 120 BCE, may be the most useful. He skillfully covers the military activities of the city-states of Sumer and Akkad Egypt, the Hittites, Mitanni (located on the steppe between the Euphrates and Tigris Rivers and the first users of spoke-wheeled chariots as a primary combat vehicle in 1600 BCE), Assyria, China, Persia, Greece, and Carthage. A fine study.

———, *Thutmose III: A Military Biography of Egypt's Greatest Warrior King* (Washington, DC: Potomac Books, 2009). According to the author, Thutmose III was an astonishingly effective commander and builder of armies: "Thutmose III was the greatest warrior king that Egypt had ever known or ever was to know in its long and varied history on the earth." According to the author, previous to his reign, Egypt was on the defensive, and parts of it were occupied by the Hyksos. Under Thutmose III, the invaders were expelled and the army was modernized with "revolutionary" chariots. Egyptian chariots incorporated three major innovations in the middle of the 15th century BCE: The axle was moved to the far rear of the carrying platform, providing speed and stability; the six-spoked wheel, with tapered spokes for smoother and quicker handling; and the U-shaped socket joint for speed and stability. The Egyptian army was divided into archers, heavy infantry with spears and swords, and axmen. Thutmose ended up expanding Egypt's imperial reach into Canaan, Syria, and Nubia. Upon his death, the empire and its military protected the Egyptian homeland for 500 years.

Richard Gabriel and Donald W. Boose Jr., *The Great Battles of Antiquity: A Strategic and Tactical Guide to the Great Battles That Shaped the Development of War* (Westport, CT: Greenwood Press, 1994). A huge, sprawling look at many battles in the ancient world, one that sees the Iron Age "spawning a warfare mode in human society and psychology" with standing armies and mass-produced Iron Age weapons. The authors point out that we think that only 6 percent of the time humans have been on Earth they have truly waged war, but "[b]y 2000 BCE, all the major elements of genuine, large scale modern warfare have been put in place by human inventive genius. Mankind had given birth to war."

Richard Gabriel and Karen Metz, *From Sumer to Rome: The Military Capabilities of Ancient Armies* (New York: Greenwood Press, 1991). From 4000 BCE to 100 AD, war increased in scope, scale, and sophistication, and the authors help to draw us into that world, lamenting as they do that there are no think tanks devoted to the study of this period because war is the "legacy of the ancients."

Yvon Garlan, *War in the Ancient World: A Social History* (New York: W. W. Norton, 1975). The author really means Greece and Rome but does some interesting analysis of the views of war by the ancients, claiming, "War is both a law of nature and a human invention," and focuses on the early assumptions that the right to start and stop was the "privilege of the organ of government which was the depository of sovereign power."

Hans Goedicke, *The Battle of Megiddo* (Baltimore, MD: Halgo, 2000). The first major recorded battle (1496 BCE) is described in a carefully researched monograph. The author says, "The towering event in the reign of Thutmose III was shaping Egypt into an imperial political state." Even though the battle with the Canaanites was a draw.

——— (ed.), *Perspectives on the Battle of Kadesh* (Baltimore, MD: Halgo, 1985). In the ongoing battle between Egyptian scholars who say Ramesses II won this battle in 1285 BCE and Hittite scholars who say Muwatallis did, Goedicke's work seems to come down to this: The battle was a draw, and although Ramesses II may have exaggerated things, the Egyptians continued to campaign in Palestine and Moab outside the Hittite sphere of influence thereafter, and the Hittites maintained the status quo ante with a free hand to deal with their problems in the north.

John Hackett (ed.), *Warfare in the Ancient World* (New York: Facts on File, 1989). An excellent, in-depth introduction to the subject, with a variety of UK experts writing on various topics, from the Assyrians to Hellenistic warfare to the Romany army in the age of Polybius. Useful maps and pictures. Would have liked more on the Hittites and the Sea Peoples, however.

Mark Healy, *The Ancient Assyrians* (Oxford, UK: Osprey, 1991). Dominating a great deal of the Fertile Crescent from Anatolia to Egypt to the Persian Gulf during the 10th to 7th centuries BCE, the Assyrians had the most massive military machine of the ancient world, with extensive chariot (from two- to four-man machines), cavalry, infantry, and siege units (plus foreign specialty troops, such as Israeli bowmen). The Assyrians were finally done in by civil war; dynastic troubles; and outside pressures from the Egyptians, Medes, and Babylonians before "Nineveh is laid waste" in 615 BCE. Many good illustrations featuring the Assyrians in their "cone" helmets.

Frank Holt, *Into the Land of Bones: Alexander the Great in Afghanistan* (Los Angeles: University of California Press, 2005). An interesting reexamination of Alexander's conquest of Bactria marred by the author's insistence of force-fitting current US efforts into a fixed pattern of failures (British, Soviet) while glossing over the ease with which Genghis Khan, Tamerlane, Babur, and the Huns took over Afghanistan.

Sa-Moon Kang, *Divine War in the Old Testament and the Ancient Near East* (New York: W. de Gruyter, 1989). A careful reading of the Bible underscores the incessant warfare, with or without divine intervention.

John Keegan, "Stone" and "Flesh," in his *A History of Warfare* (New York: Vintage, 1993), pp.77–235. The origins of warfare as seen through the eyes of one of its masters. He seems a little weak on the early periods, such as Sumer, Egypt, and Sea Peoples, however.

Lawrence Keeley, *War before Civilization: The Myth of the Peaceful Savage* (New York: Oxford University Press, 1996). Argues that wars predate recorded history and that war's iniquitousness applies to "uncivilized" as well as "civilized" peoples.

Paul Kern, *Ancient Siege Warfare* (Bloomington: Indiana University Press, 1999). Reacting against the overemphasis on field warfare, he provides an excellent antidote to such emphasis on the phalanx, chariot, and individual combat in the early period to force our attention on the sieges of Jericho, Troy, Ur, Tyre, Gaza, and Jerusalem, with their higher stakes for noncombatants. Manages to tie all this to the sieges of Leningrad and Warsaw. Finds at the ends of sieges looting ("universal") and massacres and transportations ("common"). Not a good idea to be on the losing side of a siege beginning, if not earlier, when Joshua brought down the city of Jericho (fortified from 7000 BCE) and then killed every living thing in it.

Robert G. Korkot, *The A to Z of Ancient Egyptian Warfare* (Lanham, MD: Scarecrow Press, 2010). A surprisingly useful work covering the much-neglected subject of early Egyptian warfare. Overviews of unification and civil wars as well as thousands of years of Egyptian expansion into Sinai and Nubia, Kush, and the Middle East and struggles with the Hittites, Asiatics, Sea Peoples, Libyans, Assyrians, Babylonians, Persians, Greeks, and Romans from 3100 BCE (about the time of unification) to 642 and the Arab conquest. Notes that the clash with the Libyans and Sea Peoples around 1207 BCE is the first chronicle of a major sea battle on record.

Jim Lacey, "The Persian Fallacy," *Military History* (July 2012), pp. 42–51. Argues that the Greeks were fortunate that they defeated the Persians and they were dead wrong that the Persians lacked courage and were "weak, degenerate and corrupt," despite what Herodotus and others said. Lacey blames the Persians' Greek difficulties on their failure to adapt to the heavy infantry phalanx (other military historians have simply blamed poor leadership and a failure to adjust to the battlefield realities).

Stuart Legg, *The Barbarians of Asia* (New York: Douhet, 1970). Overview with some interesting insights concerning the societies that underpinned the war-making potential of them as they came out of their respective heartlands.

Alan B. Lloyd (ed.), *Battle in Antiquity* (London: Duckworth, 1996). A stimulating set of essays on Homeric battle; hoplite warfare; the making of the Macedonian army; homosexuality and warfare in Greece; and my favorite, "Battle in Ancient Egypt," covering the lengthy period from the Proto Dynastic era to that of the New Kingdom (3200–1070 BCE).

J. G. Macqueen, *The Hittites* (London: Thames and Hudson, 1986). Well illustrated with a good chapter on "Warfare and Defense" for these stalwarts who ruled Anatolia from 1500 BCE to 750 BCE. Unlike the Egyptians, who used chariots as mobile firing platforms, the Hittites seem to have used them more as heavy assault forces armed not with bows as much as with spears and three-man crews.

Ira Meistrich, "War's Cradle: The Birthplace of Civilization Is Also the Home of Culture's Nemesis," *MHQ: The Quarterly Journal of Military History* 17, no.3 (Spring 2005), pp. 84–93. Meistrich ties organized warfare and first true armies with the rise of civilization in the confluence of the Tigris and Euphrates Rivers in the third millennium BCE.

Susan Niditch, *War in the Hebrew Bible: A Study in the Ethics of Violence* (New York: Oxford University Press, 1993). Lots of ethical and moral dimensions of a savage set of wars fought by the Israelites.

A. T. Olmstead, *History of the Persian Empire* (Achaemenid Period) (Chicago: University of Chicago Press, 1948). A lengthy and detailed description of the history of that empire, which at its height stretched from the Indus to Thrace and from Armenia to Egypt and included a vast collection of peoples, all ruled by the king of kings and his well-organized satraps ("King's Ear"). The author describes a huge and largely effective armed force, beginning with the Ten Thousand Immortals and including a navy of Phoenicians, Egyptians, Anatolians, and Greeks. Believes that Marathon and Salamis were mere checks "to the Persian frontier advance" (the Battle of Mycale in 479 BCE was the real defeat) and blames Persian losses not on the abilities of the Greeks but on "timid, incompetent or disloyal allied commanders." Take that, Herodotus.

Keith Otterbein, *How War Began* (College Station: Texas A&M University Press, 2004). Pushes the beginning of war far back into the mists of time, even before agriculture.

C. J. Peers, *Ancient Chinese Armies 1500–200 BC* (Oxford, UK: Osprey, 1990). With accent on swords, bows, and spears, as well as a number of key battles throughout the period, especially those of the Chin, this volume gives a sense of the vast sweep of warfare during 1,300 years of Chinese history. Weak on the role of chariots, however.

David Perlmutter, *Visions of War: Picturing Warfare from the Stone Age to the Cyber Age* (New York: St. Martin's, 1999). See especially "Origins" and "Primitive War."

Ed Phillips, *The Royal Hordes: Nomad Peoples of the Steppe* (London: Thames and Hudson, 1965). Many peoples coming from the same place, driven by the same desire: to make war and prosper.

Simon Price and Peter Thonemann, *The Birth of Classical Europe: A History from Troy to Augustine* (New York: Viking, 2010). Precious little on the strictly military—and they do not credit the Sea Peoples with much of anything, decrying their "monocausal" nature—but there is quite a bit about the Aegean and other worlds, fragmenting around 1200 BCE. Lots on culture, religion, and trade among the Minoans, Mycenaeans, Hittites, Greeks, Egyptians, Romans, and Babylonians. I did learn that a hectare is the size of a soccer pitch, or two football fields, and that two and a half hectares make one acre, so all was not lost.

James B. Pritchard (ed.), *The Harper Concise Atlas of the Bible* (New York: HarperCollins, 1991). A glorious compendium of maps detailing the military campaigns of the various pharaohs, the Hyksos, Sea Peoples, Romans, and others. A very useful companion from which to study the history of the region.

Kurt Raaflaub and Nathan Rosenstein (eds.), *War and Society in the Ancient and Medieval Worlds* (Cambridge, MA: Harvard University Press, 1999). A cross-cultural and diffuse study ranging from China to Egypt to Japan and the Achaemenid Persian Empire and then on to the medieval period. Puts war in a cross-cultural and historical context, seeing war as a rational and pragmatic institution. For the topic of early warfare, there are good chapters by Robin Yates on "Early China," Andrea Gnirs on "Ancient Egypt," and Pierre Briant on "The Achaemenid Empire."

Charles Rollin, *The Ancient History of the Egyptians, Carthaginians, Assyrians, Babylonians, Medes and Persians, Macedonians and Grecians*, 8 vols. (Philadelphia: W. W. Woodward, 1825). An early, extensive, and all-encompassing sweep of history seen through the author's lens.

Nancy K. Sandars, *The Sea Peoples: Warriors of the Ancient Mediterranean 1250–1150 BC* (London: Thames and Hudson, 1978). Outstanding and groundbreaking. A classic.

Ralph Sawyer, *Ancient Chinese Warfare* (New York: Basic Books, 2011). A vast array of information about ancient fortifications of China. The myths and legends of early warfare there, citadels, dagger-axes, archery, and chariots are all covered.

A. H. Sayce, *The Hittites: The Story of a Forgotten Empire* (New York: Fleming Revell, 1888). The section on "The Hittite Empire" traces that empire through its monument throughout Asia Minor and the Levant. Argues that its empire was not like the Roman Empire, with centralized authority firmly in place, but required constant force to subdue and resubdue the conquered provinces.

Neil A. Silberman, "The Coming of the Sea Peoples," *MHQ: The Quarterly Journal of Military History* 10, no. 2 (1998), pp. 6–13. A provocative but convincing account of how the chariot nobility in many countries came to be defeated by infantry armed with iron swords.

Anthony J. Spalinger, *War in Ancient Egypt: The New Kingdom* (London: Blackwell, 2005). Extremely detailed account of how war was conducted in this era—including tactics, strategy, size of camps, weapons, and water and food requirements for animals of war, as well as a careful examination of many battles, including the Thutmose III and Ramesses II campaigns of Megiddo (1479 BCE) and Kadesh (1294 BCE).

David Ussishkin, *The Conquest of Lachish by Sennacherib* (Tel Aviv: Institute of Archeology, 1982). A detailed and well-illustrated account of the king of Assyria's conquest of Lachish in Judea in 701 BCE. Many biblical references buttress the text and findings. Demonstrates the Assyrian siege-making capacity, including siege rams, siege machines, battering rams, and long stone-and-earthen ramps to overcome walls. They were the masters of the art form for centuries.

Hans van Wees, *Status Warriors: War, Violence and Society in Homer and History* (Amsterdam: J. C. Giebeng, 1992). Senses the shift between heroic, single-champion combat and the later, much more effective, Greek phalanx. This simple dichotomy has been challenged by those who see echoes of the phalanx even in Homer.

H. T. Wallinga, *Ships and Sea-Power before the Great Persian War: The Ancestry of the Ancient Trimere* (Leiden, Netherlands: E. J. Brill, 1993). Useful insights into the battleship of the ancient era and how it was employed in battle and in trade.

Terence Wise, *Ancient Armies of the Middle East* (Oxford, UK: Osprey, 2010). A useful introduction, one with helpful illustrations, covering a wide range of military entities from the Sumerians to the Egyptians to the Sea Peoples (Philistines and Teukroi).

Serge Yalichev, *Mercenaries of the Ancient World* (London: Constable, 1997). The earliest mercenaries—Amorites, Gutigs, Libyans, Nubians, Bedouins, and Greeks—are all from agriculturally unproductive regions and served a variety of empires, including the Sumerians, Egyptians, Israelites, Hittites, Assyrians, Persians, Greeks, Romans, Carthaginians, and many others. They appear to have been reliable, loyal (at least when paid), and professional. King David, for example, hired Philistine and Canaanite mercenaries as he expanded his kingdom. The author concludes, “Mercenary soldiering is—like prostitution—an extremely ancient pastime, perhaps about as ancient as organized warfare itself.”

Chapter Eleven

Classical Infantry

Greek

Useful topics here include the rise of the Greek city-states, the development of the phalanx and classical infantry warfare, Alexander the Great, the Macedonian Empire, city-states of commerce, and force projection. It should be noted that some scholars, such as Robert L. O'Connell, argue that the phalanx was actually first depicted on a Sumerian victory monument, the Stele of the Vultures, carved around 2500 BCE.

J. K. Anderson, *Military Thought and Practice in the Age of Xenophon* (Los Angeles: University of California Press, 1970). Basically points out that "[o]pen and honest and honorable" warfare leads to great slaughter when the phalanx breaks down, but there was no need to do more than temporarily occupy agricultural land at harvest time. Could have used an overarching conclusion.

Flavius Arrianus, *The Life of Alexander the Great* (London: Folio Society, 1958). This second-century-CE account by a Greek citizen of Rome who was also a Roman imperial government official. Concentrating on Alexander's military affairs, this is a rich, luminous, and ultimately enriching account, accenting, as he does, the life and times of Alexander as well as his early vision of an association of nations with their own cultures but ruled by a king. Arrianus is wise to advise us of Alexander's lifelong efforts with both the infantry phalanx and cavalry for a "suppleness of movement" in battle.

Stringfellow Barr, *The Will of Zeus: A History of Greece from the Origins of Hellenic Cultures to the Death of Alexander* (New York: J. B. Lippincott,

1961). Some interesting chapters include "A Race of Iron" and "Alexander's World Polis." Readable, with superb maps and good illustrations. Doesn't really focus on warfare but gives the context of that warfare.

A. B. Bosworth, *Conquest and Empire: The Reign of Alexander the Great* (Cambridge: Cambridge University Press, 1986). This scholarly campaign narrative focuses heavily on the "gaining" of the empire of Alexander from 336 to 323 BCE and how it eventually functioned under Alexander for the short time he ruled it. Lots of intrigue around him on the campaign trail and back in Macedonia and Greece.

Pierre Briant, *Darius in the Shadow of Alexander* (Cambridge, MA: Harvard University Press, 2015). The author, an international expert on all things Achaemenid, lays out how lucky Alexander was to have as an enemy a "devoted husband and father perhaps, but a weak king."

Andrew Burn, *Persia and the Greeks: The Defense of the West c. 546–478 B.C.* (London: Edward Arnold, 1962). Highly descriptive and illuminating, it covers the rise of Persia, with Darius leading the empire into Greece and Xerxes trying to finish the job, only to come a cropper at Salamis and Plataea. Burn rightly regrets the 50-year gap between Herodotus and Thucydides and wishes someone had provided us with a firsthand account of how and why the Greeks fell to fighting among themselves again.

Paul Cartledge, *After Thermopylae: The Oath of Plataea and the End to the Greco-Persian Wars* (London: Oxford University Press, 2013). An erudite and perhaps overly pedantic work that makes the important case that, in the words of George Cawkwell, "[i]t was at Plataea, not at Salamis, that the new satrapy was lost." Claims Plataea in 479 was more important because it ended Persian land dominance (70,000 to 100,000 Persians against an almost equal number of Greeks, mostly Spartans). Plataea is northwest of Athens; Salamis, northeast.

———, *Ancient Greece: A History in Eleven Cities* (London: Oxford University Press, 2009). Jam-packed with interesting historical nuggets on a polis-by-polis basis. From Cnossos and Mycenae in prehistory to Argos and Sparta in the Dark Ages and archaic period (to 500 BCE) to Athens and Syracuse in the classical period (500–330 BCE) and Alexandria in the Hellenistic period (330 BCE on), this work provides a tour de horizon. Thebes, for example, is well represented by a discussion of the Theban Sacred Band, an "elite strike force composed of 150 homosexual couples," which seems to have aroused the Spartans (and others) to enmity, even though the Spartans (and others) "had made homosexual relations between an adult male and an adolescent boy an integral part of their educational system."

Jack Cassin-Scott, *The Greek and Persian Wars 500–323 BC* (Oxford, UK: Osprey, 1977). Good illustrations of various units of the Persian Empire and the Greek city-states. Not much on how the Persians were able to sustain their large and successful empire for so long.

George Cawkwell, *The Greek Wars: The Failure of Persia* (London: Oxford University Press, 2005). From the 6th-century-BCE expansion of Cyrus the Great to the defeat of Darius III by Alexander, the history of Greek–Persian warfare is more explained, the author believes, by the failure of Persia than by the heroism of the Greeks, and only the military power of Macedonia eventually destroyed Persia as a threat.

Peter Connolly, *Greece and Rome at War* (Englewood Cliffs, NJ: Prentice Hall, 1981). A marvelous, lavishly illustrated (with diagrams, maps, and photos) look at the battles of Marathon, Salamis, Plataea, Cannae, Zama, Cynoscephalae, and many others. Excellent depictions of the history of the Greek and Roman helmets (and many other equipment of war). A very worthwhile study. I wish the author had started earlier during the time of the Sea Peoples! Very useful photos of the various battle sites.

R. D'Amato and A. Salimbeti, *Bronze Age Greek Warrior 1600–1100 BC* (Oxford, UK: Osprey, 2011). A well-illustrated and -documented account of the Greeks before and during the coming of the Sea Peoples, a good overview of the Mycenaean period before the Dark Age. Interesting descriptions of various chariots and spear types as well as equipment variations over this 500-year period.

Hans Delbruck, *Warfare in Antiquity* (Lincoln: University of Nebraska Press, 1990). Volume 1 in his *The Art of War*. A sweeping, inclusive study of the Persians, the Greeks, the Romans, the Carthaginians, the Macedonians, and their interactions over a millennium. Contains valuable studies of the major battles (reading in reverse chronological order), such as Pharsalus, Zama, Cannae, Gaugamela, Issus, Plataea, Salamis, and Marathon.

Jacqueline de Romilly, *Thucydides and the Athenian Imperialism* (New York: Barnes and Noble, 1963). How Athenian imperialism started out as a good thing and ended up as a bad thing.

Donald Engels, *Alexander the Great and the Logistics of the Macedonian Army* (Los Angeles: University of California Press, 1978). A truly amazing effort to replicate the true needs of this most marvelous of supply systems, the author makes the case for the need of 11,000 pack animals to carry the supplies for his army for *four* days. The author concludes, "Supply was indeed the basis for Alexander's strategy."

Robin Fox, *Alexander the Great* (London: Penguin, 1973). Deeply grounded in the facts and myths surrounding Alexander, the author provides an elegant, engaging, and ultimately illuminating look at a complex man who conquered a great deal of the known world but could not conquer his own inner demons, concluding, "He had the romantic's sharpness and cruel indifference to life."

Richard A. Gabriel, "The Genius of Philip II," *Military History* 25, no. 6 (2009), pp. 39–43. Without Philip II and his innovations, there would never have been an Alexander the Great, argues the author, who points to Philip II's improvements in strategy, tactics, logistics, vision, and long war to unify the Greek city-states, a previously impossible task. The Battle of Chaeronea in 338 BCE unified by force a Greece under Macedonia and set the stage for the subsequent attack on Persia led by Phillip II's son Alexander.

———, *Philip of Macedonia: Greater Than Alexander* (Washington, DC: Potomac Books, 2010). Convincingly, the author indicates that the long-pike phalanx, with its greater flexibility, and the long-lance shock-cavalry arrangement for use at the *Schwerpunkt*, when coupled with the Macedonian Empire's supply system and its national recruitment and replacement system, enabled Alexander to succeed. "In fact, Philip's legacy was so superior that without it, there would have been no Alexander the Great."

Yvon Garlan, *War in the Ancient World: A Social History* (New York: Norton, 1975). Greece and Rome at war are examined. Garlan claims, "War is both a law of nature and a human invention," yet he seems to do very little with actual warfare during preclassical military history.

Michael Grant, *The Rise of the Greeks* (New York: Charles Scribner's Sons, 1987). The "Early Greeks" of the 13th century BCE morphed into those we know in Athens, Sparta, Crete, Corinth, Spina, Gela, Cyrene, and so on, to 700 poleis across the Mediterranean world, each with their own military strength. Aided by the "hoplite revolution," whose short description is worth noting.

Peter Green, *Alexander the Great* (New York: Praeger, 1970). A breezy, nicely illustrated volume with good maps, a lot of illustrations, and some fine descriptions of the battles of Chaeronea, Issus, Gaugamela, and Jhelum.

———, *The Greco-Persian Wars* (Los Angeles: University of California Press, 1969). Starting with the notion that "[m]odern Europe owes nothing to the Achaemenids," the author gives a thoroughly enjoyable and often insightful account of the invasions of Darius and Xerxes. Points out that the Achaemenid Persians were very successful at warfare before and after their Greek failures. Places a lot of emphasis on the role of Themistocles, who in 493 was elected archon of Athens and led a shaky coalition to victory. Good detail on Xerxes's expedition crossing the Dardanelles, huge supply problems, need for Phoenician ships, and captains. Vivid account and an amusing one at that. Says Xerxes was right to withdraw after Salamis due to the danger of the Greek fleet attacking Dardanelles's bridges and supply ships. No final treaty to this war for 30 years.

Nicolas Grguric, *The Mycenaeans c. 1650–1100 BC* (Oxford, UK: Osprey, 2005). A straightforward account of the Mycenaeans (also known as Achaeans, Danaans, and Argives), who, in the period before the fall of Troy and the arrival of the Sea Peoples, dominated the western Aegean as well as Crete (replacing the dominant Minoans after a huge earthquake at Knossos). Principle weapons: 12-foot spears, swords, figure-8 and other shields, boar's-tusk helmets, regular bows and composites, and two-man/two-horse chariots. Heavy infantry, light infantry, skirmishers, archers, and charioteers gave them a variety of formations and battlefield deployments. No phalanx formations as yet. For a further exploration of these aspects, see also R. D'Amato and A. Salimbeti, *Bronze Age Greek Warrior 1600–1100 BC* reviewed previously.

John R. Hale, *Lords of the Sea: The Epic Story of the Athenian Navy and the Birth of Democracy* (New York: Viking, 2009). A marvelous read—well illustrated with excellent maps—by a writer of clear prose who finds Greek ships emblems of liberty and democracy as well as imperial ambition. Focuses on the 120-foot-long triremes, with their 170 rowers who could make the trireme go 10 knots an hour, and notes how frequently they had to be beached in order to stave off sea worms. The rowers were, as with the Greek hoplites, not slaves but free men fighting for causes they believed in. Hale claims that the Gold Age of Athens (480–322 BCE) and the "Greek Miracle" came after the fleet, not before, in large part because of the hectorings and orations of Themistocles. I had not realized how skillfully Philip and Alexander of Macedonia used the Athenian fleet to assist them in their eventual Asian conquests.

Nicolas Hammond, *Philip of Macedonia* (London: Duckworth, 1994). Philip's military revolution (7-foot spear into 12- to 15-foot pike, adding depth to the phalanx and making his infantry pikemen with phalanx training, more use of the wedge, and more flexible hoplite formations and integral cavalry arm) and his unification of the "Greek community" with him as hegemon after the Battle of Chaeronea (338 BCE) gave Alexander a military of 200,000 infantry and 15,000 cavalry upon which to draw for his campaigns. Says Hammond, "It was Philip as leader of the Greeks who made the Hellenistic Age possible."

Victor Hanson (ed.), *Hoplites: The Classic Greek Battle Experience* (London: Routledge, 1991). A variety of looks at Greek infantry during its heyday. Again, there is great emphasis on discipline and free association, as well as short, sharp battles with clear victors.

——— (ed.), *Makers of Ancient Strategy: From the Persian Wars to the Fall of Rome* (Princeton, NJ: Princeton University Press, 2010). A collection of essays on a variety of subjects, including Peter Heather's "Frontier Defense and the Later Roman Empire," David Berkey's "Why Fortifications Endure: A Case Study of the Walls of Athens during the Classical Period," and John Lee's "Urban Warfare in the Classical Greek World," illuminate.

———, *A War Like No Other* (New York: Random House, 2005). The Peloponnesian War was both tragic and vicious, with plagues and massacres and ethnic cleansing (especially by Athens), and both sides put to death each other's diplomats. The "Greek Miracle" continues to offer ironic discoveries.

———, *The Western Way of War: Infantry Battle in Classical Greece* (New York: Knopf, 1989). Argues that free men, fighting under the leaders they choose and relying on discipline and better equipment, emerge triumphant against others lacking those attributes. And the Greek experience proves it.

Waldemar Hekel and Ryan Jones, *Macedonian Warriors* (Oxford, UK: Osprey, 2006). Straightforward, with excellent emphasis on the Macedonian 18-foot-long spear (sometimes 24 feet long), which had to be held with two hands and necessitated a smaller "hanging" shield. The Macedonian army under Alexander marched 20,000 miles.

Herodotus, *The Histories* (Baltimore, MD: Penguin Books, 1954). Writing in the fifth century BCE, this wandering Greek was the "first Greek and first European to use prose as the medium of a work of art," writes Aubrey de Selincourt, and the "great traveler" gives us exciting pictures of both Greece and the Persian invasions.

———, *The Persian War*, translated by William Sheperd (Cambridge: Cambridge University Press,1982). This shorter and more manageable account leaves out some 300 pages of *The Histories* devoted to the rise and expansion of the Persian Empire and is thus more accessible for those seeking just the wars with the Greeks. Interestingly, Herodotus puts the blame for the Persian rout at Plataea on their equipment, saying, "What they suffered from most of all was their lack of armour," for they were lightly armed soldiers battling against hoplites.

Tom Holland, *Persian Fire: The First World Empire and the Battle for the West* (New York: Random House, 2005). Very useful maps on pages xxiv to xxv showing the Persian Empire and its Khorasan Highway and Royal Road stretching from Gandhara to Babylon to Ionia. Breezy and arresting, it covers the efforts of Darius and Xerxes to capture the "remote margin of the world"; that is, Greece with good accounts of Marathon, Thermopylae, Salamis, and Plataea. Author sees the East–West struggle as continuing over the millennia.

Homer, *The Iliad*, translated by Robert Fagles (New York: Penguin, 1998). The Greek states attack Ilium and its capital, Troy, in Asia Minor sometime during the period 1194–1184 BCE. Although gods and goddesses still interfere with human activities, *The Iliad* points to individual, heroic conflict with chariots in battle (Hector kills Patroclus and is in turn killed by the half-god Achilles) as opposed to the disciplined unit fighting of the phalanx/hoplite type, which comes considerably later in Greek history. At the same time, there is a very serviceable description of what is to come: "At last the armies clashed at one strategic point, they slammed their shields together, pike scraped pike with the grappling strength of fighters armed in bronze and their round shields pounded, boss on welded boss, and the sound of struggle roared and rocked the earth" (p. 160). Gods in constant interaction in the ebb and flow of actual battles. Will the Hectors, men of peace as well as war, always be defeated by men of war and glory, Achilles? Homer seems to suggest the answer is yes. I find it useful, also, to remember what Philip Roth says in *The Human Stain* about *The Iliad*: "You know how European literature begins? With a quarrel. All of European literature springs from a fight."

Note: See also Fagles's translation of his *The Odyssey* (New York: Penguin, 1996), which depicts the long journey, metaphorical and actual, following the war. These dimensions are of great import when looking at the decompression that always lies on the trail of warfare.

Note: See also Edward Luttwak's illuminating "Homer Inc." in *London Review of Books*, February 23, 2012, pp. 3ff. Showing off his own substantial knowledge about *The Iliad* and its worldwide appeal ("*Iliad*'s Iron Age reminiscence of Bronze Age combat"), he concludes, "*The Iliad* is an epic—the point is not to win but to gain honour by fighting not effectively but heroically."

Donald Kagan, *The Archidamian War* (Ithaca, NY: Cornell University Press, 1970). Covers in great detail the first phases of the Peloponnesian War (431–421 BCE). Good material on the Corcyran revolution.

———, *The Peloponnesian War* (New York: Viking, 2003). An exhaustive study of the first major Greek civil war. Kagan's work is outstanding and provides a clear view and understanding of what happened and why. Argues that the Greek wars with Persia, by vaulting Athens to the top of the power heap in that world, resulted in an overmighty city bent on domination. The results were most serious and far reaching.

———, *Thucydides: The Reinvention of History* (New York: Viking, 2009). Thucydides was the first historian in the West to write without reference to the intervention of the gods and, Kagan argues, the first revisionist historian.

J. E. Lendon, *Soldiers and Ghosts: A History of Battle in Classical Antiquity* (New Haven, CT: Yale University Press, 2005). Looks at the genesis, evolution, and high points of Greek and Roman warcraft, pointing out how slow the evolution really was: the hoplite form of warfare took 200 years to dominate the battlefield. Sees an ongoing interplay between literature and warfare: "The story of the evolution of ground warfare in antiquity reflects the evolution of ancient literature." From Homer through Polybius.

Madeline Miller, *The Song of Achilles* (New York: HarperCollins, 2012) (F). Putting Achilles in a homoerotic frame of mind due to his love of Patroclus, the author accounts for his success with speed (shades of John Boyd!), and the "best warrior of his generation" goes on to immortality in story and song. Achilles is not frightened of war for it was "what I was born for."

Josiah Ober, *The Rise and Fall of Classical Greece* (Princeton, NJ: Princeton University Press, 2015). See especially "The Termination of the Classical Era," the final chapter, which does an excellent job at pointing out the reasons Philip II conquered Greece. Special attention is paid to the military improvements of the 15- to 18-foot sarissa, used by both infantry and heavy cavalry, and the superiority of the Macedonian training, coinage, and use of scarce resources, including gold, silver, and land.

Steven Pressfield, *Gates of Fire* (New York: Doubleday, 1998) (F). The 480 BCE Battle of Thermopylae is depicted with excitement and verve as the 300

Spartans fight and die to hold off the Persians led by Xerxes. Courage depicted and insights into the Greek phalanx in action provided.

———, *Tides of War* (New York: Doubleday, 2000) (F)/ The novel about the life and times of Alcibiades, who, during the Peloponnesian War, fought for Athens, Sparta, and Persia and had much to say about it.

John Prevas, *Xenophon's March into the Lair of the Persian Lion* (Cambridge, MA: Da Capo Press, 2002). Gives the basics of an incredible feat of arms as the Greeks fight their way home after an interventionist campaign comes a cropper.

Mary Renault, *The Persian Boy* (New York: Pantheon, 1972) (F). Alexander's alleged bisexuality explored from the perspective of one of his imagined lovers. This forms the literary device that enables the author to chronicle his life and times.

James Romm, *Ghost on the Throne: The Death of Alexander the Great and the War for His Crown and Empire* (New York: Knopf, 2011). When Alexander died in 323 BCE after a reign of 12 years, his vast empire split and split again, until by 315 BCE it was divided among four rulers: Cassander, Ptolemy, Antigonus, and Lysimachus. The status quo ante had returned. The author tells us why.

William Shepard, *Plataea 479 BC* (Oxford, UK: Osprey, 2012). An in-depth look at this very important battle that led directly to the failure of the Persian invasion of Greece under Xerxes. Detailed with good maps, illustrations, and photographs. Some argue that Plataea was more important than Salamis.

Harry Sidebottom, *Ancient Warfare: A Very Short Introduction* (London: Oxford University Press, 2004). Short it may be, but this is a provocative, stimulating, and satisfying account of the "Western Way of War" and how its basic, "long-lived, highly adaptable, and powerful ideology" gave the Greeks and Romans and other Europeans a psychological advantage when fighting "lesser breeds without the law," even though their stereotyping was not always correct. Points out a variety of exceptions to the "Greek-Roman" uniqueness by citing findings from 350 BCE in Denmark and the 19th-century Zulus. Looks at controversies surrounding the "barbarization" of the Roman army and the "unhorsing" of the Huns. Sidebottom has also written a number of historical novels, such as his Warrior of Rome series, which are chock-full of history and his interpretation of scholarship of the ancient world.

Barry Strauss, *The Battle of Salamis* (New York: Simon and Schuster, 2004). Places great emphasis on Themistocles and his role in this great battle, which many, including Strauss, believe protected "Europe" from "Asian" conquest.

W. W. Tarn, *Alexander the Great* (Cambridge: Cambridge University Press, 1948). In fourth century BCE, this youth needs to subdue Greek states before moving into Persia. Largest army in the field (47,000) at Gaugamela. On to Turkestan, India, and others. Leisurely way home after army mutiny on the Indus. Author believes he was fortunate to die when he did: "He had made war as few have made it, it remained to be seen if he could make peace" (p. 121). Never aimed to rule the world but believed in the brotherhood of mankind, claims the author.

Thucydides, *The Peloponnesian War* (New York: Modern Library, 1982). The first true historian whose work shines brilliantly thousands of years later and illuminates our understanding of war as well as of human nature. A work of almost immeasurable value.

Matthew Trundle, *Greek Mercenaries: From the Late Archaic Period to Alexander* (London: Routledge, 2004). The author finds mercenaries playing a "central role in Greek history" and serving as a "link between citizen and political community" in a context where the Greeks had no word for *mercenary* and no sense of nationalism. Believes that Greek mercenaries fought for the Assyrians, Babylonians, Egyptians, and Persians (although his estimate of 50,000 Greek mercenaries serving Darius III seems quite high). Trundle argues that the expansion of naval warfare as well as the poverty of Greece provided pull and push in addition to the appeals of food, pay, and booty.

John Warry, *Warfare in the Classical World* (London: Salamander Books, 1980). From "Homeric and Mycenaean Warfare" to the "Coming of the Barbarians," this well-illustrated volume provides short, pithy, and helpful accounts of centuries of warfare in the Mediterranean basin.

Robin Waterfield, *Dividing the Spoils: The War for Alexander the Great's Empire* (London: Oxford University Press, 2011). Fascinating and very illuminating, this work describes in considerable detail and with penetrating analysis the almost continual warfare during the 40 years following the death of Alexander in 323 BCE. Many of his successor generals contested each other and the subject peoples, and three succeeded quite grandly, forming definite empires in the tradition of the Hittites (1430–1200 BCE), Neo-Assyrians (934–610 BCE), and Achaemenids (550–330 BCE). Seleucus (sometimes Seleucid) created one containing large chunks of the former Achaemenid Empire of Persia (1.5 million square miles and 15 million people).

Ptolemy carved out one in Greater Egypt and claimed Alexander's embalmed body (8,000 square miles and 4 million people). The Antigonids seized Macedonia and parts of Greece. All were based on military might and claimed Hellenistic kingship. For 40 years after Alexander's death, there was almost continual warfare. Very good descriptions of both land and naval warfare in the Hellenistic period.

Michael Wood, *In Search of the Trojan War* (New York: Facts on File, 1985). Lavishly illustrated and relying heavily on a wide array of archeological evidence (and some written accounts, such as Hittite references), Wood concludes that there was a Trojan War, pretty much as described by Homer, and not one Troy but a series of nine Troys, as well as that the Bronze Age merged into the Iron Age with the arrival of the Sea Peoples.

———, "The Road to Samarkand," in his *In the Footsteps of Alexander the Great* (Los Angeles: University of California Press, 1997), pp. 133–66. On the risings and guerrilla warfare in Afghanistan: "He had had to fight a war he had not anticipated" (p. 159).

Andrew Young, *The Lost Book of Alexander the Great* (Yardley, PA: Westholme, 2014). This is a "book about a book," an attempt to piece together the lost text by Ptolemy I, companion of Alexander, after he took over the Egyptian portion of his empire after Alexander's death. The author concludes, "In war, he was of singular brilliance."

Chapter Twelve

Classical Infantry

Roman

> "The Romans fought to destroy the enemy army and end its capacity ever to fight them again."—Adrian Goldsworthy, *Roman Warfare*

This chapter covers the development of the Roman state and its expansion, the Punic Wars, Scipio Africanus, and Julius Caesar. The Roman Republic versus Roman Empire, Gaius Marius, the self-contained legion and the state, classical infantry warfare, Caesar Augusta, Pax Romana, Trajan, power centralization are all possible topics for further investigation.

Also worth studying is the Roman integrative capacity (they borrowed military innovation wherever found), their strategic sense that war could solve most any problem, and their development of the Roman legion, which superseded the Greek phalanx as the dominant formation of the age. Dwight Zimmerman, for example, notes that the Roman legion was a formation so supple that military historian Lynn Montross compared it to a human hand: "The legion was capable either of closing up solidly like a fist, or of feeling out the enemy's weaknesses with the effect of exploring fingers."

Note: Adrian Goldsworthy estimates that, at its height, the Roman army consisted of 30 legions, between 350,000 and 375,000 soldiers, or 1 soldier for every 130 civilians (*How Rome Fell*, p. 59).

Ancient Warfare (periodical). All you could ever want to know about the military of ancient times.

Marcus Aurelius, *Meditations* (Oxford: Oxford University Press, 1990). Roman emperor from 161 to 180 CE, Aurelius and his words of wisdom have echoed over the years. Focuses on the individual and adjusting to whatever life thrusts at you. A great deal of "practical" philosophy and advice: "Where a man can live, there he can also live well."

Alvin Bernstein, "The Strategy of a Warrior-State: Rome and the Wars against Carthage, 264–201 BC," in *The Making of Strategy: Rulers, States, and War*, edited by Williamson Murray, MacGregor Knox, and Alvin Bernstein (Cambridge: Cambridge University Press, 1994), pp. 56–84. Excellent maps and clear prose make this chapter most useful.

Gregory Bolich, "Terrorism in the Ancient Roman World," *Quarterly Journal of Military History* (Spring 2006), pp. 52–59. There were a lot of uprisings, and not surprisingly, Rome dealt harshly with the great majority of them.

Thomas Burns, *A History of the Ostro-Goths* (Bloomington: Indiana University Press, 1984). This volume is of note in pointing out how long and consistent the interaction between the "barbarians" and the Romans was long before the Romans reached the Rhine. Particularly useful is the "Warriors and Military Systems" chapter and the portrait of Theodoric, "The Greatest Germanic leader of his day." Theodoric, sent by the emperor in Constantinople to regain northern Italy, did so but then kept it for himself and the short-lived Ostro-Gothic Kingdom.

Julius Caesar, *The Civil War* (London: Oxford University Press, 1997). Of course this conflict is presented from Caesar's point of view, but it nevertheless offers very useful insights into the actual warfare of the time, in particular the allegiance of the legions in the era of the late republic. The legions were comprised of citizen soldiers, not professionals, and their loyalty to their commander depended on many things, including their perceptions of the rightness of his, and hence their, cause, as well as his treatment of them and their opportunity for past, present, and future shared booty. After Pompey defeated Caesar (Battle of Dyrrhachium in 48 BCE), for example, the legions followed Caesar despite the defeat, and he—and they—emerged triumphant at the subsequent Battle of Pharsalus (48 BCE) despite being outnumbered two to one.

———, *The Conquest of Gaul* (Baltimore, MD: Penguin Books, 1951). Caesar spent nine years of his life away from Rome, campaigning in Gaul and its frontiers (58–50 BCE). He had two marvelous gifts as a commander. The first was the ability to inspire his troops; the other, an unusual proclivity for thinking quickly and correctly when things were going wrong. The conquests of what are now Belgium, France, and Switzerland and the invasions of Germany and Britain gave him many opportunities to employ both skills.

———, *Caesar's Commentaries: On the Gallic War and on the Civil War* (El Paso, TX: El Paso Norte Press, 2005). Caesar always seemed to know how to take the main chance in life and in battle. He was also a pretty good self-promoter.

Averil Cameron, *The Later Roman Empire AD 284–430* (Cambridge, MA: Harvard University Press, 1993). See especially chapter 9, "Military Affairs, Barbarians and the Late Roman Army." Makes the point that the eastern portion of the empire made considerable strides, but these were eventually undercut by the rise of Islamic rule.

Brian Campbell, *War and Society in Imperial Rome 31 BC to AD 284* (London: Routledge, 2002). Provides a good interplay among the social, political, and economic forces that saw Rome become the dominant superpower in the Mediterranean but eventually to be overcome by forces from beyond its borders.

A. S. Esmonde Cleary, *The Ending of Roman Britain* (Savage, MD: Barnes and Noble Books, 1990). Makes a sturdy case that, between 300 AD and 500 AD, there was a rather abrupt change from Roman to non-Roman Britain and then an interlude of 50 years of non-Roman Britain before Saxon England began to develop. Points out that Britain had a Roman period lasting 250 years. Very good maps and illustrations of both Hadrian's Wall and the protocastles of "The Forts of the Saxon Shore." The author believes the collapse was "sudden and total" as well as "nasty, brutish and short."

Ross Cowan, *Roman Legionary: 58 BC to AD 69* (Oxford, UK: Osprey, 2003). Gives good visuals of what the Romans must have looked like as they went into battle during this period.

Gregory Daly, *Cannae: The Experience of Battle in the Second Punic War* (London: Routledge, 2002). Stating that "[h]umility is absolutely necessary when analyzing the nature of battle in antiquity," the author looks at the 216 BCE Battle of Cannae in which upward of 50,000 Romans and Italians and 8,000 Gauls, Spaniards, and Carthaginians lost their lives. Relying on such ancient sources as Polybius, Plutarch, and Livy, as well as the modern insights of John Keegan, Victor Hanson, and Adrian Goldsworthy, he reproduces a feeling for the thirst, dust, din, vomit, sweat, and blood, all mixed in with fear and triumph and the ongoing killing of the battle qua battle that set what was perhaps a one-day record for battle deaths.

Stephen Dando-Collins, *Nero's Killing Machine: The True Story of Rome's Remarkable Fourteenth Legion* (New York: John Wiley and Son, 2005). Despite its lurid title, thought up no doubt by a marketing maven at John Wiley, this is actually a good, readable account of Rome's 14th Legion, raised by Caesar but wiped out to a man in 54 BCE by the Belgics. It was reformed with new recruits and would fight for the better part of five centuries (finishing up in 410 AD) in Gaul, North Africa, and Germany, indeed all along the Rhine and Danube Rivers, and it was instrumental in crushing the Celtic uprising of Queen Boudicca in Britain. Emperor Nero labeled them Martia Victrix ("Victorious and Blessed by Mars") after they crushed the Iceni rebellion under Queen Boudicca in 60 CE. The author believes they were justified in saying at the end of their operational life, "Good-bye and applaud us."

Hans Delbruck, *The Barbarian Invasions* (Lincoln: University of Nebraska Press, 1990). Often considered the first "modern" historian, Delbruck wrote a four-book *History of the Art of War*. This volume focuses on the conflict between the Romans and the Germans and gives good accounts of many battles and the gradual decline of Roman military prowess and the beginnings of the medieval system. A Prussian's Prussian, his academic star and standing went into some decline when, after predicting the victory of Germany in World War I, he then had to explain its defeat.

Philip de Souza, Waldermar Heckel and Lloyd Llewellyn-Jones, *The Greeks at War: From Athens to Alexander* (Oxford, UK: Osprey, 2000). Strong narrative covering the period with useful photos of archeological and artistic evidence. See "The Wars of the Successors 323–301 BC" for the little-known fact that, in these wars, the amount of their men the successors to Alexander killed was more than Alexander did in all of his battles.

Cassius Dio, *Roman History* (Cambridge, MA: Harvard University Press, 1924). For a long time, an important overview. Still a useful source, although much new material has been uncovered since this was written. However, this earlier work shows the long shadow of Roman culture and warfare.

Peter Ellis, *Caesar's Invasion of Britain* (New York: New York University Press, 1978). A delightful read, well-illustrated and straightforwardly informative. Caesar comes in 55 BCE with 80 ships and 10,000 troops but, after storming ashore and winning a battle, heads back to Gaul only to return the next year with 30,000 troops. His motives: loot, prestige, and getting back at the Celts for assisting their brethren in Gaul. The Romans are impressed with the Celts' scythed wheeled chariots and their "hot-blooded courage," but in the end, the Roman discipline and legionary steadfastness wins the day. But there is precious little loot and more problems with resupply, so Caesar returns to Gaul with only slaves to finance the expedition. The Romans do not come again for 97 years.

Hugh Elton, *Warfare in Roman Europe, AD 350–425* (New York: Oxford University Press, 1996). The divided empire faced perils and in-migrations, especially in the West, where Vandals, Suevi, Visigoths, and Burgundians came to stay and Britain went away for good. The author sees little technological, tactical, or organizational change during this period and says that the 476 collapse in the West was due to too much pressure; too much civil war; and, ultimately, the loss of North Africa with its money, grain, and troops. There were no structural failures, he believes.

Arther Ferrill, *The Fall of the Roman Empire: The Military Explanation* (London: Thames and Hudson, 1986). This overlooked work is quite convincing and worthy of further exploration. The author's premise is that the Roman military was not overcome by barbarian horse soldiers but by infantry decline in its own ranks: "By 410 the Roman army no longer had any special advantage, tactically over barbarian armies—simply because the Roman army had been barbarized." In the West especially, Roman shields got smaller, and they abandoned heavy armor, so that in effect, by the early fifth century, they "looked just like and fought just like the barbarians they faced." Often without their motivation.

Richard A. Gabriel, "Hannibal's Big Mistake," *Military History* (November 2011), pp. 64ff. Gabriel makes the claim that Hannibal should have moved on Rome following his big initial victories in northern Italy, including the Battle of Lake Trasimene (217 BCE), rather than go marauding around Italy for several more years, winning tactical victories but not ever defeating Rome.

Edward Gibbon, *The History of the Decline and Fall of the Roman Empire*, 3 vols., edited by Betty Radice (London: Folio Society, 1983). Originally published as a six-volume study during the period 1776–1789, this multivolume classic has sometimes been dismissed as not adequately explaining that fall because of an overemphasis on the role played by Christianity (the "religion of Constantine achieved, in less than a century, the final conquest of the Roman Empire") and for slighting the subsequent accomplishments of Byzantium (which went on for another 1,000 years in various forms), but Gibbon's work is so much more than just a history of Rome. Rather, with Rome as the centerpiece, it is really an in-depth exploration of the Mongols, Tartars, Arabs, Byzantines, and many Germanic and other tribes and their interaction with Rome over time. As such, it is a useful introduction to these many interactions.

Adrian Goldsworthy, *Caesar: The Life of a Colossus* (New Haven, CT: Yale University Press, 2006). An excellent overall account of the life and times of Caesar.

———, *How Rome Fell: The Death of a Superpower* (New Haven, CT: Yale University Press, 2010). Failure to remain strong on the Danube probably was as big a reason as any that the Romans contributed to their own decline. The allegiance of specific legions to their commanders rather than to the state, which is typified by Caesar's XIV Legion, was also a factor.

———, "Julius Caesar and the General as State," in *Makers of Ancient Strategy*, edited by Victor Hanson (Princeton, NJ: Princeton University Press, 2010), pp. 206–26. The militarization of politics followed his seizure of power.

———, *Roman Warfare* (London: Cassell, 2000). Well written and shows the conquest of Italy and the defeat of Carthage and the Greeks, leading to "World Conquest" and the Pax Romana before "Collapse in the West, Recovery in the East."

Michael Grant, *The Ancient Historians* (New York: Charles Scribner's Sons, 1970). Herodotus, Thucydides, Xenophon, and Polybius from the Greeks and Caesar, Livy, and Tacitus from the Romans. Livy is featured, writing 3 or 4 books a year for 40 years, for a total of 142, on the history of Rome. He marvels most at the Pax Augusta. For good reason.

———, *The Armies of the Caesars* (New York: Charles Scribner's Sons, 1974). A fine and highly readable account of the Roman army from the days of the republic (107–31 BCE) to the ending period of "Anarchy, Reconstruction and Collapse" (235–476). Points out that the very institution that made the empire possible also made it inherently unstable at the political center as Rome oscillated between civil war and tyranny. The Roman army was a "constant peril to its imperial master." Useful examinations of the way Sulla, Julius Caesar, Augustus, Trajan, Hadrian, and others dealt with that institution.

Guy Halsall, *Barbarian Migrations and the Roman West 376–568* (Cambridge: Cambridge University Press, 2007). The rise of various kingdoms in the Roman West, Vandals in Africa, Visigoths in Gaul and Spain, Burgundi in Gaul and Clovis, and the triumph of the Merovingians all were made possible by a "loss of a monopoly over political legitimacy" and argues cogently that the "fall of the Roman Empire produced the barbarian migrations, not vice versa."

Victor Hanson (ed.), *Makers of Ancient Strategy: From the Persian Wars to the Fall of Rome* (Princeton, NJ: Princeton University Press, 2010). A stimulating collection of essays on a variety of subjects, as the title implies.

William Harris, *War and Imperialism in Republican Rome 327–70 BC* (Oxford, UK: Oxford University Press, 1979). Concludes as it begins, stating, "For a war against some enemy or other, with some 'justification' or other, the Romans expected and intended about every year." They liked war, not just for glory and territory and economic advantages, but also, as Polybius notes, power for its own sake. Harris declares that, for the Romans, war was almost a "biological necessity."

B. H. Liddell Hart, *Scipio Africanus: Greater Than Napoleon* (Boston: Little, Brown, 1928). Maybe that is stretching it a bit, but Scipio was the right general at the right place before and at the time the Battle of Zama was fought. His magnanimity toward Hannibal after the battle was not matched by the Roman senate. Nor was their treatment of him much better. I liked Hart's summation: "Scipio's claim to eternal fame is that he was the staff, not the whip of Rome and of the world."

Richard Haywood, *The Myth of Rome's Fall* (New York: Thomas Cromwell, 1958). Reiterates the fact that only the Roman Empire in the West was overrun and that the loss of North Africa to the Vandals was the "sorest single blow which the Empire suffered during all this troubled time" but points out that North Africa was recaptured by Justinian in 527 and ruled by Constantinople for 100 years until overrun by the forces of Islam.

Peter Heather, *The Fall of the Roman Empire: A New History of Rome and the Barbarians* (New York: Oxford University Press, 2005). Quite an engaging revision, with lots of emphasis on the various "barbarian" groups and their interplay rather than on what the Romans did or didn't do at any given juncture. A worthwhile read.

Tom Holland, *Rubicon: The Triumph and Tragedy of the Roman Republic* (Boston: Little, Brown, 2003). In 49 BCE, the 13th Legion crossed the Rubicon after Caesar took 800 cities and subdued 300 tribes in Gaul. This would lead to a great tragedy in the mind of the author: "A thousand years of civic self-government were [*sic*] brought to an end." Except for the chapter "World War," there isn't much military material here, although Pompey's premature flight from Rome confirms that "all the big mistakes are made early."

Michael Howard, *The Fall of the Roman Empire: A Reappraisal* (Radnor, PA: Annenberg School Press, 1976). This very erudite and useful volume is in honor of Gibbon's earlier work. Focuses on 13 defects, including "Failure of the armies to perform," "Failure to grant citizenship to the Visigoths and other Germans," and the "Christian other world approach in this world."

Benjamin Isaac, *The Limits of Empire: The Roman Empire in the East* (New York: Oxford University Press, 1990). Why the Romans never truly conquered the Persians. Some interesting analysis on why and how Persia was able to prevail.

Simon James, *Rome and the Sword: How Warriors and Weapons Shaped Roman History* (London: Thames and Hudson, 2011). The martial aspects of Roman history as well as their various wars are covered around the symbol and reality of the Roman sword. Good maps on pages 160 and 161 and pictures of the evolution of the sword on page 31. Highlights the Roman's integrative and absorptive strengths. "The Roman world that Jesus knew was largely created, maintained and ultimately destroyed by the sword." A fine read.

Josephus, *The Jewish War* (75 AD) (Cambridge, MA: Harvard University Press, 1997). Sustained ruthlessness in a titanic struggle as described by an eyewitness. See also Harry Sidebottom's novel about siege warfare in the east against the Persians with the Roman garrison under attack: *Fire in the East* (New York: Overlook Press, 2009). Romanize German Marcus Claudius Ballista goes to Arete on the Euphrates and holds off the Sassanid Persians—at least for a time. The description of siege and countersiege warfare is quite good.

J. F. Lazenby, *Hannibal's War: A Military History of the Second Punic War* (Norman: University of Oklahoma Press, 1998). The first scholarly work devoted solely to this war since the 19th century, it relies on Spanish, Italian, German, and British sources and scholarship and provides a fine account. Unfortunately, the photographs and maps are most prosaic, in black and white and often blurry. He gives Hannibal high marks for consistently excellent generalship but also credits the Roman refusal to accept defeat as "their finest hour." Lazenby also states that this war did not end the republic but rather simply sped up and intensified the movement of the Roman political system toward empire and its ultimate conquest of 53 different peoples.

J. E. Lendon, *Soldiers and Ghosts: A History of Battle in Classical Antiquity* (New Haven, CT: Yale University Press, 2005). Highlights the importance of tradition in warfare during the 500 BC–400 AD period, declaring that there were only "subtle shifts of emphasis" and few great technological changes. He asserts, for example, that there were more technological changes in the 1910–1940 period than in the entire 800-year span he covers. Echoing Messer (see later this chapter), the author also believes that Roman discipline has been overpraised as a cause of Rome's success, saying rather it was links that bound individual soldiers to the traditions of society and the military that carried the day. In the end, he feels, it was the "commanders, not the soldiers who had lost the legacy of Roman virtues" by the time the empire fell.

Neil Litten, *The Defeat of Rome in the East* (Havertown, PA: Casemate, 2008). Persia was not to be trifled with, as many emperors, including Julian, found out!

Livy, *The History of Rome: Books 1–5*, translated by Valerie M. Warrior (Indianapolis: Hackett, 2006). Livy, originally from Cisalpine Gaul, lived from 64 or 59 BCE to 17 CE, wrote a 28-book history of Rome. This compendium concentrates on the foundation of Rome and its early history and gives a good description of the beginnings of Roman military prowess. Warrior points out that Mars was originally an italic god of vegetation and associated with war but eventually assimilated the Greek war god Ares and became the father of Romulus and Remus. Given Rome's penchant for incessant warfare for the next centuries, this seems like a sensible parentage to claim.

Edward Luttwak, *The Grand Strategy of the Byzantine Empire* (Cambridge, MA: Harvard University Press, 2009). Superb, amazingly detailed account of how the eastern portion of the Roman Empire held out for so long. One of those books that truly amazes for its depth, range, breadth, and valuable original insights.

———, *Grand Strategy of the Roman Empire: From the First Century A.D. to the Third* (Baltimore, MD: Johns Hopkins University Press, 1979). Excellent overview that gives the reader a much-needed sense of time, distance, and scope of the empire, even in decline.

Simon MacDowall, *Adrianople AD 378: The Goths Crush Rome's Legions* (Oxford, UK: Osprey, 2002). A quick study with excellent illustrations giving readers the why and wherefore of the battle the Romans decried, "No battle in our history except Cannae was such a massacre." The Goths, pushed by the Huns, came into Roman territory seeking protection. A whole people on the move across the Danube into Dacia and Trace were then treated badly as subjects, revolted, and caused great harm. Alan and Greuthung cavalry proved decisive, flanking the legions and, as at Cannae, pushing the legionnaires into a smaller and smaller space for slaughter. Food shortages and several years of warfare drove the Goths back, but under Alaric they would revive and sack Rome in 410.

Susan Mattern, *Rome and the Enemy: Imperial Strategy in the Principate* (Berkeley: University of California Press, 1999). Although claiming that there was more bungling and luck to Rome's success than expert military strategy, the author nevertheless points out the Roman commitment to "honor," which "was maintained by conquest, terror and retaliation." Caesar's use of "fear, dignity and revenge" is thus an illustration of the importance of Roman "dignity."

Brian McGing, *Polybius' Histories* (London: Oxford University Press, 2010). This marvelous study in the Oxford Approaches to Classical Literature series provides the reader with a very good overview of Polybius's 40-volume study of the rise of Rome. Like Xenophon—but unlike Thucydides—he has no problem inserting himself into the events that altered and illuminated the times in the second-century-BCE period. Of considerable importance for our purposes is book 18, in which Polybius gives the reasons for Roman success over the Greeks, and in the process (the Battle of Cynoscephalae against the Macedonians had been fought and won in 167 BCE), he highlights the advantages of the Roman infantry sword formation over the Greek phalanx: The phalanx is only invincible in very specific circumstances where flexibility on the part of opponents is not possible, requiring as it does a "ground that is plain and naked and free from obstacles of every kind." He goes on,

> The Roman order, on the contrary is never attended even on such occasions with any disadvantage. Among the Romans every single soldier, when he is once armed and ready for service, is alike fitted to engage in any time or place, or on any appearance of the enemy, and preserves always the same power and the same capacity in action whether in separate companies or man to man. As the parts, therefore, in the Roman order of battle are so much better contrived for use than those in the other, so success in action must also be greater in the one than the other.

See also the additional review of Polybius himself in this chapter.

Ronald Mellor, *The Roman Historians* (London: Routledge, 1999). Livy, Tacitus, Sallust, and other Roman historians are explicated in useful fashion as the author sheds light on how and why the Romans wrote the history they did.

William S. Messer, "Mutiny in the Roman Army," *Classical Philology* 15, no. 2 (April 1920), pp. 158–75. Argues that "contrary to the usual convention, mutiny and insubordination were surprisingly prevalent in the Roman Army." This, he explains, was true regardless of the size of Roman armies, their ethnic composition, or the time frame studied. Roman success was based on the quality of the individual soldier as well as the fact that the "Romans were the only people of antiquity who seriously and studiously made the science of war their main business over a period covering centuries."

Richard Miles, *Carthage Must Be Destroyed: The Rise and Fall of an Ancient Mediterranean Civilization* (New York: Penguin, 2010). For 700 years before its 146 BCE destruction by Scipio Aemilianus, Carthage showed its Phoenician heritage (from Tyre) as its captains sailed from Britain to Gabon. It made colonies out of Sardinia, Sicily, and Ibiza and at the time of the First Punic War (264 BCE) had "undisputed command of the sea." Razed in real life and in memory claims the author. This is a superb piece of historical writing and enriches the reader.

Sam Moorhead and Davie Stuttard, *AD 410: The Year That Shook Rome* (Los Angeles: J. Paul Getty Museum, 2010). Lavishly illustrated account of the events leading up to Alaric's and the Visigoths' sack of Rome. Alaric offered generous terms that were refused prior to the sack, and he died afterward, following a failed attempt to sail to North Africa, from which Rome imported much grain and olive oil.

Adrian Murdock, *Rome's Greatest Defeat: Massacre in the Teutoberg Forest* (Gloucestershire, UK: Sutton, 2006). The savage defeat of the Romans by the Germanic tribes in 9 CE resulted in Rome setting its northern border as the Rhine, not beyond.

David Nicolle, *Rome's Enemies (5): The Desert Frontier* (Oxford, UK: Osprey, 1991). Often overlooked are Rome's adversaries in North Africa, Arabia, Palmyra, and Nubia, among others. This volume covers them.

Robert O'Connell, "Carthage's Road to War," *MHQ: The Quarterly Journal of Military History* (Summer 2001), pp. 14–23. Even though Carthage started the Second Punic War, the Romans bear a huge responsibility for its occurrence.

———, *The Ghosts of Cannae: Hannibal and the Darkest Hour of the Roman Republic* (New York: Random House, 2010). Until I read this book, I had not realized that the survivors of Cannae were disgraced and sent to Sicily, becoming "ghost soldiers," and that Scipio gave them a chance at redemption before and during the Battle of Zama and the final defeat of Hannibal. Roman losses at Cannae, estimated at 50,000 in 1 day, were among the highest ever recorded on any battlefield. The author also makes the important point that many generals have subsequently tried to imitate Hannibal in a "perfect battle of annihilation," but few have ever succeeded.

A. M. D. Parker, *The Roman Legions* (Oxford: Oxford University Press, 1922). A useful overview of the various periods of Roman military history—pre-Marian, Marian, Caesar and Pompey, Augustus, post-Augustus, and so on. Somewhat episodic to our more modern eye but, still, provides a great deal of valuable information.

Michael Pitassi, *The Navies of Rome* (Woodbridge, UK: Boydell Press, 2009). From its inception during the First Punic War (753–264 BCE) to the "renewal and decline" (285–476 CE), the Roman navies played an important, even vital, role in achieving and preserving Roman hegemony. Pitassi firmly declares, "To build her Empire, Rome had to be in control of the sea," and he notes that, when she lost that power, the "Western Roman State quickly ceased to exist." This is a traditional military history work with many useful side-boxes on such topics as "Sundials" and "Trade with India."

Plutarch, *The Lives of the Noble Grecians and Romans* (New York: Modern Library, 1985). Fascinating descriptions of all the Greeks and Romans you have heard about from a contemporary of many.

Polybius, *The Rise of the Roman Empire* (New York: Penguin Books, 1980). Near-contemporary account (40 volumes) of how the Roman Empire rose from humble beginnings by a Greek who was brought to Rome as a hostage in 167 BCE after the Romans defeated the Macedonians in the battle of Cynoscephalae. Cited over and over throughout history, so if he got anything wrong, then it got passed on. Polybius remains our best source for Roman warfare; he saw it firsthand and wrote convincingly about it.

John Rich and Graham Shipley, *War and Society in the Roman World* (London: Routledge, 1993). Rome increasingly became a slave-owning society whose continued prosperity required constant conquest and continuous integration of subject peoples into the empire.

W. L. Rogers, *Greek and Roman Naval Warfare: A Study of the Strategy, Tactics, and Ship Design from Salamis (480 BC) to Actium (31 AD)* (Annapolis, MD: United States Naval Institute, 1937). A splendid book with excellent maps and illustrations. While the Athenian prowess at sea is well known, the author boasts that, of the Roman navy (claiming, among other things that Agrippa was the equal of Themistocles), "[i]n the whole course of Roman expansion, from the First Punic War to Actium, her fleet was indispensable to Rome's conquests." Offers the interesting independent variable, for example, that, at the battles of Pharsalus, Philippi, and Actium (31 CE), the losing armies were goaded into hasty action and ultimately defeated by the efforts of the opposing navy. Also, he reminds that the Battle of Actium (south of Corcyra) began a 300-year period of Roman control and economic unity of the Mediterranean. A must-read for this topic.

Philip Sabin, "The Face of Roman Battle," *Journal of Roman Studies* 90 (2000), pp. 1–17. What it was like to serve in and command the Roman legions.

Stacy Schiff, *Cleopatra: A Life* (New York: Little, Brown, 2010). Schiff has provided a most-engaging analysis of the world of Rome seen from the vantage point of Egypt and its last empress. Of considerable interest is her description of the battle of Actium. Why did Antony and Cleopatra go to Greece and then throw away their initial advantages (19 legions and 500 ships) with inactivity? Why did Antony, a long-proven commander, choose to fight at sea when he could have fought on land, where he was by far the abler leader? Did Cleopatra and Antony mean to flee rather than fight? And what terrible dishonor to leave most of the legions behind—those he took with him refused his prizes when they reached Alexandria. This is a splendid book and portal into the age of the Roman Empire just as it was dawning.

Harry Sidebottom, *Warriors of Rome* series (New York: Penguin Books, various) (F). This series of novels by an Oxford don contains some exciting yarns embedded in a great wealth of historical detail (sometimes too overwhelming). A pleasant way to pick up useful information about the Roman military (tactics, campaigns, and war) in a variety of eras. As of 2012, there were *Fire in the East*, *King of Kings*, *Lion of the Sun*, *The Caspian Gates*, and *The Wolves of the North*.

Chester Starr, *The Roman Imperial Navy, 31 B.C.–A.D. 324* (Ithaca, NY: Cornell University Press, 1941). Starting with the great imperial fleets of Misenum and Ravenna as well as the independent squadrons, this work ends with the 324 CE disappearance of the Augustan navy, superseded by the eastern Roman Empire at Constantinople.

Barry Strauss, *The Spartacus War* (New York: Simon and Schuster, 2009). A sprightly and free-flowing account of the revolt of Spartacus the gladiator and his slave rebellion, which threatened Rome for two years, destroying nine Roman armies before being itself crushed by Marcus Licinius Crassus in 71 BCE. To make sure everybody got the message that slave revolts were not to be tolerated, Crassus had the 6,000 followers of Spartacus crucified on the way from Capua to Rome. The author ends by making the somewhat extravagant claim that the Spartacus rebellion made it more likely that Romans, faced with such disorder, would accept the end of the republic and the rise to emperorhood of Octavius Augustus.

Tacitus, *Complete Works of Tacitus* (New York: Modern Library, 1942). Tacitus, who lived from 56 to 117 CE, was a legion commander, and his *The Annals* and *The History* contain some of the only contemporary descriptions of Roman legions from that era. Unfortunately, from a military perspective, one has to wade through hundreds of pages of Roman leaders, mutinies, slave insurrections, family backgrounds, and a great deal of extraneous material for our purposes. His editor in this work, Moses Hadas, points out that, to the Romans of that period, a "historian" was also a "literary figure." His reference to Christians and "Christ" during the rule of Nero and Tiberius (*The Annals*, book 18, p. 44) is one of the first historical mentions from the period. Another is found in Josephus.

Richard Tada, "Trajan's Last Conquest," *Military History* (November 2007), pp. 44–49. Trajan conquered Dacia (northwest of Thrace) and then made it to the Persian Gulf in 116, defeating the Parthians along the way and moving a fleet from the Euphrates to the Tigris Rivers before heading back to North Africa and Cyprus in 117 to put down another Jewish revolt (although not one in Judea this time).

Flavius Vegetius, *On Roman Military Matters* (St. Petersburg, FL: Red and Black, 2008). First appearing in 420 CE, this work highlighted a recall to Roman greatness based on the old-fashioned discipline and training of the early empire's legions. There are many suggestions here, including teaching recruits to swim and how to deal with elephants (a variety, including letting them go through the first ranks or using slingers against their handlers), as well as extolling the virtues of entrenched camps constructed on a nightly basis. "Troops are not to be led to battle unless confident of success" seems relevant for all armies. Often studied in the Middle Ages by those who practiced or studied the art of war.

Virgil (Publius Vergilius Maro), *The Aeneid*, translated by Robert Fagles (New York: Penguin, 2006). Seeking to provide "Homeric luster" to the founding of Rome, Virgil took his Trojan hero from Trop to Thrace to Delos to Crete to Actium to Sicily to Carthage to Sicily and hence on to Italy. The gods still intervene with humans, and Aeneus is securely in with the heroes. Much Roman myth-making in this wonderful poem.

Graham Webster, *The Roman Imperial Army of the First and Second Centuries AD* (Totowa, NJ: A and C, 1998). A quick look at the Roman Empire in the centuries before its decline in the West. Considerable description of interest.

Peter Wells, *The Battle That Stopped Rome: Emperor Augustus, Arminius, and the Slaughter of the Legions in the Teutoburg Forest* (New York: W. W. Norton, 2003). In 9 CE, three Roman legions under Varius were savagely defeated and wiped out by German tribes led by Arminius, who had learned Roman tactics and weapons firsthand while serving with them. Very important battle, for although punitive Roman expeditions killed a number of Germans in later raids, by 17 CE, Tiberius discontinued the Roman campaigns along the lower Rhine, and the "Rhine became a highly fortified frontier of the Empire."

Peter Wilcox, *Rome's Enemies (1): Germanics and Dacians* (Oxford, UK: Osprey, 1982). Good maps showing migrations and various peoples who filtered into and subsequently invaded much of the Roman Empire, east as well as west.

———, *Rome's Enemies (2): Gallic and British Celts* (Oxford, UK: Osprey, 1985). Persistent enemies they were, and widespread as well as persistently on the move into what is now Spain, Britain, France, and Italy. Quite good on weapons.

Greg Woolf, *Rome: An Empire's Story* (London: Oxford University Press, 2012). A comprehensive one-volume analysis of the rise and fall of the Roman Empire and the various sociological, political, environmental, and economic aspects of it, including an effort to place it in the context of other empires. Depressingly weak, however, on the actual military sinews and historical dimensions of its warcraft. The chapter on "War" gets a mere 18 pages.

Chapter Thirteen

The Horsemen Cometh

The Centrality of Mounted Warcraft

In a 9,000-mile arc from England to Persia, Arabia to India, China and Japan to Korea, from 500 AD until 1500 AD, there was feudal warfare in the core with an accent on cavalry, most often heavy cavalry (following in the cataphract tradition). From the widespread use in the 9th century of the stirrup and high saddle, larger horses, and chain to plate armor (in Europe) in the 13th century, warfare changed its character considerably.

Both horse and rider were armored, and swords and lances were the primary weapons, with battle axes also in use. Very expensive warriors needed a great deal of training and very special war horses, which required fodder (not just grazing), special trainers, blacksmiths, and stables that were often better equipped than peasant huts.

Some writers make something out of an interesting feature, calling it a "bow line." Horsemen in Europe and West Africa generally didn't use bows much, while in Egypt, China, Byzantium, and India and in Muslim armies, they did. The closer to the core areas of the steppes, the higher the likelihood of the use of bows by mounted archers. In any case, out of the "sea of grass" in the Eurasian heartland would come wave upon wave of invaders, fanning out in all directions with enormous consequences for all concerned.

Susan Wise Bauer, *The History of the Medieval World: From the Conversion of Constantine to the First Crusade* (New York: W. W. Norton, 2010). This fine work is an excellent introduction to this period and on a worldwide basis. Parallel histories of Near East, Europe, Asia, and even a little Africa. Constant warfare depicted.

Brian Carey, Joshua Allfree, and John Cairns, *Warfare in the Medieval World* (London: Pen and Sword, 2006). Covers warfare from the fall of the Roman Empire to the military revolution of the 16th century. Very good on the rise of heavy-horse cavalry and its domination of centuries of warfare regardless of location until the coming of the lighter, more mobile horsemen from the steppes. Good short sketches of various battles, such as Dara, Tricameron, Manzikert, Tours, Liegnitz, Ain Jalut, and Agincourt, among others.

Antony Karasulas, *Mounted Archers of the Steppe 600 BC–AD 1300* (Oxford, UK: Osprey, 2010). A good introduction to the vast sweep of the horsemen and their impact on the surrounding peoples and modes of warfare.

V. Vuksic and Z. Grbasic, *Cavalry: The History of a Fighting Elite: 650 BC–AD 1914* (London: Cassell, 1993). A gorgeous, even glorious, work dedicated to the horsemen. From the Assyrians of the seventh century BCE to the Polish cavalry of World War I, many horse archers and cavalrymen are displayed lovingly and in Technicolor.

PART I: THE MOUNTED HORSEMEN OF THE STEPPES

Following the 1700 BCE expansion of the steppe people, with the Hyksos moving into Egypt, the Kassites into Mesopotamia, and the Hittites into Asia Minor and Syria, and the 1500 BCE movement of the Aryans, whose warfare revolved around the chariot, into India (through Afghanistan) by 700 BCE, the mounted bowmen followed them—such peoples as the Scythians, Avars, Goths, Magyars, Sarmatians, Heruli, Alans, Avars, Cumins, Huns, and Hsiung Nu (Northern Horse Barbarians)—into China. The Hiung Nu (Eastern Horse Barbarians into Manchuria), Ephthalites, Juan-Juan, and others, plus various Turkic peoples, such as the Bulgars, Uigaurs, and Khazars, also moved into Europe, Persia, and China in waves and set in motion thousands of years of conflict with the existing, more settled peoples.

The Eurasian steppes, known as the "Sea of Grass," were the center of the earth for the horse peoples. Antony Karasulas (reviewed in this chapter) gives us an excellent geoportrait of this more or less "continuous expanse of grassland." Bounded in the north by the thick forests of the taiga and in the south by mountains and deserts, the grass steppes run for 5,000 miles (averaging 600 miles in width), from western China to the plains of Hungary, and extend into such territories as eastern Anatolia. An amazing sweep of territory across two continents in the heartland of the world. Harry Sidebottom puts the following words into the mouth of the Roman eunuch Mastabates in his *The Wolves of the North* to sum up the views of the settled peoples who encountered those from the steppes: "The grasslands are a terrible place, inhabited by terrible people."

Of course, the "terrible people" did not consider themselves as such, and we should look at their success in warfare and indeed marvel at the longevity of their successes. Consider also what a long run they had, for it was not until 1771 that the Russians finally annexed the Khanate of the Crimea and not until 1912 that the Manchu Dynasty in China was finally overthrown. Considering that the steppe archers began their remarkable run in the 600 BCE era, they were a truly formidable set of warrior people with a long-running successful history.

Note: In terms of language, the steppe people fall into at least three distinct groups: the Turkic peoples (Huns, Seljuks, Cummins, Uigaurs, Kazakhs, Uzbeks, and Khazars), the Mongol peoples (Mongols, Naimans, Merkits, Tartars, and Avars), and the Indo/Aryan peoples (Scythians, Sarmatians, Alani, Toxolani, and Bulgars). The Magyars were a Finno-Ugric language people.

The horses of the steppes were, of course, the sine qua non for warfare. Each warrior needed three to four horses per man, and a warrior could cover 40 to 60 miles per day riding geldings and mares (stallions were much harder to tame and control in battle) with no horseshoes (a later European invention), using a recurved, composite bow of wood, sinew, and horn. Remember that stirrups were not seen until the Avars and the sixth century AD, making the accomplishments of the horsemen of the steppes in the earlier period all the more remarkable.

The history of the steppe people, for millennia then, has been the history of the Eurasian heartland.

David Anthony, *The Horse, the Wheel and Language: How Bronze-Age Riders from the Eurasian Steppes Shaped the Modern World* (Princeton, NJ: Princeton University Press, 2007). An illuminating set of hypotheses that are a nice break from normal Eurocentric accounts of "development."

Thomas J. Barfield, *The Perilous Frontier: Nomadic Empires and China* (Oxford: Oxford University Press, 1989). From the third-century-BCE Hsiung-nu Empire to the Manchu takeover, China has had warfare with the steppe people for 2,000 years. Barfield notes the "surprising durability of imperial organization in the Eurasian steppe." No wonder the Chinese don't want the Uigaurs and others in Xinjiang to get stirred up even today!

Louis A. DiMarco, *A History of the Military Horse and Rider* (Yardley, PA: Westholme, 2008). A long, long process, sometimes helped by saddles, pommels, stirrups, and other mechanical aspects we often take for granted.

Nick Field, *The Hun: Scourge of God AD 475–565* (Oxford, UK: Osprey, 2006). Well-researched and in-depth look at these quintessential horse archers. Good pictorial representations of them and excellent background on their equipment (high-backed and -fronted saddle but no stirrups) and battle tactics (rained down arrows before attacking and shooting directly). Probably originated as a people to the west of China's Great Wall as one of the feared "northern barbarians."

Rene Grousset, *The Empire of the Steppes* (New Brunswick, NJ: Rutgers University Press, 1970). Scholarly research on the rise presenting the Mongols by a noted French author and specialist—although, if truth be told, he is more interested in their culture than their means of conducting war.

Peter Heather, *Empires and Barbarians: The Fall of Rome and the Birth of Europe* (London: Oxford University Press, 2010). Reexamines the migrations and interactions of many peoples, including the Goths, Huns, Avars, Magyars, Slavs, many Germanic tribes, Vikings, and other peoples who came for "wealth collection" and to escape the wealth collection of others.

Erik Hildinger, *Warriors of the Steppe: A Military History of Central Asia, 500 B.C. to 1700 A.D.* (New York: Sarpedon, 1997). A very clear and readable introduction to a most fascinating subject. A good place to start making contact with the warriors of the Eurasian land mass and their preeminent place in history.

Antony Karasulas, *Mounted Archers of the Steppe 600 BC–AD 1300* (Oxford, UK: Osprey, 2002). A clear, concise account of the mounted horsemen, true pastoralists armed primarily with the bow, who came in successive waves out of the 5,000-mile-long "Sea of Grass" that averaged 600 miles in width and propelled peoples west (and east into China) for almost 2,000 years. Because they had no permanent homes, the author asserts, the journey was always what mattered. He also cautions that, because so much of the mounted archers' histories were written by the sedentary peoples on whom they preyed, we need to keep an open mind about what they actually did and why.

Ed Phillips, *The Royal Hordes: Nomad Peoples of the Steppe* (London: Thames and Hudson, 1965). Castles eventually were quite helpful in stemming the tide of the steppe people in Eastern Europe because they were difficult to breach without a lengthy siege, and many, if not most, steppe incursions were raids, not attempts at conquest.

PART II: EUROPEAN HEAVY HORSE: FEUDALISM AND KNIGHTHOOD, POWER DIFFUSION AND ITS CONSEQUENCES, AND LATER POWER CENTRALIZATION

While some scholars prefer to use the term *feudalism* principally for the phenomenon in Europe, other military scholars see a pattern of global feudalism, whereby the warcraft of Europe, the African savanna kingdoms, Byzantium, the Muslim world, and Rajput India, as well as Ming and Han China, which revolved around the use of the heavily armed horse nobility, tended to produce centrifugal forces that resulted in similar power diffusions all across the Eurasian core.

The heavy-horse nobility in all these places eventually developed a code or ideology to idealize their forms of warfare and culture: European chivalry, Muslim *futuwa*, Byzantine chivalry, Indian Rajputs, Japanese Bushido, and Chinese Hsia, among others. It is always important when studying warfare during this period to remember Maurice Keen's wise words: "Chivalry is a word that came to denote the code and culture of a martial estate which regarded war as its hereditary profession." The horse nobles were central to it and to the institutionalization of perpetual warfare.

For European feudal warfare, see the rise of the Franks and Normans, feudalism in medieval England and France, and the changes produced by the Battle of Hastings in 1066. The Crusades represent an interesting example of force projection of European medieval warfare and should be studied in the context of John French's characterization of Urban II's call in 1095 for the liberation of the Holy Land as "salvation for slaughter."

Some patterns of battle repeat themselves, as in Crecy (1346), Poitiers (1356), and Agincourt (1415). Also note the eventual end of the horsemen and the cannon as "tax collector" as the medieval period wound down. Interestingly, there are enduring "medieval aspects" even in the contemporary period. There are, for example, some quite amazing parallels in medieval warfare to the 2002 American campaign in Afghanistan in conjunction with the Northern Alliance, including the use of soldiers on horseback.

Christopher Allmand, *The Hundred Year War: England and France at War c. 1300–c. 1450* (Cambridge: Cambridge University Press, 1988). "Generally speaking, the Middle Ages accepted war with fatalism. It was part of a divine plan." Allmand sees the war in a variety of ways and especially through the optics of a wider civil war of royal centralization (Paris versus Aquitaine, Normandy, Brittany, and Burgundy) and the rise of national identity in France and England. He makes the worthwhile point that the great victories of the English at Crecy, Poitiers, and Agincourt were all fought as they were leaving France, not when they arrived to conquer. In terms of weapons, he sees the English having ratios of 7 or even 10 to 1 of archers over men of arms by the end of the period under review.

———, (ed.) *Society at War: The Experience of England and France during the Hundred Years War* (New York: Barnes and Noble, 1973). Many useful snippets from primary sources covering attitudes, armies, "enticements" and war, and the civilian population. Interesting on looting: "But the principle reamed the same: he who took a risk in war, whether by paying an army or by hazarding his person in battle, stood to gain part of the benefits which an army was expected to derive from war."

Dorsey Armstrong, "Weapons and Warfare" and "Revolts, Uprisings, and Wars," in her *The Medieval World* (Great Courses), lectures 34 and 35. She wisely focuses on the excesses of European warfare inflicted on the poor peasants of many lands.

Thomas Asbridge, *The Greatest Knight* (New York: HarperCollins, 2014). The life and times of William Marshal, advisor to and warrior for five English rulers, is illuminated, along with the process of war in the early 12th century. Based on a largely unknown 13th-century manuscript, the *History of William Marshal*, this work provides some useful insights: *chevauchées* were designed to deprive opponents of such resources as rents and to goad opponents to leave fortifications; "tournaments" were actually huge affairs running over miles and featuring captures and ransoms of horses and men, underscoring the notion that "[y]ou are what you have got and no more than that." Marshal also led the royalists to a huge victory at Lincoln in 1217, when they defeated the army of the invading French.

Bernard S. Bachrach, *Armies and Politics in the Early Medieval West* (Norfolk, UK: Variorum, 1993). Shines some important light on the interplay between the military and political strivings all across Europe and provides insights into the process by which they became intertwined.

Macomb Barber, *The Crusader States* (New Haven, CT: Yale University Press, 2012). Perhaps the definitive account of how these European feudal forces, energized in 1095 by the cry "Deus Le Volt," took their war-making efforts to the Holy Land and held out until 1191, establishing their presence in the kingdom of Jerusalem, the principality of Antioch, and the counties of Tripoli and Edessa. The marvel is not that they were eventually overcome but that the "Franks" lasted as long as they did in this European exercise of power projection.

Juliet Barker, *Conquest: The English Kingdom of France 1417–1450* (Cambridge, MA: Harvard University Press, 2012). If you are only going to read one book about the way medieval warfare was conducted in Europe, this is it. Sieges, *chevauchées*, dynastic underreach and overreach, alliances and treachery, crosses and double-crosses, the English and their alliance with the Burgundians (southwest of Paris) and the Armagnacs—it's all here. In the aftermath of Agincourt in 1415, Henry V gets Charles VI to make him his heir, but he dies before being able to consolidate the kingdom, although his young son Henry VI is finally crowned king of France in 1431 (the only person ever to hold the English and French thrones). Covering the last 40 years of the Hundred Years' War, this admirable volume captures the flavor of the constantly shifting political and military landscape, including the brief career of Jehanne d'Arc, the Pucelle. But the author maintains that the battle of Patay overshadows the lifting of the siege of Orleans in importance, and the loss of Paris in 1437 effectively dooms the English kingdom. At the end, the English were driven out of Normandy, ending a long chapter of history.

Note: This fascinating period is also covered in Philippa Gregory's *The Lady of the Rivers* (New York: Simon and Schuster, 2011) (F), a novel that views the events of the end of the English kingdom of France and those of the War of the Roses (*The Cousins War*) through the eyes of Jacquetta of Luxembourg, first wife of John of Lancaster, First Duke of Bedford, who held the English kingdom for Henry VI. Jacquetta's daughter, Elizabeth Woodville, by her second husband, Richard Woodville, First Earl Rivers, would eventually marry Edward IV in secret.

Thomas Basbridge, *The Crusades: The Authoritative History of the War for the Holy Land* (New York: HarperCollins, 2010). This work is both authoritative and illuminating. Does a pretty good job of showing the Muslim military side of things as well as the usual European aspects.

Matthew Bennett, *Agincourt 1415: Triumph against the Odds* (Oxford, UK: Osprey, 1991). In the Campaign series, this account is fairly straightforward, with good illustrations. Two different theories about English positioning—those of A. H. Burne and J. Bradbury—are presented. I tend to side with Bradbury's formation, but readers can choose for themselves. I wish that the extensive showings of English and French heraldry could have been in color, however.

——— (ed.), *The Medieval World at War* (London: Thames and Hudson, 2009). Lavishly illustrated, this fine work covers 1,000 years of history with excellent essays and maps. Integrates the European military activities and actions within the broader contexts of the Vikings, Magyars, Mongols, Mamluks, and Ottomans. Numerous key battles explored as well. An outstanding addition to our understanding of this period. Most illuminating is James Howard-Johnson's "Byzantium against Persia and Islam" for its emphasis on the former's "ultra-elastic form of defense." This is a gorgeous book, well presented, and a wonderful introduction to medieval warfare across the globe (with the exception of Africa and the kingdoms of Ghana, Mali, and Songhay).

Constance Brittain Bouchard, *Strong of Body, Brave and Noble: Chivalry and Society in Medieval France* (Ithaca, NY: Cornell University Press, 1998). Although there is precious little in this work about actual fighting, it does provide some interesting analysis of chivalric codes, always cautioning the reader not to retrofit this High Middle Ages phenomenon with the earlier 11th and 12th centuries and to avoid too many generalizations and simplifications.

Jim Bradbury, *The Medieval Siege* (Woodbridge, UK: Boydell Press, 1992). But once their opponents got behind stone walls, the mounted horsemen needed a long time to destroy them through starvation and disease as much as by armed conflict.

Alfred Burne, *The Crecy War* (London: Oxford University Press, 1955). A good look at both Crecy and Poitiers: careful, descriptive, providing simple but very readable battle maps, and a good examination of relevant sources, both French and English. In describing what happened in a variety of battles, the author concludes correctly, "But in those days considerations of Chivalry were held to be of greater account than those of strategy." This explains a lot, especially on the part of the French, where command and control was almost always a problem.

Norman F. Canton, *Inventing the Middle Ages* (New York: William Morrow, 1991). A delightful investigation of how various individuals, groups, and schools retrofitted their views about what the Middle Ages were all about on the 11th to 15th centuries and how the "Middle Ages" were thus "conceptually created." Such interpretations as those by the "Nazi Twins," "French Jews," "Oxford Fantasists," and "American Pie" are all explained with insight and some humor. Author sees the major differences between the dark period of 600 to 1000 and the later one as being due to increased population, agricultural productivity, and more centralized political authority in a variety of places.

Brian Carey, *Warfare in the Medieval World* (Barnsley, UK: Pen and Sword, 2006). A marvelously detailed and excellently illustrated and diagramed work revolving around the primacy of heavy-horse cavalry during the High

Middle Ages (1000–1300) and dramatic changes to that supremacy from 1300 to 1500, whether from the light cavalry of the Mongols or the more balanced combined arms of Prince Maurice of Nassau, who urged "shot over shock" through the use of light infantry and linear formations, as well as the rise of the Swiss battle square and the return of heavy infantry. Best illustrated of all medieval battles. Puts heavy emphasis on three items: the stirrup and ends-raised saddle (which enabled the couching of the lance, the armor, and especially the 44-inch-long sword, which he calls the centerpiece). Points out that Charlemagne was involved in an "unprecedented" 54 campaigns during his 46-year reign.

Christine Carpenter, *The War of the Roses: Politics and the Constitution of England c. 1437–1509* (Cambridge: Cambridge University Press, 1997). The Lancastrians and Yorkish forces finance the sinews of war with more modern methods, claims the author, and the English kings of this period could not tap the country's resources more effectively than previous monarchs.

Roger Collins, *Charlemagne* (Toronto: University of Toronto Press, 1998). Charles the Great took eighth-century Francia and institutionalized the conquest of the Saxons (across the Elbe), the defeat of the Lombards (in Italy), the checking of the Avars (in the Danube Valley), and the Danes (North Sea), but the author focuses on the "decomposition" of his empire after his death, contrasting its decline with the survival of the Byzantine Empire until 1453.

Philippe Contamine, *War in the Middle Ages* (Oxford: Basil Blackwell, 1984). Interesting section on how the trebuchet and balance siege equipment (as well as the torsion-utilized catapults) were gradually replaced around 1500 AD by artillery due, in part, to the falling price of powder and the replacement of stones with cast-iron balls. Regarded as a classic.

Stephen Cooper, *Agincourt: Myth and Reality 1415–2015* (Barnsley, UK: Pen and Sword, 2014). Traveling though the "fog of uncertainty," the author examines many revisionist theories about the battle, finding the French courageous but foolish and not totally dominated during the three-hour battle (dented helmet/crown of Henry V) and seeing the battle as less a *chevauchée* than an invasion. He believes that there were 8,000 English and Welsh versus 12,000 French. The archer stakes were a first, and not much really changed strategically after the battle. He concludes, "No king of England ever led an army into that province again."

Bernard Cornwell, *1356* (New York: Harper, 2013) (F). While not his best novel by any means, this work contains some excellent descriptions of a medieval battle, in this case, Poitiers (1356), with a fine command of the various "battles" led by various commanders, the important role of loot and taking of prisoners (and not seeking either too early in the battle), and above all the continuing role of discipline even in medieval melees.

R. H. C. Davis, *The Medieval Warhorse: Origin, Development and Redevelopment* (Woodstock, NY: Thames and Hudson, 1989). A lot of breeding, a lot of training, and a lot of care and feeding were required. They needed to be big and strong and tall and aggressive for the medieval form of heavy-horse warfare, but by the middle of the 17th century, the demand went up for faster, lighter, more agile horses—simply a "different type of horse."

Hans Delbruck, *Medieval Warfare* (Lincoln: University of Nebraska Press, 1990). A classic, in-depth look at war during the medieval period. Book 3 in his four-volume look at warfare over the ages.

Christine de Pizan, *The Book of Deeds of Arms and of Chivalry* (reviewed in chapter 3).

Kelly Devries, *A Cumulative Bibliography of Medieval Military History and Technology* (Leiden, Netherlands: Brill, 2002). A good book for those who are really into the medieval period and its military.

John W. Eadie, "The Development of Roman Mailed Cavalry," *Journal of Roman Studies* 57, nos. 1–2 (1967), pp. 161–73. Surprisingly, the Romans did not make much use of the cataphract heavy-horse cavalry until very late in their hegemonic period.

Sarah Foot, *Aethelstan: The First King of England* (New Haven, CT: Yale University Press, 2011). Argues that the "Unknown King" was the first English king to rule all the Anglo-Saxon kingdoms, and his 15-year reign from 924 to 939 "constitutes one of the most significant periods in the history of England before the Norman conquest. The grandson of Alfred the Great, king of Wessex, defeated the Scots, Picts, Northmen, Dublin Norse, Northumbrians, and East Anglians. Unfortunately, not much on his warcraft or even his means of making war.

Jean Froissart, "The Battle of Crecy, 26 August 1346," in *The Book of War*, ed. John Keegan (New York: Penguin Books, 2000), pp. 46–52. How the English defeated the French in this pivotal battle and how little the French learned from their defeat.

George Goodwin, *Fatal Colours: The Battle of Towton 1461* (London: Weidenfeld and Nicolson, 2011). Pushing the end of the medieval period, this work captures the most savage battle ever fought on English soil, where 75,000 (Lancastrian and Yorkish) fought with a "complete lack of mercy" and upward of 25,000 died out of a population of 3 million. Illuminates the violence of civil war, as well.

Christopher Gravett, *Norman Knight 950–1204* (Oxford, UK: Osprey, 1993). An excellent, well-illustrated compendium of Norman weapons and fighting skills. Amazing how successful the small (25–30) *conrois* unit was and how important the sword was to the Normans over such a long period of time. Endless practice and fighting gave them an advantage over most foes.

Philippa Gregory, *The White Queen* (New York: Simon and Schuster, 2009) (F). The dynastic struggles that were such an intrinsic part of the Hundred Years' War are seen through the eyes and actions of the rulers and would-be rulers.

Tom Holland, *The Forge of Christendom: The End of Days and the Epic Rise of the West* (New York: Anchor Books, 2010). Although I did not buy his basic thesis that the failure of Christ to return in 1000 stimulated the rise of the West and the edit in 1072 at Canossa by Pope Gregory VII declaring the "heavens are the Lord's heavens—but the earth He has given to the sons of man" led to modernity, I found this a rollicking, fast-moving, and interesting read. It is especially good at underlining the incessant warfare that was central to the rise and fall of various peoples: the Saxons, Wends, Capetians, Franks, Normans, Wessex, and of course the "Saracens." A nice historical overview of war in the Middle Ages.

David Howarth, *1066: The Year of the Conquest* (London: Penguin Books, 1981). Pages 166 to 188 contain the best description of the actual battle (and why William ended up there) that I have read. Good map showing the Norman, French, and Breton positions (7,000) versus the Saxons (8,000) on their hill. In this "battle of attrition," both sides called on God ("God's help" for the Normans, "God Almighty" for the Saxons). The author notes that the Saxon army was not differentiated (a "throng"), while the Norman was divided into archers, infantry, and cavalry. Remarks that after William's victory there was a "baffling silence" across England. Well worth reading.

William Chester Jordan, *Europe in the High Middle Ages* (New York: Viking, 2003). Looking at the period 1000–1300, the author examines Europe from a variety of vantage points, including religions, culture, politics, and agrarian society. Precious little on the military. More's the pity because Jordan might have found the changes in military activities and power centers illuminating.

Maurice Keen, *Chivalry* (New Haven, CT: Yale University Press, 1984). How this code of conduct was supposed to work, how it actually worked, how infrequently its precepts held sway when confronted with the reality of the battlefield, and yet how the idea of it permeated so extensively the lives of various courts.

——— (ed.), *Medieval Warfare* (Oxford: Oxford University Press, 1999). A variety of essays on various aspects of medieval warfare. Particularly useful are the ones on fortifications and sieges, arms, armor and horses, and mercenaries.

H. W. Koch, *Medieval Warfare* (New York: Barnes and Noble, 1995). Large, lavishly illustrated with many color plates, and written in a lively fashion. Good on a number of specialized topics, especially "The Condottieri and Mercenary Corps," "The Hussites," and "Cities in Warfare."

Sean McGlynn, *By Sword and Fire: Cruelty and Atrocity in Medieval Warfare* (London: Weidenfeld and Nicolson, 2008). Surely Chaucer's 14th-century gentle knight is an ironic inversion. No "gentled knight" would have been worth much on a medieval battlefield, says the author as he focuses on the "breathtaking brutality" of war in the medieval period, when people were boiled alive, burned (taking 45 minutes), drawn and quartered, impaled, and generally treated poorly. Massacres of civilians and prisoners were widespread. In Accra in 1191, Richard the Lion-Hearted, for example, beheaded 2,500 prisoner Muslims, and Charlemagne killed 4,500 hostage Saxons in a single day at Verdun in 782.

Rosamond McKitterick, *Charlemagne: The Formation of a European Identity* (Cambridge: Cambridge University Press, 2008). Charlemagne ruled a "Greater Frankish Kingdom," including Saxony, Bavaria, and Lombardi from 768–814. He was a towering figure and spent most of his rule on military campaigns. Regrettably, the author spends virtually no time on his military instrumentalities of statecraft, let alone on the military campaigns themselves. Quite a pity because the rest of the work shows admirable depth and insight. Charlemagne traveled thousands of miles every year, bringing his person and his government to the various parts of his empire.

Ken Mondschein, *The Knightly Art of Battle* (Los Angeles: Getty, 2011). A wondrous reproduction of the early-15th-century illustrated fencing book by Fiore dei Libori. Superb illustrations detailing the "how to" of wrestling; self-defense; swordplay; and hafted weapons, including spears and pole axes. Shows how the male upper classes of the Middle Ages "defined themselves by the profession of arms." Should be required reading and viewing for anyone interested in the warfare of the Middle Ages.

R. I. Moore, *The Formation of a Persecuting Society: Power and Deviance in Western Europe 950–1250* (Oxford, UK: Basil Blackwell, 1987). The devil was at work trying to subvert the Christian order of Western Europe and bring chaos, so the "others" had to be marginalized and persecuted, including heretics, Jews, witches, lepers, and homosexuals, in order to save the "community of the faithful." Ugly business, that.

Tim Newark, *The Barbarians: Warriors and Wars of the Dark Ages* (Poole, UK: Blandford Press, 1985). Demonized by the "civilized world," the Huns, Germans, Visigoths, Vandals, Ostrogoths, Franks, Moors, Arabs, Danes, Swedes, Avars, Magyars, and Mongols threatened the European core for 1,000 years from the 4th century to the 13th. They had a great deal more influence over contemporary and later warfare than often noted.

Richard Newhall, *The English Conquest of Normandy, 1416–1424: A Study in Fifteenth Century Warfare* (New Haven, CT: Yale University Press, 1929). Puts Crecy, Poitiers, and even Agincourt in much-needed perspective, showing how Henry V's linear use of time paid dividends as English seized a line of fortifications and reduced the individual cities at leisure. Largest armies: 15,000. Very good account of military organization and "provision and material."

David Nicolle, *The Age of Charlemagne* (Oxford, UK: Osprey, 1984). An incisive look at the military of what would become the largest kingdom in Europe, from which feudalism emerged. For readers not familiar with the stirrup controversy, the author spends considerable time fleshing out his case that the early Carolingian horsemen did *not* use stirrups, although the contemporary Avar, Lombard, Persians, and Turks seem to have been using metal stirrups. Good map of campaigns, showing Charlemagne almost constantly at war. "Spears were the most common and cheapest weapons for all warriors," but the cost of chain mail and other armor was already making heavy-horse cavalry an elite in warfare.

———, *European Medieval Tactics*, 2 vols. (Oxford, UK: Osprey, 2011). Lavishly illustrated and a huge, sprawling, but never overwhelming look at the European "fall and rise of cavalry" and "new infantry, new weapons." An excellent overview of military change across time and space.

———, *Hungary and the Fall of Eastern Europe 1000–1568* (Oxford, UK: Osprey, 2006). The Magyars came from the steppes onto the plains of Hungary and adopted European-style feudalism over time, becoming Christian in 1000. By the 16th century, they were defeated by the Ottoman Turks (Battle of Mohacs in 1526) and then partitioned between the Ottomans and the Habsburgs in 1568.

A. V. B. Norman, *The Medieval Soldier* (New York: Thomas Crowell, 1971). The origins of feudalism and chivalry are covered in this work. It is especially good on the contributions of the Lombards, Franks, Vikings, and Saxons to war in medieval Europe.

Charles Oman, *History of the Art of War in the Middle Ages*, 2 vols. (New York: Stackpole Books, 1999). The art of war across a vast sweep of time, with lots of detail on mercenaries, gunpowder, the Crusades, and famous battles. Ends on the note that the Middle Ages end in the 15th century when "[w]ar was no longer an occupation in which feudal chivalry found its glory and the rest of society its ruin." Very useful sections on the Byzantine Empire and its battles.

Sharon Kay Penman, *Lionheart* (New York: G. P. Putnam, 2011) (F). A well-written novel about the Third Crusade and the convoluted interactions between the Saracens and the Franks and within the Frankish world. Quite an engaging read.

———, *When Christ and His Saints Slept* (New York: Ballantine Books, 1995) (F). Excellent novel giving a sense of war in the Middle Ages and how humans carried it out on so many occasions that it is possible to lose track of some of them.

Edouard Perroy, *The Hundred Years War* (Oxford: Oxford University Press, 1951). Long a standard, still a classic, this study by a French historian sees the war as "one of the many solvents of the medieval political order" and blames France's "recurrent disorder" and the centrifugal forces unleashed by the lords of Anjou, Armagnac, Burgundy, and Aquitaine as enabling the war to go on through four generations. Yet, the author maintains, out of the war, the modern nations of England and France were born.

Jonathan Phillips, *Holy Warriors: A Modern History of the Crusades* (New York: Random House, 2009). A very useful and well-organized overview of the Crusades and how they were fought on both sides.

Michael Prestwich, *Armies and Warfare in the Middle Ages: The English Experience* (New Haven, CT: Yale University Press, 1996). Well illustrated and wide ranging, this volume explores the differences (absence of regiments, no career paths for officers, and no standing armies) and similarities (efforts needed to dominate, the importance of logistics, and the centrality of preparation) between the wars of then and now. Asserts that there was a decline in the use of mercenaries after the Magna Carta was signed in 1215, but demand rose again during the Hundred Years' War.

Clifford J. Rogers, *War Cruel and Sharp: English Strategy under Edward III 1327–1360* (Woodbridge, UK: Boydell Press, 2000). Sees the nation-state slowly evolving out of the Hundred Years' War and, most importantly, sees strategy in Edward III's reducing "much of France to smoke and cinders" in a series of chevauchées, whereas many previous experts, such as Charles Oman and Basil Liddel Hart, saw only "purposeless parades." In one very interesting chapter, Edward III, seeking "to make an end to the war by battle," leads his chevauchée to Crecy after landing in Normandy; moving on Paris and refusing to be trapped beyond the Seine and then the Somme, in 1346 he finally makes the battle he always wanted. Then it was on to Calais and a siege that made the town English for a long, long time.

Nigel Saul, *For Honour and Fame: Chivalry in England, 1066–1500* (London: Bodley Head, 2011). Looking at a "world at once glamorous and violent, alluring and yet elusive," the author sees chivalry as arriving with the Normans (who he claims were far less savage than the Anglo-Saxons) in 1066. Coming with the French knight on horseback, the chevalier, chivalry became a "value system and behavior code of the secular aristocratic elite of the Middle Ages" and died out gradually as the knights became men of the king and eventually paid by the political center. Saul's chapter, "The Face of Chivalric War," is especially good.

Walter Carruthers Sellar and Robert Julian Yeatman, *1066 and All That* (London: Folio Society, 1930). A most amusing account of British history as seen through the eyes of the English.

Desmond Seward, *Henry V: The Scourge of God* (New York: Viking, 1988). Makes a good case that the reign of Henry V saw a "Norman conquest in reverse" of much of France, but despite—or because of—the success at Agincourt in 1415 and 30 years of chevauchée and land seizure, his reign set in motion a decline of English power in France. By 1450, that power had terminally eroded. The medieval period, he claims, was over.

Hugh D. Soar, *The Crooked Stick: A History of the Longbow* (Yardley, PA: Westholme, 2005). A paean to the weapon that won a lot of battles for the Welsh and the English but that was not adopted by very many other militaries because of the long time needed to train the archers.

Richard Suskind, *Cross and Crescent: The Story of the Crusades* (New York: Norton, 1967). Cleverly illustrated by Victor Lazzaro, this book is written at a deceptively basic level, for it contains, in a few short pages, a great deal of important information about the Crusades; medieval life; and the interaction between and among Europeans, Byzantines, and the "Saracens," as they were known to the Europeans at the time.

Brenda Rickman Vantrease, *The Illuminator* (New York: St. Martin's Press, 2005) (F). The medieval world seen through the eyes of a vulnerable, but not unresourceful, woman. Good glimpses of daily life that went on in the shadow of seemingly ceaseless warfare.

J. F. Verbruggen, "The Rise of the Foot Soldiers," in his *The Art of Warfare in Western Europe during the Middle Ages from the Eighth Century to 1340* (New York: North Holland, 1997), pp. 99–184. Note the *exceptional* battles where infantry overcame heavy-horse formations: Courtrai (Flemish) in 1302, Bannockburn (Scots) in 1314, and Morgarten (Swiss) in 1315. Very rarely did infantry best heavy-horse cavalry during this period.

E. Vinauer (ed.), *The Works of Sir Thomas Malory*, 3 vols. (Oxford: Oxford University Press, 1947). These "loved deeply and darkly understood" Arthurian romances were originally published as *Morte Dartur* in 1485. Malory retrofitted existing French chivalric tales to the swamps of Wessex and affixed them anachronistically to society of the age. This was somewhat ironic, for even as Malory was writing to celebrate the wonders of knighthood, the medieval knight was being undercut by pikes and "fiery weapons." This three-volume set is not for the faint of heart.

Chris Wickham, *Framing the Early Middle Ages: Europe and the Mediterranean 400–800* (Oxford: Oxford University Press, 2006). Far ranging and encyclopedic, seeking a broad synthesis by looking at different regions—from Denmark and Ireland to Egypt—and coming up with the key socioeconomic-political features. Precious little on military matters, let alone battles and technology. A disappointment in that regard.

———, *The Inheritance of Rome: Illuminating the Dark Ages 400–1000* (New York: Viking, 2009). Wickham makes the important point that there was a gradual transition from Rome to the "Dark Ages" and that many structures, political and legal, stayed in place. A major difference between the Arab and Byzantine worlds and Europe during this period was the lack of a comprehensive tax system in the latter and a state military paid for by taxes in the former. A book of astonishing depth and breadth and one that challenges many of the assumptions others have made about what happened after Rome fell. For example, he stresses the previously undercredited importance of the Vandal invasion of the North African granaries in 439 and the subsequent breaking of the Carthage-Rome tax spine.

Michael Wood, *In Search of the Dark Ages* (New York: Facts on File, 1987). By focusing on such historical figures as Boudicca, Alfred the Great, Eric Bloodax, Ethelred the Unready, and William the Conqueror, the author gives good insights into medieval warfare and sheds considerable light on important battles. I liked his chapter on Eric Bloodax, the son of the Norwegian king Harold Fine Hair, who was invited to become king of Northumbria (centered in York) and who, after his death, a saga was composed that ended with him being greeted in Valhalla by Odin himself: "Hail to you Eric, be welcome here and come into the hall, gallant king." Actually, his brief reign signaled the beginning of the end of the success of the Northmen in England, unless, of course, one remembers William the Conqueror, who was but a transplanted Northman himself.

Nicholas Wright, *Knights and Peasants: The Hundred Year's War in the French Countryside* (Woodbridge, UK: Boydell, 1998). Peasants were targets and suffered greatly, but they also resisted and revolted enough to make history less clear cut. Wright does conclude, "Medieval non-combatants suffered greatly."

Note: Beginning with the Viking sack of the monastery of Lindisfarne off the northeast coast of England in 793 through the Norman conquest of England in 1066, the Scandinavian seafarers and their descendants often substituted shallow draft ships and infantry for horse-borne warriors. There are many useful works on the military, political, and economic impact of the Vikings on medieval Europe, including the Volga basin and the Mediterranean. See chapter 14.

PART III: BYZANTIUM

This amazing story needs to be more widely known. The "Byzantine" or "Eastern Roman Empire" (it called itself neither) lasted 800 years longer than the Western Roman Empire and did so against a series of challenges as great as, if not greater than, those faced by Rome. How it was able to survive is an important tale in its own right but offers many contemporary lessons on survival in a hostile international environment.

Anthony Brogna, *The Generalship of Belisarius* (Ft. Belvoir, VA: Defense Technical Information Center, 1995). A master's thesis at the US Army

Command and General Staff College. Taken primarily from Procopius's *History of the Wars*, this work gives a good introduction to why Belisarius may have been the outstanding general of the sixth century CE and well beyond, winning battles (such as Ad Decium and Daras) on three continents (against a variety of foes, such as the Vandals and the Goths) and retiring undefeated. Said to have a magnificent coup d'oeil like Napoleon.

Lars Brownworth, *Lost to the West: The Forgotten Byzantine Empire That Rescued Western Civilization* (New York: Three Rivers Press, 2009). Who would have thought 1,000 years of history could be told so engagingly and illuminatingly? A fine read, with a vast array of colorful characters, including the iconoclasts; Irene, the first female emperor of the Roman Empire to rule in her own right; Theodosius II; Maxentius; Diocletian; Basis II (whose empire in 1025 was grand again); and Alexius I Comnenus. The description of Constantine laying out the city that would bear his name for so many centuries is itself worth the price of admission, as is the reaction to the sacking of the city by the Crusaders: "Let the powers of Islam come, they thought. Better to be ruled by an infidel than those heretics who made a mockery of Christ." Overall, a great read. The appendix includes a time line of all the emperors of Constantinople.

J. B. Bury, *History of the Later Roman Empire: From the Death of Theodosius I to the Death of Justinian* (New York: Dover, 1958). The splitting of the Roman Empire and how the longer-lasting part settled in the east and how Christianity came to play an important role irrespective of its, and the empire's, bifurcation.

Brian Carey, *Road to Manzikert: Byzantine and Islamic Warfare 527–1071* (Barnsley, UK: Pen and Sword, 2012). Like his previous books on warfare in the ancient and medieval periods, Carey analyzes battles and war-making with excellent maps and clear descriptions. From the battle of Dara (against the Sassanian Persian Empire) in 530 through the rise of Islam and its territorial expansion to the arrival of the Seljuk Turks and their success in 1071), this work illuminates and entertains. A fine read.

Roger Crowley, *1453: The Holy War for Constantinople and the Clash of Island and the West* (New York: Hyperion, 2005). The Greek-speaking inhabitants of Constantinople (whom we have come to know as the Byzantines, a word first used by Europeans in 1853, 400 years after the fall of Constantinople) are overcome by massive cannons, a substantial fleet, and the determination of Mehmet, but we need to remember that the city fell a full 800 years after Rome was sacked. With the capture of Constantinople, the Ottoman Empire had become a "superpower of the age" and would remain so for more than 200 years.

Timothy Dawson, *Byzantine Cavalryman c. 900–1200* (Oxford, UK: Osprey, 2009). A most useful account of Byzantine cavalry over time, with special focus on the tactics and strategy of their use and especially the interplay among their skirmishers, mounted bowmen and heavy-horse lancers. Good map on page 15 showing the extent of the Eastern Roman Empire in 1025 before that "Terrible Day" of Manzikert in 1071.

Lawrence Fauber, *Narses Hammer of the Goths: The Life and Times of Narses the Eunuch* (New York: St. Martin's Press, 1990). An amazing account of a truly amazing war leader who in his 70s and 80s reconquered Italy for Byzantium, ruling that province from 554 to 567. Winner of the battles of Taginae, Mons Lactarius, and Casilinum (Capua); vanquisher of the Ostrogoths and Alamanni Franks, and subduer of the Lombards.

John Haldon, *The Byzantine Wars: Battles and Campaigns of the Byzantine Era* (Charleston, SC: Tempus, 2001). Should be studied by all those seeking to understand how this "island" state lasted as long as it did. Valuable descriptions of battles won and battles lost. Don't be put off by the battle-after-battle progression. In 1,000 years, you are bound to lose some and win some, and Byzantium did both in great numbers, yet it always prided itself on avoiding war and buying off opponents as well. Its survival for so long in such a strategic location gets, and keeps, the author's attention.

———, *Byzantium at War AD 600–1453* (Oxford, UK: Osprey, 2002). A "World around War" as the ongoing, ceaseless effort to buy off, scare off, and defeat would-be predators went on apace. Good maps about contraction and expansion. "Byzantium was, in a sense, always at war. . . . [I]t always had an enemy or a potential enemy on one front or another."

Victor Hanson, "Byzantium at the Brink: The Fireman Flavius Belisarius—AD 527–59," in his *The Savior Generals* (London: Bloomsbury Press, 2013), pp. 49–95. The resurgence of Byzantium ("A Civilization in Crisis") under Belisarius, good on his counterinsurgency and general war tactics against the Vandals, Goths, and Persians. See more detail on his successful battles Dara, Ad Decimum, and Tricameron elsewhere, but this is a nice, easy-to-read introduction to the man and the times.

Jonathan Harris, *Byzantium and the Crusades* (London: Hambledon and London, 2002). See especially "The Search for Security." Shows how the empire, under pressure in Armenia, the Danube, southern Italy, and the Dalmatian coast, tried to cope. Ironically, it was not the defeat at the Battle of Manzikert in 1071 that did the damage but rather the dynastic instability that followed. Using Varangians and other western mercenaries at the end of the 11th century helped stabilize things, but the various Christian Crusades did not. Many of the Crusaders, for example, took umbrage at Emperor Alexios warning the Egyptians they were coming! Useful descriptions of various ways the Byzantines survived the Fourth Crusade and outlived the Outré Mer kingdoms in the Levant as well. Thus the Byzantine Empire was "one of the most successful and long-lived bodies politic, lasting for over a thousand years" (330–1453). Few polities can make such a claim.

Ian Heath, *Byzantine Armies 886–1118* (Oxford, UK: Osprey, 1979). Focusing on Byzantium during the 10th and 11th centuries' high point of power (up to the disaster at Manzikert in 1071), this work outlines the role of the elite Tagmata units stationed close by, the almost 50 military "Themes" or districts, the role of the Varangian Guard ("[a]xe bearing barbarians" from Scandinavia), and other mercenaries. Interesting depictions of swords, spears, and other weapons, albeit a bit weak on the cataphracts.

———, *Byzantine Armies AD 1118–1461* (Oxford, UK: Osprey, 1995). Follow-on to his earlier work, this one takes the story of Byzantine armies through the fall of Constantinople in 1453 and the overwhelming of the final remnant on the Black Sea, the "empire" of Trebizond in 1461 (after two failed attempts). This volume is clearly written, well-illustrated, and full of fascinating information. For example, it lists (and illustrates) many of the mercenary troops employed by the Byzantine Empire over its last 300-plus years: Alans, Albanians, Armenians, Bulgarians, Burgundians, Catalans, Cretans, Cumans, Englishmen, Georgians, Hungarians, "Latins" (Italians, Germans, Spanish, and Franks or Kelts), Mongols, Patzinaks, Russians, Scandinavians, Serbs, Turks, Uzes (a Turkish people), and Vlachs (Wallachians). The decline of the Byzantine navy turns out to be a major factor in the fall of Constantinople.

Erik Hildinger, "Belisarius' Bid for Rome," *Military History* (October 1999), pp. 31–37. This brief, oddly named (Belisarius took Rome in 536 and held it against two sieges by the ousted Ostrogoths) article provides very useful insights into the sixth-century Byzantine army, with its reliance on heavy cavalry that used both bows and lances. Shows Belisarius at his flexible, strategic best. After he was recalled by the ever-jealous Justinian, the Byzantines lost Rome but continued to hold enclaves in southern Italy, Sicily, Sardinia, Corsica, and southern Spain.

Benjamin Isaac, *The Limits of Empire: The Roman Army in the East* (Oxford, UK: Clarendon Press, 1990). Limits of time, space, and will more than ability, claims the author. And he is convincing.

Leo VI, *The Taktika of Leo VI*, text, translation, and commentary by George Dennis (Washington, DC: Dumbarton Oaks Research Library and Collection, 2010). An immensely valuable and comprehensive work summarizing the collected war wisdom of Byzantium (or, as Leo would more likely have called it, "of Rome") compiled for Leo VI. Called the "Tactical Constitutions," the volume brings together contemporary and previously written wisdom on leadership, baggage trains, armament and training, equipment for infantry and cavalry, siege engines, Saracen and naval warfare, and especially the most illuminating "The Day of Battle." Quite a summation and one that helps to explain the long-term success of the eastern branch of the Roman Empire.

Edward Luttwak, *The Grand Strategy of the Byzantine Empire* (Cambridge, MA: Harvard University Press, 2010). A classic, comprehensive, and grand undertaking. A must-read, well-written book, one that illuminates and stimulates at the same time. Many stimulating and challenging descriptions, analyses, and accounts.

Ramsay MacMullen, *Constantine* (New York: Dial Press, 1969). Third century AD, a time of troubles and trials and Roman emperors' focus on "restoring" and "reestablishing," as Roman "pride in war" had given way to "inadequacy in war." Rome is no longer the center of things; various outbreaks and migrations demand constant attention. The east is more settled and richer due to trade and a higher tax base. Constantine fights along the Rhine and in Spain, Gaul, and Italy before becoming the first among equal claimants and then "Maximus." Converts to Christianity (and picks up Mithraism and the sun god along the way) and thus becomes the first Christian to sit on the Roman throne. Defeats his rival in the east and sets up a new city on the Bosporus. He has a real talent for administration and proves it before dying in 337 AD.

Christopher Marshall, *Warfare in the Latin East* (Cambridge: Cambridge University Press, 1991). The amazing thing about the Crusader states seems to be how long they lasted and how much their survival depended on Muslim disunity. Some valid warnings and comfort for the West at present can be read into both aspects.

Maurice (Byzantine emperor), *Maurice's Strategikon: A Handbook of Byzantine Military Strategy*, translated by G. T. Dennis (Philadelphia: University of Pennsylvania Press, 1984). Provides incredible insights into warfare of the era and the relationship between Byzantium and their steppe nomads and other opponents. A very meaningful and clearly written book. Written between 575 and 628 after the Justinian revival, this work shows the transition away from infantry to armored cavalry armed with both bows and arrows and lances (stirrups now appearing). Comprehensive look at training; formations; and role of the *tagma*, or companies (300 men). Accent on cavalry two- or three-line attacks for concentration. How to handle plunder and use "Great hunts" for preparation. Enemies: Persians, Lombards, Avars, and Slavs at a time when the Byzantine Empire encompassed North Africa, the Balkans, Palestine, Turkey, and Armenia (contested with Persia).

Donald Nicol, *Constantinople 1453* (Oxford, UK: Osprey, 2005). Very good analysis of why the Ottoman Turks finally succeeded and why they succeeded just when they did. Fine maps and the whole work give the reader many worthwhile insights into the conduct of the battle from both sides.

———, *The Last Centuries of Byzantium 1261–1453* (Cambridge: Cambridge University Press, 1993). Does an excellent job at capturing what happened and why and provides many useful insights into the Byzantium thinking.

J. J. Norwich, *Byzantium: The Decline and Fall* (New York: Viking, 1995). Begins after the Battle of Manzikert (1071) with "The Rise of Alexius" and traces the pressures on the empire of the Normans, the Angevins, the Catalans, and the Ottomans, among many others. Interesting account of how Byzantium was saved by Tamerlane, who in 1402 defeated the Ottomans. The author points out that Byzantium gets no respect despite having survived (and often prospering) for 1,123 years (from 330 to 1453). Designating it "very successful" and "very civilized," he quotes Robert Byron's "Triple Fusion," for its Roman body, its Greek mind, and its "oriental mystical" soul.

David Potter, *Constantine: The Emperor* (London: Oxford University Press, 2013). A wordy effort, big on context and theology but woefully weak on the force of arms that got this chap his throne and the military might that would keep him there as he formed a presence in the area, a presence that would last 1,000 years, based on "God, the army, marriage and the bureaucracy."

Warren Treadgold, *A History of the Byzantine State and Society* (Stanford, CA: Stanford University Press, 1997). An expert on the Byzantine army, the author gives an overview to the long history and the sharp end of the empire. Not always an easy read.

Speros Vryonis Jr., *Byzantium and Europe* (New York: Harcourt, Brace and World, 1967). A once-over, light account. Of most use is the chapter on "The Victory of the Military," which suggests that, under pressure from the Seljuks, Patzinaks, and Normans, Byzantium turned more and more to the use of mercenaries, who "began to victimize the empire they had been hired to protect."

Mark Whittow, *The Making of Byzantium, 600–1025* (Berkeley: University of California Press, 1996). Solid study with useful insights into how it came to be and how it stood for so long.

PART IV: AFRICAN HEAVY HORSE: POWER CENTRALIZATION AND POWER DIFFUSION

Another often overlooked subject when the history of warfare is discussed is the phenomenon of African heavy horse. This section covers a number of important aspects of heavy-horse battlefield domination in Africa, including how the salt and gold trades and control of them led to the great war empires of the savanna. These empires, known to us as Ghana, Mali, and Songhai, all had a central component of heavy-horse cavalry as an intrinsic dimension of their war-making tool kit.

Later, the limits of heavy-horse feudalism were shown when they tried to penetrate the African rain-forest kingdoms and the Moroccans acquired firearms to use against the heavy horse arrays. The end of heavy-horse feudalism happened across a broad front and was caused not only by the Moroccan invasions but also by the rise of the Asante and other forest-kingdom peoples who acquired guns from the European transatlantic slave trade and were thus able to break out onto the savannas of northwest Africa and defeat their traditional opponents.

Molefi Asante, *The History of Africa: The Quest for Eternal Harmony* (New York: Routledge, 2007). Well maybe, but more like the quest for eternal warfare to any student of Mars.

Lester Brooks, *Great Civilizations of Ancient Africa* (New York: Four Winds Press, 1971). See especially his chapters on Ghana, Mali, and Songhai (pp. 111–85) for useful background on these empires founded on control of the salt and gold trades.

J. I. Dibua, "Sudanic Kingdoms of West Africa," in *Africa*, vol. 1: *African History before 1885*, edited by Toyin Falol (Cambridge: Cambridge University Press, 1980), pp. 137–60. A very useful analysis of the great kingdoms of Ghana, Mali, and Songhai and what they had in common on a space-time continuum that lasted for 1,000 years.

Nehemia Levtzion, *Ancient Ghana and Mali* (New York: Africana, 1980). The rise of the Sudanic kingdoms based on their control of the salt and gold trades. Armored horsemen dominate the military and protect the caravans.

Patricia and Frederick McKissack, *The Royal Kingdoms of Ghana, Mali, and Songhay: Life in Medieval Africa* (New York: Henry Holt, 1994). Compared to life in medieval Europe, there was a great deal more wealth and a much larger and far more centralized military establishment for these three kingdoms.

Richard Reid, *Warfare in African History* (Cambridge: Cambridge University Press, 2013). Takes a most useful overview of continent-wide warfare, not from discrete or even overlapping perspectives, but from such broad themes as "The Contours of Violence" and "Transformations in Violence."

S. P. Reyna, *Wars without End* (Hanover, NH: University Press of New England, 1990). Africa was never a peaceful place and still isn't. More than 4 million people have died in and around the Congo alone during the past 30 years. Sad tale, really.

Joseph Smalldone, *Warfare in the Sokoto Caliphate* (Cambridge: Cambridge University Press, 1977). The Hausa jihads of the 19th century show the ebb and flow of the horsemen prior to the European invasion.

John K. Thornton, *Warfare in Atlantic Africa 1500–1800* (London: UCL Press, 1999). Useful excursions into the modes of warfare in northwestern Africa. For most African ethnic units, there was a terrible Hobbesian choice when the Europeans and their own neighbors showed up wanting slaves for trade goods. Their choices were stark and morally horrible: join in the slave trade as sellers or join in the slave trade as victims.

PART V: ARAB LIGHT(ER) HORSE: FLOW AND EBB, EBB AND FLOW

Here we find the melding of religion as an ideology for conquest, the new horsemen eschewing the early forms of heavy horse, success and failure and the continuing ripples of battle spreading out from the Middle East. The Arab Muslims had military religious leaders, who in turn had soldiers who paid no taxes and were supported by the cities, which paid for their upkeep. As the

Berbers, Persians, and Turks (and later the slave Mamluk soldiers) became Muslim, they greatly expanded their empire until the Umayyad and Abbasid Caliphates ruled from the Atlantic Ocean to the borders of Tang China. Their principle mode of warfare was the horse archer. They used lighter, smaller, rounder shields than the Europeans and prided themselves on mobility, feigned retreats, and the use of speed to trap their opponents. Superb horsemen and dedicated fighters, they did have problems with the heavily armed, heavy horse of the Franks in many situations but on balance were masters of their domain elsewhere.

Note: Muhammad's flight from Mecca to Yathrib (later called Madinat an-Nabi, the City of the Prophet, according to Tom Howard), which took place in 622, begins the calendar of Islam. Thus, for Muslims, 622 became *Anno Heirae* (AH), the year of the *Hijra* AH.

Bruce Ware Allen, "The Siege That Made Suleiman Magnificent," *MHQ: The Quarterly Journal of Military History* (Autumn 2006), pp. 6–21. Very interesting examination of this fascinating campaign.

Karen Armstrong, *Muhammad* (New York: HarperCollins, 1992). Straightforward analysis of the Prophet, who not only founded a religion but also brought about a society and state based on that religion.

Thomas Basbridge, "The Response of Islam," in his *The Crusades: The Authoritative History of the War for the Holy Land* (New York: HarperCollins, 2010), pp. 225–366. Reaction to European penetrations.

Patricia Crone, *From Arabian Tribes to Islamic Empire* (Ashgate, UK: Variorum, 2008). Especially interesting is Crone's discussion of the non-Arab but Muslim *mawali* (clients), "renegades" who served in the army and were paid like regular Muslim army members, with annual stipends in cash and monthly allocations in kind, albeit at a lower rate during the first several centuries after the conquest. See especially "The Pay of Client Soldiers in the Umayyad Period."

Patricia Crone and Michael Cook, *Hagarism: The Making of the Islamic World* (London: Cambridge University Press, 1997). Well worth a long look into yet another dimension of Islam and how it, like all the world's major religions, splintered into sects.

Roger Crowley, *1453: The Holy War for Constantinople and the Clash of Islam and the West* (New York: Hyperion, 2005). Amazing how long Byzantium held out against the Muslim tide but how important for the history of the region that it fell. Even at the very end, it was a close-run thing.

Fred Donner, *The Early Islamic Conquests* (Princeton, NJ: Princeton University Press, 1981). A step-by-step analysis of the rise of the Muslim state by professional conquest. "Few events in human history have transformed the face of such a large part of the globe as rapidly and as decidedly as did the expansion of early Islam and the conquest by Muslims of much of the ancient world."

Hammond Historical Atlas of the World, H-15, "Islam and Christianity: 622–700 A.D." and "Islam and Christianity: 799–900 A.D."; H-24, "Europe in 1559"; and H-25, "Europe in 1648."

Albert Hourani, *A History of the Arab Peoples* (Cambridge: Harvard University Press, 1991). Extensively covers the social, economic, and religious impact of Islam and its fractures over the ages and suggests its initial expansionary success as being due in part to its location between the Byzantine and Sassanian (Iraq) Empires. Regrettably weak on the military aspects of that expansion.

Tom Howard, *In the Shadow of the Sword: The Birth of Islam and the Rise of the Global Arab Empire* (New York: Doubleday, 2012). A challenging and even exciting book filled with statements one remembers ("Only in the Qur'an had the awful purity of the divine revelation been properly preserved. Every last word of it, every last syllable, every last letter came directly from God and from God alone."). The author claims that, far from enjoying this awful clarity and directness, the first several hundred years of Islam are based on the "barest shreds of shreds" or the "delusory shimmering of mirages." He asserts that the first true biographies of Muhammad were created 200 years after his death and that the references to figures in Judaism and Christianity require us to accept that fact that Muhammad was influenced by contemporary religious knowledge. Mecca was then simply a backwater of a backwater. Surprisingly enough, this chap seems to have escaped a fatwa with his name on it, at least to date. Sees the "awesome sweep of their dominions" as being helped immeasurably by the existing declines of the Persians and Byzantines (Romans) and struggles between them *and* by the powerful war image of an almighty Allah, not one tainted by humanness, as in the teachings of Jesus. Islam plus the sword plus relentless light horsemen became a religious and military revolution to a vast area, including the Persian Empire. Turned back only at Tours and Byzantium, as this richly presented and deeply researched volume describes.

Efraim Karsh, *Islamic Imperialism* (New Haven, CT: Yale University Press, 2006). There is a long history here and one not necessarily politically correct but a real one nevertheless. A good source on the history of the region seen from the point of empire expansion and contraction.

Hugh Kennedy, *The Armies of the Caliphs: Military and Society in the Early Islamic State* (London: Routledge, 2001). Kennedy makes the point that Arab and later Muslim (Berber, Persian, Turkish) armies were not better led, nor were they better equipped than their adversaries, but they often had greater mobility and higher morale. Yes, their "spectacular" successes were "among the most dramatic and far-reaching in history." For three centuries after the death of the Prophet in 632, the author claims there were no significant military technological advances. Some interesting parallels can be found today in the "Long War" between the societies of the West and the radical Salafists.

———, *The Great Arab Conquests* (New York: Da Capo Press, 2007). A larger empire than that of Rome, carved out in half the time, all in the name of Allah and carried out swiftly by lightly armed men on fast horses, using small armies, never more than 20,000 men in size. Author takes the reader though the conquest of Syria, Palestine, Iraq, Egypt, Iran, the Maghreb, Transoxania, Tukharistan, and the Sind and gives credit to religious motivation, high-quality leadership, self-confidence, and the ease of submission allowed subject peoples.

Malcolm Cameron Lyons, *Saladin: The Politics of the Holy War* (Cambridge: Cambridge University Press, 1982). Contains a most useful map of "Indo-Arabia" to put Saladin's accomplishments in some perspective. He was quite an amazing leader, ruthless when he had to be, compassionate when he wanted to be, and politically astute much of the time. Quite ironic how, being a Kurd, he was chosen by Saddam Hussein as a hero; indeed, he claimed to be the "Saladin" of modern times.

David Nicolle, *The Armies of Islam 7–11th Centuries* (Oxford, UK: Osprey, 1982). Useful overview of Islam's conquest of Arabs, Persians, and Turks and the resulting fusion of military equipment and tactics.

———, *Armies of the Muslim Conquest* (Oxford, UK: Osprey, 1993). Provides a good chronology and offers some valuable insights. At the end of the Great Expansion and the Abbasid revolution by 800 CE, for example, Islam's territory was larger than Rome at its height and ran from northern Spain to the Indus River, from the Horn of Africa to the Caspian Sea. Light horsemen with bows, spears, and swords made the Muslim armies formidable and supple. The West was fortunate to have Muslim infighting to help stem this flood tide. Contains a very useful analysis of will in battle from Mu'tasism, the Abbasid Caliph, in the early ninth century:

> For the common soldier to confront an enemy champion at sword point is a hard and excellent act, but less so than the ignorant suppose . . . for if there was nothing to balance the horrors of going into battle, a man could always choose inaction instead of action. . . . Courage can be inspired by many things—anger, alcohol, stupidity, inexperience or youthful enthusiasm. It may be due to innate bloodthirstiness, jealousy, hatred of foreigners or ambition. It can result from hardheartedness or mercy, generosity or meanness, fear of punishment or a sense of resignation. It can also be produced by religion, though a soldier driven solely by religious feelings will not go into battle unless he is also inspired by one of these other motives.

———, *The Moors: The Islamic West 7th–15th Centuries AD* (Oxford, UK: Osprey, 2001). An overview of the forces of Islam that swept across North Africa and into Spain, ruling parts of it until 1492, when the Nasrid Emirate of Granada finally fell. Does a good job of depicting the 700 years of Muslim rule, especially the flowering of civilization at Cordova (mid-8th to 11th centuries). Interesting description of the frontier society that developed between Islam and Christianity in northern Spain.

———, *Yarmuk AD 636: The Muslim Conquest of Syria* (Oxford, UK: Osprey, 1994). The Arabs capture the Byzantine province of Syria with serious repercussions for the empire but growing success for the Arabs in this six-day battle.

Anthony Pagden, *Worlds at War: The 2500-Year Struggle between East and West* (New York: Random House, 2008). Islam's western thrusts from the 17th century onward mirror those of Persia before. East and West trade imperial desires and designs as they have for millennia—with no end in sight.

Russ Rodgers, *The Generalship of Muhammad: Battles and Campaigns of the Prophet of Allah* (Gainesville: University Press of Florida, 2012). An extremely valuable look at the rise of Islam through the generalship of Muhammad. A stunning story, well told. Asserts that Muhammad's declaration that only Allah was to be worshiped was truly revolutionary, asking individuals to give up idols *and* family/clan loyalties. He also points out how small the percentage of Muslims was, even in Medina in the beginning, and how small were the margins of victory at Badr (p. 623) and the Trench (p. 627) and of defeat at Uhud (p. 624). There is a fascinating chart on page 160 showing the growth in the use of horses in force projection (as opposed to camels) in the expansion of the Muslim world, from 2 at Badr to 10,000 at

Tabuk (p. 630). Muslim "light horse" became the rule and would rule in vast stretches of the Middle East and North Africa. Muhammad was to win three major battles and five sieges, losing one battle (Uhud) and failing at one siege (al-Ta'if). Rodgers also asserts, "Muhammad's campaigns came with a hefty price tag, much greater than many would assess today." But, says the author, "In this way, despite any weakness exhibited, Muhammad fulfilled the most critical dictum of warfare: he compelled his enemies to do his will." This is a most useful book for anyone seeking to understand the military rise of Islam.

Stephen Shoemaker, *The Death of a Prophet: The End of Muhammad's Life and the Beginning of Islam* (Philadelphia: University of Pennsylvania Press, 2012). Finds sourcing for the first century and indeed the origins of the Quran "profoundly ahistorical" and shot full of Muslim midrash. He does fix Muhammad firmly in the eschatological prophet tradition and points out that, until "Believers" became "Muslim," there was considerable coexistence among them, Jews and Christians and against polytheism. Would that it have lasted that way. A bit weak on the military aspects, so interested students need to consult Rodgers reviewed previously.

R. C. Smail, *Crusading Warfare 1097–1193* (Cambridge: Cambridge University Press, 1956). See especially "Muslim Armies," pp. 64–87, in which he suggests, "The Turks were quicker and more flexible in manoeuver than the Franks. This was ascribed to the pace and agility of their horses and the lightness of their weapons." They fired arrows from afar before using swords and lances up close, and they feigned retreats and were very coordinated in using their superior mobility.

Diane Stanley, *Saladin: Noble Prince of Islam* (New York: HarperCollins, 2002). A praiseful biography of the Kurd who was one of Islam's finest warriors. Perhaps a bit too praiseful in spots for the true historical record but nevertheless a good introduction to this important figure in Middle East history.

PART VI: JAPANESE AND CHINESE HORSEMEN

The Japanese experience is only somewhat contemporaneous (1200–1877) with the European medieval experience, but it had many similarities during the heyday of the samurai and way of Bushido. At the same time, the Tokugawa Peace (1600–1867) is without precedence in the West. The warhorse first came to Japan after their disastrous defeat in 400 AD at the hands of the Koreans. As William Farris puts it, "Equestrian skills, the birthright of every samurai, originated with the hated and feared Koreans." By the sixth century AD, the mounted, armored archer with a composite bow had become the central military figure in warfare and stayed so until the rise of the pike-armed peasant infantry in the 14th century.

The Chinese experience was different in that, while heavy cavalry played an important role from time to time and in place to place, infantry armed with crossbows were often the preferred military arm, and the heavy horse knight never dominated the way it did in the West, although by the fifth and sixth centuries CE, the Hans fielded heavily mailed cataphracts. Also, scholars indicate that Chinese armies rarely exceeded 50,000 men and that, when analyzing the Han Dynasty period, it is important to remember that the Han state did not cover the same land mass as "China" does today, lacking as it did in the period 300–900 AD Tibet, Xinjiang, Quinghai, most of Manchuria, and Yunan.

Thomas. J. Barfield, *The Perilous Frontier: Nomadic Empires and China* (Oxford: Oxford University Press, 1989). From the third-century-BCE Hsiung-nu Empire to the Manchu takeover, China has had warfare with the steppe people for 2,000 years.

Anthony Bryant, *Sekigahara 1600* (Oxford, UK: Osprey, 2005). A good overview of the feudal clash in which Tokugawa Ieyasu emerged victorious and united Japan in the footsteps of Toyotomi Hideyoshi. Good map on page 19 shows the medieval clan structure. The use of "fiery weapons" is still in its infancy.

Thomas Cleary, *Training the Samurai Mind: A Bushido Sourcebook* (Boston: Shambhala, 2008). The warrior mind-set that resisted guns and gunpowder and fought World War II with a very "thin-gruel" strategy.

Thomas Conlan, *Weapons and Fighting Techniques of the Samurai Warrior 1200–1877* (New York: Amber Books, 2008). Useful examination of these subjects by a former Bowdoin professor now at Princeton.

William Wayne Farris, *Heavenly Warriors: The Evolution of Japan's Military 500–1300 AD* (Cambridge: Harvard University Press, 1992). Samurais and shoguns, codes of honor, banditry, and resistance to military change.

Karl Friday, *Hired Swords: The Rise of Private Warrior Power in Early* Japan (Stanford, CA: Stanford University Press, 1992). Looks at the Heian period in the seventh to eighth centuries, which the author claims, is an important transition to the medieval Japanese of the Age of the Samurai.

David A. Graff, *Medieval Chinese Warfare 300–900 AD* (London: Routledge, 2002). Infantry with crossbows and cavalry came to grow in importance, eventually replacing the earlier domination by chariots, which were the force projection weapons of choice for centuries.

Cameron Hurst, *Armed Martial Arts of Japan: Swordsmanship and Archery* (New Haven, CT: Yale University Press, 1998). Practice, practice, practice. Why firearms were such a sociological threat as well as a military one to the ruling warrior class.

Charles Shiro Inouye, *Evanescence and Form: An Introduction to Japanese Culture* (New York: Palgrave Macmillan, 2008). A good book to get a flavor for differences and similarities between Japanese culture and others.

Harold Kleinschmidt (ed.), *Warfare in Japan* (London: Ashgate, 2007). Wide ranging and comprehensive, with 19 essays on such diverse topics as "Formation of Bushi Bands" to "Chinese Influences on Japanese Siege Warfare." Graduate level.

John Kuehn, *The Military History of Japan* (Santa Barbara, CA: Praeger, 2014). An engaging and useful overview of warfare in Japan from the Age of the Samurai to World War II.

Mitsuo Kure, *Samurai: An Illustrated History* (Boston: Tuttle, 2002). How they trained and fought and what they believed.

Takao Mukoh, *The Hagakure: A Code to the Way of the Samurai* (Tokyo: Hokusido Press, 1980). The official Bushido text of the Samurai.

Inazo Nitobe, *Bushido: The Soul of Japan* (Rutland, VT: Charles Tuttle, 1969). A noble code that got the Japanese into terrible trouble during World War II.

C. J. Peers, *Imperial Chinese Armies 200 BC–AD 589* (Oxford, UK: Osprey, 1995). The expansions of the western and eastern Han and others are chronicled here. Principle weapons include crossbows and spears, heavy lances, bows, and swords. Good illustrations of Chinese chariots and true lancers. Comprehensive look at 3,500 years of one civilization's war-making.

David Robinson, *Empire's Twilight: Northeast Asia under the Mongols* (Cambridge, MA: Harvard University Press, 2009). The 14th century sees the rise of the Ming dynasty and the Mongols superseded in Korea, Manchuria, and the area around Beijing.

Stephen Turnbull, *Battles of the Samurai* (London: Arms and Armor Press, 1987) The author examines nine battles, from Kurikara (1183) to Sekigaha (1600), providing useful descriptions and analysis.

———, *Nagashino 1575: Slaughter at the Barricades* (Oxford, UK: Osprey, 2000). How muskets and volley fire came to Japan and why the Samurai banned both. This diminished receptivity to innovation cost the Japanese a great deal in the 19th century, but they then moved inexorably into the modern modes of warfare with a vengeance and by 1905 defeated Russia.

———, *The Samurai Invasion of Korea 1592–98* (Oxford, UK: Osprey, 2008). Argues that the invasion of Korea by Toyotomi Hideyoshi was a continuation of his successful effort to unite Japan, and his attempt to project power into Korea and China arose out of a need to keep his samurai occupied and personal desire to keep winning. Good map on page 25 shows the Japanese invasion routes in 1592. Toyotomi considered the invasions a "success" because, although he sent 160,000 soldiers, they brought back 185,738 Korean and 29,014 Chinese heads plus 50,000 captives. The Korean guerrillas, turtle ships, and Chinese assistance eventually sent the Japanese home, however. The Japanese invasions weakened both Korea and the Ming dynasty, paving the way for the eventual takeover of both by the Manchus.

Royall Tyler (trans.), *The Tale of the Heike* (New York: Viking, 2012). Called a "seminal masterpiece" and "Japan's great martial epic," this work provides an in-depth look at culture and war in 12th-century Japan as the Heike and their rivals, the Genji, fight it out. It would help the samurai class to create the bushido code, claims the translator.

Arthur Waldron, "Chinese Strategy from the Fourteenth to the Seventeenth Centuries," in *The Making of Strategy: Rulers, States and War* edited by Williamson Murray, MacGregor Knox, and Alvin Bernstein, (Cambridge: Cambridge University Press, 1994), pp. 85–114. The history of Chinese military strategy is clearly outlined and can be read with considerable profit.

———, *The Great Wall of China: From History to Myth* (Cambridge: Cambridge University Press, 1990). At 1,500 miles long and built primarily between 246 and 209 BCE, the Great Wall has fascinated and intrigued since it was built. Like the French Maginot Line of World War II, it has been of little use in dealing with real threats, as when the Manchus ousted the Mings. Now a symbol of Chinese resurgence, it is also an image of oppression and despotic rule, claims the author. It mostly regulated commerce rather than providing an impenetrable barrier to invaders.

Erik Zurcher, *The Buddhist Conquest of China* (Leiden, Netherlands: Brill, 1972). An absorbing tale of how one of the most peaceful religions conquered so many Chinese. How do you make the tenets of Buddhism serve Mars? A most difficult but apparently not impossible task, says the author.

PART VII: PARTHIAN AND SASSANIAN HORSEMEN

The heavily armed cavalry of Iranshahr, known as the "heroes of steel," conform to the models of heavy-horse cavalry described previously, with thick armor plate; a long, heavy lance; and a one-piece helmet. The cataphracts served long and well in the East.

Tom Howard, "Iranshahr," in his *In the Shadow of the Sword: The Birth of Islam and the Rise of the Global Arab Empire* (New York: Doubleday, 2012), pp. 61–122. A sprightly overview of the dynastic and religious underpinning of Persian heavy horse.

Gareth Sampson, *The Defeat of Rome in the East* (Philadelphia: Casemate, 2008). How and why Rome had such a difficult time at the eastern end of its territories.

Peter Wilcox, *Rome's Enemies 3: Parthian and Sassanid Persians* (Oxford, UK: Osprey, 1986). Quite an eye-opening read if one is interested in cataphracts, for Rome long had difficulty on its eastern borders. The Medo-Persians routed by Alexander and their successor states gave way to the Scythian clan of the Parni known as the Parthians from northern Persia. These Parthians ruled for 400 years and were overcome by their own internal dynastic rivalries and the rise of the Persian Sassanids. The feudal levies of both Parthians and Sassanid Persians display excellent examples of heavily mailed men and horses and a form of feudalism that is quite recognizable. Great artist renditions of heavily mailed Parthian cataphracts of the first century BCE and the later-sixth-century Sassanian *clibanarius*.

PART VIII: MONGOL AND LATER TURKIC ASIAN HORSEMEN

The Mongols set the gold standard for many aspects of warfare: mobility; sustained ruthlessness; and success in battle due to careful planning, clever strategy, and the principle of mass times velocity. Their conquest of much of Eurasia and their successful forays across Europe and Asia provide a most fertile ground for the study of warfare; so, too, do their failures to conqueror Japan and Vietnam and their inability to sustain the empire in the face of dynastic rivalries. Still, whenever war and warcraft are studied, the Mongols need to be at the heart of those investigations, for between Rome and the 19th century, they were probably the most "modern" of all war-making machines. The works of this section are but an introductory sample.

Thomas T. Allser, *Mongol Imperialism: The Policies of the Grand Qan Mongke in China, Russia, and the Islamic Lands, 1251–1259* (Berkeley: University of California Press, 1987). Examines the extremely impressive extractive capability of the Mongols, who were able to draft en masse from both nomadic tribesmen and sedentary auxiliaries from China, Persia, Russia, and central Asia. Their ability in this regard was not, according to the author, equaled until the time of Napoleon or exceeded until World War I.

Reuven Amitai-Preiss, *Mongols and Mamluks: The Mamluk-Ilkhanid War, 1260–1281* (Cambridge: Cambridge University Press, 1995). Fascinating study of this little-studied phenomenon.

———, *The Mongols in the Islamic Lands* (Cornwall, UK: TJ International, 2007). The Mongols did quite well in most situations, but we remember the exceptions, such as Ayaan Jalut, more clearly.

Zahiruddin Muhammad Babur, *The Baburnama: Memoires of Babur, Prince and Emperor*, translated by Wheeler M. Thackston (London: Oxford University Press, 1996). The life and times of Babur in his own words, with digressions on nature, geography, history, and peoples, as well as all the battles he fought. Quite a read from, and about, quite a man.

Gyorgy Balas and Karoly Szelenyi, *The Magyars: The Birth of a European Nation* (Budapest: Corvina, 1989). A lavish, color-image-laden attempt to bring us the story of the Magyars, the "last wave of Turkic conquerors to reach Europe." Starting in the Urals and going to the Don and Volga river basins before hitting the Carpathians and moving out onto the plains of Hungary in the 10th century, the Magyars became Christian and settled after their defeat at the Battle of Augsburg in 955.

James Chambers, *The Devil's Horsemen* (New York: Atheneum, 1985). The Mongols in all their glory are presented in straightforward fashion. Some useful descriptions of wars and warriors.

Sam Djang, *Genghis Khan: The World Conqueror*, 2 vols. (New York: New Horizon Books, 2011) (F). A long, loving portrait of Temujin, who propelled himself onto the world stage with great power and force. Although told as a novel, it reads like a history book, with a surprising lack of excitement. Most useful in depicting the harsh life that spawned the Mongol (and Turkic) expansion.

Richard A. Gabriel, *Subotai the Valiant: Genghis Khan's Greatest General* (Westport, CT: Praeger, 2004). Should be required reading at anybody's military academy. He was one of the greatest generals of all time, regardless of culture. Subotai made use of the major Mongol advantages: speed, maneuverability, and surprise. Their armies could cover hundreds of miles in a few days; arrive at a designated point from a variety of columns; and, using trickery, subterfuge, and feints, keep opponents off guard.

Nicholas Hooper and Matthew Bennett, "Thirteenth Century Eastern Europe and the Mongols," in their *Cambridge Illustrated Atlas of Warfare: The Middle Ages* (Cambridge: Cambridge University Press, 1995), pp. 60–63. The Mongols turn back from their ongoing conquest of Europe only because the khan dies.

Colin Iggulden, The Mongol Series, including *Genghis: Birth of an Empire*, *Genghis: Lords of the Bow*, *Genghis: Bones of the Hills*, *Khan: Empire of Silver*, and *Conqueror: A Novel of Kublai Khan* (New York: Delacorte Press, 2011–) (F). Marvelous blending of fiction and fact, an absorbing way to learn about the Mongols and their tumens, which conquered far and wide under truly amazing leaders. Too bad all history couldn't be absorbed this pleasantly.

Peter Jackson, *The Mongols and the West: 1221–1410* (London: Longmans, 2005). A straightforward account of the Mongols who came and almost stayed. Interesting interplays between the newly arrived and existing leaders.

Paul Kahn (ed. and trans.), *The Secret History of the Mongols: The Origins of Chingis Khan* (Boston: Cheng and Tsui, 1984). This 13th-century text documents the life of Chingis Khan and provides a superior look into Mongol culture and military strategy. See especially how the "largest empire in world history" was created.

Judith Kolbas, *The Mongols in Iran* (New York: Routledge, 2006). An episode that lasted longer than most think and a much-understudied topic. Well worth further investigation.

Beatrice Manz, *The Rise and Rule of Tamerlane* (Cambridge: Cambridge University Press, 1989). His career from 1382 to 1405 is covered in-depth, with emphasis on his rise in a nomadic culture and his consistent decisions to rule the "sown" areas as well as the steppe. The sown areas turn out to be much easier to control once conquered. His empire fell away due to this principle when the nomadic base could not unify. Using his Turkic name, Temur, the author shows his empire and how it grew and how it was governed.

Justin Marozzi, *Tamerlane: Sword of Islam, Conqueror of the World* (New York: Da Capo Press, 2004). Tamerlane is in the tradition of Genghis Khan, but with additional Turkmen overlays. Tamerlane, or Temur, was an amazing conqueror for any age and any culture, as Cairo, Constantinople, Delhi, Bagdad, Kabul, and Tiflis all fell to him. After defeating the Ottomans, he turned to attack the Mings in China instead of Europe, the author feels, because Europe was so poor: "No jewels, no jihad." He died during that brutally cold winter after he had decided to make war on China. Tamerlane was a superb tactician as well as strategist. The Temurid Empire was short lived, however, although his great, great, great grandson Babur conquered India and founded the Mughal dynasty there.

Timothy May, *The Mongol Art of War* (Yardley, PA: Westholme, 2007). Gives credit where credit is due and outlines how "modern" the Mongol army really was. Quite an engaging read.

Frank McLynn, *Genghis Khan: His Conquests, His Empire, His Legacy* (Boston: Da Capo Press, 2015). Holistic, wide ranging, and based on many sources—Arab, Chinese, and Russian, as well as traditional ones. Provides the reader with a vast amount of information and some important frameworks for analysis. Accents the tremendous revolution (especially between 1206 and 1209) Genghis made by creating a common law (the Great Yasa), careers open to talent, and the integrative function of "if you act like a Mongol and obey like a Mongol, you are a Mongol" (as with Chaka the Zulu). This resulted in a true state that outlasted Genghis by generations and functioned even under less-gifted khans. Great use of captured prisoners to attack the next city and "soak up" casualties, thus preserving Mongol soldiers. Took 23 years to conquer Jin China. Author also argues he was the greatest conquer in history (his empire was 12 million square miles, the size of North America, while the Roman Empire at its height was only half the size of the United States) and most importantly of all created so much of the empire's substance and structure himself. Subedei and Jebe's ride around Russia was "history's longest cavalry raid" and resulted in seven major victories. Also believes that no commander in history ever spawned more outstanding lieutenants, arguing the Muqali, Jebe, and Subedei were true military geniuses. If you are only going to read one book about the Mongols, this should that work.

David Morgan, *The Mongols* (London: Basil Blackwell, 1986). Comprehensive overview of the history of the Mongols with a good introduction to the sources of knowledge about them. But honestly concludes that the "study of the Mongol Empire is in its infancy."

Rhoads Murphey, *Ottoman Warfare, 1500–1700* (New Brunswick, NJ: Rutgers University Press, 1999). Looking at the "material restraints" of Ottoman warfare as well as their methods, the author finds that during this period, at least, the Ottomans were ahead of the Europeans in terms of the centralized means for financing and administering war, and they were simply "too big to fight" for any other single power.

David Nicolle, *The Age of Tamerlane* (Oxford, UK: Osprey, 1990). A very interesting and useful contrast with the Osprey accounts of the Mongols, particularly notable are the significant advances in armor by the Timurid period. Tamerlane, or Timur, was also much closer to the armies of Genghis than the contemporary Muslim ones. Not sure why Tamerlane gets so little historical coverage, for he was a very successful captain of war and conquered a vast territory and many peoples from the Black Sea to China.

———, *The Mamluks 1250–1517* (Oxford, UK: Osprey, 1993). By focusing on the Mamluk Sultanate based in Cairo but founded by earlier Mamluks or slave soldiers, the author and illustrator give a focus and cohesion to the subject within the broader context of slave soldiers (from Central Asia; Anatolia; and, later, Africa), who became more than slaves by rising in the ranks and eventually taking over. The procurement and training system was also used by Seljuk and Ottoman Turks. A variety of weapons were employed, from ordinary bows to lances, from crossbows to swords and later matchlocks. Good map on page 4 of the sultanate at its apogee, circa 1450.

———, *The Mongol Warlords* (London: Brockhampton Press, 1990). A fine introduction to the careers of Genghis Khan, Kublai Khan, Hulegu, and Tamerlane. Finely illustrated, with useful sections on Mongol armies and warfare. Interesting note on how the Mongols encountered Vietnamese war elephants on their failed invasions southward.

Tamara Talbot Rice, *The Seljuks in Asia Minor* (New York: Praeger, 1961). Covers the spread of the Seljuks from Anatolia into the Middle East (1071–1300), including Persia, Syria, and Iraq, before being supplanted by the Ottoman Turks.

David Robinson, *Empire's Twilight: North East Asia under the Mongols* (Cambridge, MA: Harvard University Asian Center, 2009). The "apocalyptic millenarianism rooted in Buddhism" of the Red Turban wars contributed greatly to the fall of the Mongol Yuan Empire and the rise of the Chinese Ming dynasty with battle fought over 250,00 square miles.

Jason Mason Smith, "Ayn Jalut: Mamluk Success or Mongol Failure?" *Harvard Journal of Asiatic Studies* 44, no. 2 (December 1984), pp. 314–20. A little of both, the author feels, for the Mongols were at the far reaches of their dynastic tether and not necessarily at their best, but the Mamluks were to rise higher still as time went by. Blames much of it on the demands of Mongol horses (they needed 2.8 million pounds of grain per day and 1.5 million gallons of water but were inferior to larger Mamluk mounts and their attendant supply system) and arrows (Mongol warriors made their own; Mamluks had centralized production). Mamluks were war slaves purchased from the steppe peoples by various Turkic groups. Highly trained in warfare, they eventually took over the sultanates of Egypt and Syria and waged war in their own right.

Colin Thubror, *Shadow of the Silk Road* (New York: HarperCollins, 2007). The author travels along its path and finds echoes of its earlier glory. Good pictures give the reader a sense of place where these events occurred and continue to occur.

S. R. Turnbull, *Genghis Khan and the Mongol Conquests 1190–1400* (Oxford, UK: Osprey, 2003). Interesting details found here on some normally overlooked campaigns, including those against Burma, Cambodia, and Thailand. Useful map of the attempted invasions of Japan plus some little-known indication of the impact of the Mongol conquests; for example, the value of herring dropped precipitously in Western Europe due to the fall of Novgorod.

———, *The Mongols* (Oxford, UK: Osprey, 1980). A good introduction to the warcraft of some truly amazing warriors. "The history of the Mongol armies is a catalogue of superlatives. No armies in history have ever won so many battles or conquered so many territories." Good on Mongol equipment (their composite bows were more powerful than the Welsh longbow) and their success in both sieges and open warfare.

Mesut Uyar and Edward Erickson, *A Military History of the Ottomans: From Osman to Ataturk* (Santa Barbara, CA: ABC/CLIO, 2009). Arguing that the Ottoman military was the "backbone of the empire throughout its life span," the authors boldly assert, "As the world's dominant military machine from 1300 to the mid-1700's, the Ottoman Empire led the way in military institutions, organizations structures, technology and tactics." Very detailed although not battle centric, the work points to the importance of the Janissary corps and its instrumental use of musket-bearing infantry well ahead of the European militaries of the period after the initial rise of the Turks on highly mobile horse archers. They also note that the Ottoman armies had better organization, rations, medical and sanitary arrangements, and supplies generally than your European, African, or Asian contemporaries. Positioned at the nexus between the late-medieval and early-modern periods, the Ottoman military is well worth studying in its entirety. Quite good on the classical period of Ottoman ascendancy, 1451–1606.

Jack Weatherfield, *Genghis Khan and the Making of the Modern World* (New York: Crown, 2004). Author gets a bit carried away but in a good cause. For sure, Genghis and the Mongols spread around a lot of their DNA. Lots of good material, not just on the Mongols and their battle tactics, but also on the spirit of their armies and their desire to win battles regardless of what had to be done, and if trickery could be employed, then so much the better.

David Woolman, "Primitive Warriors or Not, the Hardy Mongols of the 12th and 13th Centuries Used the Most Advanced of Tactics," *Military History* (October 1995), pp. 12–18. Why I call the Mongols "modern" in their military tactics and strategy as well as their organization.

PART IX: INDIAN AND SOUTH ASIAN HORSEMEN

Although South Asian armies employed a variety of combined arms, the use of cavalry and war elephants is well documented.

Charles Allen, *Ashoka: The Search for India's Lost Emperor* (New York: Overlook Press, 2012). By tracing the reign of Ashoka from the height of his empire in 250 BCE (stretching from south India to the Himalayas and Kandahar), the author eruditely traces the subsequent history of the region through its monuments and archeological remains. A good example of history changing and rechanging the story it is telling in any given era. Needed to bring in the role of Kautilya in the rise of Ashoka, however.

Babur, emperor of Hindustan, *The Baur-Nama in English*, translated by Annette Beverage (London: Luzac, 1922). A vast compendium of statements attributed to Babur and those who came after him, describing in great detail his 16th-century battles to rule Kabul, Hindustan, Samarkand, and Garghana. Includes a workable history of its compilation. The Mughal Empire probably reached its high point during the reign of Babur's grandson Akbar

(1556–1608), when trade, administration, armament, and social-religious liberalism became a potent brew.

Simon Digby, *War Horse and Elephant in the Delhi Sultanate* (Oxford: Oxford University Press, 1971). The steppe warriors, in this case, Mughals, used horses and composite bows to win their major battles, although by the 15th century, cannons and muskets were also in use.

Jos L. Gommans, *Mughal Warfare 1500–1700* (London: Routledge, 2002). It was an "infantry" revolution, not a gunpowder revolution, that ended the Mongol (and other mounted archers) habitual success in warfare over time: the bayonet, flintlock, and prefab paper cartridge were all important but so, too, were light field pieces and the discipline of the infantry square.

Jos L. Gommans and Dirk H. A. Kolff (eds.), *Warfare and Weaponry in South Asia 1000–1800* (Oxford: Oxford University Press, 2001). Good introduction to the role weapons played when embraced by different cultures in South Asia, showing that the weapons were often the independent variable militarily but depended on the degree to which a particular culture would accept the new weapons and their impact on society as well as the military.

Peter Jackson, *The Delhi Sultanate: A Political and Military History* (Cambridge: Cambridge University Press, 1999). Five Delhi-based sultanates of Turkic origin in medieval India (lasting between 1206 and 1526) are covered extensively by the presentation of reams of names and doings but unfortunately precious little military analysis. We do learn that the Delhi sultanate rulers often used the Mongol winter-hunt model for keeping troops sharp and appear to have been able to field 400,000 horses. They managed to defeat the Mongols on many occasions, although the sultanate would eventually be conquered by the Mughal emperor Babur in 1526.

R. K. Nehra, *Hinduism and Its Military Ethics* (Frankfort, IL: Lancer, 2010). An overview of the interplay between Hinduism ("the most ancient and complex religion of humankind") and military activity from the Mauryan period to the Muslim invasions through the time of the Rajputs, the Marathas, and then the British invasions, as well as independence, and beyond. Mostly description.

Kaushik Roy, *From Hydaspes to Kargili: A History of Warfare in India from 325 BC to AD 199* (New Delhi: Manohar, 2004). See especially "The Chaturanga Armies of India, 500 BC–AD 600" and "Cavalry and Chivalry: The Men on Horseback, 700–1520." The author has some interesting things to say about the differences between the Rajputs and the "Turks" (nomads of

Asia who adopted Islam). The Turks had more and better horses, iron stirrups (so they could stand up in the saddle and fire arrows), cross- and composite bows, and better discipline and tactics, including a heavy cavalry reserve to be used at the crucial stage of the battle. All of these obviated Rajput advantages in numbers, quality of swords, and emphasis on personal valor. Interestingly, Roy also claims that, in 1225, the Mongols came to the Indus River and turned back due to the heat and muggy climate. After the Turks took over and formed the Delhi sultanates, they used a professional army (featuring a Mamluk core), that trained year round using the Mongol-style "great hunts." Eventually the Delhi sultanate suffered from the Mongols cutting off their supplies of horses and gunpowder from China and from the rise in use of gunpowder weapons by the southern Indian states. It was eventually overcome by Babur and the Mongols in the 1526 Battle of Panipat.

———, *India's Historic Battles: From Alexander the Great to Kargil* (Delhi: Permanent Black, 2003). Heavily skewed to the post-Panipat (1526) era but does have some useful information on Jhelum (326 BCE) and the Second Battle of Tarain (1192). Happily, the author fights the twin scourges of economic and demographic determinism and rightly takes issue with Victor Hanson's strangely overhyped notion that decisive battles only began with the Greeks. Roy simply points to the Aryan Battle of Dagarajna in 1900 BCE and calls it good!

Gurcham Singh Sandhu, *A Military History of Medieval India* (New Delhi: Vision Books, 2006). Giving the reader a sense of the sweep of 5,000 years of Indian military history, the author examines such battles as Khanna, Gogra, Sirhind, and Tukaroi. His son-in-law may have been dragooned into doing the drawings.

Harjeet Singh, *Cannons versus Elephants: The Battles of Panipat* (New Delhi: Pentagon Press, 2011). See especially "Cannons in the Indian Medieval Period," pp. 175–90.

Vincent Smith, *Akbar: The Great Mogul 1542–1605* (Delhi: S. Chandam, 1966). Focusing on his ancestry (Turkic and Mongol from his father and Persian from his mother but "no Indian blood"), the author traces the rise of the conqueror of much of India in considerable detail. Most useful is his chapter on "Wars and Annexation." The grandson of Babur, the lord of Kabul who swept into northern India in 1526, Akbar did his grandfather proud.

Douglas Streusand, *The Formation of the Mughal Empire* (New Delhi: Oxford University Press, 1989). Mughal martial superiority accounts for the small number of battles during their rule, and the author notes that the Mughal mounted horsemen from the steppes with composite bows (which could shoot six arrows in the time it took a musketeer to shoot once) rather easily overcame the elephant- and infantry-based system of the previous Indian style of warfare. Iron discipline also was a key component.

H. N. Verma, *Decisive Battles of India through the Ages* (Delhi: GIP Books, 1994). From Dasarajna in 3100 BC to Gujarat (1948) and a lot in between, this introduction gives a feel for the ebb and flow of battle across the subcontinent.

B. N. S. Yadva, *Society and Culture in North India in the Twelfth Century* (Allahabad, India: Central Book Depot, 1973). Yadva makes a strong case for the ideal of "chivalry" in Indian warfare with ancient precedents but acknowledges that, as in the West, "chivalry" and its code of ethics was often swept away by the true nature of warfare.

Chapter Fourteen

The Vikings

> "Praise not the day until evening has come; a woman until she is burnt; a sword until it is tried; a maiden until she is married; ice until it has been crossed; beer until it has been drunk."—Viking proverb

In and of themselves, the Vikings are not necessarily worth a major section, but in terms of force projection and their successive reincarnations, such as the Normans, who would go on to conquer England, Sicily, Italy, and even Byzantium, they are worth an inclusion here.

Jessie L. Byock, *Viking Age Iceland* (New York: Penguin, 2001). Medieval life from 10th to 13th centuries in Iceland and a "headless polity" that converted relatively peacefully to Christianity in 1000, with the church quickly integrated into the chieftainship array already present. Interesting material on blood feuds and vendettas.

Michael Crichton, *Eaters of the Dead* (New York: Bantam Books, 1977) (F). Using the trope of a 10th-century Muslim account of encounters with the Northmen in their various war-making diaspora, the author captures many dimensions of their warriorhood.

Ian Heath, *The Vikings* (Oxford, UK: Osprey, 1985). An extensive compendium showing weapons, ships, and maps of the Viking era. Accent on the battle formation known as the shield wall, a variation of the phalanx of the Greeks, which, when combined with the Viking longship with its shallow draft and swift arrival and ferocity in battle, made them such intrepid warriors for so long and in so many places (from northern Europe, where they were known as the Norsemen, to Spain, where they were known as Al-

Madjus, or "heathen wizards," and Byzantium, where they were known as Varangol; they cut a wide swath).

Gwyn Jones, *The Norse Atlantic Saga* (London: Oxford University Press, 1986). Shows why the smarter of the Vikings went south and took over Normandy and Sicily and England instead of going into the foggy cold mists of the North Atlantic. Contains the important "Greenlanders' Saga" and "Erik the Red's Saga," as well as "Karlsefni's Voyage to Greenland."

M. K. Lawson, *Canut: The Danes in England in the Early Eleventh Century* (London: Longman, 1993). England had been a rich source of plunder for several centuries, but Danish kings Swegen Forkbeard (1013) and Canut (1016) want more. This work puts the Danish interlude in England in the broader context of an arc running from Scandinavia to the Elbe.

G. A. Loud, *The Age of Robert Guiscard: Southern Italy and the Norman Conquest* (London: Longman, 2000). The movement of some Normans to Sicily and southern Italy at the beginning of the 11th century was to produce "Robert the Cunning" or "Robert the Weasel," who turned out to be the "greatest warlord in Latin Christendom," challenging Muslims in Sicily, the Byzantine Empire, the Holy Roman Empire, and the papacy. Shows that force projection is not always as important as leadership projection, and although the duke of Apulia was subsequently almost forgotten, his story should be retold.

John Norwich, *The Normans in Sicily* (London: Penguin, 1970). An all-encompassing account of the Normans (1016–1194), including graphic accounts of the "Kingdom in the Sun." Shows how the Normans defeated Lombards, Saracens, Greek Byzantines, the Holy Roman Emperor, and the people. Robert Guiscard of the House of Hauteville is finally done in—along with 500 of his knights—by typhoid in 1085. Quite a saga.

Philip Parker, *The Northmen's Fury: A History of the Viking World* (London: Johnathan Cape, 2014). One of the most readable accounts of the period and the far-flung military expeditions of the Vikings. Brings together a lot of scholarship from a variety of sources.

Alexios G. C Savvides, *Byzantio-Normannica: The Norman Capture of Italy (to A.D. 1081) and the First Two Invasions in Byzantium (A.D. 1081–1085 and 1107–1108)* (Leuven, Belgium: Peeters, 2007). A most useful list of rulers and bibliography chronicling Norman ascendancy in Sicily and Italy in the 11th century, precursors to the Norman invasion of Byzantine Greek lands in the late 11th and early 12th centuries.

Donald Scragg, *The Return of the Vikings: The Battle of Maldon* (Stroud, UK: Tempus, 2006). After the Danes are checked by Alfred the Great and Wessex flourishes, the Danes renew their raids and depredations and win in 991, as 90 Viking ships landed troops in Essex. Thereafter, the Normans would replace the Danes as the biggest threat to Anglo-Saxon England. Good map on page 14 of Viking depredations of the time.

Terence Wise, *Saxon, Viking and Norman* (Oxford, UK: Osprey, 1979). This is quite an amazing account of how much havoc the Saxons, Vikings, Frisians, Normans, Jute, and Angles, among others, were able to wreak with swords, spears, axes, and shields (including the distinctive "Norman" teardrop version), spreading out from the North Sea to Russia, Sicily, Byzantium, Spain, Wales, Ireland, England, and France from the 6th to the 11th centuries.

Harriet Harvey Wood, *The Battle of Hastings: The Fall of Anglo-Saxon England* (London: Atlantic Books, 2008). Quite disappointing, with endless dynastic descriptions and no diagrams of the battle or maps showing the various political entities of the time. Seemed overwhelmed with the "loss" of Anglo-Saxon Wessex, as if the Anglo-Saxons themselves hadn't shown up to dominate the Celts. "William the Lucky Bastard" was once removed from Vikingdom, having learned in Normandy how to build castles and harry the countryside until it surrendered.

Chapter Fifteen

The Gunpowder and Discipline Revolution and the Expansion of the Gunpowder Empires

> "Though a simple mixture, no substance has ever probably given rise to so extensive a literature or to so much death and destruction."—O. F. G. Hogg

> "Without these munitions and artillery one cannot preserve any state, nor defend it, nor attack the enemy."—Gabor Agoston, quoting minutes from the Venetian Senate, June 16, 1489

> "Whosoever commands the sea commands the trade; whosoever commands the trade of the World commands the Riches of the World and consequently, the World itself."—Sir Walter Raleigh

Although gunpowder was invented much earlier in China and used in warfare (as rockets, bombs, grenades, and a component in Greek fire) for hundreds of years, by both Asian and Muslim armies, it was not until after 1429 that Europeans made extensive use of "corning" (which turned it into more usable crystals that did not break down into sulfur, charcoal, and saltpeter) and other refinements that an arms race in Europe produced a number of new "fiery" weapons, such as the matchlock or harquebus, and truly transformed warfare globally. Also, useable gunpowder depended on finding the right amount of saltpeter (75 percent) to go with sulfur (15 percent) and charcoal (10 percent) to get the necessary big bang, and discovering that balance and proportion took a surprisingly long time to achieve.

What is very important, however, is that the Gunpowder Revolution meant more than eventual European military supremacy over much of the world. It meant that settled, agricultural people no long had to fear the steppe people because, with pikemen and harquebuses and cannons, they could defeat them, and as the Chinese and Russians could eventually conquer the steppes themselves. The Gunpowder Revolution led to gunpowder empires; that is, those non-Europeans who followed the revolution faithfully—the Ottoman Empire, the Safavid (Persian) Empire, the Mughal Empire, the Muscovite (Russian) Empire, and later the Manchu Empire (last of the steppe people who overthrew the Ming dynasty with a lot of help from the Mings themselves)—and reigned supreme thereafter. Gunpowder gave the settled peoples their first major advantage over the steppe people in a thousand years, and they made the most of the opportunity.

Social and political dynamics also were affected by the spread in use of gunpowder. Much greater centralization of political authority was now possible for a variety of reasons, including the inability of nobles to afford either the new fortresses (castles couldn't stand up against cannon) or the cannons and mass-produced firearms (which required foundries, powder houses, stockpiles of gunpowder and ammunition, cannonballs, etc.), so that the power of the political center grew at the expense of local noble authority. Henceforth, in most situations, only the state could afford major wars.

In terms of the overseas projection of power, it was the marrying of the cannon to the ship and especially the development of broadside firing and the various technologies that accompanied it that enabled some countries to enforce their wills on distant societies and peoples. European dynasties seem to have been more adaptive in this regard and more clearly saw the growing importance of cannons on ships.

All this emphasis on gunpowder and its ultimate effects on warfare should not be allowed to overshadow the ongoing revolution in military affairs that occurred in Europe from 1550 to 1650 and that gave the Europeans a set of advantages in dealing with each other and the rest of the world. Some, such as Michael Roberts, believe the revolution began in Holland (Maurice of Nassau) and especially Sweden (Gustavus Adolphus), while Geoffrey Parker focuses more on the Habsburg Empire as the radiating incubator; but both agree that revived patterns of discipline were as important as the invention of gunpowder, and initially, at least, the pike was as essential as the gun in reducing the power of the heavy-horse knight. As medieval levees became standing armies and dynastic armies eventually became national ones, the world would never be the same.

In this regard, different peoples responded differently to both the Gunpowder Revolution *and* the appeal and demands of the gunpowder empires. Some of these variables are presented in the readings here:

Funso Afolayan, "Kingdoms of West Africa: Benin, Oyo and Ashante" in *The Middle Age of African History*, edited by Roland Oliver (London: Oxford University Press, 1967), pp. 161–90. How the slave trade and the introduction of firearms enabled some forest kingdoms to wax powerful, while others, especially savanna kingdoms, who had dominated them earlier were overtaken by the harsh realities of changed power configurations.

Gabor Agoston, *Guns for the Sultan: Military Power and the Weapons Industry in the Ottoman Empire* (Cambridge: Cambridge University Press, 2005). Agoston argues that "channels of military acculturation" remained open in the Ottoman Empire and it was economic decline and weakening of the central administration that made it difficult for them to keep up with the West in terms of "fiery weapons."

Thomas Arnold, *Renaissance at War* (London: Cassell, 2001). "A Gunpowder Revolution?" "The Cross versus Crescent," "Dueling Kings," and "Faith versus Faith" are all treated here. Ottomans and Christian clash but not as much as Christians against Christians.

David Ayalon, *Gunpowder and Firearms in the Mamluk Kingdom* (London: Frank Cass, 1978). As with many other "horse nobles," sociopsychological antagonism to firearms prevented much-needed receptivity to innovation, contributing to their defeat in the early 19th century by the Ottoman Turks, who had accepted the Gunpowder Revolution. Ayalon speaks of an "unbridgeable gulf" that caused the social and military elite of the Mamluks to repudiate these "fiery" weapons.

Jack Beeching, *The Galleys of Lepanto* (New York: Charles Scribner's Sons, 1982). Despite the title, this book is mostly about the extensive backstory leading up to the battle. Marvelous portraits of Emperor Charles V; Don John; Don Carlos; Selim; and others, such as Maria la Bailadora, a woman who fought at Lepanto as a man, killed her Turk, and was then officially admitted into the ranks. Charles V is described as a "glutton whose motto was 'I came, I saw, God conquered.'" The pope apparently gave him a dispensation to eat whenever and whatever he wished, including before confession and during Lent!

Jeremy Black, "Dynasty Forged by Fire," *MHQ: The Journal of Military History Quarterly* (Spring 2006), pp. 34–43. A must-read on the Gunpowder Revolution and its origins. How the Battle of Pavia (1525) illustrates the changed military balance in Europe, beginning with the Habsburgs.

———, *European Warfare, 1494–1660* (London: Routledge, 2002). Black challenges the Geoffrey Parker "revolution" in a military affairs thesis arguing that 200 years is a long time but ends with us seeing what he calls a "set of military revolutions."

——— (ed.), *The Thirty Years' War*, 2nd ed. (London: Routledge, 1987). A series of essays covering many aspects of this Hobbesian struggle of all against all in Germany of the 17th century. A struggle of enormous import to both the future of war and the decimation of Germany: "Until 1939, the Thirty Years' War remained by far the most traumatic period in the history of Germany. The loss of people was proportionally greater than in World War II." Europe had witnessed an apocalypse.

John Brewer, *The Sinews of Power, War, Money and the English State 1688–1783* (New York: Alfred F. Knopf, 1989). How the Gunpowder Revolution and the gunpowder empire it spawned presaged the rise of state power in order to pay for the new warfare and its demands, as well as the complex intertwining of war and taxes.

Kenneth Chase, *Firearms: A Global History to 1700* (Cambridge: Cambridge University Press, 2003). A long and convoluted history it is, with true firearms dating from 1100 in China and later, the early1300s, in Europe, but it is not until the 1500s that were they decisive in warfare and then only when combined with pikemen (later bayonets) and drill discipline. It was their ease of learning to operate, not their lethality, that eventually made firearms superior to the recurved or longbow.

Carlo M. Cipolla, *Guns, Sails and Empires: Technological Innovation and the Early Phases of European Expansion 1400–1700* (New York: Pantheon Books, 1965). The author, an inveterate pacifist, writes, "I was forced, by overwhelming evidence, to recognize, against my tastes and inclinations, the importance of guns and sails." How the Atlantic overcame the Mediterranean with "gunned ships" is covered in considerable and useful detail.

Weston F. Cook Jr., *The Hundred Year War for Morocco: Gunpowder and the Military Revolution in the Early Modern Muslim World* (Boulder, CO: Westview Press, 1994). The military revolution paradigm holds up well in the Moroccans' struggle against the Portuguese, Spanish, and Ottoman Turks as the Sa'adian gunpowder army proves decisive during the 1460s–1560s struggle. Morocco stays independent until the 1842 French takeover (although a gunpowder army eventually breaks the back of Sa'adian absolutism). By contrast, leaders in Algiers, Libya, and Tunis do not adopt the gunpowder and military revolution and are taken over by the Ottomans.

Julian Corbett, *Drake and the Tudor Navy: With a History of the Rise of England as a Maritime Power*, 2 vols. (London: Longmans, Green, 1898). A very readable account of the beginning of the gunpowder empires, in this case, England under Elizabeth, although the author gives credit to Henry VIII for starting the country toward maritime rather than continental power. Drake was accused of the "Idolatry of Neptune," but his voyage around the world (1578–1571) in the *Golden Hind* electrified the nation and made him rich and Queen Elizabeth richer. The 16th century saw the transition from galley to galleasses to galleons as "high-sea endurance and broadside fire (not fire end and forward) came into vogue," reconciling sea endurance with free movement and great striking power compared with the Mediterranean galley. The author suggests that England and Holland seized upon the galleon "greedily."

Roger Crowley, *Empires of the Sea: The Siege of Malta, the Battle of Lepanto, and the Contest for the Center of the World* (New York: Random House, 2009). A very engaging study shot through with the Turkish ambiguity toward fiery weapons, which only manifested itself over time. How power configurations in the Mediterranean world might have turned out is covered and covered well.

Kelly De Vries, *Infantry Warfare in the Early Fourteenth Century: Discipline, Tactics and Technology* (Woodbridge, UK: Boydell Press, 1996). They were only waiting for the pike and the musket to usurp the 1,000-year dominance of the armed horseman. Took a while.

Felipe Fernandez-Armesto, *1492: The Year the World Began* (New York: Harper, 2010). Here is a well-known historian who takes an important year and shows us what was happening all around the world at that time and in the process tries to explain the coming rise of Europe. He certainly could have used the Template of Mars to explain some of it. For example, he sees the changes in Eurasia coming from the rise of a revitalized Russia but without recognizing the military advantage the musket-armed Russians now had against the people of the steppe. Also, he notes the failure of China, which he describes as a "global superpower . . . bigger and richer than all its possible competitors combined" but without seeing the importance—as did the Europeans—of melding the cannon to the ship for force projection. To his credit, the author does mention that, when Ahuitzotl became Aztec paramount and dedicated a new temple at Tenochtitlan, he sacrificed 20,000 captives, no doubt laying the seeds for some Native Americans to welcome the eventual arrival of the Spaniards, however naïve that welcome turned out to be.

John Francis Guilmartin Jr., *Gunpowder and Galleys: Changing Technology and Mediterranean Warfare at Sea in the Sixteenth Century* (Cambridge: Cambridge University Press, 1974). Challenges some assumptions about how and why gunpowder changed warfare and how long it took to do so.

Bert S. Hall, *Weapons and Warfare in Renaissance Europe: Gunpowder, Technology and Tactics* (Baltimore, MD: Johns Hopkins University Press, 1997). Concentrates on Western European warfare and sees heavy-horse cavalry as less important than previously assumed, hence initially gunpowder was not crucial. Pikemen had to face the terrifying charges of cavalry, but if they could, then they could change the nature of warfare. The 15th and 16th centuries saw the rise of pikemen and harquebusiers together, and in the 17th century, the bayonet replaced the pike.

Lisa Hull, "How to Capture a Castle," *MHQ: The Quarterly Journal of Military History* (Spring 2004), pp. 34–45. Very readable account of the specifics involved. An excellent introduction to this sometimes-overlooked phenomenon in the history of warfare (albeit sometimes overemphasized in romantic literature).

Colin Imber, *The Ottoman Empire: The Structure of Power, 1300–1650* (New York: Palgrave Macmillan, 2002). The Ottomans' conflicted relationship with gunpowder and weapons.

John Lamphear (ed.), *African Military History* (Hampshire, UK: Ashgate, 2007). This set of wide-ranging essays, primarily from the *Journal of African History*, gives good insights into the arrival of guns into a variety of societies, including West Africa, Ethiopia, and Madagascar.

Brian Lavery, *Empire of the Seas: How the Navy Forged the Modern World* (London: Conway, 2009). Lavishly illustrated, this work traces the rise of the British navy from the defeat of the Spanish armada and the age of "Tudor Galleons" to World War I and the age of "Steam, Steel and Shellfire." I did not realize that the British navy shrunk from 1,009 ships and 145,000 men in 1813 to 179 ships and 23,000 sailors in 1826.

Ignacio and Ivan Lopez, *The Spanish Tercios 1536–1704* (Oxford, UK: Osprey, 2012). Because of the Spanish terrain, infantry played an important role during the Middle Ages, and the Spanish were among the first to successfully adapt to the changes brought about by the "fiery weapons." The tercios were formations from 1536 that initially featured one-third pikemen, one-third swordsmen, and one-third gunners (later variations changed the composition) organized in squadrons or "squares." Very effective and depended on superior discipline and small-unit cohesion.

Collin Martin and Geoffrey Parker, *The Spanish Armada* (Manchester, UK: Manchester University Press, 1988). Erudite, fun to read, and offering quite important insights into the military revolution on land (duke of Palma in the

Spanish Netherlands) and sea (Spanish armada confronts English ships, which marry heavy cannon to nimble ships), as European powers practice on each other before taking their military show globally.

William H. McNeill, *The Age of Gunpowder Empires 1450–1800* (Washington, DC: American Historical Association, 1989). Corned gunpowder spread throughout the globe, making winners and losers along the way.

David Nicolle, *The Age of Tamerlane* (Oxford, UK: Osprey, 1990). A very interesting and useful contrast with the Osprey accounts of the Mongols. Particularly notable are the significant advances in armor by the Timurid period. Tamerlane, or Timur, was also much closer to the armies of Genghis than the contemporary Muslim ones. Don't know why Tamerlane gets so little historical coverage, for he was a very successful captain of war and conquered a vast territory and many peoples from the Black Sea to China.

———, *Armies of the Ottoman Turks 1300–1774* (Oxford, UK: Osprey, 1983). Captures the rise and subsequent decline of these most expansive of Turks and their ongoing ambivalence with gunpowder weapons. Author makes an important point on how the Ottomans did experience some major military reforms during the 17th and 18th centuries, enabling the "sick man of Europe" to survive for another century and a half.

———, *The Mamluks 1250–1517* (Oxford, UK: Osprey, 1993). By focusing on the Mamluk sultanate based in Cairo but founded by earlier Mamluks, or slave soldiers, the author and illustrator give a focus and cohesion to the subject within the broader context of slave soldiers (from Central Asia; Anatolia; and later, Africa) who became more than slaves by rising in the ranks and eventually taking over. Their procurement and training system was also used by the Seljuks and Ottoman Turks. A variety of weapons were employed from ordinary bows to lances, from crossbows to swords and, later, matchlocks. Good map on page 4 of the sultanate at its apogee, circa 1450.

Peter Padfield, *Maritime Supremacy and the Opening of the Western Mind* (Woodstock, NY: Overlook Press, 2000). How the sea made the Enlightenment and the world as we know it. Quite an outstanding thesis and history used to back it up can teach us a lot about force projection.

Geoffrey Parker, *The Military Revolution: Military Innovation and the Rise of the West 1500–1800* (Cambridge: Cambridge University Press, 1988). A true classic. One of the most far-reaching assessments of the period and one showing considerable in-depth knowledge concerning changes in warfare that persisted over time and across space. Even those who disagree cannot ignore the arguments Parker makes.

———, *Warfare: The Rise of the West* (Cambridge: Cambridge University Press, 1995). The West conquered most of the globe because it most fully followed the Gunpowder Revolution and its countries practiced mightily on each other before venturing out into the rest of the world.

——— (ed.), "The Western Way of War," in his *The Cambridge Illustrated History of Warfare* (Cambridge: Cambridge University Press, 1995). Discipline and technology trump numbers, but you have to practice a lot.

J. R. Partingon, *A History of Greek Fire and Gunpowder* (New York: Barnes and Noble, 1960). A very valuable work that traces the use of incendiaries with great detail from the time of the Assyrian Empire (ninth century BCE) through the use of Greek fire (and its antidote, vinegar) until the Gunpowder Revolution. Includes a meticulous examination of gunpowder ratios from saltpeter (7 parts), sulfur (5 parts), and charcoal (5 parts) to the modern high explosive (75 parts to 10 parts to 15 parts). An amazing work.

Robert Payne and Nikita Romanoff, *Ivan the Terrible* (New York: Thomas Crowell, 1975). Ivan IV facilitates subsequent Russian expansion by defeating the Kazan khanate in the middle of the 16th century, setting the stage for further Russian expansion into the Black Sea and Baltic Sea regions and eventually Siberia. The authors conclude, however, that "[h]is greatest gift to the Russian people was his own death."

Peter C. Perdue, *China Marches West: The Qing Conquest of Central Eurasia* (Cambridge, MA: Harvard University Press, 2005). "Cannons on Camels" as the expanding Russian and Chinese Manchu Quing Empires squeeze central Asia between them, as Mongol Zughars, Manchu Qings, and Muscovite Russians vie for power in the region, setting boundaries, many of which lasted until the collapse of the Soviet Union in 1991.

I. A. Thompson, *War and Government in Habsburg Spain 1560–1620* (London: University of London, Athlone Press, 1976). The Habsburgs were almost constantly at war during this period, as Charles V's indirect administration of the military gave way to Philip II's direct administration and three major theaters held sway: the Low Countries (the Dutch Revolt), the Mediterranean (forces of Islam), and the Atlantic (England and France).

J. F. Verbruggen, "The Rise of the Foot Soldiers," in his *The Art of Warfare in Western Europe during the Middle Ages from the Eighth Century to 1340*

(New York: North Holland, 1997), pp. 99–184. How infantry replaced the cavalry of the Middle Ages as the dominant military formation.

Jenny West, *Gunpowder, Government and War in Mid-Eighteenth Century* (Woodbridge, UK: Boydell Press, 1991). Argues that the Gunpowder Revolution was also a revolution in state power and resulted in far-reaching consequences for both state and warfare.

Chapter Sixteen

Era of the European Great Captains

As warfare changed in this era, there developed a pattern that was eventually exhibited by all successful armies and continued until the end of the 19th century. Dominic Lieven sums that pattern up most nicely, writing in *Russia against Napoleon*,

> To win battles in this era required infantry trained to deliver rapid volleys and to move in formation across the battlefield. The troops must be able to shift between column, line and square according to circumstances and to do so rapidly and in good order. The infantry need to be supported by mobile artillery and by cavalry trained to charge home in massed formation to exploit any wavering by the enemy. Though all this sounds simple, amidst the terrors of the battlefield it was anything but. To achieve this, an army required good training, a strong core of veterans, and experienced officers and NCOs. Behind the army there had to stand a state and a society capable of providing reliable officers and of paying the large sums needed for men, arms, food and equipment. The main European armies achieved this and so did the British in India.

The Great Captains, almost regardless of society or culture, saw the advantages in this style of warfare and imprinted their own personalities upon it. Later, as seen in another chapter 20, the French Revolution was, through the institution of the levy en masse, able to provide a great expansion of the size of the armed forces, but the previous tactical requirement remained in force. Russell Weigley goes even further in his *The Age of Battles*, arguing that, between the Thirty Years' War's Battle of Breitenfeld (1831) and the Battle of Waterloo (1815), there was indeed an ongoing search for the decisive battle. And the Great Captains were at the center of that endeavor. In the process, they changed the nature of warfare.

Note: See other Great Captains explored in this chapter. As Lieven points out, where adoption of this pattern was absent—such as with the Ottoman

Turks—the country involved was at a great disadvantage when fighting states that employed the new pattern.

Jeremy Black, *Warfare in the Eighteenth Century* (London: Cassell, 1999). Broad ranging, with excellent maps and illustrations. Covers "Wars without Europeans," "Europeans versus Non-Europeans," "The American War of Independence," and "The Coming of Revolution."

Tim Blanning, *The Pursuit of Glory: Europe 1648–1815* (New York: Viking, 2007). A holistic and overarching history of the period in which the nations and states of Europe found their power and sought glory and riches all over the world after practicing on each other. Covers a broad range of subjects and developments beyond just the military. The military developments of this age are found in part IV, "War and Peace" (pp. 531–674). Blanning is particularly good on summing up the French revolutionary and Napoleonic wars. He is not, however, a fan of either.

John Childs, *Warfare in the Seventeenth Century* (London: Cassell, 2001). Childs claims, "Most seventeenth-century campaigns were 'affairs of post and ambuscades,' involving small garrisons. . . . Full-scale battles and sieges were infrequent." He is not a big proponent of the "military revolution." Still, Childs provides useful chapters, including "The Thirty Years War," "The Rise of the Standing Army," and "The Wars of Louis XIV."

Christopher Duffy, *The Military Experience in the Age of Reason* (New York: Atheneum, 1988). The revolution is well under way and practiced by the Great Captains of Europe. Good on land war in the middle of the 18th century. See N. A. Rogers in this chapter for an analysis of war at sea during this period.

Geoff Mortimer, *Wallenstein: The Enigma of the Thirty Years War* (New York: Palgrave Macmillan, 2010). A sympathetic study, one of only four in English, the others having been written in 1837, 1938, and 1976, respectively. Argues that Albrecht Wallenstein was a grand strategist who twice saved the Holy Roman Empire only to be assassinated on orders of the emperor for being an overmighty and overreaching vassal. A count of Bohemia who lost his estates when the Protestants took over, he was a true warrior who sought peace, and Mortimer claims that he was not as overly ambitious as previously described.

Note: Friedrich Schiller was also fascinated by the personality of Wallenstein and his role during the Thirty Years' War. See his "dramatic poem" "Wallenstein," in *The Robbers and Wallenstein* (Baltimore, MD: Penguin Books, 1979), pp. 161ff. Writing in 1799, Schiller saw this "great man" as one "betrayed by his delusions of historical destiny" and takes some artistic liberty with his story.

Geoffrey Parker (ed.), *Global Crisis: War, Climate Change and Catastrophe in the Seventeenth Century* (New Haven, CT: Yale University Press, 2013). A vast, sweeping study and overview of the 1600s, concluding that an amazing confluence of no sunspots, global cooling, famine, disease, and war led to the most horrendous century on record. Encyclopedic in scope.

———, *The Thirty Year's War* (London: Routledge, 1987). A comprehensive reevaluation of the Thirty Years' War. Thick and expansive, the various chapters illuminate, especially "On Total War" and "The Indecisive War 1618–1629." Some interesting conclusions: there was an "unprecedented concentration of men at arms" (200,000), the majority of whom were volunteers; the 13-foot pikes and 5-foot matchlock muskets, along with cavalry pistols, swords, and especially extensive use of cannons, made war more expensive due to the new scale of activity. At the end, however, the work concludes that a "mere handful" of "determined figures" were the primary reason this war was fought for so long. At the end, a new international state system was in place, along with a decline in confessional politics.

David B. Ralston, *Importing the European Army: The Introduction of European Military Techniques and Institutions into the Extra-European World, 1600–1914* (Chicago: University of Chicago Press, 1990). Chronicles the "irrepressible spread of European ways" in warfare in Muscovy, the Ottoman Empire, China, and Japan, especially after the Meiji Revolution and the arrival of Admiral Perry. Argues that Swiss pikemen were as vital as portable firearms to the Gunpowder Revolution.

N. A. Rodger, *The Wooden World: An Anatomy of the Georgian Navy* (New York: Norton, 1996). How the European navies were simply a world apart as they revolutionized the globe with their military and naval technology. Establishing the gunpowder empires required new ships and new technologies to make them possible.

Michael Rogers, "The Military Revolution 1560–1660," lecture delivered before the Queen's University of Belfast, 1956. This provocative and profound lecture captures the essence of the changes that came about during this seminal century: the end of the effectiveness of heavy cavalry, the defining role of pikes, the ability of cannons to reduce castles, the reintroduction of discipline, drill by units and the cohort system with officer command and control, a lineal pattern of attack and defense with mobile artillery, and uniforms to make "king's men" out of mercenaries. He concludes,

> By 1660 the modern art of war had come to birth. Mass armies, strict discipline, absolute submergence of the individual had already arrived and the conjoined ascendancy of financial power and applied sciences was already established in all its malignity; the use of propaganda, psychological warfare, and terrorism as military weapons was already familiar to theorists as well as to commanders in the field, and the last remaining qualms as to the religious and ethical legitimacy of war seems to have been stilled. The road lay open, broad and straight, to the abyss of the twentieth century. (p. 32)

Well said.

Desmond Seward, *Prince of the Renaissance: The Golden Life of Francois I* (New York: Macmillan, 1973). The House of Valois in full bloom under Francois I (1494–1547). Despite his mother's warnings, he goes into Italy and annexes Milan only to later be challenged by Holy Roman Emperor Charles V, who decisively defeats Francois at the famous Battle of Pavia (1525) and indeed captures him. A good description of Pavia, one of the first real situations in which the "fiery weapons" were decisive. Seward says categorically, "The battle was decided by the arquebus [harquebus]," although cannon played a role as well. Eight thousand French and their allies lost their lives compared to 700 for the Imperialists.

John Stoye, *The Siege of Vienna: The Last Great Trial between Cross and Crescent* (New York: Pegasus, 2000). The 1683 siege that lasted from July to September was lifted by a relieving army led by Charles of Lorraine; the king of Poland, John III Sobieski; and contingents of Saxons, Bavarians, and Franconians, as well as forces from Habsburg emperor Leopold II. The Ottoman, Tartar, and Magyars were defeated, and this battle marks the start of the Ottoman rollback. The Ottoman grand vizier was executed for his failure to win the battle as the Sultan Mehmet IV cut his losses early. Sobieski ended up with the majority of the spoils, including the flag of the Prophet Muhammad.

C. V. Wedgewood, *The Thirty Year's War* (London: Jonathan Cape, 1938). Long a standard, although written right before World War II, it tends to blame Germany for the war, declaring, "Germany's tragedy was essentially her own," and arguing that a majority of people in Germany and in Europe wanted no war, let alone a 30-years one. Wedgewood believes that, while casualty figures are wildly inflated for the period, nevertheless there was a very widespread breakdown of the social order, and the resulting disintegration of society was not to be easily remedied.

Russell F. Weigley, *The Age of Battles: The Quest for Decisive Warfare from Breitenfeld to Waterloo* (Bloomington: Indiana University Press, 1991).

From Gustavus Adolphus, Maurice of Nassau, and the birth (or rebirth) of military professionalism through Louis XIV and the fortresses of Sebastien Vauban and through Frederick the Great, the author notes the "recalcitrant indecisiveness of war" (p. 537) and concludes, "The larger the objective of war, the more acute the frustrations of indecisiveness." An important work, well worth reading.

Andrew Wheatcroft, *The Enemy at the Gate: Habsburgs, Ottomans and the Battle for Europe* (New York: Basic Books, 2008). Interesting discussion of the Ottomans' sanitary, quiet, and well-supplied war camp of the Turks in the 1683 campaign of Sultan Mehmet IV against the Vienna of Holy Roman Emperor Leopold II. He believes that pikes (12 feet long) were one of the keys to the Habsburg victory (as were infantry with bayonets); another was cavalry armed with firearms. Vienna is saved by a relief column led by Charles of Lorraine and the Polish army under King John Sobieski, whose heavy cavalry (3,000) and light cavalry (13,000) carried the day. The defeat of the Ottomans and their Tartar allies led to 16 more years of warfare and the reconquest of Hungary as well as the death of the Turkish commander Vizier Kara Mustafa.

James Williamson, *The Age of Drake* (New York: World, 1965). Makes a strong case that the Great Captains Walter Raleigh, John Hawkins, Francis Drake, and others of the Elizabethan Age, operating as privateers and even pirates but also as agents of the crown, set the stage for England's successful defense against the Spanish armada of 1585 and eventual force projection around the world.

Sam Willis, *The Fighting Tereraire* (London: Quercus, 2009). Taking Joseph Turner's extremely popular painting *The Fighting Tereraire Tugged to Her Last Berth to Be Broken Up 1838* as his point of departure (and the several ships that held that name), the author explores the naval world of the 18th century, showing some of the reasons England proved superior to some of its European rivals (France, Denmark, and Spain) in the age of sail. Contains some excellent accounts of the exhausting nature of naval gunnery during the period.

Peter Wilson, *The Thirty Years War: Europe's Tragedy* (Cambridge, MA: Harvard University Press, 2010). A comprehensive, even exhaustive, study of the war in all its dimensions: political, religions, economic, and social. Surprisingly weak as to the revolution in military affairs during the period but gives good material on the "great—and not so great—captains" of the era. Points out that political scientists are very positive about the Treaty of Westphalia, which concluded it because it set up the international system still in place today, but historians are more negative because the treaty really didn't solve any problems for the Holy Roman Empire, and on balance, the results of the war were destructive, degrading, and futile. Wilson tries to put that destruction into some kind of comparative framework, claiming 20 percent of the empire's prewar population was killed (5 million), compared with 5.5 percent for Europe in World War I and 6 percent in World War II (p. 787). He disagrees with Wedgewood (reviewed earlier) concerning the extent of the casualties.

John B. Wolf, *The Emergence of the Great Powers 1685–1715* (New York: Harper and Brothers, 1951). Looks at how all the Great Captains and their service to states resulted in the expansion of power and led to the "World War of 1683–1699." Sees the bayonet replacing the pike and the musket, mobile artillery, and new fortresses as keys to military change in this era. The chapter on the economic demands of war is quite good. He believes that war during this period lies halfway between medieval warfare and modern industrialized warfare, and shortages of funds always threatened to start or at least cut the caloric intake of Mars.

MAURICE OF NASSAU (1567–1625)

This son of William of Orange searched antiquity to then accent light infantry and combined arms with a strong belief in "shot over shock" and linear formations. He invented volley and countermarch fire in Europe (the Japanese did the same in Asia) and instituted the "Dutch formation," which was more linear (only 10 ranks deep) than the contemporary Spanish heavy, more concentrated formations (15–25 ranks deep). The smaller "battalions" of 550 men (with 250 pikes and 300 firearms) offered considerable tactical flexibility over the larger Spanish 1,500-man tercios.

He also helped organize the rebellion against Spain and excelled in sieges, infantry, and cavalry strategy and tactics. He made discipline and training the sine qua non for all military matters, instituting a professionalism for officers that was very innovative at the time. His successes enabled Holland to develop behind secure borders and keep its freedom, and his musket and pike formations spread throughout Europe, albeit slowly. Overall, the Dutch revolt against the Spanish Habsburgs lasted from 1568 until 1609. In terms of tactics, the Maurician reforms took a mercenary army or a medieval levy and turned it into a professional one.

Note: Michael Howard points out that one of Maurice's pupils, Jacob Delagardie, became military instructor to Gustavus Adolphus before he became king. Many of Adolphus's innovations were based on those of Maurice. Maurice should be studied far more than he is.

Pieter Geyl, *The Revolt of the Netherlands 1555–1609* (New York: Barnes and Noble, 1958). This Dutch scholar provides an in-depth look at the life and times of Maurice of Nassau and the many interlocking events and developments against which were set his military reforms. Richly developed, this work is perhaps the definite history of the period. See especially "The Conquests of Maurice" (pp. 217–52).

Geoffrey Parker, *The Dutch Revolt* (Ithaca, NY: Cornell University Press, 1977). An in-depth look at the revolt in the Spanish Netherlands, which ended up with the Dutch Republic. Good coverage of the ebb and flow of battle during the war.

GUSTAVUS ADOLPHUS (1594–1632)

The Swedish "Lion of the North" strode about in the 16th century, reintroduced some Roman dimensions back into warfare, played a big part in the Thirty Years' War, and died at the Battle of Lutzen. More importantly, he is one of the first rulers in Europe to merge the Gunpowder Revolution and the Military Revolution with the State-Building Revolution to extract from the governed society the necessities for success in warfare. He developed standard equipment (as for cannons and cannonballs), organized pay and supply systems, and responded vigorously to the increased scale of war. Sometimes called the "Father of Modern Warfare," one of his many students was Napoleon Bonaparte. He left Sweden the dominant power in the Baltic and was a master of "command and control" in battle.

Michael Roberts, in particular, is most fulsome in his praise of Gustavus Adolphus for having truly changed the nature of warfare during the Thirty Years' War, including emphasis on linear formations; mobile artillery; Roman use of NCOs for command and control on the local, disciplined level; drilling by unit year round (replacing tournaments and individual skill acquisition); uniforms for making "king's men" out of mercenaries; and, above all, a revolution in strategy, making "big war" on many fronts, with strategic battle plans in many areas simultaneously. Noting his emphasis on integrated combined arms and shock value of firepower and speed, Adolphus wrote, "All the wars of Europe are now blended into one."

Richard Brzezinski and Richard Hook, *The Army of Gustavus Adolphus: Book Two, Cavalry* (London: Osprey, 1993). Takes issue with Roberts on uniforms and a few other points but accepts his important accent on Gustavus's "national" army and his heavy accent on fiery weapons (both artillery-blasting infantry and cavalry and infantry firing in big volleys). Good diagram of combined arms formation (p. 33) and some useful descriptions of cavalry units (cuirassier—heavy armor and pistols; dragoons—mounted infantry; harquebusiers—armed with harquebuses to protect cuirassiers and skirmish; and lancers—out of fashion).

T. A. Dodge, *Gustavus Adolphus*, 2 vols. (Boston: Houghton Mifflin, 1895). He made a great deal out of highly mobile light artillery, inventing a gun that could be moved by a couple of horses or men. He used artillery to outshoot infantry, increasing the ratio of cannons to men from 1:1,000 to 6:1,000. This massive—almost-900-page work—is not only about Adolphus but also about the art of war from the Middle Ages on and covers some of the campaigns of Marlborough, Turenne, Conde, Eugene, and Charles XII (who Dodge finds lacking in "discretion"). Believes that "Gustavus Adolphus has fairly earned the title of Father of the Modern Art of War" for his successful meshing of prince, state, and population and for his mixed used of foot, cavalry, and artillery.

Philip J. Haythornthwaite, "Gustavus Adolphus—Lion of the North," in his *Invincible Generals* (Bloomington: Indiana University Press, 1992), pp. 7–49. Gustavus Adolphus, as Swedish king Gustavus II, led a military revolution in the 17th century, using pikemen in the center, musketeers on the flanks, and light cavalry armed with pistols and swords as shock troops once the musketeers had decimated the opposing pikemen and his pikemen had pushed forward. All arms were supported by both highly mobile light artillery and heavy standardized artillery (Adolphus reduced 26 types of cannons to 3).

G. A. Henty, *The Lion of the North* (London: Blacky and Son, 1995) (F). A charming and cleverly illustrated introduction to the life and times of Gustavus Adolphus and the first half of the Thirty Years' War written with fictitious but helpful dialogue. Covers the period up to the death of Wallenstein with skill and dispatch and teaches many good virtues, such as courage, daring, and military service.

James Lacey and Williamson Murray, "Breitenfeld: The Creation of Modern War 1631" in their *Moment of Battle: The Twenty Clashes That Changed the World* (New York: Bantam, 2013), pp. 162–81. These authors place Gustavus Adolphus as the very center of the changes that made for modern war. Excellent maps and descriptions of all the alterations to the battle arrangements of the day credited to Gustavus Adolphus. Good introduction to the battle and this Great Captain.

C. A. La Croix, *Gustavus Adolphus* (Sand Springs, OK: Grace and Trust Books, 2002). A short, loving portrait of the Lion of the North, heavy on his religious motivations and good Christian character. According to the author, he was "[o]ne of the greatest characters of modern history and one of the purest of the Reformation" and under "divine protection," at least until he was killed during the Battle of Lutzen in 1632. But this sacrifice was not in vain, for "in losing it (his life) he delivered nations from the attempt to eliminate Protestantism entirely" by his efforts during the Thirty Years' War.

Michael Roberts, *Gustavus Adolphus and the Rise of Sweden* (London: English Universities Press, 1973). In-depth and covers a great deal on both showing how Adolphus's system of rewarding soldiers with royal land at the end of their 20 years' military service integrated them into the nation and deferred monetary payment in that way. Also, Adolphus was instrumental in paying soldiers in cash on the march to reduce looting and ill discipline during the Thirty Years' War.

THE DUKE OF MARLBOROUGH (1650–1722)

Regarded by some as something of a treacherous English swine, switching sides and loyalties depending on which way the English throne was swinging or giving him honors or gold or both, he was a very important military figure. Others highlight his role as marvelous captain, one who never lost a major battle and left a magnificent estate, as well as his descendant Winston Churchill, to posterity. Churchill was master of both movement and fire, and his strategic brilliance came from being able to see the whole of the war, the whole of the enemy, and the whole of the battlefield in his mind's eye. His victories at Blenheim, Ramillies, Oudenarde, and Malplaquet remain classics on how to win decisive battles. Perhaps we should judge his career on strictly military grounds, taking C. T. Atkins's caveat in *Marlborough and the Rise of the British Army*, "It was not an age of high standards of morality, public or private," at face value.

C. T. Atkinson, *Marlborough and the Rise of the British Army* (New York: C. P. Putnam's Sons, 1921). A thorough examination of Marlborough and "his peculiar excellency" as a leader, strategist, administrator, organizer, and tactician (in particular, his use of infantry fire discipline and fire control, making full use of the infantry socket bayonet—the pike having been eliminated in 1706—and the shock action of cavalry). Deals quite effectively with the criticisms of Marlborough from Jonathan Swift, Thomas Macaulay, and E. M. Thackeray, ending up agreeing with Captain Robert Parker's eulogy, "he never fought a battle which he did not gain, nor laid siege to a town that he did not take." As for his seeming venality, Atkinson overrides it with a quote from Henry Bolingbroke: "He was so great a man that I forgot his errors."

David Chandler, *The Art of Warfare in the Age of Marlborough (1688–1748)* (New York: Sarpedon, 1995). Holistic and well organized, with functional sections, such as "The Foot," "The Horse," and "The Engineering Services." Gives much-needed acclaim to the lead engineers of the era, Sebastian le Presete de Vauban, Menno van Coehoorn, and George Rimpler.

———, *Marlborough as Military Commander* (London: Penguin, 1973). A very positive but still insightful account and analysis of Marlborough's life and career. "A general who never rode off a battlefield except as a victor" in 10 consecutive major campaigns, Marlborough was nevertheless a "lifetime intriguer" and a "profound dissembler" with a ruthless drive who was "undoubtedly an opportunist" and one who, with a lack of remorse, burned 400 villages during a 1704 ravaging of Bavaria in the name of "cruel necessity." Nevertheless, "England has never produced a better soldier," claims the author.

Winston S. Churchill, *The Life of Marlborough*, 4 vols. (London: George G. Harrap, 1933). Proud is the author, and detailed is the presentation. An excellent place to start one's study of this warrior prince. Not only was Winston very proud of his heritage, but also, during World War II, he very much fancied himself a master strategist the equal of his illustrious ancestor. Wild Churchillian strategic lunges like those into Norway and Greece and the Dieppe raid suggest that this was not always the case.

Virginia Cowles, *The Great Marlborough and His Duchess* (New York: Macmillan, 1983). A praiseful account, in-depth and novel-like in spots, as masses of details of romance and spite are covered. There is some good discussion of things diplomatic, political, and military as they pertain to Duke John, but at heart, this is more the story of Sarah Churchill, a "fearless, outspoken, liberty-loving woman two centuries ahead of her time."

Eric Niderost, "Stressed to the Breaking Point," *Military History* (February 1988), pp. 42–49. A useful analysis of the Battle of Blenheim, one that credits Marlborough with institutionalizing "rolling volley fire" and checking the ongoing ambitions of Louis XIV.

George Malcolm Thomas, *The First Churchill* (New York: William Morrow, 1980). The life of John Churchill, the first duke of Marlborough, is examined, primarily from the point of English and continental politics. He is described as having "cuckolded one king, betrayed another and had become the distrusted confidante of another," making himself enormously rich and successful in the process. Churchill was a complex personality, says Thomas, possessed of "military acumen, even genius."

Peter Verney, *The Battle of Blenheim* (New York: Hippocrene Books, 1976). The author makes a strong claim that this was the battle that checked the military ambitions of Louis XIV for the first time in 40 years, as well as made the duke of Marlborough's military bones.

Note: "The First Churchills," a 12-part *Masterpiece Theater* miniseries, gives "Jack" the benefit of the doubt, albeit in interesting fashion, although the "Duchess" steals the show.

CHARLES XII (1682–1718)

Another Swedish captain who ends up losing to the Russians and others after many brilliant victories. Horrendous losses by the Swedes, which ultimately take 25 percent of all fighting-age males by the end of his reign. He instituted "universal" military service by assigning each village in his realm a quota and enforcing it. Charles XII is reported to have abstained from enjoying alcohol or women, leaving him quite free to wage war for much of his adult life.

R. Nisbet Bain, *Charles XII and the Collapse of the Swedish Empire 1682–1719* (New York: G. P. Putnam's Sons, 1895). Shows imperial overreach can undo great generalship. A lot of campaigns far from home over a long period of time would wear out anyone and his or her army.

Peter Englund, *The Battle That Shook Europe: Poltava and the Birth of the Russian Empire* (London: I. B. Tauris, 2003). Explains how and why this battle in 1709 was such a crushing defeat for the Swedes and their Cossack allies during the Great Northern War, when Russia, Denmark, Saxony, and Poland combined against Sweden. Astonishingly high Swedish casualties totaled nearly 60 percent. Some survivors did not return home for more than 30 years after being trapped at the Dnieper River.

R. M. Hatton, *Charles XII of Sweden* (London: Weidenfeld and Nicolson, 1968). An important contribution to the evolution of warfare was made by his village-by-village draft, the use of mobile artillery, and extremely good training. Small Swedish armies defeated much larger ones all across Europe. Very good on the Great Northern War and Charles's victories in Denmark,

Saxony, Poland, and Norway and initially against the Russians. But if you go to a map and see where Poltava is (southern Ukraine), where the Swedes were finally defeated after Charles XII was wounded, you can understand the term *imperial overreach* quite well. Charles managed to survive in Turkey, making it back to Poland and Sweden before being killed in battle in Norway—a huge hole was made in his head, and scholars continue to argue about what caused it. Doesn't really matter, I suppose, but it's a hell of a hole when you see the photograph in this book.

Jonathan W. Jordan, "Tipping Point at Poltava," *Military History* (June/July 2009), pp. 50–57. A short but very worthwhile introduction to a very important battle that showed how Russia was on the rise and Sweden had tried to go a geopolitical bridge too far. But by all means, study the little map above the June 28, 1709, battlefield illustration, and see just how far Poltava is from Stockholm. This location was a long, long way from Sweden.

Michael Roberts (ed.), *Sweden's Age of Greatness 1632–1718* (New York: St. Martin's Press, 1973). A collection of solid essays putting the accomplishments of Charles XII in historical perspective by giving considerable credit to Gustavus Adolphus and Charles X as a strong monarchy and an efficient administration made for a solution to Sweden's previous security problems. Sweden's empire expanded into Saxony, Denmark, Norway, and Pomerania as "war sustained war." But Swedish victories spurred Peter the Great to modernize the Russian army, and by 1721, only Sweden, Finland, and bits of Pomerania remained under the Swedish Empire.

Francois Marie Arouet de Voltaire, *The History of Charles XII King of Sweden* (London: Folio Society, 1976). This reprint of Voltaire's 1731 work focuses on the Great Captain theme and sees the Battle of Poltava as a terrible disaster: "Charles XII had lost in one day the fruit of nine years labours and nearly a hundred battles." He provides vivid "inside information" about Charles XII's three-and-a-half years as a "guest" of the Sultan and his eventual return to Sweden via Hungary, Moravia, Germany, Austria, and Bavaria, where he fought again, only to be killed in battle at the age of 36½.

LOUIS XIV (1638–1715)

While not strictly a Great Captain in the sense of being a battlefield general, he nevertheless greatly changed the nature of warfare during his era through the raising of a professional standing army of 400,000 and the creation of a series of fortresses designed for both offensive and defensive purposes on the frontiers. Louis XIV instituted a military administration that was a great leap forward, creating a secretary of state for military affairs (1643), a country-wide magazine system, and an outstanding engineering corps. He also pushed the boundaries of the French state to their present limits, adding territory in Artois, Lorraine, Alsace, Franche-Comté, and Provence, and then defended them successfully.

Michael Howard also points out that Louis XIV's creation of a civil bureaucracy to administer the army, crown control of procurement, and supervision of regiments and military districts through a series of inspector generals resulted in a French army that was the "most remarkable instrument of state power that Europe had yet seen." And Geoffrey Parker underscores his truly martial nature, indicating that he participated in more than 20 sieges from 1650 to 1692. Young Louis started at age 12.

John Childs, *Warfare in the Seventeenth Century* (London: Cassel, 2001). In the 1600s, through the Thirty Years' War and the campaigns of Louis XIV, war shifted from late-medieval to the dawn of modern warfare, including the rise in importance of paid soldiers and new equipment.

James Falkner, "Father of the Fortress: Vauban, France's Master Engineer of the 17th Century, Designed Nearly 100 Fortresses for Louis XIV—and Defined the Form for the Ages," *Military History* (September 2012), pp 34–45. An amazing article, with truly astonishing maps and diagrams. You will never look at a "fort" again without thinking of Vauban's masterpieces and his "fence of iron" after reading this article.

———, *Marshal Vauban and the Defense of Louis XIV's France* (South Yorkshire, UK: Pen and Sword, 2011). Engineer-in-chief to Louis XIV and general of fortifications, he applied "reason" to warfare in order to secure the boarders of France with his "fence of iron" of 180 fortresses, citadels, and forts, a fence that eventually thwarted Marlborough. He was also a master of fortress destruction, thus contributing both offensive and defensive dimensions of warfare. Falkner's section on siege warfare is the best I have yet encountered.

FREDERICK THE GREAT (1712–1786)

Famous for reintroducing Roman-style discipline to his armies and for the massive use of artillery. He also set up horse artillery and trained it to move almost as fast as regular cavalry. Very mobile with light three- and six-pound guns, horse artillery used limbers, to which the horses and cannon were attached. Horse artillery would run *ahead* of the infantry, stop outside musket range, and fire into opponent's infantry. Voltaire once said, "Prussia is not a country which has an army, but an army which has a country." For his part, Frederick the Great came to see the advantage of speed in battle: "The faster the attack, the fewer the soldiers it costs. By making your battle short you will deprive it of the time to rob you of many soldiers. The soldier who is led in this manner will gain confidence in you and expose himself gladly to all danger." Also very good at flanking maneuvers and feints, as well as thinking and acting in the middle of his battles.

Robert Asprey, *Frederick the Great: The Magnificent Enigma* (New York: Backinprint, 2007). A wide-ranging and perceptive biography of a genius at war (and other things). Very much an in-depth account of him and his times.

Christopher Duffy, *The Army of Frederick the Great* (New York: Hippocrene Books, 1974). The author is a tad harsh on the "old devil" Hun and rightly gives credit to Frederick's father, Frederick William, for his important innovations in terms of the Prussian armament industry, its recruiting patterns, and especially his protection of capital from people. Frederick the Great's innovation included harsh Roman-style discipline, sustained ruthlessness, and an eventual replacement of "cold-steel" bayonet charges with sheer firepower and maneuver.

David Fraser, *Frederick the Great: King of Prussia* (New York: Fromm International, 2001). While not great on the military aspects of Frederick the Great, except in the chapters "Education of a Prince," "Quadrille to Distant Gunfire," and "The Last Campaign," this work provides depth, balance, and detail and does emphasize his correct emphasis on supply, discipline, and overall efficiency in warfare and kingship.

Jay Luvaas (ed.), *Frederick the Great on the Art of War* (New York: Free Press, 1966). A wide-ranging look at the thought of Frederick the Great, whose *Military Instructions of the King of Prussia for His Generals* is a classic of 18th-century war, waxes eloquent on tactics, military change, and the anatomy of battle, declaring, "[B]attles decide the fate of nations." Very valuable insights into the mind of one of the truly Great Captains.

Theodor Schieden, *Frederick the Great* (London: Longman, 2000). The "Soldier King" believed that "war could only be learned by studying the great examples in history." The author concludes that, for the Prussian army under Frederick, the "wars it fought had only a single political goal: to preserve and enlarge the state." This, he feels, was the source of Frederick's greatness but also shows his limits militarily.

Chapter Seventeen

Warfare on the Periphery and the Expansion of the European Gunpowder Empires

While the advances in military technology and the expansion of the European empires as they fanned out across the globe by land and by sea brought enormous and often calamitous impacts on the peoples found in their path, most of that warfare took place on the periphery of meaningful military change. From the military point of view, virtually all advances in military technology and warfare generally grew out of discoveries and innovations of the peoples of the core, that is Europe, Eurasia, Asia, and Sudanic Africa. The rest of Africa, the Americas, Oceania, extreme northern Europe, and Asia remained on the periphery of both the world's trade and warfare patterns. Therefore, an interesting question arises: In order to keep pace with improvements in warfare technology, does a polity have to be on trade routes, engaged in constant warfare, or both?

Also, what develops over time is the expansion of the European powers out of Europe but always in competition with *other* European powers. This is why the European-built forts along the African coast have guns trained out to sea, not inland. European rivalry eventually spawned a global colonial race.

As World War I and World War II would demonstrate, at the end of the day, the fate of the European empires were decided by both the winners on the battlefield *and*, upon their exhaustion, such new players as the United States and the Soviet Union, which rose to global power as Europeans fought each other. Materials on nationalist movements and the end of the empires follow chapter 28.

What follows are some of the works that capture the interplay between the military core(s) and the peripheries into which the European gunpowder empires expanded.

Fred Anderson, *Crucible of War, The Seven Year's War and the Fate of Empire in British North America, 1754–1766* (New York: Random House, 2000). Shows what happens when the core states, such as England and France, fought on the periphery.

Jeremy Black, *Western Warfare, 1775–1882* (Bloomington: Indiana University Press, 2001). Looks to the American and French Revolutions, as well as the wars of Napoleon, as setting the stage for new naval technology and strategic imperial outreach and the "varied character of military modernity" involved in the great expansion of European power force-projecting worldwide.

C. R. Boxer, *The Dutch Seaborne Empire 1600–1800* (New York: Alfred A. Knopf, 1965). The United Provinces were born in war (an 80-year struggle), expanded in war overseas, and ultimately lost most of their empire to war (fighting the English, Portuguese, French, and Spanish, as well as the native peoples along the way). These merchant adventurers traded with Japan, China, India, and established colonies and trading posts in North (New Amsterdam) and South America (Suriname, etc.), as well as Africa (Cape Coast and Cape Colony), Formosa, and the Dutch East Indies. Their example highlights the great shift in world power (1450–1700) from the rest of the world to Europe thanks to the marriage of the cannon and the ship and the exported sense of reality: "There are no Ten Commandments south of the equator." "The real world was the Dutch world."

———, *The Portuguese Seaborne Empire 1415–1825* (New York: Alfred A. Knopf, 1969). Using the themes of Guinea gold, slaves, and spices and leading to a very interesting chapter entitled "Soldiers, Settlers and Vagabonds," Boxer underscores (in contrast with the Spanish Empire) the "conspicuous military element." Despite the poverty of its military strategy ("headlong charges only"), the reliance on African slaves, and the Portuguese military's "notorious indiscipline," Boxer believes that the Portuguese sense of superiority, that they were God's chosen people, resulted in a "nationalism of exceptional durability and toughness."

M. Kathryn Brown and Travis W. Stanton, (eds.), *Ancient Mesoamerican Warfare* (Lanham, MD: Rowman and Littlefield, 2003). Armed only with spears and atlatls, clubs, and darts until much later, when bows arrived from Asia (they did not arrive in the Yucatan until 800 AD despite having arrived in Alaska during the first century AD). These immigrants lacked bows, horses, and firearms. Nevertheless, "[w]arfare was woven deeply into the social fabric of most Formative and all Classic and Post Classic Mesoamerican civilizations."

Eliot A. Cohen, *Conquered into Liberty* (New York: Free Press, 2011). The Albany–Montreal corridor was the "Great Warpath" and conditioned the fighting among French, British, Canadians, Americans, and Indians. Woodland fighting conditioned warfare for two centuries from 1600s through the War of 1812.

George A. Collier, Revato I. Rosaldo, and John D. Wirth (eds.), *The Inca and Aztec States 1400–1800* (New York: Academic Press, 1982). Flower wars and capturing humans for sacrifice left the Incas and Aztecs ill-prepared for Europeans, who practiced sustained ruthlessness and liked their human sacrifices to be in combat or perhaps directly after it.

Dietrich Geyer, *Russian Imperialism* (New Haven, CT: Yale University Press, 1987). Points out how much of Russian expansion was "borrowed imperialism" and driven by Russian industrialization. The author traces the expansion into the Balkans, Central Asia, the Caucasus, Turkistan, Turkmenistan, and then East Asia, where the Russians bumped into Japanese imperialism and got a bloody nose for their trouble in 1905. Others have termed it the "saltwater fallacy," when authors overlook the fact of Russian imperialism because it was conducted primarily on land and involved contiguous territory.

Ross Hassig, *Aztec Warfare: Imperial Expansion and Political Control* (Norman: University of Oklahoma Press, 1988). How the Aztecs captured an empire that disintegrated quickly under pressure from the Spanish.

———, *War and Society in Ancient Mesoamerica* (Berkeley: University of California Press, 1992). Good material on the Mayans as well as others, showing anything but a peaceful people living happily in the tropics. Bloodletting was central to the culture on many levels.

Ashley Jackson, *Distant Drums: The Role of Colonies in British Imperial Warfare* (Brighton, UK: Sussex Academic Press, 2010). Imperial warfare on both the periphery and in some key core areas required allies, minions, and manpower drawn from the colonies. From the Falkland Islands to Malta to Ceylon to Singapore and Mauritius, the "martial character of colonies" and the "importance of the imperial monarchy are traced."

Kwasi Kwarten, *Ghosts of Empire: Britain's Legacies in the Modern World* (New York: Public Affairs, 2011). The author sees the British Empire as unique, a "series of improvisations" marked by individualism, elitism, snobbery, and the "sheer audacity of its self-belief." Kwarten makes the point that that British Empire was never really administered in the field by the upper-class elite but rather by the working-upper-middle class, albeit with some upper-elite boneheads (Randolph Churchill comes to mind) often in charge at the colonial office. His five case studies are illuminating: (1) selling Gulab Sigh Kashmir for $50 million in today's money, an "almost whimsical" gesture with continuing ramifications for its majority Muslim population; (2) creating Iraq, where the British Hashemite kings were a "disaster"; (3) seizing upper Burma, truly a "white elephant," even though the British were forced to raze whole villages in order to conquer it in, "[o]ne of the most ruthless episodes in the British Imperial story," while, after WWII, the situation led to a power vacuum and one balls up after another; (4) occupying the Sudan, which was an artificial construct in which grandiosity and the Nile combined to create a poisonous racial and ethnic brew, as well as an almost unending civil war; (5) creating Nigeria, where the "Center Cannot Hold" and never could in this totally artificial creation as Lord Lugard's "Dual Mandate" of making money and doing good soon became a world-class fiasco of ethnic strife, civil war, and corruption rivaling any anywhere; and (6) intervening in Hong Kong, where the British wanted to insure a market for Indian opium and pedaled an ossified, hierarchical, snobbish society that sold out any local democrats. It is hard to fault his central conclusion that, on balance, the British Empire was a destabilizing force or, at least, a force that created as many problems as it solved. Certainly not in the examples he selects.

Douglas Leach, *Flintlock and Tomahawk* (New York: Macmillan, 1958). King Philip's War as an object lesson in core/periphery imbalance. Very little from the periphery ever worked its way back into the core as meaningful innovation.

Patrick M. Malone, *The Skulking Way of War: Technology and Tactics among the New England Indians* (Baltimore, MD: Johns Hopkins University Press, 1993). Outside the core, military technology lagged for many reasons, and from a Mars perspective, the Native Americans "deserved" to be at a disadvantage.

Robert Massie, *Catherine the Great* (New York: Random House, 2012). Fascinating woman, fascinating account of her life. The expansion of Russia into Poland and to the Black Sea set the stage for further expansion in the 19th century, all the way to the Pacific. Interesting dynamic of the backward portions of Russia and their spectrum flow into new lands. Also good on the political machinations of Frederick the Great, France, England, and Austria in the Seven Years' War and beyond.

Matt Matsuda, *Pacific Worlds: A History of Seas, People and Cultures* (Cambridge: Cambridge University Press, 2012). "Civilization without a Center" is analyzed and its "many Oceanian Worlds" water-connected and put in perspective. Interesting how the pullback of Chinese imperial power in the 15th century led to a Hobbesian pirates-and-raiders eruption all over the Asian periphery. Good on the arrival of the European colonial powers.

William H. McNeill, *The Rise of the West: A History of the Human Community* (Chicago: University of Chicago Press, 1963). From the "Era of Middle Eastern Dominance to 500 BC" to "Eurasian Cultural Balance 500 BC–1500 AD" to the "Era of Western Dominance 1500 AD to Present," the author finds the expansion of the West to be the defining movement in human history, hence the subtitle. Gunpowder and the Industrial Revolution sure helped.

James Onley, *The Arabian Frontier of the British Raj: Merchants, Rulers, and the British in the Nineteenth-Century Gulf* (London: Oxford University Press, 2007). The Gulf residency system enables Great Britain to influence much of the politics on the Arabian littoral (Aden, Oman, and Trucial Oman, as well as the more distant British Somaliland), reinforcing 2,000 years of history with India and the Arabian Peninsula in close interaction.

Douglas Peers (ed.), *Warfare and Empire* (Aldershot, UK: Ashgate, 1997). A series of essays looks at European firearm penetration of the Ottoman Empire, the Americas, India, Africa, Java, and Japan. Most interesting is the chapter by J. E. Inikori, "The Import of Firearms into West Africa 1750–1807" (pp. 245–74), in which he provides hard evidence in support of his contention that, during this period, 1,615,309 firearms were imported! This averages 283,00 to 390,000 per year (p. 267). Astonishing support for the title of this section. Argues that the gun-slave thesis is far more important than the gun-hunting or gun-protection-of-crops approaches. Makes sense.

Nathaniel Philbrick, *The Last Stand: Custer, Sitting Bull, and the Battle of the Little Bighorn* (New York: Viking, 2010). A very good summary of a battle that has fascinated the American people ever since it occurred in June 1876. Six hundred fifty of Custer's troops were overcome by thousands of Lakota, Cheyenne, and others in a battle they initiated, breaking all the sensible rules of war in the process. With hindsight, Custer should have taken along his Maxim guns, but he believed they would only slow him and his men down, which, when you think about it, would have had a salutary effect on his campaign just by itself.

Claudio Saunt, *West of the Revolution* (New York: W. W. Norton, 2014). A fascinating look at the rest of North America in 1776, showing the very tenuous holds of the Spanish in the Southwest and California, Russia in Alaska, and the many changes within Native American cultures of the era, including the movement of the Lakota into the Black Hills in the year the Declaration of Independence was signed. Gives a good sense of the early European dependence on Native American goodwill.

Yuri Semyonov, *Siberia: Its Conquest and Development* (Baltimore, MD: Helicon Press, 1963). A Russian version of why Russian expansion was good for the native peoples it overwhelmed. A good reminder of the "saltwater fallacy," which points out that just because colonial expansion did not take place across the ocean does not mean it did not take place. Interesting side aspect depicted here: The czar and the grand duke sell Alaska to the United States because they want to concentrate on Russian Asia, and the author concludes, "It is permissible to say that Russian America paid for Russian Asia."

C. P. Stacey, *Quebec, 1959: The Siege and the Battle* (Toronto: Macmillan, 1959). How one gunpowder empire bested another for an entire continent and, in the process, ironically sowed the seeds for its own loss of much of it by removing the colonies' existential threat. British imperial power waxes as Wolfe defeats Montcalm, but France would get its revenge during the American Revolution. Author believes Wolfe was both a good commander and lucky, while Montcalm did pretty well with what he had but was not a great commander.

George M. Thomson, *Sir Francis Drake* (New York: William Morrow, 1972). Don't be put off by the sometimes flowery writing or the conversational style. This work captures much that is relevant in the competition of the gunpowder empires. The raw piracy backed by the city of London bankers and the queen drives the expansion of the English, as they challenge the formidable Spanish Empire, culminating in the defeat of the Spanish armada of 1588 (incidentally the royal portion of the armada actually did rather well, losing 7 ships compared to 51 for the rest of the fleet). Ironically, the ordinary sailors of both nations paid a tremendous price for their service from the battles, disease, and stingy governments. Drake, says the author, "shared with all the sea rovers: Columbus, Leif Erikson, Ulysses himself. . . . The thrill of escape from queens, statesmen, domestic concerns, and the cramping letter of the law, into the infectious anarchy of the sea."

Carl Waldman, *Atlas of the North American Indians* (New York: Facts on File, 1985). An excellent resource with maps, charts, and useful illustrations, such as the spread of the horse on the Great Plains from 1600 to 1770. Did notice that the Niantics actually occupied territory to the west of their included sphere, but on balance, most maps are clear and helpful.

Note: Warfare during colonization and decolonization does not enjoy a separate section in this annotated bibliography. This is due in part to the notion that, in terms of the history of warfare, very little innovation, or indeed change in war-making techniques and weapons, occurred during the colonization or decolonization process. Instead, a number of works dealing with revolutions in the former European colonies are reviewed in the next chapter, "Revolution and Civil War." Still, there are some recent, stimulating works that can provide the interested reader with insights into the nature of colonial formation and deformation. Among the most enjoyable is:

Robert Wilkinson-Latham, *The Sudan Campaigns 1881–1898* (Oxford: Osprey Books, 1976). A quick overview of how one gunpowder empire (the British) turned away another (the French) as they both scrambled to take over what was left of Africa. The British campaigns against the Mahdi and his successor also resonate.

Chapter Eighteen

Revolution and Civil War

"Every revolution begins as a movement of liberation but ends in tyranny."—Ignazio Silone, *Bread and Wine*

"The patricians started the revolution, the plebeians finished it."—Francois Rene Chateaubriand, "*Essai Sur Les Revolutions*"

Revolution is such a debased term (think of the "revolution in hair coloring") that it is often difficult to appreciate the deep-seated socioeconomic-political upheaval accompanied by the widespread violence that a true revolution entails. It is also very difficult not to read the literature and come away with a prorevolution bias in terms of outcomes because most accounts are by successful revolutionaries *or* studies of successful revolution. In fact, truly successful revolutions are few and far between, and in the works cited in this chapter, emphasis is placed on the most important, the ones that became models for others, especially the American, Russian, French, and Chinese.

Revolutions are great incubators of conflict, not just in the societies in which they occur, but also in neighboring countries, whether attempts are made to spread the revolutions to them or the neighboring states attempt to destroy the revolution. In the process, other states get involved on the side of the revolutionaries or on the side of the counterrevolutionaries. Starting with the revolution on Corcyra in 427 BCE, which drew in first Corinth, then Athens, and then Sparta, or the American Revolution, which drew in the French and Spanish, or the later French and Russian Revolutions, many produced a major reaction among many of the states of Europe. Revolutions can produce great changes in warfare, as well as the French levy en mass, which eventually led to the democratization of warfare throughout the continent of Europe and eventually across the world.

The human dimension of revolution and the mind-set of revolutionaries is well-captured in the words of Sergei Nechaev in his *The Catechism of a Revolutionary*:

> The revolutionist is a doomed man. He has no private interest, no affairs, sentiments, ties, property nor even a name of his own. His entire being is devoured by one purpose, one thought, one passion—the revolution. Heart and soul, not merely by word but by deed, he has severed every link with the social order and with the entire civilized world: with its laws, good manners, conventions and morality of that world. He is its merciless enemy and continues to inhabit it with only one purpose—to destroy it.

Note: For additional works on the American and other revolutions, especially their insurgency and counterinsurgency dimensions, please see chapter 8.

Fred Anderson, *Crucible of War: The Seven Year's War and the Fate of Empire in British North America 1754–1766* (New York: Random House, 2000). This great military conflagration of the 18th century (the "first world war") left Great Britain supreme in North America but ironically removed the threat of the French and Native Americans from the British colonies and thus led to one of the key conditions for the American Revolution.

Keith Michael Baker, *The Old Regime and the French Revolution* (Chicago: University of Chicago Press, 1987). Especially good on "Revolutionary Politics" and "The Evolution of the Terror." A grim reminder that revolutions can assume a life of their own.

Gregor Benton, *Mountain Fires: The Red Army's Three-Year War in South China, 1934–1938* (Los Angeles: University of California Press, 1992). A much-needed, long-overdue book looking at the "other" Red Army in China, with its "southern guerrillas and its urban submarines." The author argues that, as a "symbol of patience and stoical endurance," the southern comrades deserve great credit and recognition, not just for their contribution to the war effort (tying down up to 180 divisions of Nationalist troops at some points), but also for their "polycentrism and regional diversity," both of which were submerged by the mythology of the northern Great March, with its centralization, homogeneity, and rise to power of those who would rule China after the revolution. A very important look at the Chinese Revolution.

Crane Brinton, *A Decade of Revolution 1789–1799* (New York: Harper and Row, 1934). A classic study, not just of the French Revolution, but also of the notion that nationalism is a civic religion, with all the attractions and detriments of that phenomena. Captures the "dram of the day" quite well.

Brinton provides a good lesson for all those who advocate revolution in a time or a place: "The Revolution did not bring freedom of thought."

Ron Chernow, *Washington: A Life* (New York: Penguin, 2010). What a masterwork! Intimate and intricate details and much vital analysis. No matter how much you know about Washington and the American Revolution, you will be amazed at the findings in this volume: how close the British came (several times at Brandywine) to winning the war; how fragile the American army was (many times); how important, even decisive, the French contribution was; how General Howe had Washington close to the end but lacked the killer instinct; how Washington was fixated on getting the French to attack New York (when the revolution would be won in the South); and what an important role disease (small pox) played in Cornwallis's decision to capitulate. Interesting sidebar: A number of Washington's slaves were freed by the British. He got eight of them back at the end of the war, but the British saved thousands and brought them to Canada as freed people.

Nikolai Chernyshevsky, *What Is to Be Done*? (*Shto Delat*?) (Ithaca, NY: Cornell University Press, 1989) (F). A utopian novel of bold feminists and stalwart revolutionaries by a 19th-century Russian who, after Marx and Engels (claims Ian Frazier), was the "writer who had the most influence on the generation that made the Russian Revolution," including Lenin (who may have seen himself as Rakhmetov in search of his feminist counterpart Vera Pavlovna). It seems to have had the impact of discrediting moderate reformers and idealizing the more radical "totally rational revolutionaries," even as it convinced supporters of the status quo to hold firm.

Joseph Conrad, *Under Western Eyes* (Buffalo, NY: Broadview Press, 2010) (F). A reprint of the 1904 novel by the Polish (later, a naturalized British) writer, one that captures both revolutionary fervor and counterrevolutionary suppression before the 1905 Russian Revolution. Unusual slant in that the author is equally critical of the "atrocious" nature of both the revolutionaries and those who repress them.

Charles Dickens, *A Tale of Two Cities* (London: Folio Society, 1985) (F). The essence of revolutionary change is truly captured in this melodramatic tale: "It was the best of times, it was the worst of times, it was the Age of Wisdom, it was the Age of Foolishness, it was the epoch of Belief, it was the season of Doubt, it was the spring of hope, it was the winter of despair, we had everything before us, we had nothing before us." The turbulence; the casting up; and the pulling down of careers, lives, classes, and structures—it's all here.

Frank Dikotter, *Mao's Great Famine* (New York: William Morrow, 2010). The author of seven previous books on China gets access to Chinese Communist Party records and archives and concludes, "Between 1958 and 1962, China descended into hell," when "at least 45 million people died unnecessarily" as a result of the Great Leap Forward. In that time frame, more damage was done to China's property and infrastructure than by all the bombing of World War II. According to party records, 2.5 million people were tortured or summarily killed. After 1962, of course, would come Mao's Cultural Revolution, "which would cover up his unmatched crimes." This book should be required reading for any serious person interested in politics, power, ideology, and intellectual gullibility.

William Doyle, *The Oxford History of the French Revolution* (Oxford: Oxford University Press, 1989). Illuminating account of how the central government has to make a series of blunders in order to give the revolutionaries a chance of success. The French ancient regime did more than most to assist them. Good on the Thermidorian Reaction of 1794–1795, which he terms "burial of the Revolution."

Laurent Dubois, *Avengers of the New World: The Story of the Haitian Revolution* (Cambridge, MA: Harvard University Press, 2004). Provides an in-depth look—and largely proves—its basic proposition, namely, the "revolution, which began as a challenge to French imperial authority by colonial whites, but it soon became a battle over racial inequality and then over the existence of slavery itself." The slave revolt started in 1791, but the abolition of slavery wasn't actually recognized in France until 1794. Thus it was the creation of a new nation, not simply a revolt against colonial rule, guided by the American and French Revolutions that preceded it.

Jonathan Dull, *The French Navy and American Independence* (Princeton, NJ: Princeton University Press, 1975). See especially "De Grasse's Campaign" (pp. 238–49). Gives Admiral De Grasse high marks for going to the Chesapeake rather than the Hudson, for winning the Battle of the Virginia Capes (September 5, 1781), and for bringing siege artillery to win the battle of Yorktown.

Frederick Engels, *Revolution and Counter Revolution in Germany* (Beijing: Foreign Language Press, 1977). His *New York Daily Tribune* reports from 1851 to 1852 shed light on the events of 1848 and success of the counterrevolutionaries thereafter. Sees the real revolutionary class as the working classes of the towns. As both an observer and an activist, he has a keen, if ideologically filtered, eye.

John King Fairbank, *The Great Chinese Revolution: 1800–1985* (New York: Harper and Row, 1986). Very useful in showing Mao's cleverness as well as his deviousness and putting both in historical perspective. Shows how revolutionaries ultimately depend on the counterrevolutionaries to make bad decisions or to fail to carry out good ones.

David Fischer, *Washington's Crossing* (London: Oxford University Press, 2004). This "vivid portrait" of a "pivotal moment in American history" is well-captured as the author declares, "No single day in history was more decisive for the creation of the United States than Christmas 1776." Battered by successive defeats (New York to Philadelphia) and seemingly endless retreats, Washington's crossing of the Delaware to attack the British Hessians at Trenton (2,400 colonials versus 1,500 Hessians); followed by his movement back across the Delaware and coming back to Trenton on January 2, 1777, to win another victory, this time over Cornwallis; and then his subsequent avoidance of a set-piece battle there by moving on to another victory at Princeton (when a sudden freeze made movement possible after a January thaw). These victories broke the back of British momentum and led to the "Forging War" of 1777, in which the American militia and regulars forced the British back to their bases and gained the support of many in the New Jersey countryside, as well as making it clear internationally that the British had not finally defeated the insurgents. Very clear maps, the best I have seen on this campaign.

Alexandra Fuller, *Scribbling the Cat: Travels with an African Soldier* (New York: Penguin, 2004). Alexandra meets K, a former member of the Rhodesian Light Infantry who is rather unrepentant about his roles in "hunting gooks" and "killing mundts," including civilians. The chap is obviously suffering from posttraumatic stress disorder, but it is still a bit hard to feel too sympathetic. The encompassing corrosive effects of war are writ large in this tale of an African rural revolution.

David P. Geggus and Norman Fiering (eds.), *The World of the Haitian Revolution* (Bloomington: Indiana University Press, 2009). A wide-ranging collection of essays covering the revolution, its precursors, and its aftermath. Too often, Haiti is overlooked in the examinations of revolutions.

Pieter Geyl, *The Revolution of the Netherlands 1555–1609* (New York: Barnes and Noble, 1958). The "Revolution in Netherlands Civilization" and the new nation it produced went on for an 80-year war. Now that was a long march! Another dimension seldom highlighted by revolutionary theorists.

Nadine Gordimer, *July's People* (New York: Penguin Books, 1981) (F). In every revolution and counterrevolution, there is a moment when, in Antonio Gramsci's words, the "old is dying and the new cannot be born; in this interregnum there arises a great diversity of morbid symptoms." Gordimer captures the dynamics of this twilight zone where those not committed to the revolution or to its suppression find themselves threatened and becalmed, confused and in great danger. South Africa is the setting.

Colin D. Heaton, "Chetnik Mountain Guerrilla," *Military History* (July 2005), pp. 30–36. If you like your guerrilla wars complicated, try former Yugoslavia during World War II, when German soldiers, the Royal Yugoslav Liberation Army, Croatian Fascists, and Marshal Tito's Yugoslav Communists all fought each other over their visions for the future and their down drags from the past.

Alistair Horne, *A Savage War of Peace, Algeria, 1954–1962* (New York: NRRB Books, 2006). The Algerian Revolution seen quite convincingly from both sides. A classic examination of the upheaval of change and the turmoil it brings. Well worth reading.

Hugh Howey, *Wool* (New York: Scribner's, 2013) (F). All you ever need to know about revolution is contained in this futuristic novel. Revolutions require (1) inequities in the political, economic, and social systems of an entity; (2) knowledge of those entities; (3) alternative models; (4) a spark or series of sparks to focus attention; (5) revolutionarily minded leaders; and (6) a call to arms. To be successful, the revolutionaries have to make a number of crucial decisions correctly and have the counterrevolutionaries make major mistakes. And then, voilà, the silos are run by new people who restructure the polity, the society, and the economy. One can apply these basic tenets to all revolutions.

C. L. R. James, *The Black Jacobins: Toussaint L'Ouverture and the San Domingo Revolution* (New York: Vintage Books, 1989). In 1979, Haiti supplied two-thirds of the overseas trade of France. The Slave revolt, which started in 1791, went on for 12 years. Eventually the revolution triumphed over French troops of the Bourbons *and* Napoleon, the Spanish, and a British force of 60,000, and Haiti gained its independence in 1803. In between, there was a violent, brutal "war of extermination" of mutual savagery, rare even for revolutionary violence.

Arthur Koestler, *Darkness at Noon* (New York: Macmillan, 1941) (F). The classic tale of the revolution devouring its own children. Rubashov, the party leader during the Russian Civil War, ends up being executed for being a party leader. The capriciousness of the Soviet system is exposed here, but the tale is timeless.

W. Bruce Lincoln, *Passage through Armageddon: The Russians in War and Revolution 1914–1918* (New York: Simon and Schuster, 1986). A highly readable description of the gradual and then sudden collapse of czarist power and the rise of the revolutionaries, although it does not carry the civil war to its conclusion.

———, *Red Victory: A History of the Russian Civil War* (New York: Simon and Schuster, 1989). A straightforward account of the Russian Revolution. Good material on the White armies, why they lost, and why they didn't have to lose but did.

Jack London, "The Scarlett Plague," in *A Treasury of Doctor Stories*, edited by Noah Fabricant and Heinz Werner (New York: Grosset and Dunlap, 1946), pp. 265–315. One of the essential features of revolution is the upwelling of talent from the lower social and economic strata. See, for example, "The Chauffer Tribe" in London's work about a revolution, albeit one caused by a plague. There is a great deal to be learned about the true nature of revolution by reading this seemingly superficial tale.

Piers Mackesy, *The War for America 1775–1783* (Cambridge, MA: Harvard University Press, 1965). Examines how the "world got turned upside down." Points out that the French-American alliance was not only key but also that it almost fell apart at several points and the British were pressed all over the globe (including Gibraltar while this was going on). Also, had the British "destroyed the rebel armies of the rebellion they might well have succeeded."

Andre Malraux, *Man's Fate* (New York: Simon and Schuster, 1934) (F). An existential novel about the failed Communist uprising in Shanghai in 1927, when Mao and the Chinese Communists attempted to follow the Leninist notion of urban proletariat-based revolution, only to have it crushed. After this, Mao transformed Communist doctrine to make the rural peasants the primary revolutionary cadre.

Karl Marx, *The Civil War in France* (Moscow: Foreign Language, 1952). Looks at the Paris Commune of 1970 and its eventual extinguishment, echoing the judgment of Engels: "What does the Dictatorship of the Proletariat look like?" "Look at the Paris Commune." Of course the commune was suppressed (as are most uprisings and insurgencies, so we shouldn't be surprised that this one failed), but Marx concludes that the "martyrs will be enshrined in the great heart of the working classes." Perhaps. It could happen; at least it did for Horst Wessel.

Robert Middlekauff, *The Glorious Cause: The American Revolution 1763–1789* (London: Oxford University Press, 2005). Holistic, comprehensive, vivid, and full of insights and tales of ordinary people caught up in the battles and movements. Good military chapters on "The War of Posts," The War of Maneuver," and "Inside the Campaigns," concluding, "Some 200,000 men carried arms in the Continental Army and state militias" and suffered the highest percentage of casualties of any American war, save the Civil War.

Rana Mitter, *A Bitter Revolution: China's Struggle with the Modern World* (Oxford: Oxford University Press, 2004). Chronicles the making of a new China and the human sacrifice it cost. Doesn't spare Mao in its account of the Cultural Revolution, quoting his aphorism, "A revolution is not a dinner party. It cannot be so kind, so refined." It wasn't. Although I've often thought it a pity that families of his victims didn't get their hands on him and his.

George Orwell, *Homage to Catalonia* (New York: Harcourt Brace Jovanovich, 1952). Although Lionel Trilling perhaps goes too far in calling this "one of the most important documents of our time," there is a universal quality to its description of the Spanish Civil War that transcends that contest and provides us with a useful glimpse into all revolutions. Orwell's enthusiasm for Communism dims as the revolution devours its children in 1937 Barcelona and by crushing the Trotskyite Partido Obrero de Unificación Marxista (POUM), and the Anarchists, the Communists, "made sure the Revolution never happened." Honest and painful disillusionment for Orwell, who was wounded in the fighting and barely escaped with the help of the British consulate.

Peter Paret, *Internal War and Pacification: The Vendee, 1789–1797* (Princeton, NJ: Princeton University Press, 1961). An early voice crying out for the study of insurgency and counterinsurgency. The author uses the French counterrevolution as an example of a counterinsurgency strategy that worked. Very important work illustrating how the revolution eventually must become a counterinsurgency and devour its own children.

Marifeli Perez-Stable, *The Cuban Revolution: Origins, Course and Legacy* (Oxford: Oxford University Press, 1993). Highly sympathetic to the revolution but still questions whether at the end of the day which legacy, Castro the revolutionary or Castro the caudillo, will endure.

Nathaniel Philbrick, *Bunker Hill: A City, A Siege, A Revolution* (New York: Viking, 2013). A book of almost excruciating detail about the beginning of a revolution in Boston, the hows, the whys, and the might-have-beens.

Kevin Phillips, *1775: A Good Year for Revolution* (New York: Viking, 2012). A somewhat prolix account of why 1775 was more than just an incubator of revolution but also a real beginning. Claims the American militia were much better armed and drilled than previously thought and played a major part along with the grassroots patriot committees. Author makes a firm case for his thesis.

Richard Pipes, *The Russian Revolution* (New York: Alfred Knopf, 1990). Regarded as the classic, for good reasons. This is an excellent, hardheaded work providing a most useful overview of all the dimensions of that revolution, written with the keen mind of a realist. A must-read.

Jeremy Podkin, *You Are All Free: The Haitian Revolution and the Abolition of Slavery* (Cambridge: Cambridge University Press, 2010). A most useful argument that the antislavery message was in the end the central element in the Haitian Revolution and sees the June 20, 1793, burning of La Ville Du Cap Francais as a "true turning point in the struggle for freedom."

Jack Rakove, *Revolutionaries: A New History of the Invention of America* (Boston: Houghton Mifflin Harcourt, 2010). Heavily lardered with stories of the Founding Fathers and the constitutional background of the formation of the creation we know as "America."

Mike Rapport, *1848: Year of Revolution* (New York: Basic Books, 2010). An excruciatingly detailed book covering the flowering of revolution during the spring of 1848 and the eventual success of counterrevolution later in the year. Covering events in France, Austria, Prussia, Italy, Hungary, and Croatia, the author declares, "The Springtime of Peoples crystallized and sharpened national differences," and the brief upsurge in liberties eventually came to flower, albeit long after the horror of World Wars I and II, reaching fruition only in 1989 with the fall of Communism in Eastern Europe. The revolutions of 1848 then were thus a "seed-plot of history." Written by the director of the master's program in Revolution and Counterrevolution at the University of Sterling.

Thomas W. Robinson (ed.), *The Cultural Revolution in China* (Berkeley: University of California Press, 1971). A bridge too far, even for most revolutionaries, as the Red Guards ultimately run amok. Lacks a firm conclusion as to the costs of the upheaval.

John Shy, *A People Numerous and Armed: Reflections on the Military Struggle for American Independence* (New York: Oxford University Press, 1976). A much-needed look at the American Revolution in insurgency/counterinsurgency terms. The author writes convincingly that the local patriot militias were the "sand in the gears of pacification" and that the British thought law and order was the key, drastically underestimated their foes, spread their forces too thinly, and did "search and destroy" rather than pacification for much of the war. Most Americans, he argues, were afraid of revolution but over time underwent a "social process of political education."

Julia Sweig, *Inside the Cuban Revolution* (Cambridge, MA: Harvard University Press, 2002). Ties Fidel Castrol squarely inside the existing Cuban urban underground leadership, which was more collective in nature than previously assumed.

Nguyen Khac Vien, *Tradition and Revolution in Vietnam* (Berkeley, CA: Indochina Research Center, 1974). Written before the ultimate triumph of the North, this work provides a highly personal perspective in which social solidarity and relentless determination shine through.

Harry M. Ward, *The War for Independence and the Transformation of American Society* (London: University College London Press, 1999). The war for independence played a major role in creating the new nation, as wartime developments reinvented the body politic and transformed society, as "victory confirmed the moral character of the patriots."

Richard Wolin, *The Wind from the East* (Princeton, NJ: Princeton University Press, 2010). A very engaging and valuable study of how the dictator and mass murderer Mao became a cult hero to French intellectuals, who anointed "revolutionary violence" as "justifiable homicide." A cautionary tale, and one more indication of the need to suspect French intellectuals as arbiters of this or any other age. A must-read.

W. J. Wood, *Battles of the Revolutionary War 1775–1781* (New York: Da Capo Press, 1990). An excellent collection of battles well researched and well presented, including Quebec, Trenton, Oriskany, Kings Mountain (the biggest British victory of the war, with 5,000 prisoners, 6,000 muskets, and 391 cannon at a cost of 76 killed), Guilford Courthouse, and especially the Chesapeake Capes (quoting Harold Larrabee, decision at the Chesapeake: "On the afternoon of September 5th, 1781, of the Capes of Virginia, two and a half hours of cannonade between warships of the British and French navies determined the outcome of the American Revolution"). The author gives one a sense of the battle itself, the strategy, and impact of the outcome in very manageable doses. So many authors who write on this subject could benefit from his close reading of the relevant sources. Some interesting factoids: The French shipped 100,000 muskets and socket bayonets to the insurgents; there was great diversity in the tactics used (the Redcoats did not march in easily shot ranks in many instances); the British were virtually totally dependent on seaports, but even after Yorktown, they still held New York, Savannah, Wilmington, and Charlestown; and finally, although the war was decided in 1781, the British government didn't accept that outcome until two years later.

Zhang Xianliang, *Half of Man Is Woman* (New York: W. W. Norton, 1986). The author, who spent 20 years in jail, labor camps, and state farms from 1966 to 1986, writes poignantly about his life, with the great irony that life in the camps was safer and more predictable than life outside during the chaos and upheavals of the Cultural Revolution, which sought "to stir up positive events." Longing for love and unable to find it, either in the camps or in ideology, he concludes, "I curse you and I love you, demonic land and demonic woman." Good portrayal of the internal dynamics of revolutionary indoctrination turned on its head.

———, *Mimosa and Other Stories* (Beijing: Panda Books, 1985) (F). "Rightist" Chang's first day out of the reeducation camp in 1961 is a good one, until the rats eat his two buns instead of being eaten themselves. Still, he has crawled out from under a "heap of corpses" in the reeducation camp, and life must go on, even as he reads *Das Capital*, his only book because the writings of Mao are forbidden (because prisoners might learn how to overthrow their masters)! Good-hearted Mimosa provides some comfort before he moves on toward further rehabilitation.

Chapter Nineteen

The Real United States Is Born

For most works dealing with the many aspects of the American Revolution, see chapter 18. Here are contained some of the books that deal with the transition of the United States into a longer-lasting version of itself and the wars (such as that of 1812 and the Mexican War) that shaped its future so cogently. There are also some other works on the American Revolution in this section, which help us to understand its military formulation and various other elements of the United States of America and how it evolved.

The post-Revolutionary period was to determine if the United States was to live or die. In this regard, certainly Allan Millett and Peter Maslowski are correct in their *For the Common Defense: A Military History of the United States of America* (New York: Free Press, 1984): "The post-Revolutionary era, which was one of serious peril for the infant republic, necessitated the development of a military policy that reconciled ideological concerns for liberty with military effectiveness." They also point out that, for 100 years before the War of 1812, those living in what was to become the United States were invariably enmeshed in the wars of Europe and that, 100 years after that war, the Americans would not be so entangled. For his part, Donald Hickey in his *Glorious Victory* (Baltimore, MD: Johns Hopkins University Press, 2015), pp. 6–32, calls the war, and especially the American victory at New Orleans, "Completing the Revolution"

AMERICAN MILITARY HISTORY (GENERAL)

David A. Clary, *George Washington's First War: His Early Military Adventures 1753–1758* (New York: Simon and Schuster, 2011). Or misadventures. It is easy to see why the British weren't keen on giving our hero a queen's commission. Still, this on-the-job training would later pay off, as would his marriage to a wealthy widow. Although in 1775, when he was offered the post of supreme commander of the continental forces, the author concludes, "He was correct that he was not prepared for this new responsibility."

Department of the Army, *American Military History 1607–1953* (Washington, DC: US Government Printing Office, 1956). Quite an interesting, even charming, once-over of the American military's checked career in the service of Mars.

Vincent J. Esposito, *The West Point Atlas of American Wars*, 2 vols., vol. 1: *1689–1900*; vol. 2: *1900–1953* (New York: Praeger, 1959). This superbly designed and extremely well-illustrated (the maps make the wars come alive) should be required reading for anyone interested in the American way of war and the ground on which it was fought.

Don Higginbotham, *The War of American Independence: Military Attitudes, Policies and Practice, 1763–1789* (New York: Macmillan, 1971). The chapters on "Military Origins of the Revolution," "The Continental Army," and "Defeat and Victory in the South" are well-documented. Makes considerable fanfare about the successes of General Nathanael Greene and his military and political activities in the Carolinas. Accents the extent to which the rise of the American militias, the removal of the French and Indian threat after 1863, and the demands of housing British soldiers during a time of "profound peace" all led to the revolutionary impulse. Stresses how important British military and strategic incompetence aided those seeking independence.

Robert Leckie, *The Wars of America* (New York: Harper and Row, 1968). Covering the colonial wars through Vietnam (another colonial war but, of course, in reverse), this is a workmanlike account backed by considerable research.

Robert Middlekauff, *The Glorious Cause: The American Revolution 1763–1789* (London: Oxford University Press, 2005). States the case clearly: "And throughout the long struggle, the passion of the people and their moral strength played a far more important part than in any eighteenth century war before the French Revolution."

Allan Millett and Peter Maslowski, *For the Common Defense: A Military History of the United States of America* (New York: Free Press, 1984). Military institutions, relations with civil authorities, policies, and warcraft from the American Revolution to the end of the Vietnam War and beyond—this is the definitive account of the subject.

Walter Millis (ed.), *American Military Thought* (New York: Bobbs-Merrill, 1966). A wide-ranging collection, from Benjamin Franklin's calls for a volunteer militia (1747) to Robert McNamara's "perils of the nuclear age," this work scans the likes of George Washington, Thomas Paine, James Monroe, Henry Halleck, William Tecumseh Sherman, U. S. Grant, Alfred Mahan, Harry Truman, Douglas MacArthur, and Dwight Eisenhower. Millis concludes, however, "We have produced no Clausewitzes."

Robert Sobel, *For Want of a Nail . . . If Burgoyne Had Won at Saratoga* (New York: Macmillan, 1973). A captivating alternative history of North America if Gentleman Johnny Burgoyne had won the battle and gone on to be a conciliatory governor of the colonies. For without French intervention, the author does not believe the Americans would have prevailed, and the Confederation of North America would have included Mexico and Canada, all under English guidance to the benefit of all.

WAR OF 1812

> "Many people say that our American Republic was born on the Fourth day of July, 1776, in Philadelphia. This is not true. It was only begotten then. It was never confirmed until the 8th of January last."—Captain Henry Garland at New Orleans, February 1815

Whether the War of 1812 was a "civil war," as some writers listed here suggest, or simply a continuation of the American Revolution or simply a war necessary to safeguard that revolution, the War of 1812 is an intrinsic part of the formation of what would become the continental power known as the United States. In light of all these dimensions, it is very interesting to examine the might-have-beens if the British had won the War of 1812.

Charles Brooks, *The Siege of New Orleans* (Seattle: University of Washington Press, 1961). Quite a readable and easily understood account of what happened, including a lot of interesting information on the British troops, where they came from, and how they got to the Mississippi Delta to besiege this yellow fever–infested city of 18,000. Of the slaughter of redcoats, General Andrew Jackson said, "It is magnificent but it is not war." Brooks concludes with some finality, "For the last time, England and America had embraced in violence on the battlefield."

Samuel Carter, *Blaze of Glory: The Fight for New Orleans, 1814–1815* (New York: St. Martin's Press, 1971). Quoting Samuel Eliot Morison that the "Battle of New Orleans . . . wiped out all previous American defeats, ending the 'Second War of Independence' in a blaze of glory," Carter gives a sprightly account of the run-up to the battle, the battle itself, and its aftermath. British losses were extremely high: 3,326 killed and wounded, compared to 13 killed and 39 wounded for the Americans. The Highlanders, for example, lost all but 132 of the 1,000 men with which they started, a higher casualty rate than suffered by the Light Brigade when it charged the Russian guns at Balaklava.

George C. Daughan, *1812: The Navy's War* (New York: Basic Books, 2011). Smooth reading and navy centric, this account gives the reader an overview of the northern, Chesapeake, British Isles, Caribbean, and Mediterranean theaters of operations and concludes that the United States was fortunate to end up on the winning side. Also points to the twin outcomes of better relations with Great Britain and the strengthening of democratic principles by the push of returning veterans for dropping the property qualifications for voting. Readers will want to study the drawing of "The Sails of a Square-Rigged Ship" in the map section, as the author uses a considerable array of nautical terms and assumes familiarity with the various sails on a fully rigged ship. Despite having only 20 ships when the war started, the US Navy and privateers would eventually seize 800 British ships. The United States was fortunate that Great Britain was fighting Napoleon at the same time.

Donald Hickey, *The War of 1812: A Forgotten Conflict* (Urbana: University of Illinois Press, 1989). This most perceptive work belongs here because the author correctly sees the War of 1812 as one of those hinges of history, as the "Second War of Independence" morphs into the myth of manifest destiny. He writes, "According to the emerging myth the United States had won the war as well as the peace," instead of acknowledging that in reality that it had "barely escaped dismemberment and disunion." Good synopses of the campaigns of 1812, 1813, and 1814. Although the United States had low casualties, 528,000 Americans were engaged in the war. That's a lot of people thrashing around with weapons. Often unsupervised.

Alan Taylor, *The Civil War of 1812* (New York: Alfred A. Knopf, 2010). Author makes the case that between the northern United States and "Upper Canada," the war of 1812 was basically a civil war in which the United States played so poorly that its military efforts were a total failure. In fact, ill-trained and ill-equipped American troops "proved more adept at looting than at fighting." With 7.7 million Americans and 300,000 Canadians, the imbalance on paper was considerable, but the British armed forces were greatly improved over those who had lost the war against the colonies. Additionally, they made superb use of their Native American allies, terrorizing the American forces. "Having failed to conquer Canada or compel British maritime concessions, the Republicans redefined national survival as victory."

Ironically, the peace favored the Americans, constrained the British, and left the Indians at the mercy of the United States.

THE MEXICAN WAR

Perhaps no war in American history, except for the Vietnam War, has been as vilified and condemned as the Mexican War of 1848. Yet it is one of the most important wars in American history in terms of its outcomes. The addition of Texas, California, New Mexico, and Arizona to the territory of the fledgling republic was to have enormous consequences, both in terms of additional territory, as well as the concomitant shrinking of Mexico to a more manageable area and form (from the point of view of the United States). The Mexican War was also to serve as an important training ground for a generation of military leaders who would play such crucial roles in the subsequent Civil War. The additional territories and their slave-or-free status would in turn also help to lead to that conflagration.

However one looks at the Mexican War, it is an important war and one with profound positive and negative consequences, regardless of how one feels about manifest destiny and the United States going to war for material gain. Ulysses S. Grant famously called it the "most unjust war ever waged by a stronger against a weaker nation," but it was one of the most profitable for the stronger country, and its task was made very easy by the policy, diplomatic, and military failures of the weaker one. It was a war on the periphery to be sure, but it was of great long-term significance for the American Civil War that was to come. The Mexican War deserves much greater focus and attention than it is usually accorded.

K. Jack Bauer, *The Mexican War 1846–1848* (New York: Macmillan, 1994). One in the series the Wars of the United States, edited by Louis Morton, this volume looks at the war by making some comparisons with the contemporary one in Vietnam (something of a stretch). It is a straightforward account with easy-to-read maps and interesting commentary. Sees the immediate results of this decisive victory by the United States as "largely negative," except for the acquisition of California. Quite a big exception, that.

Charles L. Dufour, *The Mexican War: A Compact History 1846–1848* (New York: Hawthorn Books, 1968). A readable, straightforward account of what happened and when, on the way to the "Halls of Montezuma." One of the few works featuring a comprehensive master map showing the broad scope of US operations from San Francisco to San Diego, Las Vegas to Sante Fe, Chihuahua, Corpus Christi, Veracruz, Mexico City, and the two Monterreys.

John S. D. Eisenhower, *"So Far from God": The U.S. War with Mexico 1846–1848* (New York: Random House, 1989). There is a lot in this very useful volume. It reads easily, and the reader gets a sense of the military action from the excellent maps and surprisingly useful photos. Eisenhower puts the war in perspective, quoting Professor Smith as saying, "The hostilities were deliberately precipitated by the will of Mexico," and the war actually had the highest death rate of any US war—13 percent of its 104,556 soldiers died. The author is high on Generals Scott and Taylor and more than a little ambivalent toward President Polk, although he argues that Scott's landing at Veracruz was the "largest amphibious invasion yet attempted in history," with 6,750 men, 50 ships, and 140 flat-bottomed boats.

Note: Actually, the Athenians were supposed to have brought 260 ships and 27,000 men to Syracuse in 415 BCE, while the Mongols in 1274 and 1281 tried to conquer Japan with 140,000 troops and more than 600 (1274) and 3,000(1281) ships, and William the Conqueror in 1066 brought 7,000 troops to England in 600 ships, to name a few larger efforts. By contrast, the Anglo-American Normandy invasion involved 5,000 ships, 8,000 aircraft, and 130,000 men.

David S. Heidler and Jeanne T. Heidler, *The Mexican War* (Westport, CT: Greenwood Press, 2006). A breezy, easy-to-read description of the events and personalities leading up to the Mexican War. Precious little on the military aspects but some acumen shown with regard to Polk's steering a middle course to achieve the peace, with some American leaders wanting all of Mexico swallowed and others still opposing the war in the first place.

Robert W. Johannsen, *To the Halls of Montezuma* (Oxford: Oxford University Press, 1985). An interesting counterpoint to many of the other studies of the Mexican War that focus on territorial expansion, the extension of slavery, and the problems the results of the war would provide. Johannsen looks at many contemporary sources and sees America' "first foreign war" as a patriotic enhancer of what it meant to be an American, a strong message to European observers of American military process, and especially highlighting the success of volunteer armed forces, as well as an event that fueled a heroic war literature. It was also the first war with newspaper correspondents.

Robert Merry, *A Country of Vast Designs: James K. Polk, the Mexican War and the Conquest of the American Continent* (New York: Random House, 2010). The author makes a strong case for the importance of the presidency of James K. Polk and shows how his personal agenda came to pass in four short years from 1844 to 1848. The 600,000 square miles of new territory came at a cost of 13,780 American and 25,000 Mexican lives. What if the Mexico of today still included Texas, California, New Mexico, and Arizona?

And what if the United States had decided to incorporate all of Mexico in 1848 and tried to make that a reality? The actual differences between the Democrats and Whigs of this era remain confusing to many, and the parties seem as much driven by personalities and politics as by ideology or program.

Otis Singletary, *The Mexican War* (Chicago: University of Chicago Press, 1968). Describing it as a "melodrama of the irritation and internal conflicts that have continued to plague American democracy at war," this work is a breezy introduction to the various military campaigns: General Zachary Taylor in northern Mexico, Colonel Stephen Kearny in New Mexico, John C. Fremont and Commodore Robert Stockton in California, and General Winfield Scott taking Veracruz and Mexico City. The sketch maps are useful, as is the analysis of the "hidden warfare" within the American military, warfare that is not confined to the oft-cited badmouthing and backbiting between Taylor and Scott.

Steven E. Woodworth, *Manifest Destinies: America's Westward Expansion and the Road to the Civil War* (New York: Alfred A. Knopf, 2010). Some excellent introductions to (and easy-to-read maps of) the various battles that determined the outcome of the war: Buena Vista, Monterrey, San Pasqual, Cerro Gordo, Churubusco, and Chapultepec. Sees great irony in the fact that the annexation of Texas, California, and New Mexico was a "culmination of the nation's hope [that] brought the country to the brink of the conflict it long had dreaded."

Chapter Twenty

Further Democratization of War

The French Revolution and the Napoleonic Wars Spread Nationalism to Prussia, Austria, and Especially Russia

One of the most momentous developments in the history of war was the rise of nationalism in the 18th and 19th centuries. Mass identification with the nation enabled the state to mobilize the nation for many purposes, including war. This happened perhaps earliest and most decisively with the French Revolution, where nationalism became a civic religion. The leaders of the French Revolution, and later Napoleon, were able to use this new wellspring of dedication to create huge mass armies and, in the case of Napoleon, to create them over and over again as the French Empire expanded. Yet ironically in terms of the politics of nationalism, "what one has done, another can imitate," and the French nationalism eventually produced equally powerful—and countervailing—nationalisms in Spain, Germany, and Russia.

In any case, the rise of nationalism was to have a disastrous impact, not only on the frequency and lethality of warfare, but also on the ability of governments to stop wars once they have begun. We are still today living with the pernicious aspects of the melding of warfare and nationalism. What happened after the French Revolution and the wars of Napoleon were the great expansions of both war and the state—with profound consequences, it should be added. The literature in this chapter contains a variety of explorations and explanations of these phenomena.

What changed in warfare was the dedication of more and more people to involve themselves in warfare. The French levy en mass mobilized hundreds of thousands of soldiers, and the new doctrine of mass attacks altered the way wars were fought. As Lazare Carnot, the organizer of the French revolutionary armies put it, "No more maneuvers, no more military art but fire, steel and patriotism" coming together to produce what Michael Howard was to call "sheer ferocity in war." War was becoming total, a conflict "not of armies but of populations." And it would continue: The Russians were able to field an army of 600,000 during this period; by 1870, the North German Confederation would field 1.2 million men against France; and by 1914, Germany would field 3.4 million against France and Russia, according to Howard.

Napoleon's contribution to the history of warfare is considerable. He saw the importance of artillery and its utility in smashing the forces of his opponents and the need to concentrate at the decisive point and to use his reserves to exploit that decisive point. He was also to greatly emphasize velocity (and see its ability to overcome mass) and the need to disperse and then combine his forces over long distances, enabling them to travel along parallel lines of attack and recombine at the point of battle.

M. S. Anderson, "Towards a New World: Nationalism and People's Armies 1740–1789," in his *War and Society in Europe of the Old Regime, 1618–1789* (New York: St. Martin's Press, 1988), pp. 157–206. As for national armies, "[i]n 1789, their day was about to begin."

David Bell, "Napoleon's Total War: Napoleon's Suppression of Spain's 1808–1814 'Guerrilla War' of Independence Was Something New under the Sun: A War against Everyone," *Military History* (April 2007), pp. 38–47.

Geoffrey Best, *War and Society in Revolutionary Europe 1770–1870* (New York: St. Martin's Press, 1982). Best makes a strong case for the resulting "nationalization and militarization of society," as well as focusing on the "most revolutionary" dimensions of warfighting: steam, railways, and rifles.

Scott Bowden, *Napoleon and Austerlitz* (Chicago: Emperor's Press, 1997). Extravagant, elegant, and lovingly crafted, this work provides the most detailed account of Napoleon's 1805 masterpiece. As he himself put it, "First one must see. . . . Then, the vulnerable point in the enemy's dispositions having been determined, an all-out attack was loosed against it. The secret was in the timing. . . . the transition from the defense to the offense is one of the most delicate operations in war."

David G. Chandler, *Austerlitz 1805: Battle of the Three Emperors* (Oxford, UK: Osprey, 1990). An in-depth look at this, one of Napoleon's true masterpieces, where he envisioned the entire battle the night before. This account is most useful for the series of maps that show the complexity of movement as it occurred. Most useful.

———, *The Campaigns of Napoleon* (New York: Macmillan, 1966). Almost 1,100 pages of well-researched, carefully presented material on his battles. Very clear and useful maps, with worthwhile appendixes, including some on the organization of Napoleon's various armies. A must-read both for the clarity of its presentation and for an understanding of its conclusion: "But the shade of Napoleon engendered an irresistibility for many years after his death in 1821. It lingers yet." Chandler's summation also may rankle many but rings true: "Napoleon will always be regarded as a soldier of genius and the creator of modern Europe."

Note: For readers interested in the purported relationship between Napoleon's love interests and his campaigns, see the insightful series by Thames Television, *Napoleon and Love* (1974), presented in three discs.

Carl von Clausewitz, *1812 In Russia* (Hattiesburg, MS: Academic International, 1970). Reprint of the 1834 edition. The author, who was an eyewitness, insists that Napoleon had a much better chance than traditionally noted and was very much in character, as "Napoleon owed everything to his boldness of determination."

Julian Corbett, *The Campaign of Trafalgar*, 2 vols. (London: Longmans, Green, 1919). Napoleon and the Spanish are whipped in this classic naval battle by Pitt, Nelson, and Collingwood, among others, as the author makes us understand that 50,000 men were involved in the overall campaign that ultimately culminated in the huge victory at Trafalgar. This holistic, overarching study does not diminish the contribution of Nelson (still the "greatest of admirals") but does put his accomplishments in a broader perspective.

Bernard Cornwell, *Waterloo: The History of Four Days, Three Armies, and Three Battles* (New York: Harper, 2015). The duke of Wellington was fond of saying it was impossible to tell the story of a battle, but Cornwell proves this wrong. His account is clear, accurate, easy to follow, and impressive in both its sweep and detail, and the reader is swept along with the various *Schwerpunkts* (possible and actual) appearing properly and in perspective. Very useful description of the "lines and squares" that were so tactically relevant and strategically defining (see pp. 93–94). A magnificent achievement.

Christian Duffy, *Austerlitz 1805* (London: Seeley Service, 1977). Chapter 8, "A Storm of Fury and Bloodshed," gives detailed accounts of the battle and relatively clear maps. Napoleon pronounced at the beginning, "And now, gentlemen, let's go and do great things." They did, and after the battle, the emperor dispensed brandy and blankets to his wounded and pronounced,

"And it will enough for you to say, 'I was at the Battle of Austerlitz,' and they will reply, 'There stands a hero.'" This stunning victory deprived Austria of 2.5 million of its subjects, bought off Prussia with Hanover, and smashed a large Russian army, but in the end, the battle prompted all three to revamp their militaries, with the result that France eventually lost.

J. F. C. Fuller, *Sir John Moore's System of Training* (London: Hutchinson, 1925). Fuller gives Sir John Moore great credit for the outcome of British training at Waterloo and other prominent battle sites: "Waterloo Won on the Training Ground of Shorncliffe," which makes more sense than it being won on the "playing fields of Eton," as the great duke always claimed. Excellent focus on discipline and small-unit cohesion. Sir John Moore would serve in Corsica, St. Lucia, Ireland, Holland, Egypt, Sweden, and Portugal and died at Corunna in Spain.

John Gill, *1809: Thunder on the Danube*, 2 vols. (London: Frontline Books, 2009). The definitive account of these campaigns in Italy, Germany, and Austria with Napoleon in full battle cry and the Hapsburgs on the defensive from beginning to end. Great depth and valuable insights, illustrated with excellent maps.

F. R. Gridge and Roger Bullen, *The Great Powers and the European States System, 1815–1914* (New York: Longman, 1980). An engaging look at the intricate alliance system that developed over the century and that ultimately resulted in the inexorable march toward World War I. Seeking to protect themselves, the nations of Europe created the very conditions where a slide toward war became uncontrollable and popular identification with "my country right or wrong" grew during this period.

B. H. Liddell Hart (ed.), *The Letters of Private Wheeler* (Boston: Houghton Mifflin, 1951). "A unique find," says an eyewitness of the Napoleon era in the Peninsula campaign and Waterloo, through this true son of the 1st Battalion, King's Own Yorkshire Light Infantry. Fresh and illuminating.

Henry Kissinger, *A World Restored: Metternich, Castlereagh and the Problem of Peace 1812–1822* (Boston: Houghton Mifflin, 1957). Statesmen stave off chaos and stifle revolution for a while. This blend of history and politics focuses on the nature of statesmanship as seen through the careers of Viscount Castlereagh and Prince C. W. N. L. Metternich. The structure of Europe before the Napoleonic wars is reestablished by their actions and intrigues.

Dominic Lieven, *Russia against Napoleon: The True Story of the Campaigns of War and Peace* (New York: Viking, 2010). An astonishingly good and welcome addition to the literature of the period. Makes a strong claim that Napoleon was actually defeated during 1812–1814 by Russian military achievement and the skillful and overarching diplomacy and military strategy of Alexander I and his closest advisors. Outstandingly revelatory on the Russian march from Moscow to Paris and the campaigns in Germany, Austria, and France, which made victory for the Grand Coalition possible. Rightly emphasizes the importance of the Russian supply system and Russian horsepower and gives substantial credit to the important, even crucial, role played by the Russians during the campaigns of 1813 and 1814. Points out that the British and Russians, like the French, were "expansionist and predatory empires" and gives great homage to the "courage, decency and self-sacrifice of the ordinary Russian soldiers." Also claims that the Russian army's march from Vilna to Moscow and then back across Europe to Paris was the "longest campaign in European history."

John A. Lynn, *The Bayonets of the Republic: Motivation and Tactics in the Army of Revolutionary France, 1791–94* (Urbana: University of Illinois Press, 1984). Élan certainly, the cult of the bayonet, of course, but the nation's gratitude toward its soldiery turns out to be a key variable, along with the revolutionary soldiers eventually learning discipline and the tactics of the square. An in-depth look at what Napoleon inherited.

Piers Mackesy, *The War in the Mediterranean: 1803–1810* (Cambridge, MA: Harvard University Press, 1957). The Peace of Amiens (1802) concluded the struggle against revolutionary France, but the fight with imperial France was just beginning. As Napoleon moved into the Mediterranean, the English counters, blockading Toulon (the "key to military action"). The result of British activity was that the "enemy's commerce quickly vanished from the sea."

John Masefield, *Sea Life in Nelson's Time* (Annapolis, MD: Naval Institute Proceedings, 2002). The 18th-century author's work is reproduced to provide chilling insights into life at sea, where the punishment for many offenses was truly draconian with cat o'nine tails flogging, keel hauling, and hanging. "Three hundred lashes were very frequently given" (even for theft!), and if the victim died during their administration, his corpse got the remaining lashes. Good diagrams, text, and models, with superior detail on all manner of cannons.

Hermann Maurice, Comte de Saxe, *Reveries on the Art of War* (New York: Dover Books, 2007). A classic work of military philosophy that was reported to have had a great impact on Napoleon. Maurice wrote, for example, "It is not the big armies that win battles; it is the good ones." He was, and remains, right.

Andrew Roberts, *Napoleon: A Life* (New York: Viking, 2014). A truly magisterial work, overarching in its scope and yet grounded in minute detail, having the benefit of more of Napoleon's own letters than any previous biographer. Calls him the "founder of modern France and one of the great conquerors of history" and points out that more opponents declared war on him than he on them and that he fought 60 battles and lost only 7 (Acre, Aspern-Essling, Leipzig, La Rothiere, Laon, Arcis, and Waterloo). The duke of Wellington, when asked who was the greatest captain of the age, would reply, "In this age, in past ages, in any age, Napoleon" (p. xxxv). Roberts remains surprised that the French lost Waterloo with inexplicably bad battles by most of the French commanders, including Napoleon, who "[v]ery much deserved to lose it." He does not believe that Napoleon had to lose the Russian campaign, but at 1,800 miles from Paris, Borodino was the largest battle in the history of Europe since Roman times and would remain the largest until the first battle of the Marne in 1914. Lauds Napoleon's blending of speed, maneuver, convergence, a genius for topography, and an acute sense of timing but points out that he never understood naval warfare and instituted the disastrous Continental System, and only he could make the unforced errors that would pull him down from the pinnacle of his empire in 1810, the largest since Rome and larger than Charlemagne's. A magnificent work.

Stephen T. Ross, *Quest for Victory: French Military Strategy 1792–1799* (New York: A. S. Barnes, 1973). Great power politics dissected in an era when France, Austria, Prussia, and Russia all fielded armies of 130,000 to 200,000. Useful in showing how the First French Republic actually produced greater success than previous regimes and set the stage for Napoleon's efforts.

Gunther E. Rothenberg, *Napoleon's Great Adversaries: The Archduke Charles and the Austrian Army 1792–1814* (Bloomington: Indiana University Press, 1982). Despite many Austrian defeats, the author argues that the restoration of the Hapsburg monarchy led to the Austrians fielding an army that ultimately became the largest in the final campaigns against Napoleon. I would personally not go that far, preferring to put Alexander II and the Russian army in that position, but Rothenberg makes his case with skill and dispatch.

Samuel Scott, *The Response of the Royal Army to the French Revolution: The Role and Development of the Line Army 1787–93* (Oxford: Oxford University Press, 1978). Looking at the French Royal Army as a social institution responding to the deep-seated military and societal changes of the French Revolution, Scott sees the army (along with the revolutionary levees) as ultimately saving the French Revolution from its foreign foes before it was disestablished as a distinct army in February 1793.

Charles Sedalia, *Napoleon's Wars: An International History, 1803–1815* (New York: Viking, 2007). The author presents not just a biography of Napoleon nor a simple military history but also a "Pan-European" analysis of the period. Interesting in its own right, the study also posits the impact on the nature of warfare as a result: 40 major battles fought, wars that went on for a long time, so large had the armies become that they had to be organized as divisions within corps, and he believes, the Napoleonic wars cost more lives in Europe than any time until World War I.

Harry Smith, *The Autobiography of Sir Harry Smith 1787–1819* (London: Constable, 1999). Vivid, exciting eyewitness account to the British and Spanish campaign in Spain against Napoleon and includes the great romance he shared with his wife Juana. The battles of Salamanca and Villoria are described, and he also served in the English forces at the Battle of New Orleans and Waterloo.

Leo Tolstoy, *War and Peace* (New York: Thomas Y. Crowell, 1899) (F). See the review in chapter 40. Wonderful on battle depiction and as a critique of Napoleon and why he lost.

Chapter Twenty-One

The American Civil War

Fratricide in Living Color

This is the defining war in American history, although sometimes it appears that not all the issues from it are yet totally settled. From the point of view of warfare, however, it was but one civil war among the many chronicled in human history going back to Thucydides's account of the revolt on the island of Corcyra. The American Civil War did make use of some "modern" elements, of course: the use of watches to coordinate attacks (Lee could have used more of these on the second day of Gettysburg), railroads to move troops quickly to the battlefield (the South won the First Battle of Bull Run by bringing up forces to tip the balance), balloons for aerial observations, ironclad ships, and many other innovations.

In terms of strategy and tactics, the American Civil War, however, remains one of the more fertile and productive grounds for these analyses, and the endless debates about leaders, battles, and turning points are indeed the essence of the study of the malleability (or lack thereof) of history.

This chapter presents a small sample of the works available on this vital subject. Highly recommended as places to start are U. S. Grant's amazingly good memoir *Personal Memoirs* and the magnificent sweep of James McPherson's *Battle Cry of Freedom* (both reviewed here).

Richard Beringer, Herman Hattaway, Archer Jones, and William Still Jr., *Why the South Lost the Civil War* (Athens: University of Georgia Press, 1986). "More united in defeat and the subsequent Reconstruction than in their brief independence," Confederates seem to forge a distinctive unity after the war that they did not have before or during it. The authors point to the rebellion in Paraguay for a contrast with people who really wanted their independence.

Deanne Blanton and Lauren Cook, *They Fought Like Demons: Women Soldiers in the Civil War* (New York: Vintage Books, 2002). An in-depth look at the broader (many more than the normally quoted 250–400) range of women who fought in the Civil War, beginning with First Bull Run; ending with Appomattox; and including the Peninsula campaign, Second Bull Run, Antietam, Fredericksburg, Chancellorsville, Gettysburg (including Pickett's Charge), Vicksburg, the Wilderness, Cold Harbor, Red River, Shiloh, Petersburg, and Sherman's March to the Sea. Also examines the many motivations for enlisting, including patriotism; love of freedom; desire to be with one's spouse, father, or brother; spirit of adventure; economic advancement; and desire to escape the confines of womanhood in the mid-19th century.

Dora Costa and Matthew Kahn, *Heroes and Cowards: The Social Face of War* (Princeton, NJ: Princeton University Press, 2008). Analyzing notes, diaries, and records of 40,000 Civil War soldiers, the authors underscore the importance of community in small-group cohesion, whether in training, battle, or prisoner-of-war situations, leading them, albeit reluctantly, to question some contemporary notions of the "benefit of community diversity."

Stephen Crane, *The Red Badge of Courage* (New York: Pocket Books, 1945) (F). See review in chapter 40.

Thomas J. Evans and James M. Moyer, *Mosby's Confederacy: A Guide to the Roads and Sites of Colonel John Singleton Mosby* (Shippensburg, PA: White Mane, 1991). Being the account of "The Grey Ghost," John Singleton Mosby, a guerrilla operator and commando who led 800 men and made the Union furious because the 800 seemed to cover the ground of 10,000. Lavishly illustrated, this work covers the locations, routes, and actual buildings where "The Grey Ghost" rode.

Amanda Foreman, *A World on Fire: Britain's Crucial Role in the American Civil War* (New York: Random House, 2010). A clever interweaving of British interest in and action during the American Civil War. Striking is Sherman's claim in 1864: "The British mark on everything on the battlefield," including muskets, cartridges, caps, and shells. Suggests that the "British Navy and Confederate courage" might have carried the day, but the last chance for the necessary diplomatic recognition was in 1863 while Lee was riding high and would have required the freeing of the slaves at that time.

Gary Gallagher, *The Confederate War* (Cambridge, MA: Harvard University Press, 1997). Why popular will, nationalism, and military strategy could not stave off defeat and why a "new slaveholding republic was not created." The author would like to know why the non-slave-holding yeomen, who "formed a solid majority," fought so hard for it.

Adam Goodheart, *1861: The Civil War Awakening* (New York: Alfred A. Knopf, 2011). The run-up to the war and how attitudes changed, hardened, and morphed into different forms, this work examines who supported the dissolution of the Republic and who did not and why and the "courage and heroism beyond the battlefields." Looking at such diverse participants as Ellsworth the Zouave, three slaves who fled to Fort Monroe, and a great comet, the author skillfully weaves his narrative to provide true understanding. Also makes the point I have not seen elsewhere that, at the time of the outbreak of the Civil War, the value of slaves held in the South was approximately $2 billion, or more than the nation's factories and railroads combined. Incidentally, the price/value of slaves fell precipitously after the war broke out.

Ulysses S. Grant, *Personal Memoirs* (New York: J. J. Little, 1885). An outstandingly honest, illuminating, and refreshing memoir, the standard against which all military autobiographies must be measured. A must-read for anyone wishing to understand the American Civil War or what it takes to be a great general or write a great autobiography—and he wrote this one as he was dying of cancer and desperately trying to earn some money for his family.

Note: For a recent and highly readable account of Grant's life and time, see H. W. Brands, *The Man Who Saved the Union: Ulysses Grant in War and Peace* (New York: Doubleday, 2012). Fast-paced yet with considerable depth where it matters and chockfull with useful quotes, this is a fine addition to our literature on the man who showed Lincoln how to win the war and then did it.

Mark Grimsley, *The Hard Hand of War: Union Military Policy toward Southern Civilians 1861–1865* (Cambridge: Cambridge University Press, 1995). Sustained ruthlessness trumps "conciliation and its challenges," as the war begins in earnest and emancipation is the party of the "Hard Way." A Vermont solider marauding in the Shenandoah Valley with Sheridan sums up its efficacy: "They see the grim determination of the North and they begin to feel that to hold out longer is to fight against inevitable destiny."

Allen Guelzo, *Gettysburg: The Last Invasion* (New York: Knopf, 2013). Perhaps the best overall book on this battle. It boosts Howard, depresses

Chamberlain, excuses Stuart, gives Lee more of a chance, punishes Meade severely, and sees John Reynolds as the one who sprung Lee's trap too soon and he could never regain its possibilities. A very understandable series of maps gives good insights into the three days' fighting. A superb work.

Edward Hagerman, *The American Civil War and the Origins of Modern Warfare* (Bloomington: Indiana University Press, 1988). In this straightforward, essentially descriptive account, the author lists a number of precursors of modern warfare, including the use of railroad, the emergence of trench warfare after 1863, the tactics of attrition, the strategy of exhaustion, and the need for massive logistical tails.

Herman Hattaway and Archer Jones, *How the North Won: A Military History of the Civil War* (Urbana: University of Illinois Press, 1983). A superbly clear military history with magically simple but very useful maps on the very pages where the battles are described. Classic thumbnail sketches of military figures: Robert E. Lee says of the 33-year-old John Bell Hood, "All lion, none of the fox." Very useful appendix, "An Introduction to the Study of Military Operations."

Doranne Jacobson, *The Civil War in Art* (New York: Smithmark, 1991). A lavishly illustrated work capturing the efforts of, among others, Winslow Homer, Matthew Brady, and Currier and Ives. The Currier and Ives lithograph of the Battle of Antietam, 1862, gives a particularly graphic depiction of the fighting around Burnside's Bridge.

William B. Jordan, *Red Diamond Regiment: The 17th Maine Infantry 1862–1865* (Shippensburg, PA: White Mane, 1996). Another Maine regiment pleads for recognition—with good cause—and gets it here.

John Kegan, *The American Civil War: A Military History* (New York: Knopf, 2009). The military master takes a turn at a one-volume work, somewhat derivative but still worth a quick read.

Edward G. Longacre, *Joshua Chamberlain: The Soldier and the Man* (Conshohocken, PA: Combined, 1999). A praiseful account of the hero of Little Round Top and beyond. Doesn't seem to realize the small number of men commanded at a far corner of the battle that would rage on for eight more hours elsewhere.

Karl Marx and Frederick Engels, *The Civil War in the U.S.* (New York: International, 1937). Dispatches to the *New York Daily Tribune* and the *Vienna Die Press*, as well as correspondence between the two authors. Foresaw that "Negro slavery would not long outlast the Civil War," and both men saw the "germ of future revolution" established as the "slave oligarchy" was overcome by the "bourgeois republic."

James M. McPherson, *Battle Cry of Freedom: The Civil War Era* (New York: Oxford University Press, 2003). A true classic and a wonderful introduction to the American Civil War, with all its tragedy and triumph. Reviewing a truly staggering amount of material that covers the entire Civil War era and always writing clearly and cogently, McPherson has created a tour de force that captures the war that killed more Americans than all its other wars combined. Insightful on "The Revolution of 1860" and "The Counter-Revolution of 1861." This is actually volume 6 in the Oxford History of the United States.

———, *Drawn with the Sword: Reflections on the American Civil War* (Oxford: Oxford University Press, 1966). A collection of provocative essays discussing a wide range of subjects, including *Uncle Tom's Cabin* (which sold 3 million copies), with Uncle Tom the strongest and veritable Christ figure who dies at the hand of Simon Legree. Argues that the War of Southern Aggression was the reason the South lost; "[m]ilitary defeat caused loss of will, not vice versa," and the North won because it had Grant, Sherman, and Sheridan and the South did not. Lee, he feels, was a great tactical genius and a motivator of men par excellence, but he lacked true strategic vision.

———, *Embattled Rebel: Jefferson Davis as Commander in Chief* (New York: Penguin, 2014). A largely derivative account but one that makes a number of important points: Davis, a West Point graduate, (1) actually had a pretty good grasp of the strategic situation; (2) was sick and debilitated beyond telling during the entire war; and (3) had insoluble problems of manpower, Confederate governors, arms, and the need to arm slaves. "If a negro is fit to be a soldier, he is not fit to be a slave." This largely sympathetic biography ends, however, by saying that, by April 1865, "Davis had gone from a state of unreality to one of fantasy" (p. 241).

Harry W. Pfanz, *Gettysburg: Culp's Hill and Cemetery Hill* (Chapel Hill: University of North Carolina Press, 1993). The third book in the trilogy. Pfanz has given readers a minute-by-minute account of the Battle of Gettysburg as this and the other two volumes of the trilogy cover the battle from beginning to end. A valuable resource.

———, *Gettysburg: The First Day* (Chapel Hill: University of North Carolina Press, 2001). The first book in the trilogy. You are there.

———, *Gettysburg: The Second Day* (Chapel Hill: University of North Carolina Press, 1987). The second book in the trilogy. You are still there.

Fletcher Pratt, *A Short History of the Civil War* (New York: Pocket Books, 1935). One of those books read in adolescence that lingers with you despite the flood of subsequent literature. Breezy and opinionated but telling, nevertheless.

John Pullen, *Joshua Chamberlain: A Hero's Life and Legacy* (Mechanicsburg, PA: Stackpole Books, 1999). Covers the career of one of Bowdoin's most famous heroes—ever.

———, *The Twentieth Maine* (Dayton, OH: Morningside House, 1997). The standard and revered account of the unit that lies at the heart of the book *Killer Angels* and the film *Gettysburg*.

George Rable, *Fredericksburg! Fredericksburg!* (Chapel Hill: University of North Carolina Press, 2002). An epic work that finally gives Fredericksburg its due, not just as a misguided campaign that ended in slaughter, but also as a thoughtful look at the battle from the points of view of the common soldiers and its leaders. Uses the chant of Union troops on the third day of Gettysburg to show how deeply the battle had been etched in the consciousness of the Army of the Potomac and how sweet the payback was. A number of Confederates who were at Fredericksburg "felt the attacks at Fredericksburg had set a standard for valor and persistence that even Pickett's charge had not matched."

Michael Shaara, *A Chain of Thunder* (New York: Ballantine Books, 2014) (F). A somewhat wooden account of the siege of Vicksburg but useful to the understanding of that battle from inside the minds of the commanders, such as Grant and Sherman versus Pemberton and Johnson. Cave-dwelling civilians add realism.

———, *Killer Angels* (New York: Ballantine, 1975) (F). The novel that vaulted Joshua Lawrence Chamberlain and the Little Round Top portion of the Battle of Gettysburg high into the public domain—so high that it is often difficult to bring it back down to earth.

John Y. Simon (ed.), *The Papers of Ulysses S. Grant*, vols. 1–2 (Carbondale: Southern Illinois University Press, 1984). Extensive accounts of a great deal of what he wrote. Well worth a perusal.

Richard Slotkin, *The Long Road to Antietam: How the Civil War Became a Revolution* (New York: W. W. Norton, 2012). An illuminating and carefully done study of the intertwined nature of Lincoln, his cabinet, and McClellan, as well as Davis and Lee, and argues how revolutionary the Emancipation Proclamation (and the reception of freed black males into the armed service of the United States; eventually 180,000 would serve) really was—in addition to being the largest expropriation of private property between Henry VIII's seizures of Catholic land during the Reformation and the Soviet nationalization of all factories and farms after the revolution. An excellent and clearly written work.

Mark Smith, *The Smell of Battle, the Taste of Siege* (London: Oxford University Press, 2014). A self-described "sensory history of the Civil War" shows how the ears, nose, eyes, and stomachs all perceived the realities of Charleston, Bull Run, Vicksburg, and Gettysburg, among others. Quite a change of pace from normal books about war.

Harriett Beecher Stowe, *Uncle Tom's Cabin* (New York: Modern Library, 1938) (F). The poignant tale written in 1852 that enflamed the public of the North and West against the institution of slavery. If you haven't read it recently, do so. Even at the distance of time and space and epoch, there is a powerful wrath-inducing impact that would make virtually anyone oppose slavery. It helps to explain the power of will as a component in the North's war-making.

Noah Trudeau, *Like Men of War: Black Troops in the Civil War 1862–1865* (New York: Little, Brown, 1998). Around 179,000 African Americans served the U. S. armed forces during the Civil War. Ten thousand were killed in action, and nearly 30,000 died of infection and disease. Most, but not all, were former slaves.

———, *Southern Storm: Sherman's March to the Sea* (New York: Harper, 2008). A minute-by-minute account of the blue tornado that ripped through the Deep South and showed that the Union would win the war and sooner rather than later. Detailed and generally a fine description. Harsh on Sherman over Negros (perhaps warranted) and on Howard (surely not; Sherman always sent freed slaves to Howard). Found the entire march well-organized but flexible. Showed the poor quality of the Confederate military response (but that could not have been expected by Union leaders). A great and daring feat of arms.

Bel Irvin Wiley, *The Life of Johnny Reb: The Common Soldiers of the Confederacy* (Baton Rouge: Louisiana State University Press, 1978). An updated version of the 1943 edition that was way ahead of its time in its accent on "common folk" and "little people." Many letters speak of measles, diarrhea, venereal disease, dysentery, typhoid, small pox, and rations of "bad beef and corn bread" and mule. Argues that the preservation of slavery and racial issues as well as states' rights and adventure played a big role in motivation. Lots of fine-grain description of life day to day.

Chapter Twenty-Two

Civil Wars in Perspective

> "For a really complete, consuming, unreserved effort at slaughter and destruction, there is nothing to match a civil war."—Ramsay MacMillan, *Constantine*

> "There is nothing unhappier than a civil war, for the conquered are destroyed by and the conquerors destroy, their friends."—Dionysius of Halicarnassus, *Antiquities of Rome*

> "Civil blood makes civil hands unclean."—Shakespeare, *Romeo and Juliet*

> "A foreign war is a scratch on the arm; a civil war is an ulcer which devours the vitals of the nation."—Victor Hugo, *Ninety Three*

From Thucydides's description of the revolt on Corcyra (Corfu) in 473 BCE, we know that civil wars have almost always been savage and shot through with heinous fratricide and worse. In this chapter are some books to give the interested student insight into the particular dimensions of civil wars when savagery takes on a special character. The reader should always remember that, at base, civil wars often involve both revolutions and insurgencies as well as counterrevolutions, and their topics are included here with that understanding.

These are some of the books that do not fit easily under other topics, such as "The American Civil War." The Spanish Civil War (1936) is a case in point, for most studies put it in the context of foreshadowing World War II and the alliances of Italy and Germany versus the United States, Great Britain, France, and the Soviet Union, or simply in the clash of ideologies of Fascism versus Democracy and Communism. Yet it, like the Greek Civil War of 1944–1949, should be viewed in the same way as our study of the civil war on Corcyra: an outbreak of fratricide demonstrating the essence of civil wars everywhere throughout time and space. In the case of China, while we have focused much of our bibliographical attention on its "modern" civil war in chapter 18, we have also included some sources here from the Taiping rebellion period, which some scholars call a Chinese civil war as well.

An interesting concept to explore: Why did some revolutions lead directly to civil wars (as the Russian, Chinese, French, and Irish) and others did not (US)?

SOME ILLUSTRATIVE EXAMPLES

Chinese Civil War (1927–1950)

Stephen Platt, *Autumn in the Heavenly Kingdom* (New York: Knopf, 2012). The author makes a compelling case that the so-called Taiping rebellion of the 19th century (early 1850s–1864) was actually a major civil war (contemporary with the American Civil War), one that engulfed the entire country and took 20 million lives (out of a population of 400 million). It was at once a civil war, an international war, and a religious war. Li Xiuchen, the "Heavenly King," and his cousin Hong Rengan, the "Shield King," fight the Manchu Qing dynasty on behalf of the Chinese and their religious brethren, who accept Li Xiuchen as the "Little Brother" of Jesus and the "son" of God. Many missionaries saw the rebellion as a good thing because it would bring Christianity (albeit a modified form of it) to China. Hong Rengan believed that, after victory, the pernicious aspects of Taiping Christianity would be corrected. The British and French managed to fight both the Quing (Beijing) and the Taiping (Shanghai) before settling on siding with the Quing. Massive looting by British and French troops along the way and many horrendous massacres by both Quing and Taiping (in the name of Confucius and Jesus, respectively). The autodidact General Zeng Guofan managed to save the Quing after an extended period of on-the-job training. His motto was "Love the People," and he was thus an available sage for Mao's plagiarism of the concept. The Taiping rebels were hurt by infighting (the "Shield King" against the "Brave King," etc.). The British were caught in the middle of two civil wars, the Chinese (both directly and via India and opium for silk) and the American (raw cotton in, finished goods out), both being major trading partners. They ended up supporting the Quing by providing arms and with actual fighting, using their gunboats (which the Chinese end up referring to as the "Vampire Fleet"). Quite a tale. Well told.

Harold M. Tanner, *The Battle for Manchuria and the Fate of China: Siping, 1946* (Bloomington: Indiana University Press, 2013). Big set-piece battle won by the Nationalists during March–May 1946. But in part because of the peace efforts of General George Marshall and Chiang's strategy of "First

South and then North," it was not decisive because there was no follow-up across the Songhua River. The Communists thus had a chance to regroup and eventually took over all of Manchuria by 1949. The author concludes, however, that the presence or the absence of Marshall was "utterly insignificant," and the battle, on balance, was not determinative one way or the other. This work really needed more, bigger, and bolder maps

English Civil Wars (1642–1661)

C. H. Firth, *Cromwell's Army: A History of the English Solider during the Civil Wars, the Commonwealth and the Protectorate* (London: Methuen, 1912). This is a marvelous work. It begins modestly enough, with the author stating, "A civilian who undertakes to write the history of an army courts many perils, and cannot hope to escape them all." But then the work soars, for he proceeds to look at all the military aspects of the English Civil War, a phenomenon that he terms the "Great Rebellion." At its heart lies a simple but enduring and ultimately proven hypothesis: "The history of the Civil War is the history of the evolution of an efficient army out of a chaos."

As he explores his hypothesis, Firth skillfully explores the history of the Tudor and Elizabethan armies in-depth, judging them as close to useless, more worshipers of Bacchus than Mars. While during this period and the early years of Stuart rule a number of English officers went abroad as mercenaries and thus had good experience, the rank and file was often decimated by disease and battle, leaving no true core of well-trained soldiers. Thus when the English Civil War broke out in 1642, "neither party had an army, neither party had an advantage." The key to the war, Firth argues, was the race to create such an army. He details this process with skill and dispatch, ultimately deciding that Oliver Cromwell's "New Model Army" created after 1645 eventually won the day and the struggle.

But this work is so much more than a most useful examination of the war in question. In addition, it opens up a truly magnificent examination of all aspects of war and war change in the mid-17th century. Providing the reader with in-depth looks at the role of discipline, armament, artillery, sieges, pay, commissariats, old soldiers, and religion in the army, the author details the evolutionary and revolutionary changes in the army itself, as the English followed many of the previous military trends on the continent.

Billmen and bowmen are replaced by pikemen and musketeers, the advantage shifts from cavalry (where it had stayed for much of the medieval period and beyond) to the infantry, with the ratios changing to match, cavalry falling to one to two (and indeed one to three during sieges). The work also underscores the difficulties in maintaining the size of armies; the need for impressments to augment volunteers; and the fact that, as the war wore on, the size of the armies both sides could bring to a particular battle dwindled.

Anyone interested in military history, let alone the English Civil War or 17th-century warfare, will find this work not only invaluable but also a wonderfully interesting and even entertaining read. It is nothing less than a treasure trove of military insights.

French Civil War (1871)

Alistair Horne, *The Fall of Paris: The Siege and the Commune 1870–1* (New York: Macmillan, 1965). The defeat of France during the Franco-Prussian War resulted in a violent uprising and even more violent repression of the Paris Commune—with 20,000 to 25,000 people killed in the process, as 120,000 French troops first surrounded and then entered Paris to stop the repression of the commune. The author gives many specific contemporary insights and points out that the Franco-Prussian War was the "most disastrous war in the long course of French history," with a huge "shame at capitulation." Claims also that the failure of the commune made the Russian Revolution possible by showing the mistakes revolutionaries make when they don't have a holistic plan.

Greek Civil War (1946–1949)

Kevin Andrews, *The Flight of Ikaros* (Boston: Houghton Mifflin, 1959). American archeologist wanders around Greece during 1947–1949. He runs into lots of Greeks who have bitter memories and thoughts about their civil war and the struggle between the Communist EAM/ELAS forces and those of the royalist EDES. When a saucy villager blames the war on America, he rightly—and sprightly—retorts, "You have been fighting civil wars in Greece since the beginning of history. . . . There's a nice description in Thucydides." Well said, but still, that night I don't believe he was served the best retsina in the area.

Andres Gerolymatos, *Red Acropolis, Black Terror* (New York: Basic Books, 2004). "Pogroms of the Left" led to "White Terror," which led to a civil war that cost 158,000 dead—11,000 in the Greek Army and 38,000 among the Greek insurgents, as well as civilian casualties. Communists were defeated when they moved from guerrilla operations to conventional ones, claims the author. The Greek counterinsurgency remains a striking example of the extent to which insurgencies can be defeated, although some drew the wrong conclusions perhaps. President Truman said, "Korea is the Greece of the Far East," although he may have simply meant there was a need to draw a line in the sand. Stalin apparently kept his agreement with Churchill, who had Greece as a 90 percent Allied sphere of influence, and the Soviets actually did very little to help the Greek Communists.

Mark Mazower (ed.), *After the War Was Over: Reconstructing the Family, Nation and State in Greece, 1943–1960* (Princeton, NJ: Princeton University Press, 2000). A wide-ranging collection of essays dealing with the Greek Civil War and its aftermath. Most interesting is the section of "The Civil War in Evrytania" (pp. 184–209) set in the central, mountainous, poor, and rural area. Examining this microcosm helps one to understand the wide variety of forces at work. Thucydides would have loved it, and the inhabitants of Corcyra during the 427 BCE revolution would feel right at home 2,438 years later.

———, *Inside Hitler's Greece: The Experience of Occupation 1941–1944* (New Haven, CT: Yale University Press, 1993). Shows how starvation, repression, murder, and a variety of insurgencies led to there being "two armed camps," one promoting revolution, one fearing it when the Germans left, setting the stage for the civil war that would inexorably follow.

Edgar O'Balance, *The Greek Civil War: 1944–1949* (New York: Praeger, 1966). A military man looks at the Greek Communists' three attempts to take over Greece: the first, during the German occupation of 1943–1944; the second, the abortive coup of October 1944, which the British suppressed; and the third, during the civil war of 1946–1949. The author believes that the Greek Communists (KKE) and their armed forces failed because they followed the Chinese revolutionary model. The Greek National Army itself reorganized in midwar, with the United States providing advanced arms, and the British and Americans, guidance, but ultimately, the Communists were defeated when they tried to go from guerrilla warfare to regular warfare.

George Psychoundakis, *The Cretan Runner: His Story of the German Occupation* (London: John Murray, 1955). Although this war memoir is primarily about the Greek resistance to the German occupation of Crete during World War II, I found it useful to catch glimpses of the forthcoming Greek Civil War in the competing guerrilla bands of the EOK and the ELAS. The bitter resistance on Crete and its brutal suppression led to the Germans only wanting to surrender to the British as the war wound down. They did not always get their wish.

C. M. Woodhouse, *The Struggle for Greece 1941–1949* (London: Hurst, 2002). The commander of the Allied military mission to the Greek guerrillas in German-occupied Greece from 1943 to 1944 sets the stage for the three-cornered struggle for control of Greece. He concludes that, while there was considerable importance in US and British aid and ultimately "all depended on leadership and morale," Stalin's letting down the Greek Communists was among the most important elements in the failure of the Communists to attain their goals.

Russian Civil War (1917–1922)

Benson Bobrick, *East of the Sun* (New York: Poseidon Press, 1992). The expansion of the Russian Empire into the far reaches of the Asian continent and the exploitation of the native peoples shows in stunning detail the "salt-water fallacy," which suggests that colonies have to be overseas to truly be colonies. Siberia, added to the Russian Empire, would become many things, including the home of many of the most infamous gulags of our time.

Stéphane Courtois, Nicolas Werth, Jean-Louis Panné, Andrzej Paczkowski, Karel Bartošek, and Jean-Louis Margolin, *The Black Book of Communism: Crimes, Terror, Repression*, translated by Jonathan Murphy and Mark Kramer (Cambridge, MA: Harvard University Press, 1999). This monumental 818-page tome covers all the evil done in the name of Communism, and the set of essays leaves us gasping with one central question: Did Communism in various forms kill 60 million or 90 million people? This is a most sobering portrait of humankind's inhumanity to itself. The various authors conclude that Communism was a vast and atrocious criminal enterprise. It is illuminating how many intellectual supporters of Communism even today act as if this book had never been written.

Christopher Dobson and John Miller, *The Day They Almost Bombed Moscow: The Allied War in Russia 1918–1920* (New York: Atheneum, 1986). Allied intervention in the Russian Civil War was sporadic, unhelpful, and ultimately doomed.

Frazer Hunt, "I Captured Vladivostok," in *Men at War*, edited by Ernest Hemingway (New York: Houghton Mifflin, 1942), pp. 966ff. A lighthearted "I was there" account, quite an interesting tangent.

Oleg Khlevniuk, *Stalin* (New Haven, CT: Yale University Press, 2015). See especially the first two chapters, "Before the Revolution" and "In Lenin's Shadow," for valuable insights into both the revolution and Stalin's checkered career path before and during it.

Mikhail Khvostov, *The Russian Civil War (1) The Red Army* (Oxford, UK: Osprey, 1996). Points out that the Red Guards were originally local units formed to protect factories and other key locations as early as 1905. They greatly assisted the Bolsheviks when they seized power in 1917 and were later absorbed into the Red Army in 1918. Some good pictures and unit backgrounds.

———, *The Russian Civil War (2) White Armies* (Oxford, UK: Osprey, 1996). Argues that the "term 'White Guards' (*belogvardeytsi*) appeared only after the Bolsheviks had started to call their own troops the 'Red Guards' (*krasnogvardeytsi*)." By calling their opponents "whites," the author feels, the Bolsheviks hoped to tar the anti-Communist cause with the brush of the discredited monarchy, although the "great majority of anti-Bolsheviks" did not want it reinstated. Plates show a wide range of units from those in Europe (Latvian 1st Kurzem Division and General Kornilov Shock Infantry Division) to those in the Far East (Komuch People's Army).

W. Bruce Lincoln, *Passage through Armageddon: The Russians in War and Revolution* (New York: Simon and Schuster, 1986). Reviewed chapter 18.

———, *Red Victory: A History of the Russian Civil War* (New York: Simon and Schuster, 1989). Unfortunately, the best campaign maps in this work are on inside covers. Reviewed in chapter 18.

Richard Luckett, *The White Generals: An Account of the White Movement and the Russian Civil War* (New York: Viking Press, 1971). Clear and understandable, Luckett's work brings to life the personalities of the White Generals L. G. Kornilov, A.I. Denikin, Carl Emil Mannerheim and P. N. Wrangel, along with Admiral A. A. Kolchak, and gives a convincing account of why they collectively failed to defeat the Bolsheviks.

Andrei Makine, *Requiem for the East* (London: Hodder and Stoughton, 2001). Part 3 of this work contains a powerful description of the Russian Civil War when Nikolai, after fighting for two years, deserts and travels in the Don area: "He had to avoid the Reds, whose troops he had just quit. To avoid the Whites, for whom he was still a Red. To steer clear of armed bandits, who varied their colour to suit their looting."

Evan Mawdsley, *The Russian Civil War* (New York: Pegasus Books, 2007). Mawdsley dates the Russian Revolution from October 1917 until the end of 1920, with 1918 being the "Year of Brest Litovsk" and 1919 being the "Year of the Whites." He sees much truth in Trotsky's statement "It is obvious that Soviet power is organized civil war against the landlords, the bourgeoisie and the Kulaks."

Simon Sebag Montefiore, *Stalin: The Court of the Red Tsar* (New York: Knopf, 2003). An exciting and even magnificent biography of Stalin, covering many dimensions of his life and times in considerable depth. Stalin remains the epitome of ruthless determination and paranoia. The author sees Stalin not just as the revolutionary Bolshevik but also fitting into the time-honored tradition of czars, for whom Russia was a personal fiefdom. Outstanding.

Richard Pipes, *The Russian Revolution* (New York: Alfred Knopf, 1990). Reviewed in chapter 18.

———, *Russia under the Bolshevik Regime* (New York: Alfred Knopf, 1993), chaps. 1–3. In-depth look at how the Bolsheviks dealt with their civil war enemies and created many new enemies, with whom they also dealt.

Edvard Radzinsky, *Stalin* (New York: Doubleday, 1996). Another outstanding biography of this revolutionary turned czar. Lots of drama here using the trop of "Koba." His ability to generate purge after purge against enemies of the right, the left, and the center; even the Communist center is legendary and rightly so.

Ted Raicer, "Reds: The Russian Revolution," *Command*, no. 54 (1998), pp. 34–51. Some worthwhile observations from the military perspective of how and why the Reds won and the Whites lost.

Nicholas Riasanovsky, *A History of Russia* (London: Oxford University Press, 1968), pt. 4. Gives a good factual account of the revolution in a straightforward fashion.

Robert Service, *Lenin* (Cambridge, MA: Belknap Harvard, 2000). An excellent biography of the man who, as much as any other, made the Russian Revolution possible.

———, *Trotsky* (Cambridge, MA: Belknap Harvard, 2009). Perhaps the second-most important man in the hierarchy of those who made the Russian Revolution. It is important to remember the author's words that, at the height of the counterrevolution in 1919, the "Soviet State covered an area little bigger than medieval Moscow."

See also other works on the Russian Civil War listed in chapter 18.

Spanish Civil War (1936–1939)

> "The war, which lasted for 989 days, has been one of the longest, bloodiest, costliest and most brutal internecine struggles of modern history."—Lt. Colonel Henry Cheadle, US military attaché, Madrid, April 1939

Antony Beevor, *The Spanish Civil War* (New York: Penguin, 2001). It is apparently difficult to write about the Spanish Civil War without taking sides, and most take the Republican side. Beevor seems more balanced and nuanced without shying away from the atrocities on both sides. Clear maps and clean prose make this a good read on a complicated subject.

James W. Cortada (ed.), *Modern Warfare in Spain: American Military Observations on the Spanish Civil War, 1936–1939* (Washington, DC: Potomac Books, 2012). Based on the observations of General Stephen Q. Fugua, US military attaché to Madrid during the Spanish Civil War, who wrote meticulous and illuminating dispatches and reports. Noting that there are more than 60,000 titles on the subject, the editor points out some important conclusions: Although 1.5 million casualties have been claimed, recent scholarship suggests a figure closer to 750,000, and the small number of Nationalist troops necessary to garrison liberated areas and Fugua's trenchant comment on the Republic's forces, "sloppiness, poor training, lack of discipline and weak command and control," were the primary reasons the Nationalists won.

Stanley Payne, *The Spanish Civil War* (Cambridge: Cambridge University Press, 2013). Careful, judicious, balanced. If you are going to read a single scholarly analysis of the Spanish Civil War, this is the one. Excellent on the "Breakdown of Democracy" and especially the "Civil War within the Civil War." Shows how mistakes by the Republicans made Franco's victory possible. Startling revelations about the bombing of Guernica: Military targets were the aim; casualties were actually 10 percent of those quoted elsewhere; and there was no targeting of civilians, but there was a lousy bombing of a bridge trying to cut off Republican retreats. Revokes the notion that Germany got to test out blitzkrieg for the war was anything but.

Stanley Payne and Jesus Palagios, *Franco: A Personal and Political Biography* (Madison: University of Wisconsin Press, 2014). A most useful account of the "most dominant figure in Spain since the time of Phillip II" and the "greatest political general in Spanish history." Particularly helpful for this section is chapter 8, "Winning the Civil War." The authors point out that Franco was the last major general to join the rebellion (he was stationed in the Canary Islands at the time) and that his 21,000-Moroccan unit became the key to victory. Clearly the Nationalists had much help from the Republicans, who were divided and often militarily inept and whose violence and overreaching led to the revolt in the first place. For his part The caudillo (Castilian for *leader*) was "methodical, organized and effective." Of the Spanish population, 1.4 percent perished, and despite Hitler's pressure, Franco did not join the Axis, both out of weakness and because Hitler wouldn't offer them French Morocco (out of Vichy sensibilities!). Incidentally, Franco wrote President Lyndon Johnson several letters imploring him not to get involved in Vietnam, and the authors claim, during World War II, no Jew was ever repatriated to Nazi Germany.

Paul Preston, *The Spanish Civil War* (New York: Harper, 2006). Provides a good account of the pro-Republican side of things and rightly includes the ongoing slaughter that followed the successful takeover of Spain by the forces of Franco, especially covering the 130,000 "judicial executions" that occurred. The "heroic determination" of the Republican forces is underscored, as is the divided nature of Spanish society prior to the formation of the Republic in 1931.

Jonathan Robb, *The Second Son* (New York: Farrar, Straus and Giroux, 2010) (F). The Spanish Civil War is the backdrop for this dark novel. One gets a whiff of the desperation, capriciousness, and often randomness of the violence, especially in a war where each town or city can be under different control and where atrocities become so common as to be ordinary.

Michael Seidman, *Republic of Egos: A Social History of the Spanish Civil War* (Madison: University of Wisconsin Press, 2002). Reacting against both the "great man" interpretation of history *and* an overreliance on class, race, and gender, the author postulates the importance of the individual and how he or she responded to the demands of the civil war as a determining factor. He believes that the failure of the Republicans to control their troops from pillaging turned many against the Republic and that the failure of the collective visions of the Spanish anarchists and Communists and the utopia of the Falangist and Fascists were undone by the "acquisitive, consumerist, and entrepreneur impulses" that many individuals exhibited during war.

Hugh Thomas, *The Spanish Civil War* (New York: Harper and Row, 1961). A long, carefully done description of the war, what caused it, and why it turned out the way it did. The battlefield accounts are straightforward and the maps clear. A fine source. I thought he does a particularly good job of looking at the balancing act the caudillo pulled off among the Falange, Catholic, and monarchist elements. Goethe, the author states, defines genius and knowing when to stop. General Franco certainly exhibited something close to this when he chose not to join the Nazis during World War II, despite the debt he owed Hitler and the Germans.

Chapter Twenty-Three

Precursors to 20th-Century War

There were a number of important changes in various warfare dimensions between the American Civil War (1861–1865) and the outbreak of World War I. These included the unification of German and Italy; the growing hypertrophy of war; and the invention, refinement, and increased lethality of such weapons as the machine gun and rapid-firing artillery, the submarine, the airplane, and dreadnaughts.

The Franco-Prussian War as well as the Anglo-Boer conflict contained within them many aspects of the war that was to come during 1914–1918. So, too, did the extensive use of the trenches and dugouts of Petersburg in the American Civil War, Port Arthur in the Russo-Japanese War, and others in the Crimean War, which looked ahead to the defensive-oriented slaughter on the Western Front during World War I. Also, and this is often overlooked by even ardent students of the period leading up to 1914, the Sino-Japanese War of 1894–1895 set in motion events and processes that would upset the previous hierarchy of power in Asia. Finally, the racial ethnic cleansing of the Hereros by the Germans in southwest Africa would stand as one of the many precursors to the racial horrors of World War II.

Note especially that the events of the Russo-Japanese War stimulated both Alfred Mahan (Russia) and Julian Corbett (Japan) and buttressed their arguments and examples. These writers, in turn, greatly influenced the development and operational strategies of the Germans (in both World War I and World War II) and Japan and the United States (in World War II). On the other hand, the quick victory of Prussia in the Franco-Prussian War exerted broad and lasting influence on those who planned for the next war: reorganization of general staffs, improvement of mobilization methods, an emphasis on offensive efforts (Napoleon III was criticized for lacking an aggressive strategy), and a continuing reliance on cavalry.

These precursors deserve our attention and can be illuminated by examining the following sources.

A. F. Becke, *An Introduction to the History of Tactics 1740–1905* (London: Hugh Rees, 1909). A highly readable, easy-to-understand, yet perspicaciously informative volume dealing with the eras of (1) Frederick the Great (1740–1905), with his emphasis on discipline, drill, and fighting power enabling him to shrink 10 ranks down to 3; (2) the French Revolution and consulate (1770–1804), in which wide recruitment and careers open to talent led to column attacks; (3) Napoleon and the Wars of Empire (1804–1815), with combined arms but desperation of cannon, cavalry, and infantry; (4) the era when "[m]en were drilled, not trained" (1815–1854), when musket fire increased its lethality from 200 to 500 yards and the French returned to a three-deep line formation; (5) the Breech Loader Era (1854–1878), with the Crimean War, the Danish War of 1864, the Prussian-Austrian War of 1866, and the Franco-Prussian War of 1870–1871; and then (6) the modern period (1878–1909), with smokeless power, magazine loading, machine guns, quick-firing artillery, and high-explosive shells. The significance of this final era was to escape the major powers of Europe until well into World War I. Yet here in 1906, it is finely laid out for all to see.

Jeremy Black, *The Age of Total War 1860–1945* (Lanham, MD: Rowman and Littlefield, 2010). Looking at the sweep of history during this period, the author believes that the American Civil War primarily looks backward to the Napoleonic era but that many other conflicts look forward to the 20th century's terrible world wars. He concludes that the hangover created by the carnage of them continues to make a general war far less likely. Good at highlighting many 19th-century "small wars" (and especially European colonization by military force), which are often overlooked in general histories. Also rightly points out that the elites of all parties did not anticipate the kind of war that developed, expecting one more similar to the previous Franco-Prussian one.

———, *Warfare in the Western World 1882–1975* (Bloomington: Indiana University Press, 2002). This follow-up or companion volume is far more pedestrian and even derivative; much better on the 19th than the 20th century.

Henri Bonnal, *Sadowa* (London: Hugh Rees, 1913). The English edition of this French general's 1894 almost-hour-by-hour analysis of the battle that in effect drove Austria out of Germany and vaulted Prussia into preeminence, with its emphasis on offensive strategic design. Accented the centrality of Helmuth Moltke the Elder and his impact on the great general staff of Prussia. Longer term, Sadowa seems fateful, leading as it does to the unification of Germany, the Franco-Prussian War, and World War I.

Noel F. Busch, *The Emperor's Sword: Japan vs. Russia in the Battle of Tsushima* (New York: Funk and Wagnalls, 1969). Looking at the battle through the career, life, and times of Admiral Heihachiro Togo gives the reader a highly descriptive and straightforward account of his overwhelming victory.

R. M. Connaughton, *The War of the Rising Sun and Tumbling Bear: A Military History of the Russo-Japanese War 1904–5* (London: Routledge, 1988). The author sees one of the root causes the Trans-Siberian Railroad's terminus in Vladivostok and its alternative branch to the warm-water port of Port Arthur and the siege warfare for that port foretelling the shape of World War I battles to come—especially with the high casualties bound to be incurred when taking a defensive position in the age of rapid-firing artillery and machine guns. Very detailed and comprehensive, if occasionally hyperbolic, such as saying the Battle of the Yalu in 1904 "ranks as one of the most important in the annals of warfare."

Byron Farwell, *Armies of the Raj: From the Mutiny to Independence 1858–1947* (New York: W. W. Norton, 1989). An engaging look at the Indian army and how it changed over this period, with fascinating vignettes on life in it. Shows the dominance of the "Great Game" in much British thinking and also the high rates of venereal disease among British soldiers (nearly half of the force in 1898). Good chapters on "Sepoy and Sowar" (privates or troopers, depending on whether infantry or cavalry) and "The Martial Races of India" (despite independence, the same recruitment patterns continue). Devastating views of Gandhi for causing communal strife and pretending it was none of his doing and the dismemberment of the Indian army at the time of independence. Also calls ridiculous the efforts to make the Great Mutiny of 1857 as a precursor to "modern" India. It was retrograde, the author says, and looking to the decrepit Mogul emperor is hardly proto-democratic.

———, *The Great Anglo-Boer War* (New York: Harper and Row, 1976). Big, sprawling coverage of this vast human drama that foretold of war in the 20th century—concentration camps, guerrilla warfare, and counterinsurgency measures of note (flying columns burned 30,000 farms and killed 3.6 million sheep and herded tens of thousands into unsanitary camps in 19th century *chevauchées*). The difficulty Great Britain had suppressing the rebellion was to give the kaiser and the German general staff hope that they could prevail against Great Britain.

Orlando Figes, *Crimea: The Last Crusade* (London: Allen Lane, 2010). Perhaps the definitive book on the Crimean War (1853–1856) from the diplomatic and historical perspectives. Points out how hugely destructive it really was: Almost a million casualties, including 100,000 French; 20,000 British (out of a force of 98,000!); 450,000 Russians (he believes it was their "crusade"); and an unknown number of civilians and Turks. Most important for warfare in the 20th century, the author strongly believes that Austria's armed neutrality during the conflict broke up the old alliance between Austria and Russia, an alliance that had upheld the existing order in Eastern Europe since the defeat of Napoleon. The Crimean War was also a precursor of World War I, as the siege of Sevastopol showed. It was the first "modern" war to use steamships, railroads, telegraphs, and industrialized production methods.

Richard C. Hall, *The Balkan Wars 1912–1913: Prelude to the First World War* (London: Routledge, 2000). This well-titled volume outlines the many precursor aspects to those wars. They were the first modern wars of Europeans, by Europeans, on other Europeans, and were quite simply the "first phase of the First World War." In particular, "Serbia's national appetite remained ravenous." These wars would project their repercussions onto the world stage for more than 100 years, including the "Second Collapse of Yugoslavia in 1991." Many weapons and strategies of modern warfare were foretold, including the mass atrocities against civilians. A very useful introduction to World War I.

Richard Hough, *The Fleet That Had to Die* (New York: Viking, 1958). The Battle of Tsushima in 1905 was the catastrophic defeat of the Russian fleet, which had sailed halfway around the world under the direction of Admiral Z. P. Rozhestvensky (who, severely wounded, would return by way of the Trans-Siberian Railway): "'Mob' is the only word to express our formations at the time." Of their 50 ships, the Russians had 146,000 tons of shipping sunk and 56,000 tons captured. Admiral Heihachiro Togo won a battle that is still studied and whose results had a profound—and disastrous—effect on Japanese strategy in World War II, as the "victory disease" took hold in a desperate effort to emulate Togo's success.

Michael Howard, *The Franco-Prussian War: The German Invasion of France 1870–71* (London: Methuen, 1979). This five-month war cost Germany 100,000 casualties but destroyed the French army, which had not kept up with the 19th-century military revolution and could no longer field a "nation at arms." This outstanding, forthright, and perceptive single-volume work builds on the 7,000 books on this subject that were already in print by 1898, concluding that Germany's "masterful and well-deserved victory was, in a profound and unforeseeable sense, a disaster: for herself, and for the entire world," as it would lead rather inexorably to World War I, for the loss of Alsace Lorrain would once again make of France a "nation at arms" and give the newly united Germany a false sense of its own power and importance in the world.

Raymond Jonas, *The Battle of Adwa* (Cambridge, MA: Harvard University Press, 2011). Very readable account of how the Italians pushed off blithely from Massawa and got their military asses handed to them by Emperor Menelik II and his wife Taytu at the Battle of Adwa in 1896 as the scramble for Africa washed over the continent. The Italians were to return briefly to subjugate Ethiopia from 1936 until the end of 1941, but they would be ousted in return by the British. The work has some outstanding descriptions of how the Italian prisoners were treated—amazingly well. Menelik's campaign—5 months and 580 miles, dwarfs Sherman's 300-mile march from Atlanta to the sea. Poignant and descriptive. I could see the blue haze over Addis and smell the pungent eucalyptus woodsmoke from the cooking fires. Ends with a powerful assertion: "Adwa reminds us that the only freedom we truly possess is the freedom we are able to defend."

Geoffrey Jukes, *The Russo-Japanese War 1904–1905* (Oxford, UK: Osprey, 2002). Clear maps, lavish illustrations, and a good overview of the clash between two powers who both wanted to dominate China. Useful descriptions of the battles of the Yalu, Port Arthur, Mukden, and Tsushima Straits. Precursor of Japanese emphasis on the "spiritual" versus the "material," a mindset that would get them into terrible trouble during World War II.

Ragne Kruger, *Good-Bye Dolly Gray: A Story of the Boer War* (London: Cassell, 1959). "Victoria's England" in full flower before this amazing precursor occurs from 1899 to 2002, with a "somber adumbration of the techniques of World War I," including barbed wire, Maxim guns, concentration camps, trench warfare, and blockhouses (an "intolerable obstruction"), as well as scorched earth, rapid-firing artillery, and repeating rifles. At the time of the surrender, there were still 22,000 Boer commandos in the field.

Jim Leeke, *Manila and Santiago: The New Steel Navy in the Spanish American War* (Annapolis, MD: Naval Institute Press, 1993). A breezy, navy-centric account of Admirals George Dewey and William T. Sampson, who smashed the Spanish fleets in the Philippines and Cuba in 1898. The author concludes, "The Navy won both battles by filling the skies with metal projectiles, like the clouds of British arrows at Agincourt." Dewey tossed 6,000, and Sampson, 10,000! The American triumph in the Spanish-American War thrust the United States into the Pacific with a vengeance, leading to new bases in Subic Bay, Cavite, and Corregidor, as well as expanded facilities at Pearl Harbor. Under President Teddy Roosevelt, the Great White Fleet of 16 US battleships would sail the world from 1907 to 1908, projecting American power into a region where it remains preeminent today (although the Chinese are likely to challenge over time).

Albert A. Nofi, *The Spanish American War 1898* (Conshohocken, PA: Combined Books, 1996). A very serviceable and readable account, a "splendid little volume" on the "splendid little war," with some very useful charts and explanations, such as "The Leap to Arms: A Case Study NY," showing the activation of units and "Naval Gunnery Effectiveness." A pleasant surprise.

G. J. A. O'Toole, *The Spanish War: An American Epic 1898* (New York: Norton, 1984). The author claims, "It was the most popular of all our wars," but adds that in its wake came Pearl Harbor, the Bay of Pigs, and the Vietnam War. Makes quite a point about the peace treaty that gave the United States Guam and Puerto Rico and threw in the Philippines (for $20 million), as well as gave Cuba its independence. Teddy Roosevelt would later say in 1909 that the Philippines will eventually "form our heel of Achilles." This work really reminds how the theme of unintended consequences sometimes does apply with a vengeance.

S. C. M. Paine, *The Sino-Japanese War of 1895–1895: Perceptions, Power and Primacy* (Cambridge: Cambridge University Press, 2005). The Chinese failed to appreciate the importance of the Industrial Revolution, while the Japanese embraced it. China did not realize how radically their position was changing in the hierarchy of Asia. Making a bold assertion that "wars matter" and that this war was a "truly seminal event," the author explains how the struggle for Korea ended up mattering so much. The Japanese would expel the Chinese now and the Russians later and take over Korea, using military force and a modified version of the "white man's burden." They landed troops in 1894 to "help" Korea put down a rebellion and, after occupying Seoul, took the king of Korea hostage and then, after declaring war, defeated the Chinese at Pyongyang and in a naval battle at the mouth of the Yalu (and at Port Arthur and later, in 1895, Weihaiwei). Ironically, it was in this era that the United States, by winning the Spanish-American War in 1898, would thrust itself into the western Pacific by taking over Wake Island, the Marianas, and the Philippines, and thus set the stage for its eventual struggles with Japan. Also ironically, it was the terrible massacres of Japanese prisoners by the Chinese that would propel the Japanese into taking retaliatory actions and set the stage (and rationalizations) for further horrors to come.

Gerhard Ritter, *The Sword and the Scepter: The Problem of Militarism in Germany*, 4 vols. (Coral Gables, FL: University of Miami Press, 1968–1973). Traces German militarism from the Prussian tradition in 1740 to the disaster of World War I. What German military and civilian leaders believed and what the militaristic strain influenced all up to and during World War I.

Tadayoshi Sakurai, *Human Bullets: A Soldier's Story of Port Arthur* (Boston: Houghton Mifflin, 1907). The Russo-Japanese War through the eyes of a Japanese lieutenant, one of the "[b]rave Japanese" for whom "[m]obilization is sweet, the happiest day we had ever experienced." This little gem foretells so much: the fanatical devotion of the Japanese soldier, their utter contempt for the Chinese ("stupid looking" and "smelly"), machine guns in static warfare, and a "rain of human bullets" that creates a "hill of corpses." Flanders's fields lie ahead and are anticipated here.

Jeremy Sarkin, *Germany's Genocide of the Herero: Kaiser Wilhelm II, His Generals and His Soldiers* (Cape Town: UCT Press, 2010). After 1904, Germany, acting under the kaiser's "secret instructions," his favorite colonial general, Lothar von Trotha, wiped out between 60,000 and 80,000 Hereros and committed additional atrocities against the Nama and Damara using "race branding" and "creating a map scraped smooth." The Germans declared that "no war can be conducted humanely against non-humans," anticipating, according to the author, the concentration camps, ethnic cleansing, and "total extermination" that were such an intrinsic part of Germany's World War II.

David J. Silbey, *The Boxer Rebellion and the Great Game in China* (New York: Hill and Wang, 2012). The old imperialists—Britain, France, and Russia—and the new imperialists—Japan and the United States—seem to treat China and the Manchus as fair game. The Boxers, reacting to both Christian missionaries and Chinese converts, rise up in 1905. Eventually they are put down by the Manchu reassertion of power, but their efforts look forward to the China of 1912 and the overthrow of the Manchus (and they are hailed as protonationalists by Sun Yat-sen and Mao Zedong). Much muddling about by the Great Powers, but they win the Battle of Tianjin and much looting, too, when they take the imperial city of Peking (now Beijing). The old and new imperialists also manage to charge the Qing dynasty $335 million for their trouble.

———, *A War of Frontier and Empire: The Philippine-American War 1899–1902* (New York: Hill and Wang, 2007). Manifest destiny, with an assist from Admiral Mahan, propelled the United States into the Philippines, as American power replaced Spanish power just as it was about to be overcome by Filipino power led by Emilio Aguinaldo. The United States went on to defeat his "Army of Liberation" in three years (although the Moros were to fight on until 1913). The Filipino Revolution was done in by a combination of slow attrition, pacification, and many Filipinos coming over to the side of the Americans.

Evan Thomas, *The War Lovers: Roosevelt, Lodge, Hearst and the Rush to Empire* (New York: Little, Brown, 2010). In 1898, America rushes into the Spanish-American War in order to become a blue-water imperial power in Cuba and the Philippines. Thomas Brackett Reed of Maine and Bowdoin College emerges as an anti-imperial hero. Build a blue-water navy, and you have to find a use for it to justify its existence is the lesson of the day. The Spanish-American War would not only put the United States on a collision course with Japan with regard to the spoils of World War I but would also set in motion a perceived rivalry that would lead to World War II in the Pacific theater.

Barbara Tuchman, *The Proud Tower: A Portrait of the World before the War 1890–1914* (New York: Macmillan, 1966). The clever title is from Edgar Allen Poe's "City in the Sea": "While from a proud tower in the town Death looks gigantically down." Looking at art, music, politics, and society, the author captures the fin de siècle mood in the European countries that would bring the world to ruin. "Bristling with new tensions," their world could not be ordered by the quality of their political leaders, ending with the apt statement "I summon the living, I mourn the dead." There would be many to mourn.

Denis and Peggy Warner, *The Tide at Sunrise: A History of the Russo-Japanese War 1904–1905* (New York: Charterhouse, 1974). A very detailed, interestingly written account of the Russo-Japanese War of 1904–1905, with such chapters as "First Pearl Harbor" and "Elusive Victory." They begin with the arrival of Admiral Perry in Japan in 1853 and the stimulus that gave to Japanese industrial, nationalist, and military revolutions. Points out that more troops were involved in the 1904 battle of Liaoyang in Manchuria than at Gettysburg, Waterloo, or Borodino.

Geoffrey Wawro, *The Austro-Prussian War: Austria's War with Prussia and Italy in 1866* (Cambridge: Cambridge University Press, 1996). Italy wanted the Habsburg provinces of Venice and South Tyrol and, according to the author, assisted the Prussians in their successful war. He asserts that, as important as the needle gun and the improved and overhauled Prussian army under General Helmuth von Moltke were, the real reason for the Austrian defeat at Koniggratz was the bumbling Austrian general Ludwig Benedek, whose blunders caused one of his staff officers to remark, "Our pilot is off his head." Austria's defeat was massive and led another to remark, "We have sunk to the level of Turkey." Italy did get Venice in the peace settlement, while France made do with Nice. The development and deployment of the "needle gun" breach-loading rifle enabled the Prussian soldiers to kill their opponents at 300 yards without having to stand to reload. By World War I, all armies would be equipped with similar weapons.

———, *Warfare and Society in Europe, 1792–1914* (London: Routledge, 2000). A good overview of the period from the French Revolution to the outbreak of World War I, highlighting the premise that a "revolution in military affairs" is only valid until copied by others. Quite useful sections on the "Militarization of Europe" and "Deadly Spiral 1911–1914."

Christopher Wilkinson-Latham, *The Boer War* (Oxford, UK: Osprey, 1977). A useful set of photographs captures the precursors to the heavy casualties of trench war, which would be such an important dimension of World War I, and the development of mobile guerrilla warfare and how to suppress it (and at what cost).

Cecil Woodham-Smith, *The Reason Why* (London: Constable, 1953). The author uses the intertwined careers of Lord Raglan, Lord Lucan, and Lord Cardigan to show the sorry state of British leadership during the Crimean War. These chaps seemed totally clueless as they blundered about. The charge of the light brigade under Lord Cardigan is put in sharp perspective. Woodham-Smith believes that the horrors of their generalship and the public's view of the war helped to change the culture of the British military. One hopes.

Chapter Twenty-Four

World War I

The Hypertrophy of War and the Triumph of the Defense

"The war that will end war."—H. G. Wells

"We are in measurable distance of a real Armageddon."—Prime Minister Herbert Asquith

"The lamps are going out all over Europe, we shall not see them lit again in our time."—Sir Edward Grey, *The Great War*

The Industrial Revolution is applied to world warfare, but the statecraft lags far behind. There is also incredible barrenness of strategy and much myth-making in the Great War. World War I should always be seen on a space-time continuum with World War II, for, in simplest terms, Germany rose up too high prior to the first and was thrust down too low prior to the second. It is thus always prudent in studying World War I to be reminded of the wise words of Michael Howard that World War I and World War II should be thought of "as a single Thirty Years War interrupted by a long truce."

In terms of battlefield advances, it is interesting to note that the basic equipment of the First World War—the repeating rifle, the airplane, the machine gun, the submarine, rapid-firing artillery, flamethrowers, and many others—is still with us today. Improved, yes; made more powerful, yes; but still the same basic elements remain in use 100 years after the "war to end all wars" broke out. But in terms of impact, of course, those who had to make war were forced to come to grips with the new realities that this war included. As Michael Howard points out, a regiment of artillery could saturate a target a few hundred yards wide over a few hours with more explosive firepower than unleashed during all the Napoleonic wars. This took some getting used to, most unfortunately for those on the ground underneath those barrages.

With regard to sources on World War I, according to William Hay, there were well over 25,000 books and articles already written about it by 1991, and most probably, there have been thousands compiled since. What follows are some of most readily available and accessible of them.

American Battle Monuments Commission, *A Guide to the American Battle Fields in Europe* (Washington, DC: Government Printing Office, 1927). Compiled by the then-colonel Dwight David Eisenhower, who spent two years meticulously traversing the battle sites, this truly magnificent work has unparalleled details of the battlefields on which Americans fought during World War I. It is accompanied by very detailed maps, many photos not available elsewhere, and clear diagrams. A treasure. Obviously Eisenhower benefited from knowing the lay of the land when he revisited it as supreme Allied commander during World War II. Useful chapters include "Services of Supply," "Sector Occupation," and "Distinguishing Insignia."

Max Arthurs, *Forgotten Voices of the Great War* (Guilford, CT: Lyons Press, 2004). Taken from the oral history of those soldiers—British, German, French, Australian, German, and American—who fought it, in their own words. From rats the size of cats to soldiers dying in excrement because their fellows were too weak to extricate them to executing two lads who, new to the front, ran away in their first attack. What an awesome war it was, the Germans taking 90,000 British and French prisoners in their last offensive in 1918, while suffering nearly a million casualties themselves.

Tony Ashworth, *Trench Warfare 1914–1918: The Live and Let Live System* (New York: Holmes and Meier, 1980). This is one terrific book with two very fine, useful illustrations of the trench warfare arrangement. Four months of action and movement and then four years of static trench warfare. Points out how much of the front and for how long a second type of warfare held sway. Not the trench warfare of the big battles and desperately costly struggles but "cushy" sections that were often quiet for long periods of time, as a tacit "live and let live" philosophy pertained to those in the field on both sides. These "truces" were illegal and unofficial but widespread, says the author, and he makes a solid case for their extent. Also talks about "chatting," which in these sectors meant sitting around delousing each other and thus promoting small-group solidarity.

Stephane Audoin-Rouzan and Annette Becker, *Understanding the Great War 14–18* (New York: Hill and Wang, 2000). These two French authors look at the war through three lenses: violence, crusade, and mourning. Quite riveting and incisive. Shows extent of civilian casualties.

Arthur Banks, *A Military Atlas of the First World War* (New York: Heinemann Educational Books, 1975). One approaches this paperback work with high hopes, and indeed, many of the maps are quite helpful in understanding specific battles (such as Third Ypres). Unfortunately, it has several important drawbacks, the most important of which is the small area covered for many battles and situations, depriving the reader of the broader context within which a specific battle is being waged. Too bad, as this distracts from the overall general usefulness of the work.

Austin Bay, *Ataturk: Lessons in Leadership for the Greatest General of the Ottoman Empire* (New York: Palgrave Macmillan, 2011). A breezy biography of General Mustafa Kemal, who was born in Salonika and, as the Ottoman Empire was crumbling, entered the military, rising through the ranks on merit. He played a vital, even crucial, role in the Turkish triumph at Gallipoli, and this book does justice to that explanation as opposed to those who focus exclusively on British and French mistakes. Out of World War I would come modern Turkey, and Atatürk was one important reason.

Rudolf Binding, *A Fatalist at War* (Boston: Houghton Mifflin, 1929). From his baptism of fire in October 1914 until the end of the war, this German officer endured the opening rounds, Passchendaele, a gas attack at Ypres (1915), the campaign in Galicia (1916), and back to the Western Front (1917) until the end of the war. Laconic, matter-of-fact, but insightful: "We have been attacking again: success nil; heavy losses." Fatalist indeed.

Edmund Blunden, *Undertones of War* (London: Oxford University Press, 1924). Put his memoir with those of Sasson, *Memoir of an Infantry Officer*; Manning, *Her Privates We*; Graves, *Goodbye to All That*; Barbusse, *Under Fire*; and Junger, *Storm of Steel*, as the best of the eyewitness accounts of trench warfare on the Western Front. Marvelous word pictures providing a "sketch of a happy battalion—happy in spite of terrible tasks and daily destruction." The war on the Western Front is a combination of "ancient siege warfare and an abattoir," and "this dark enigma scrawled in blood." The author was in both Ypres battles as well as the Somme.

Brian Bond, *War and Society in Europe 1870–1970* (New York: Oxford University Press, 1984). Wisely sees World War I and World War II on a space-time continuum. I found the chapter on "The Twenty Year Truce 1919–1939" (pp. 135–67) most illuminating. There are many valuable insights, such as that of Goring's on the "fighters who could not become debrutalized"—those who joined the Free Corps; the Nazi SA and SS; the Italian Fascist *Squadristi*; the German *Stalhelm*; and the French *Croix de Feu*, who couldn't stop fighting, and their alteregos, the leaders of France and Great Britain, who were so traumatized by World War I that they couldn't bring themselves to ever imagine getting into another fight. To quote Stanley Baldwin, "Who in Europe does not know that one more war in the west and the civilization of the ages will fall with as great a shock as that of Rome?"

Guy Chapman, *A Passionate Prodigality* (New York: Holt Rinehart and Winston, 1933). Britain faces its first major war in 100 years. No wonder they were not prepared. "We descended into primal man," as the author lets us share his experience, including rubbing one's feet with whale oil for trench foot. At the end of the war, "[w]e passed over into Germany. No trumpets sounded."

Alan Clark, *The Donkeys* (New York: William Morrow, 1962). This little gem is full of irony, including the title. The donkeys are not the British foot soldiers who slogged to their deaths in record numbers during World War I—there were more British dead in the first two hours of Loos than casualties suffered by *all* sides on D-Day during World War II. The donkeys are the leaders, starting with Sir John French and going down through virtually all others, making serious strategic decisions during the opening phase of the war. The British professional army, which was virtually wiped out by 1915 (replaced by volunteers and conscripts of the New Army), was made up of lions, the author asserts. Clark finds the professional army done in by terrible leadership, the new arms, and a failure to appreciate their lethality. Haig wrote, for example, "The machine gun is a much-overrated weapon and two per battalion is more than sufficient" and "wastage" (illness and accident). In the entire war, there were only three actual "breakouts" by the British: during the first days of Neuve Chappell and Loos in 1915 and Cambrai in 1917, and these were quickly dissolved by mismanagement and vigorous German counterattacks. A scathing indictment of British military leadership during World War I. Despite some recent revisionist views (including those of John Keegan), this earlier cut of the material still rings true.

———, *Suicide of the Empires: The Battles on the Eastern Front 1914–1918* (London: American Heritage Press, 1971). A quite breezy, interestingly illustrated quick study, most useful for its counterintuitive assertions that (1) the fragility of the Austro-Hungarian Empire, not just the problems of the Balkans, was a major driver of German strategy; (2) the Russians had at least three chances to win the war in the east, finally during the 1916 offensive of General Alexei Brusilov against the Austrians (which opened a 200-mile gap in the Austrian lines), but could not capitalize on any of them; and (3) Ludendorff panicked when Bulgaria sued for peace in 1918 and lost his nerve to continue the war.

Christopher Clark, *The Sleepwalkers: How Europe Went to War in 1914* (New York: Harper, 2013). A truly magisterial work of almost unbelievable complexity (as is its subject, especially with all its Balkan dimensions). A must-read to understand the roles played by the deep-seated phantasmagoria of a Great Serbia, the intricate cross-currents within each of the major governments, the inherent instability of the Austrian Empire (the "empire without qualities," or an "egg with two yolks"), and especially the rather startling portrait of Kaiser Wilhelm II, who seems far less bellicose and far less in control of his own government and military at the crucial junctures than previous accounts have claimed. Although it must still be said that the kaiser's previous insistence upon an England-challenging sea empire and concomitant naval race changed the complexity of Europe rather dramatically. But to Clark, he seems far less bent on war and not really wanting one. On Sunday, June 28, for example, he returns to Berlin "to take things in hand and preserve the peace of Europe." On July 30, he is aggravated and shocked to learn that, if Germany and France got involved, Britain would come into the war on their side, lamenting that his grandmother would never have allowed that! Also new to me was the blame properly ascribed to British foreign minister Edward Grey, who seems to have been totally out of his depth as the countries inch ever onward to a tragic catastrophe amid a complex shift from a multipolar world in 1882 to an essentially bipolar one in 1907, making the whole system "more opaque and unpredictable, feeding a mood of mutual distrust, even within the respective alliances." The Russians, too, come out looking worse due to their insistence on early mobilization, forcing the German hand, as do Serbia and Austria, the two countries most bent on conflict. The author also insists that popular support for the war was far outweighed in virtually all countries by astonishment and chagrin that it was happening. This is an outstanding work.

George Coppard, *With a Machine Gun to Cambrai: The Tale of a Young Tommy in Kitchener's Army 1914–1918* (London: Her Majesty's Stationary Office, 1969). Provides a vivid, unvarnished look at the life of a private (later, lance corporal) serving in the British 12th Division during World War I. Enlisting at 17, Coppard served in some of the major battles of the British army on the Western Front during 4½ years of service: Loos, Arras, the Somme, and the first tank battle at Cambrai. Wounded twice, each a "blighty one" (a nonfatal wound that got one out of battle and safely home to England), he was first shot in the foot by his buddy and later, after recovering and returning to action, hit in the thigh by a German machine gun. The author's narrative is pithy and pulls no punches. A few examples of his telling descriptions include the following:

1. On British generals: "Historians say that Haig had the confidence of his men. I very much doubt whether this was strictly true."
2. On enlisted men not getting paid on time when they got out of the trenches: "Hundreds of men mooching around in a lonely village without the price of a drink is an unhappy sight."
3. On the dead hanging on the barbed wire after an attack: "Like fish caught in a net. They hung there in grotesque positions. . . . The German faith in massed wire paid off." After describing the horrendous British casualties and their lack of overall success, he gives the Battle of the Somme to the Germans "on points."
4. On being accused that his first wound was self-inflicted: "The most unpleasant of my war experiences."
5. On getting criticized for putting his finger in the hole in his thigh after he was wounded the second time as unsanitary: "But my action stopped the flow like turning off a tap."

Coppard was only 21 when finally demobbed. This "plain soldier's tale" is a marvelous work, its authenticity unmistakable and telling.

Robert Crowly (ed.), *The Great War: Perceptions on the First World War* (New York: Random House, 2003). A fine collection of essays from *MHQ: The Quarterly Journal of Military History*, including John Keegan, "The Breaking of the Armies"; Thomas Fleming, "Iron General"; and his own "The Massacre of the Innocents." Asserting that "never before in war had there been losses like the losses of 1914" (from August to December 31, 1914), the editor points to the losses of 300,000 French; 265,000 German; and 150,000 British troops in the West and more than 750,000 German, Austrian, and Russian soldiers in the East. Still, as he depressingly points out, total losses during World War I were only one-fifth of those suffered in World War II.

Charles Emmerson, *1913* (New York: Public Affairs, 2013). An in-depth look at the capitals of Europe (plus those of other participants, including the United States, Japan, Mexico, India, and China) on the eve of World War II. Chockfull of observations and insights but presented in an oddly flat and uninspiring fashion.

A. H. Farrar-Hockley, *Death of an Army* (New York: William Morrow, 1967). An account of First Ypres in Belgium, where the British professional army held off the Germans but was pretty much shattered: "The old British

Army with its venial faults and marvelous courage had died in its defense." The Germans did not capture Calais.

———, *The Somme* (London: B. T. Batsford, 1964). A solid, sometimes stolid, account of this gigantic battle that killed wounded or caused to be taken prisoner 1.2 million British, German, and French soldiers during the long period of July to November (although the author says the battle was really decided by September). "The Long Agony," which was the just after the first battle, seemed as much a result of German attacks on Verdun as anything else. He exonerates many leaders and generals, declaring, "The Somme battle, as indeed the whole of the Great War, was ultimately the responsibility of the people of Europe and the United States, who permitted conditions to come to such a pass." Hard slogging, that!

Fritz Fischer, *War of Illusions: German Policies from 1911–1914* (New York: W. W. Norton, 1979). Makes a strong case that the ruling classes in prewar Germany made the assumption that, without war, they would lose out in Germany and Germany would lose out in the world. For this German historian, the "decisive factor" in the outbreak of the war was German illusion, and Germany's chance for success in that war was ended by September 10 (Battle of the Marne), or at the latest by November 18 (first battle of Ypres), 1914.

Ferdinand Foch, *The Memoirs of Marshal Foch*, translated by Tibently Mott (Garden City, NY: Doubleday, 1931). Leader of the Ninth Army at the First Battle of the Marne, he rose to become supreme Allied commander and orchestrator of the Second Battle of the Marne and the subsequent offensive that broke the back of German political resistance. Argued that Germany had no need to go to war, as it was going to conquer the world through its economic power without warfare.

Aaron Friedberg, *The Weary Titan: Britain and the Experience of Relative Decline* (Princeton, NJ: Princeton University Press, 1988). Decision makers in Great Britain did not realize that the country, while still militarily strong compared to its earlier force dispositions, had lost ground relative to others, especially Germany.

Craig Gibson, *Behind the Front: British Soldiers and French Civilians 1914–1918* (Cambridge: Cambridge University Press, 2014). A most understudied subject. Declaring that the "sexual history of the BEF [British Expeditionary Force] has received scant attention," the author devotes an entire chapter to "Sex" and concludes with a number of interesting observations: The Tommies found the farm and small-town girls of Flanders different from the women of Paris; there was much "homoerotic" sex among officers and a great fear of STDs; "striking oil" was a widespread euphemism; and French women were amazingly "natural" about sex "on the very fringe of what constituted the civilized world."

G. Norman Gladden, *Across the Piave: A Personal Account of the British Forces in Italy, 1917–1919* (London: Her Majesty's Stationary Office, 1971). How much better to be in action in Italy (1 in 21 casualties) versus France (5 in 9). *After* the Somme and Passchendaele campaigns, the British Expeditionary Force (BEF) goes to Italy in 1917. Gladden notices right away how much nicer it is to have siestas from one to three, when the Italians and the Austrians take a break (until the Germans mess things up). After the disaster at Caporetto (north of Trieste) on the Piave line, the force is there to make sure Italy isn't knocked out of the war. Home leave was a "gift from the gods." Two divisions of the BEF-Italy were sent back to the Western Front as the war ends. A marvelously simple—and truly charming—account, with much to teach about this little-known aspect of the war: self-effacing, insightful, and smoothly written.

———, *The Somme 1916: A Personal Account* (London: William Kimber, 1974). Although forbidden, diaries of World War I can be very illuminating, and this one is no exception, as this British private in the infantry gives us a day-by-day account of the Great War. Finally out of the line due to trench foot, he finds that, at the recovery depot in England, "[n]othing was omitted to make our lives as miserable as possible" due to a parade ground, and the colonel was bound and determined to march them back to health and hence into the front lines again.

———, *Ypres 1917: A Personal Account* (London: William Kimber, 1967). The Northumberland Fusiliers get a 19-year-old who gets 4 months of training. "God what a hell" was Passchendaele. Detailed description of 70,000 British casualties to create the "Salient" June to December 1917, then mercifully on to Italy (see previous review). What an amazing amount of war for one man to endure.

Robert Graves, *Goodbye to All That* (New York: Jonathan Cape and Harrison Smith, 1929). Autobiographical work with matter-of-fact heroism: "It's murder sir." "Of course it's murder you bloody fool," I agreed. "But there's nothing else for it, is there?" Serving in the Royal Welsh Fusiliers and severely wounded at the Battle of the Somme in 1916, Graves gives the reader a sense of the inevitability of death in the trenches and the shared dream of a wound bad enough to get you sent home. Great detail, such as describing the machine-gunners firing off belts of ammunition just to boil some water for tea. Ironic observations abound: "Aside from the wounds, gas and accidents of war, the life of the trench soldier was, for the most part, not unhealthy."

Elizabeth Greenhalgh, *The French Army and the First World War* (Cambridge: Cambridge University Press, 2014). This dense, tightly packed volume is marvelous in its scope and detail. Defending the French contribution to the war, the author concludes, "The French Army's efforts during the First World War had been stupendous." Documenting that France conscripted a higher percentage of its adult male population than either Germany or Great Britain, she also points out how exhausted France was by 1918 and yet managed to supply the Americans with the necessary aircraft, tanks, artillery, and machine guns for the final offensive. This is an impressive work.

Max Hastings, *Catastrophe 1914* (New York: Knopf, 2013). Well-researched and puts many events and actions in a broader perspective, this account of year 1 of the war points out that (1) Germany was in fact most responsible for the war, and their cause *was* morally distinguishable from those of the Allies; (2) the British generals were "brothers" only in the sense of Cain and Abel; (3) British contribution to beginning was small—the British Expeditionary Force (BEF) was 52 battalions, compared to 1,077 German, 1,108 French, and 120 Belgians (p. 201)—about 3 percent by the time of the Battle of the Marne in September; (4) most Europeans did not welcome the coming of the war; and (5) casualties in the beginning were worse than later—the French suffered 1 million casualties in 1914 and the Germans, 800,000 (p. 48). He doesn't give the British navy enough credit for the blockade but rightly underscores their "Trafalgar" complex—focusing on an all-or-nothing, gigantic, climactic battle. There is a good section on the "race to the sea" and Churchill's wild scheme to hold Antwerp with Royal Marines. By winter 1914, all the belligerents, save the British, had bombed their opponents' cities (p. 462). France lost 16.5 percent of those conscripted; Germany, 15.4 percent; Great Britain, 12.5 percent; Austria, 12.2 percent; Russia, 11.5 percent; and Italy, 10 percent (p. 353). Even worse, Serbia lost 62.5 percent of all males between 15 and 55 (p. 514). He does a very nice job of illustrating the futile struggles in the east, as Russia, Austria, and Serbia exhausted each other in a horrendous ebb and flow of military disasters.

Holger H. Herwig, *The First World War: Germany and Austria-Hungary 1914–1918* (London: Arnold, 1997). Avoiding so many of Anglo-centric conclusions about the war, he gives the reader many insights into the German strategy and strategy not often explored. Particularly helpful is his analysis of Operation Michael, the German general staff's "last card." He ends up blaming General Ludendorff for not having any strategic plan beyond "busting through" and for failing *either* to drive the British to the Channel *or* to take

Paris. By trying to do *both*, he did *neither*. Operation Michael began with 6,600 guns; 3,500 mortars; 1,000 aircraft; and 10,000 full-length trains bringing troops. Initially, it was very successful, killing and wounding 200,000 British and French and taking 90,000 prisoners. But then, Germany ended up losing 200,000 soldiers and the war, in considerable part due to the arrival of fresh American troops.

———, *The Marne, 1914: The Opening of World War I and the Battle That Changed the World* (New York: Random House, 2009). Herwig claims this battle changed the course of European history and prevented a German domination of the continent. He makes a strong case.

Adam Hochschild, *To End All Wars: A Story of Loyalty and Rebellion 1914–1918* (Boston: Houghton Mifflin Harcourt, 2011). An interesting amalgam of the actions and words of the suffragettes, antiwar activists, and conscientious objectors, along with those of the British generals and other opinion makers. A great narrative that throws up so many interesting facts illuminating the war that I could not keep track of them all. Here are but a few:

1. There are 400 British cemeteries at the Somme battlefield alone.
2. Between 1914 and 1918, some 700 million artillery shells were fired, and even today, the French find and detonate 900 tons a year.
3. The war was "astonishingly lethal for the ruling classes"—12 percent of all British soldiers died, compared with 19 percent of peers and the families of peers.
4. During the 4½ months of the Battle of the Somme, 30 million shells were fired, and the British and French suffered 700,000 casualties and captured only 7 square miles.
5. On the first day of that battle, of the 120,000 British in action, 57,000 were wounded, killed, or captured, compared with 8,000 German casualties.
6. The bombardment during the Somme put so much metal into the ground that compasses often didn't work.
7. Each month in 1915 and 1916, the Germans brought up 7 tons of barbed wire per week, while the British averaged 6 million sandbags per month, both leading to what Field Marshal John French called "this tremendous *crust* of defense."
8. "Victory no longer went to the manly, the fearless and the daring."
9. World War I was the "most deadly catastrophe to strike Europe since the pandemic of the fourteenth century," leaving 9.5 million soldiers and 12 to 13 million civilians dead, as well as arguably an additional 7 to 10 million dead from the resulting Russian Civil War.

Richard Holmes, *The Little Field-Marshal Sir John French* (London: Jonathan Cape, 1981). An attempt, which almost succeeds, in rehabilitating the reputation of the first head of the British Expeditionary Force in World War I. He would later be replaced by General Douglas Haig, who would dream up the Somme fiasco as his contribution and then somehow hang onto his job until the end of the war. John French was a small man with a great appetite for women and love for the cavalry and did a fine job during the Anglo-Boer War (especially at the battles of Elandslaagte and Klip Drift) but was in way, way over his head with the demands of the Western Front during the First (September to October 1914, the "Massacre of the Innocents") and Second Ypres (April to May 1915, when Germans used poison gas for the first time). Although this is quite a charming book, it ends up giving a devastating portrait of British decision making—both military and civilian—during the war. One ends up confused as to how the Germans managed to lose it.

Alistair Horne, *The Price of Glory: Verdun 1916* (New York: St. Martin's Press, 1963). In addition to providing a very readable account of this titanic battle, with many human-interest and first-person accounts woven into the text, the author makes a strong case that the Battle of Verdun was a turning point in the war and set the stage for the German, French, and British strategies in the run-up to World War II. Provides some amazing photos of what the battlefield looked like and why—one area, 500 by 1,000 yards of a French fortress, was hit by an opening salvo of 80,000 shells. Quite beyond comprehension.

Richard Hough, *The Great War at Sea 1914–1918* (London: Oxford University Press, 1986). Covering in extensive detail the war between the British Royal Navy and the German *Kriegsmarine*, including the Battle of Jutland, the author boldly asserts that the British blockade of Germany was decisive in winning the war and its 1,571-day effort was the "greatest naval victory in history." "It was the blockade that finally drove the Central Powers to accept defeat." Arguable but defensible.

Samuel Hynes, *The Unsubstantial Air: American Fliers in the First World War* (New York: Farrar, Straus and Giroux, 2014). Straightforward account of the US fliers during World War II. Lots of detail and much on individual pilots and their travail.

Paul Jankowski, *Verdun: The Longest Battle of the Great War* (London: Oxford University Press, 2013). A masterful attempt to unite the "battle history" school and those that have followed it, looking at wars in much broader terms. Looks at this 300-day battle, which took 300,000 French and German lives and a total of 1 million casualties and shows how little it accomplished and how little it proved, except to show the power of the defense and the horrible attrition of static warfare on the Western Front. Gives credit to the French poilu in resisting the initial attacks after more than a million shells were dropped on their position. Feels both sides slid into an inconclusive and ghastly "attrition trap," with French and Germans both justifying the ongoing battle because it was wrecking the other side's army, for "generals do not readily abandon their own enterprises." The author concludes, "The men hated Verdun as they had hated no other place in this war" (p. 188). Astonishingly, General Erich von Falkenhayn, whose brain-child the initial attack was, seems to have no consistent strategic goals in mind. Ironically, when American soldiers took over the Verdun theater in 1918, they had 3,000 artillery pieces, compared to the 1,200 the Germans started the battle with in 1916, showing the ongoing hypertrophy of war. They were about to push beyond Verdun when, fortunately, the war ended.

Ernst Junger, *The Storm of Steel: From the Diary of a German Storm Troop Officer on the Western Front* (New York: Howard Fertig, 1929). A truly astonishing work. A 19-year-old German lieutenant is wounded 20 times fighting in trench warfare, the "bloodiest, wildest, and most brutal of all warfare," and ends up loving war and concluding, more ominously for those who do not, "Germany lives and Germany shall never go under."

Paul M. Kennedy, *The Rise of Anglo-German Antagonism, 1860–1914* (Boston: Allen and Unwin, 1994). How Germany and Great Britain headed slowly, but seemingly inexorably, toward rivalry and war. Much of the slide was based on the Anglo-German naval race. A classic example of the "weapons speak."

——— (ed.), *The War Plans of the Great Powers, 1880–1914* (London: George Allen and Unwin, 1979). A collection of essays dealing with American, French, British, German, and Russia military planning and thinking in the run-up to the Great War. Interesting how the war plans set in motion a war that probably didn't need to happen and prevented it from being stopped in time. This work cries out for a concluding chapter to tie together the various disparate chapters.

Lawrence Lafore, *The Long Fuse* (New York: J. B. Lippincott, 1965). Although it is now almost 100 years later, I have seen no analysis of the run-up to World War I that adequately refutes his six true causes of the war: (1) Germany and France felt an overpowering need to support their allies; (2) military plans left very narrow windows for decision making; (3) Russia strongly believed Austria had grand designs on the Balkans; (4) Austria saw the actions and intents of Serbia as a threat "to their own existence"; (5)

military men, not diplomats, were really in charge in the run-up to the war; and (6) Austria's weakness as a state was perhaps the ultimate sine qua non for a war turning into World War I.

Cecil Lewis, *Sagittarius Rising* (New York: Harcourt, Brace, 1936). An English schoolboy at 17 rushes off to join the Royal Flying Corps ("we walked off the playing-fields and into the lines") and describes all of his adventures in lyrical prose. And what adventures they were, flying over the Somme during the entire battle in 1916, shuttling back and forth to England for night-and-day fighter defense of London. (Whoever started the myth that Guernica was the first civilian target bombed from the air?) He and his comrades fought with a true sense of individual warriorhood: "It was like the lists of the Middles Ages, the only sphere in modern warfare where a man saw his adversary and faced him in mortal combat, the only sphere where there was still chivalry and honor. If you won, it was your own bravery and skill, if you lost, it was because you had met a better man." "Pilots in 1916 lasted, on average, three weeks." The bird's-eye view of the Somme battle is itself worth the price of admission.

Nick Lloyd, *The Hundred Days: The Campaign That Ended World War I* (New York: Basic Books, 2014). Again, the American entry is said to be crucial. The Allies learned to use rolling barrages as opposed to hilarious round-the-clock pounding, the Germans were demoralized after the failure of Operation Michael, and both Hindenburg and Ludendorff were unable to adapt. Mustard and other gases caused 27 percent of American casualties (p. 104), and the American attack (500,000 men) on the St. Mihiel salient deserves much more credit than it has received, says the author. But Pershing's frontal assault in the Meuse-Argonne Battle, far from being "open warfare," reproduced the scene of the Somme, Verdun, and Passchendaele. There was a pounding by British, French, and Americans too much by mid-October, ending November 11, as Germany was close to anarchy and armies retreated pell-mell. Author argues that Foch was correct: There was no point in fighting into the heart of Germany.

General Erich Ludendorff, *The General Staff and Its Problems* (New York: Dutton, 1920). An in-depth look at the relationship between the German general staff and the imperial chancellor during World War I, strongly supported by documents of the moment. In his, Ludendorff believes said documents disprove "that a peace of understanding was possible."

Margaret MacMillan, *The War That Ended Peace: The Road to 1914* (New York: Random House, 2013). If a student could read but one book about the run-up to World War I, one could well recommend this one. Careful, judicious, balanced, and well-researched, it apportions blame fairly (Austria-Hungry and Serbia finally get their share) and points out the irony that the very fact that the European powers settled previous problems (Morocco, Balkans) without going to all-out war made them complacent in the summer of 1914. The author blames a "failure of imagination" to not see how destructive war would actually be and a "lack of courage to stand up to those who said there was no choice left but to go to war. There are always choices."

Frederic Manning (British), *The Middle Parts of Fortune: The Somme and Ancre, 1916* (New York: St. Martin's Press, 1929). Initially privately printed, this is an extraordinary account of the ordinariness of war: the tumult and the shouting but also the dreary, momentous, and soul-numbing nature of World War I, where rum rations, French plonk, and whiskey from home saved the remains of some days. In particular, the descriptions of battle in chapters 1, 16, and 17 are outstanding.

S. L. A. "Slam" Marshall, *World War I* (New York: American Heritage Press, 1971). One of America's foremost military historians of the 20th century provides a fine comprehensive description of World War I, from "When the Lights Went Out" to "Versailles." Readable, with very concise maps, this work is a good place to start for someone new to the war. Especially good at summing up, as with the Third Battle of Ypres, "244,897 slain and not one thing of importance had been achieved." One of the historians to give American forces their due, especially during the German offensive in 1918 at Chateau Thierry and Belleau Wood, pointing out that, while the US Marines get a great deal of credit, the American 23rd, 7th, and 2nd divisions also fought well. Scathing about the $200 million the United States spent ordering 4,400 light and heavy tanks: "All was sheer waste," as only 15 tanks ever reached France and those after the Armistice!

Sean McKeekin, *The Berlin-Baghdad Express: The Ottoman Empire and German's Bid for World Power* (Cambridge, MA: Harvard University Press, 2010). By focusing on Germany's efforts under the kaiser to expand German influence in the world of Islam, the author shows how German "jihad" in World War I is still with us today, despite the failure of the Germans to actually accomplish much against Great Britain in that theater during the war. Berlin wanted a master railroad running from the capital of Germany to Mumbai.

Note: During World War II, the mufti helped Himmler recruit three mostly Muslim Waffen SS divisions, the most infamous of all was the *Handschar* SS division, which killed 90 percent of the Jews in Bosnia. Upward of 100,000 men served in these units.

Martin Middlebrook, *First Day on the Somme* (London: Penguin, 2001). The author provides a balanced, nuanced, and ultimately fair assessment of July 1, 1916, the British attack on the Somme front, using the equivalent of 26 American World War II–sized divisions. Five cavalry divisions were kept in reserves for a breakthrough that never came. This first day was to be the bloodiest day in British military history, as they suffered almost 60,000 casualties (compared to Wellington's 8,458 at Waterloo). This casualty total exceeded all British losses in the Crimean, Boer, and Korean Wars combined. Middlebrook blames a too-complicated plan, five days of rather inconclusive artillery preparation, superb German defensive strength, and the fact that 60 percent of the New Army units had no previous experience in warfare. Gives Haig a bit of a pass, as he supposedly opposed the plan in the first place, and puts the sacrifice in perspective, as French general Foch says it saved Verdun and the French Army. Notes how the number of machine guns had gone from 24 per battalion in 1914 to 204 in 1916. The total battle would go on for 140 days.

William Moore, *The Thin Yellow Line* (New York: St. Martin's Press, 1973). There were 346 British soldiers executed during World War I for desertion and cowardice, plus a few for violence against superior officers. The author feels most were victims of shell shock and takes Parliament to task for "reprehensible" behavior in sweeping this reality under the proverbial rug. A small tragedy within a much larger one. Well-documented.

William Mulligan, *The Origins of the First World War* (Cambridge: Cambridge University Press, 2010). Focusing on the interrelated issues of "security and expansion" during the period from 1871 and 1914, the author finds that the war planning that featured adherence to the maxim "power trumps justice" did not make a general war "either inevitable or even probable." In fact, he concludes only, "None of the great powers wanted a general war in 1914 but they were all willing to risk it."

Michael Neiberg, *Dance of the Furies: Europe and the Outbreak of World War I* (Cambridge, MA: Harvard University Press, 2011). Argues that most "ordinary Europeans," especially those in the rural areas, did not expect nor want the war in the summer of 1914. Rudyard Kipling was one who got it right, writing on August 4, "Incidentally, Armageddon begins."

———, *The Second Battle of the Marne* (Bloomington: Indiana University Press, 2008). Clear, illuminating, and always interesting to read. Makes the point that the first Battle of the Marne in 1914 was the largest battle ever fought until that time, with a front of 160 miles and 2 million men involved. Gives the French credit for adapting themselves to the new realities of warfare in the industrial age. By 1918, however, the "Americans were eager, but woefully unprepared for the industrial, attritional war of 1918." Pershing prepared them for the 1914 battlefield; 40 percent of some units, like the 77th division, had not even fired rifles before arriving in France. Germans often referred to their Austrian partners as being "shackled to a corpse." Great Britain went through four different types of armies: (1) British Expeditionary Force (BEF) professionals; (2) BEF survivors, Indian regiments, and territorials; (3) Kitchener's "New Army," with its pals regiments; and, finally, (4) those after conscription began. Gives great credit to the Second Battle of the Marne, especially the July 18 effort as a true turning point, and gives the French army and French general Foch as well as the élan of the Americans as a reason for its success.

Joseph E. Persico, *11th Month, 11th Day, 11th Hour: Armistice Day 1918 World War I and Its Violent Climax* (New York: Random House, 2004). Incredibly, the "mayhem of the last day was no different from what had been going on for the previous 1,560 days." Amazingly, even though the Armistice was signed at 5 a.m. on November 11 and was supposed to go into effect at 11 a.m., planned attacks still took place right up until the specified end time. There were 10,944 casualties on that last day of the war, including 2,738 dead, more than the average of the previous 4 years, which was 2,250 dead and 5,000 wounded. Covering the entire war with vignettes of individual experiences of soldiers, the book makes a number of telling points, such as the British ended up occupying Mons at the end of the war, the place where they began it. In between, they suffered 700,000 casualties. Most books about World War I are sad. This is the saddest one of all. Of 16 American generals, 7 stopped their troops from attacking, 9 did not. Interesting note: Although 21,640 died in the influenza epidemic, 11 million were killed or wounded on Western Front.

Robin Prior, *Gallipoli: The End of the Myth* (New Haven, CT: Yale University Press, 2009). Finally gives the Turks credit for their steadfast defense and receptivity to innovation rather than just blaming British blundering, although, in truth, there was more than enough of that as well!

Robin Prior and Trevor Wilson, *Passchendaele: The Untold Story* (New Haven, CT: Yale University Press, 1996). A long and detailed story of a 1917 battle that is sometimes known as Third Ypres, following previous struggles in 1914 and 1915 fought in the same area, the first, a German offensive that resulted in the slaughter of the "kinder," and the second, which gained little, and all were played out on a "desolate wilderness of water-filled shell craters." The three-and-a-half month struggle, from which netted such small yardage at such a frightful cost, is covered, but at the end, strangely enough, the account itself seems somewhat lifeless and listless. The cost of the battle "plan" was 275,000 British and 200,000 German casualties, which had the British "bursting out" of the Ypres Salient in Belgium and "sweeping" to the Belgian coast. General Haig certainly had some truly costly ideas. In their 1918 offensive, the Germans swept back over the little ground gained by the previous British effort, also to no avail.

Ludwig Renn, *War* (London: Martin Secker, 1929). This modest little work by a German infantryman, once wounded and also the recipient of the Iron Cross First Class, is an illumination of the end of the war, as the German 1918 final offensive peters out and the army is pushed back amid its disintegration, the "war becoming more and more an object of suspicion," and the "bloody war's done for now. We're not going to let ourselves be killed for the sake of a few days more fighting." Yet Renn keeps it all together as his unit is the rear guard for the movement through Belgium and back to Germany, where chaos and virtual civil war awaits.

Frank Richards (Francis Woodruff), *Old Soldiers Never Die* (London: Faber and Faber, 1933). There is an air of matter-of-factness to this memoir of World War I that will haunt you long after you have read it. The author, an orphan coal miner who served in the Royal Welsh Fusiliers during the entire war, describes life in the trenches with a sangfroid that is most impressive. Four years in the trenches in an amazing procession of the major battles on the Western Front—the Marne, Loos, the Somme, Arras, Ypres—and still alive, with no major wounds and winning the Distinguished Conduct Medal (DCM) as well as the Military Medal (MM) along the way, the author writes with skill and dispatch of what he knows.

The author is not short on trenchant observations either: "To good, trained, pre-War soldiers who kept their nerve, ten men holding a trench could easily stop fifty who were trying to take it, advancing from a distance of four hundred yards." Nor does he pull punches against the war profiteers who sold rotten food to the troops: "The head of that firm should have been put against the wall and shot for the way they sharked us troops." He is also merciless to the generals who never ventured into the front lines and reporters who wrote their stories from safe positions back with those generals, wanting to give the latter a special medal for "Distinguished Lying off the Field."

This is a marvelous memoir, and the Fusiliers must have been quite a unit in which to serve. Robert Graves and Siegfried Sassoon were also in it and most likely helped and encouraged the "Old Soldier" to publish this fascinating account. The Old Soldier didn't simply fade away, as the contemporary hymn predicted; he left a lasting literary legacy—one of the best firsthand accounts of war I have ever read.

J. Howard Roberts, "Passchendaele: Mayhem in the Mud," *Military History* (December 1997), pp. 67–72. An awful, awful battle fought under the most appalling of conditions. This short article captures the true horror of trench warfare and trying to run an offensive where human flesh and will were pitted against the cold steel and explosive power of rapid-firing weapons.

Eugene Rogan, *The Fall of the Ottomans: The Great War in the Middle East* (New York: Basic Books, 2015). A wide, grand, yet granular look at the war with the Turks. Note that Turkish attacks in the Caucasus, Suez Canal, and Basra all falter (December 1914–April 1915). But would later revive at Gallipoli, where the conditions were even more miserable than on the Western Front because there were no respites and both sides suffered casualties of 500,000 of the 8 million men engaged. The British surrender at Kut (Iraq) with 13,000 men lost was the worst surrender ever, according to the author (p. 267). It was a very harsh treaty for the Ottomans, but Atatürk and others refused to accept, fighting on until 1923 and gaining back much of Anatolia.

John Rohl, *Wilhelm II: Into the Abyss of War and Exile 1900–1941* (Cambridge: Cambridge University Press, 2014). Exhaustive study of this period. War brings out demands for reform. Wilhelm goes to Spa, Belgium, to be with army during Operation Michael and "never set foot in Germany again." As the war ended, a republic was declared in Berlin, and troops dispatched to fetch him. He only narrowly escaped to Holland, where he was not expected or wanted, but the massacre of the czar and his family held sway. The Allies wanted to try him for war crimes, but the Dutch wouldn't extradite him. His wife died, and he married again, suffering several nervous breakdowns along the way. Wilhelm, never a sympathetic figure, really emerges from this portrait as viciously anti-Semitic, extremely childish, and politically naïve, as well as incredibly, even by royal standards, self-absorbed.

Erwin Rommel, *Infantry Affairs* (London: Greenwood Books, 1990). Reprint of Rommel's 1930 recollections of his World War I campaigns in France and Italy. Lots of homage paid to the god of war and some almost tender memories of the war that did not end all wars.

Dennis E. Showalter, *Tannenberg: The Clash of Empires 1914* (Washington: Brassey's, 2010). Germany's biggest victory of the war, inflicting 50,000 casualties on the Russians and capturing 90,000 of them. A straightforward, detailed, and somewhat prosaic account, although interesting in showing how the myth of the battle propelled German thinking into World War II and the various miscalculations about how easily the Russians could be defeated again. Paul von Hindenburg and Erich Ludendorff became folk heroes (although they had to go south into Silesia) with the near-collapse of the Austrians after the battle. By the time they got to the Western Front, they were to become poster children for the phrase "nothing fails like success."

Jonathan Steinberg, *Bismarck: A Life* (London: Oxford University Press, 2011). A long, comprehensive account of the man who served three Prussian kings and was in large part responsible for the unification of Germany. Although he used war for his purposes, he seems to have been smart enough, this "Iron Chancellor," to have likely avoided having Germany fight World War I. But the last kaiser would have none of his advice by the time he needed it most.

David Stevenson, *Armaments and the Coming of War: Europe 1904–1914* (Oxford: Oxford University Press, 1996). An international history of the arms race(s) of the soon-to-be belligerents and how they became more and more competitive as the arms inventories grew and grew. Author correctly blames policy makers and not the arms themselves, claiming that the war was not "inevitable." Excellent chapter on the "Great Acceleration: 1912–1913," which resulted in the Western Front seeing 1 artillery piece for each 10 yards of front. Artillery may have killed far more combatants than the more widely attributed machine gun.

———, *Cataclysm: The First World War as Political Tragedy* (New York: Basic Books, 2004). A sprawling, densely packaged study. Considerable sweep and a good introduction to the size and equipment of land and naval formations. Sees the pre-1914 era as peaceful, globalized, economically interdependent, and working well. Then the long peace collapsed quickly, he feels, by design, not accident. Sees the collapse of Russia and the arrival of 2 million Americans in 1918 as decisive. Erick Ludendorff wanted the war to end quickly so as to preserve the integrity of the German army to prevent revolution on the home front.

———, *With Our Backs to the Wall: Victory and Defeat in 1918* (Cambridge, MA: Belknap Harvard, 2011). Regrettably, most books dealing with the end of World War I focus on the arrival of 2 million American troops and/or the internal collapse of Germany after the failures of their 1918 offensives (the "Germans defeated themselves"). Stevenson does a fine job in redirecting our attention to the Allies' increasing advantages in intelligence, production, propaganda, logistics, mobility (better roads, railroads, and more horses), and the ability to blunt the German offensives with greater firepower. I had not realized how important a factor fear of Bolshevik "infection" played in having the Allies not drive deep into Germany at the end of the war, thus setting the stage for the "stab in the back" theory, as the German soldiers marched home, not as POWs and often with their weapons, thinking they had not been defeated on the battlefield but let down by those at home. The British prime minister Lloyd George called Germany a "cholera area." A somewhat dull read but one with important insights and background information for the understanding of why World War I ended when and how it did.

John Stoessinger, "The Iron Dice: World War I," in his *Why Nations Go to War* (Boston: Bedford/St. Martin's Press, 2001), pp. 3–28. Tends to blame Germany perhaps more than the overarching set of circumstances, alliances, and poor political leadership of all the powers.

Norman Stone, *The Eastern Front: 1914–1917* (New York: Penguin Books, 1975). One of the best accounts of this often overlooked portion of the "War to End All Wars." There was a great deal happening on the Russian front in that war as well as the one that followed.

Richard Stone, *The Russian Army in the Great War: The Eastern Front, 1914–1917* (Lawrence: University Press of Kansas, 2015). Now the definitive account of the subject, clearly and engagingly written, with decent maps and a strong emphasis on "continuities" of how smaller events played into larger ones. For example, getting Romania into the war on the side of Russia was a disaster, requiring badly needed Russian troops to cover the now-extended southern front. Thought-provoking.

William Kelleher Storey, *The First World War: A Concise Global History* (Lanham, MD: Rowman and Littlefield, 2009). A very readable and easy-to-understand narration of the war by highlighting technology and the environment, including geographical dimensions. The maps are clear and make the reader aware of the relative lack of movement on the Western Front compared to the wide sweeps of armies on other fronts. Highlights the fact that more than 9 million soldiers would die in the effort to resolve the conflict. The author accepts the vital role played by the million-man US armed forces. A good introduction.

Hew Strachan, *The First World War*, vol. 1, *To Arms* (London: Oxford University Press, 2001). Comprehensive and magisterial, with 1,190 pages to cover just 1914! Argues that it was a global war, even though all nations weren't involved as primary actors. Notes how important it was that the major European powers felt the outcome was a zero-sum game, as the kaiser indicated when he said, "One must win, the other must go down." The ideas of the war, Strachan argues, proclaimed the 1914 absolutes that would justify all that followed.

Jacques Tardi and Jean-Pierre Verney, *Goddamn This War!* (Seattle: Fantagraphics Books, 2013). Graphically and powerfully illustrated, this work is not a cartoon. In fact, the images on pages 75 and 76 will turn even the most ardent war lover against it. Filled with ironic commentary, it is a marvelous "soldier's-eye" view of the war with all its irony, destruction, and bewilderment.

A. J. P. Taylor, *The Struggle for Mastery in Europe, 1848–1918* (New York: Oxford University Press, 1999). Good background on what led to World War I, including the rise of nationalism in Germany and Italy. Many Europeans were better off when they were fragmented ministates.

Mark Thompson, *The White War: Life and Death on the Italian Front 1915–1919* (New York: Basic Books, 2009). A grim slaughter in the mountains and the snow of Austria and Italy. Not much accomplished strategically but bloody nevertheless. If Italy and Austria had maintained a cease-fire instead of fighting, nothing would have changed.

John Torraine, *Ordeal of Victory* (New York: J. B. Lippincott, 1963). General Alexander Haig in perspective. He carried a huge load but wasn't always as concerned about the number of his troops being slaughtered—and why—as he should have been. The ordeal of those troops was considerable.

Tim Travers, *The Killing Ground: The British Army, the Western Front and the Emergence of Modern Warfare 1900–1918* (London: Allen and Unwin, 1997). Claims that a "conservative reflex" prevented most military establishments from recognizing the power of the new weapons and/or their ability to do anything about them except pile up bodies in front of them. How the British came to believe that massed frontal assaults could prevail and what the cost was—1,750,000 casualties on the Somme and Verdun fronts, 200,000 on the Somme in the first two weeks alone.

Barbara Tuchman, *The Guns of August* (New York: Macmillan, 1962). How the nations of Europe blundered into the worst war they had ever known, some enthusiastically, all expecting a very short war. A classic.

Geoffrey Wawro, *A Mad Catastrophe* (New York: Basic Books, 2014). I truly never expected to ever read a book about World War I that made me feel sorry for the Prussians who planned for it and so enthusiastically embraced it, but this volume did. The bumbling inefficiency, wild and grandiose assumptions, the out-of-date armies, and the most stupid leadership imaginable on the part of Austria-Hungary provided that rarest of reactions. One of Wawro's chapters is entitled "Between Blunder and Stupidity," and that says it all. He concludes, having made his case, "The Great War has justly earned a dark place on our historical map, and Vienna, no less than Berlin, was the heart of darkness." Somehow the Austrians even managed to make the Serbians look good in the process, and that was a most difficult task. Truly.

John Wheeler-Bennett, *The Forgotten Peace: Brest-Litovsk March 1918* (New York: William Morrow, 1939). A "peace without indemnities or annexations" was a strange marriage of convenience that would make Bolshevism safe, destroy the German Empire, and give Hitler the raison d'être for his "Drang nach Sud-Osten," claims the author. Quite a lot of truth in this book, written 20 years after the treaty and just before World War II.

Samuel Williamson Jr., *The Politics of Grand Strategy: Britain and France Prepare for War 1904–1914* (Cambridge, MA: Harvard University Press, 1969). An in-depth and detailed look at how France and Great Britain came from near-war in 1898 to Allies, as the Entente Cordiale was greatly assisted by Germany's bellicosity and especially its emphasis on colonies and naval might.

Denis Winter, *Death's Men: Soldiers in the Great War* (London: Penguin, 1978). An excellent set of insights into the soldier's life during the Great War. Makes important points about where one was stationed and when and says mental to physical casualties were a huge 1:3 with cold, mud, vermin, poor diet, and depression making for much misery beyond the life-or-death situation itself. Rum was "sunshine administered internally."

Chapter Twenty-Five

Transition within the Continuum

The Interwar Years

Note: Other relevant topics that chronologically fall within the interwar period, such as the "Spanish Civil War," are to be found under their more generic chapters (e.g., chapter 22, "Civil Wars in Perspective").

Frederick Brown, *The Embrace of Unreason: France, 1914–1940* (New York: Alfred Knopf, 2014). This political, social, and religious overview of interwar France certainly helps to explain why the Germans thought they could defeat France in World War II. What an intellectual and moral stew!

E. H. Carr, *The Twenty Year's Crisis, 1919–1939* (New York: Faber and Faber, 1967). Written in the 1930s and published in 1939, this is an early and most serious study of international relations, illuminating such concepts as utopianism (liberalism), realism, morality, the nature of international politics, and law. Sees power as essential for maintaining world order and peace and utopianism as insufficient, however well-intentioned. His growing fascination with the Soviet Union is now somewhat overridden by events.

David Glantz, *Stumbling Colossus: The Red Army on the Eve of World War II* (Lawrence: University Press of Kansas, 1989). Looking at the Red Army through the newly released documents, the author makes a strong case for the actual efficacy of the Soviet intelligence services prior to the outbreak of the war but indicts Stalin for ignoring them. Makes the astonishing clam that, by the end of 1941, 4.5 million Russian soldiers, sailors, and airmen had become casualties, about equal to the sum of those who began the war in June 1941 (p. 260).

Adolph Hitler, *Mein Kampf* (Boston: Houghton Mifflin, 1924). How is it possible that the world's leaders and people in general didn't believe that he would do what he said he would do? Eight million copies of this book were sold. This is an astonishing work, seldom (if ever) read, and yet it was precisely the blueprint he followed when he was elected to office in 1933. He said he would destroy democracy in Weimar Germany, and he did. Among other things, he documents his personal odyssey during World War I, going to war in 1916, wounded at the Somme, gassed in 1918, and concluding that German was stabbed in the back by Jews, Communists, and their dupes. He states clearly that the German boundaries of 1914 were a "political absurdity of such proposition and consequences as to make it seem a crime," adding, "The boundaries of the year 1914 mean nothing at all for the German Future." As for the new Reich, "[i]t must set race in the center of all life. It must take care to keep it pure," and "[t]hose who are physically and mentally unhealthy and unworthy must not perpetuate their suffering in the body of their children." One must never forget that he wrote *Mein Kampf* while he was in Landsberg Fortress prison and was then released. As Churchill put it in *The Gathering Storm*, "Here was the new Koran of faith and war: turgid, verbose, shapeless but pregnant with its message."

See also Gerhard Weinberg (ed.), *Hitler's Second Book: The Unpublished Sequel to* Mein Kampf *by Adolph Hitler* (New York: Enigma Books, 2003). Dictated in 1928 but not published until 1961, this volume continues the many themes in *Mein Kampf*: the need for Germany to destroy the Treaty of Versailles, the German need for living space, an analysis of World War I, and a rambling indictment of "international Jewry."

Also, for an in-depth debunking of the myth that Hitler was a true hero volunteer who served extensively on the front lines during World War I as a member of the Bavarian List Regiment, see Thomas Weber, *Hitler's First War* (London: Oxford University Press, 2011). Weber maintains that he served as a headquarters runner and was not often or long in the front lines (spending only four days in the front lines during the battle of the Somme) and that he did *not* develop his Nazi ideology while in the trenches. He also provides the disturbing, if unsurprising, information that the Jewish members of the List Regiment were treated no differently than other Jews of Nazi Germany after he came to power.

Joseph Maiolo, *Cry Havoc: How the Arms Race Drove the World to War 1931–1941* (New York: Basic Books, 2010). An ambitious effort to reconstruct the pre–World War II world of intertwined politics, rearmament, and miscalculations that led Europe to war. Although the author rails against the forces, including the military industrial complex, that spawned the war, he actually makes a pretty good case for this element of the template with his

"Conclusion: The Race Goes On" by indicating what a flywheel the assumption "there will always be another war" provides. He carries his fears forward, through the Cold War and beyond. "The rules can change, and so can the stakes, but the race goes on."

Roger Moorhouse, *The Devils' Alliance: Hitler's Pact with Stalin 1939–1941* (New York: Basic Books, 2014). Wide-ranging and well-researched, this fascinating study looks at the pact from all sides, concluding that Hitler was more devious than Stalin, who seems, against all odds, to have been truly duped. Horrifying for all those trapped in "The Bloodlands" between Germany and the Soviet Union.

Denis Smith, *Mussolini* (New York: Knopf, 1982). This may seem like an odd location for this work, but reading about Italy during and after World War I, I began to see the rise of Mussolini (cynical switches from left to right to left appeals; his "semi-autonomous hooligan squads"; his "love of posing" and his sense of theater, such as having the Black Shirts march on Rome in 1920 *after* he'd been named to head the government; his bizarre sense of historical destiny; etc.) that made him a poster child for the interwar years. Also, I like the way he blamed Hitler and Germany for all of Italy's troubles, especially claiming that they should have stayed out of the Balkans. Of course, Hitler himself blamed Germany going into the Balkans for all his troubles, too, because Germany went in to bail out the Italians, who did so poorly in Albania and Greece.

H. G. Wells, *The Shape of Things to Come: The Ultimate Revolution* (London: Penguin Books, 1933) (F). An amazingly prescient work of fiction that covers what Wells calls the "Second World War," six years before there actually was one, and sees Germany going to war with Poland over Danzig. He also correctly saw World War I and World War II on a space-time continuum.

Chapter Twenty-Six

The Battle of Nomonhan/Khalkin Gol

The Battle of Nomonhan/Khalkin Gol, it can be argued (see chapter 9, "Applying the Template: A Battle," in *Winning at War*, by Christian Potholm [Lanham, MD: Rowman and Littlefield, 2010], pp. 183–94), changed the course of World War II in many different ways. This obscure town/river in Mongolia was to have an enormous impact on the both the victor (the Soviet Union) and the defeated (Japan), setting in motion many key trends for World War II, ultimately hurting the war fortunes of Nazi Germany and the peace fortunes of the United States.

It is one of the most understudied yet important battles in the history of modern warfare, even allowing for some post hoc ergo propter hoc argumentation. For example, had the Japanese won it and continued their expansion northwestward into Siberia, it is quite unlikely that they would have subsequently chosen to expand southeastward. Likewise, had the Japanese won and continued to pose a threat to the Soviet Union, Stalin would not have had the luxury of moving so much of his Far Eastern forces to the defense of Moscow in December 1941, forces that provided the margin for victory in that crucial battle.

Some additional readings on the subject as well as the sources of many of the footnotes of "Applying the Template: A Battle" (pp. 257–58) include:

Christopher D. Bellamy and Joseph S. Lahnstein, "The New Soviet Defensive Policy: Khalkin Gol 1939 as Case Study," *U.S. Army War College* 20, no. 3 (1990), p. 24. Interesting venue, and the article rings true on several levels.

Marshall Brement, "Khalkin Gol," *MHQ: The Quarterly Journal of Military History* 5, no. 3 (Spring 1993), pp. 80–91. Considerable emphasis on Zhukov and the Soviet side of things, a bit weak on the Japanese military aspects, but spot-on when dealing with the important ramifications for World War II.

Alvin D. Coox, *Nomonhan: Japan against Russia 1939* (Stanford, CA: Stanford University Press, 1985). The definitive account of this very important battle. Great detail on the battle and its ramifications.

John Erickson, *The Soviet High Command: A Military-Political History 1918–1941* (London: Frank Cass, 2001). Erickson did pioneering work on the Soviet side of World War II. An outstanding look at a true hinge-of-history battle.

Stuart Goldman, *Nomonhan 1939: The Red Army's Victory That Shaped World War II* (Annapolis, MD: Naval Institute Press, 2012). Goldman does a superb job at bringing the details of the battle and its truly strategic grand ramifications into harmony. He also illuminates the considerable aid that Russia supplied China (990 planes, 82 tanks, and 1,140 pieces of artillery) between September 1937 and June 1941. Provides excellent analysis of the diplomatic maneuvering prior to the outbreak of World War II, with Great Britain trying to outbid Germany for Soviet support. Argues persuasively that one reason Stalin made the pact with Germany was to secure against a two-front war with Japan. Sees the Kwantung Army in Manchuria as the major instigator of the battle, much to the chagrin of elements of the government back in Japan, and shows the Soviet army way ahead of the Japanese in terms of tanks, artillery, and counterbattery firing. Elements of the Kwantung Army (such as the 23rd Division and 7th Division were simply destroyed (76 percent and 91 percent, respectively). Ironically, now that the Japanese knew the power of the Soviet military and gave up their interest in the "Go North Strategy," "only the United States was in a position to check Japan's southward expansion." The Soviet Union obviously championed the "Go South Strategy."

B. H. Liddell Hart, *The Tanks*, vol. 1, *1914–1939* (New York: Praeger, 1959). Calling the tank the "decisive weapon of land warfare," the author traces its origins from chariots and war elephants to its World War I incarnations. Minutely detailed and illuminating. Particularly interesting in the chapter by Major E. W. Sheppard and others, "The Royal Tank Corps in India between the Wars." North West Frontier chaps were a major problem!

Saburo Hayashi, *Kogun: The Japanese Army in the Pacific War* (Washington, DC: Marine Corps Association, 1989). A sweeping study of what the Japanese army did and didn't do tactically and strategically in the Pacific theater. Quite objective. A must-read. See especially pages 14–16 on the Battle of Nomonhan. Hayaski claims that the Komatsubara Force at Nomonhan suffered 73 percent casualties, or 11,124 out of 15,140 engaged. For the author, the "Nomonhan Incident gave the Japanese Army an opportunity to realize the actual ability of the Soviet Army. Seeing is believing."

Joe Maiolo, "Colonel Ishiwara Goes to Manchuria," in his *Cry Havoc: The Arms Race and the Second World War 1931–41* (London: John Murray, 2010), pp. 23–38. Interesting and useful for understanding the background in Manchuria leading to the conditions that eventually spawned the clash between the Soviets and Japanese at Nomonhan.

Evan Maudsley, *Thunder in the East* (London: Hodder Arnold, 2005). Soviet-Japanese conflict covered with useful and voluminous detail. Shows how the Soviets overcame the weaknesses of the czarist war machine and defeated the Japanese on their own terms.

Larry W. Moses, "Soviet-Japanese Confrontation in Outer Mongolia: The Battle of Nomonhan-Khalkin Gol," *Journal of Asian History* (1967), p. 79. Excellent on various aspects, such as the relative tactical and material strengths of both sides.

Geoffrey Roberts, "Khalkin-Gol 1939," in his *Stalin's General: The Life of Georgy Zhukov* (New York: Random House, 2012), pp. 48–66. Roberts credits Zhukov with not only the strategic insights that made the Soviet victory possible but also the first Red Army victory since the Russian Civil War, which resulted in a classic double envelopment, one that should be studied more carefully.

Overall, an in-depth look at the premier Russian general of World War II, the hero of Khalkin Gol, Leningrad, Moscow, Stalingrad, Kursk, and Operation Bagration, who the author feels is on a par with Alexander Suvorov, the great czarist general of the 18th century who never lost a battle. Interesting fact: The Soviets killed 158,000 of their own men during World War II for desertion and cowardice (p. 313).

Konstantine Simonov, *Comrades in Arms* (Moscow: Moskovskii Pistatel, 1965) (F). As yet we have not acquired a copy of this, Simonov's difficult-to-access paean to General Zhukov, and his dramatic victory at Khalkin Gol was first published in 1952. But Amazon had it for $120, so there may be copies out there.

Georgi Zhukov, *Marshal Zhukov's Greatest Battles* (New York: Harper and Row, 1969). One of the Soviet Union's best generals tells his side of things. A lot of good material on his interaction with other generals and his view of the "straight ahead" strategy. Pulls some punches on Stalin, though. Wisely perhaps.

Chapter Twenty-Seven

The Triumph of the Offense

The War in Europe

Blitzkrieg, or lightning war, becomes the anecdote to the bloody stalemate battles of World War I and eventually the prototype for the last 50 years of warfare. Various authors listed in this chapter accent production and destruction, true global warfare, and the rise of the garrison state. Thinking about World War II, one should always remember the chilling (and nonblitzkrieg) message of the Soviet Union's most celebrated general Georgy Zhukov, who told General Dwight Eisenhower, "When we come to a minefield, our infantry attack exactly as if it were not there. The losses we get from personnel mines we consider equal to those we would have gotten from machine guns and artillery if the Germans had chosen to defend that particular area with strong bodies of troops instead of with mine fields."

It is also always important to keep in perspective the relative levels of force involved. By mid-1944, the British had 14 divisions in action; the Polish, 1; the Canadians, 3; the Americans, 60; and the Soviets, 480 (albeit smaller ones), according to Max Hastings in *Winston's War*, reviewed later in this chapter. The broad sweep of its battles continues to attract avid readers now 60 years after the war.

There are a number of single-volume histories of World War II that are a good place to start your study. My personal favorites include:

Antony Beevor, *The Second World War* (New York: Little, Brown, 2012). His strategic assertions (importance of Nomonhan, Eastern Front, diversions to Norway and the Balkans, etc.), and the overarching narrative and conclusions seem altogether sound. Some important insights:

1. Before World War II broke out, the French military was suffering from bad morale and depression (*le cafard*).
2. The Balkans, Norway, and North Africa efforts hurt Germany and tied down many divisions (still 37 divisions in 1944, 400,000 men in Norway alone).
3. The United States refused to give Great Britain credit with which to buy weapons until Great Britain had exhausted its own foreign exchange and gold reserves.
4. Stalin should have been shot if his rules for punishing families of POWs were followed when his own son Yakov Djugashvili was captured.
5. Unlike many other German allies, the Spanish Blue Division fought well.
6. The London blitz killed 41,000 civilians and injured 137,000 more.
7. Following the Battle of Kursk, Stalin made his only visit to the front.
8. By January 1944, Germany had lost 4.2 million men but had 9.5 million in uniform, including 2.5 million on the Eastern Front.
9. Of the 125,000 British Bomber Command, 55,572 perished, and the US Eighth Air Force suffered 26,000 dead (p. 717).
10. Operation Bagration, which ran from June to August 1944 (named after the Georgian hero of the 1812 campaign), involved 15 Russian armies, 1.6 million men, and 6,000 tanks.

Walter Boyne, *Clash of Wings: World War II in the Air* (New York: Simon and Schuster, 1994). A breezy, fast-paced history of World War II through air combat, one that can give a good, short introduction to the fighting qua fighting in both the European and Pacific theaters.

Gordon Corrigan, *The Second World War: A Military History* (New York: Thomas Dunne, 2010). A very fine account of the Fourth World War (Seven Years' War, the Napoleonic Wars, the Kaiser's War, and World War II make four wars fought across the globe). Refreshingly candid and irreverent (the 1940 Norwegian campaign was a "complete and utter cock-up"; the "miracle" of Dunkirk was a "disaster"). Excellent and useful maps. Always puts things in perspective. One of the most enjoyable books about World War II I have read, as he allows no sacred cows. Ruthless in his judgments of generals, battles, campaigns, strategies, and political leaders (he is rightfully harsh on the many harebrained and deadly schemes of Churchill, including Norway, Crete, Greece, Yugoslavia, and Dieppe, and his meddling in various campaigns, such as North Africa and Burma). Concludes that with 50 to 60 million dead (including those who died of famine during the way), World War II is the most deadly war in human history.

J. F. C. Fuller, *The Second World War 1939–1945* (New York: Duell, Sloan and Pearce, 1949). The prolific Fuller wrote one of the very first accounts of the war. By the author's own admission, he passes over the war in China, insurgencies in Europe, and a number of other topics in order to provide a quick, straightforward, but quite perceptive single-volume description of the war. There is much to argue about here—Was Hitler's biggest strategic blunder not seizing North Africa? But some insights are quite useful, as is his assertion that the Japanese relied on a five- to six-month turnaround time for ships coming from the United States to the far Pacific as strategic insulation, or "space as a shock absorber." Fuller is very negative about the way the war ended, saying, "The dropping of the two atomic bombs was a political blunder of unfathomable consequences," and argues that the draconian terms inflicted on Germany and Japan would only lead to a subsequent conflagration. This is a contrarian read that can be quite stimulating.

Martin Gilbert, *The Second World War: A Complete History* (New York: Henry Holt, 1989). Highly praised, this comprehensive and straightforward account does actually not have much battle-qua-battle description, and there are some true oddments, such as two tiny paragraphs on the Battle of Midway while the struggle for Malta seems to go on and on. Anglo-centric is perhaps the best term for this work, which, on balance, turns out to be a great deal of description and little true analysis.

B. H. Liddell Hart, *History of the Second World War* (London: Cassell, 1970). "The Captain who teaches Generals" wrote this one, and it has great maps, probably the best maps of any of the works under review in this chapter. The prose is straightforward and workmanlike, and the reader will get a good overview of the political, diplomatic, and military aspects of the war. A bit too British-centric and much too much emphasis on the North African theater (although useful is the expansive material on the campaign in the Horn of Africa). I do disagree with his concluding assessment that the Allies' insistence on unconditional surrender made Churchill's "unnecessary war" "unnecessarily prolonged." Like Fuller, he should have stopped when he was ahead.

Max Hastings, *Inferno: The World at War, 1939–1945* (New York: Alfred Knopf, 2011). Along with Roberts, *The Storm of War*, and Weinberg, *A World at War* (see reviews later in this chapter), this is among the best all-around single-volume history of World War II. Skillfully blends the military, the political, and the personal. Majestic and soaring yet personalized, with many of the quotes from actual people caught up in the cataclysmic inferno. Written poignantly and with great feeling. Because Hastings has written so extensively about other phases of the war, he seems sure and confident as he handles the big issues and questions of the period. Some interesting factoids:

1. Great Britain needed 23 million tons of supplies a year, and the failure of Hitler to appreciate the power of the U-boat was one of the major miscalculations of the war.
2. Of the Red Army's vehicles in 1945, 427,000 came from the United States.
3. The British Army suffered 30,000 desertions in Italy.
4. The 1943–1944 famine in Bengal led to 1 to 3 million deaths due to Japanese occupation of Burma and British callousness.
5. In Tonkin, 1.5 million died of starvation, and a "minimum" of 5 million in Southeast Asia perished during the Japanese occupation.
6. Twelve thousand Soviets and 80,000 Japanese died during the 1945 Soviet invasion of Manchuria.
7. Sixty million died during World War II, 27,000 people per day.
8. Of the 403,271 Soviets who were trained as tankers, 310,000 died.
9. About 210,000 Hiwis died in German uniforms.

Williamson Murray and Allan Millett, *A War to Be Won: Fighting the Second World War* (Cambridge, MA: Harvard University Press, 2000). A fine, readable, one-volume account of World War II by two preeminent military historians. Provides the scholar and the layperson with a good overview of the intertwined nature of war and politics. Judicious and careful in its presentation.

Andrew Roberts, *The Storm of War: A New History of the Second World War* (New York: HarperCollins, 2011). Probably the best single-volume study of World War II. Among its useful observations:

1. Hitler perhaps could perhaps have defeated the Soviets *or* murdered 5 million Jews but not both at the same time.
2. No occupied country did more initially to assist the Germans than did the French: "The French fought each other."
3. England imported 70 percent of its food in 1939, so the U-boat menace had the potential to starve out the country had the German navy not wasted so much effort on its capital ships.
4. The war between Germany and the Soviet Union was "to dwarf everything that had taken place in the war so far, indeed in any war in the history of mankind, before or since."
5. The Japanese, by signing a nonaggression pact with the Soviets, made their victory in front of Moscow in 1941 possible, and Hitler, by declaring war on the United States after Pearl Harbor, committed a

huge blunder; quoting Conrad Black, "His failure to secure the collaboration of the Japanese against the Soviet Union must rank as one of Hitler's most serious errors."

6. Roberts is even more assertive: "It seems an unimaginably stupid thing to have done in retrospect, a suicidally hubristic act less than six months after attacking the Soviet Union."
7. Of the 100,000 Japanese interned in the United States, 69 percent were its citizens; 31 percent were not.
8. The United States contributed 296,000 aircraft, 88,000 landing craft, and 86,333 tanks; "if Britain had provided the time and Russia the blood necessary to defeat the Axis, it was America that produced the weapons."
9. A higher percentage of German POWs died in Russia than Soviet POWs died in Germany, claims the author, citing the figures from the German Sixth Army—of 90,000 prisoners after the Battle of Stalingrad, only 9,600 returned alive (p. 343).
10. Allied bombing of Germany siphoned off tremendous amounts of manpower and 88s, among other things, plus Albert Speer acknowledged the enormous wastage caused by "dehousing" and other area bombing tactics.
11. If Hitler had been killed in the summer of 1944, Goering or Himmler might not have made the same blunders and hence prolonged the war or made it more costly.
12. At Trent Park in England, the British kept the high German officers they captured from North Africa, Italy, and later France; let them talk; and bugged their conversations. Most implicated themselves in at least the knowledge of what was being done to the Jews, Poles, Russians, and others.
13. The Soviet Operation Bagration in 1944 was 10 times the size of the Anglo-British success at Falaise.
14. The Warsaw Uprising caused five times as many casualties on the Germans as did the Arnhem "Bridge Too Far" campaign.
15. The German navy's evacuation of the Kurland Pocket and the rest of the Baltic ports rescued 4 divisions and 1.5 million civilians.
16. Total Soviet deaths were 27 million out of a worldwide total of 50 million—23,000 dead for ever day of the war (p. 557).
17. For every American who died in World War II, 6 Japanese, 11 Germans, and 92 Russians lost their lives (p. 603).

This is a splendid book.

Gerhard L. Weinberg, *A World at Arms: A Global History of World War II*, 2nd ed. (Cambridge: Cambridge University Press, 2005). This 1,100-page description of the Second World War is thoughtful and provides one of the best introductions to the subject. Very good on weapons and their impact, as well as a broad-gauge look at "home fronts," especially in German- and Japanese-occupied territories. Notes that the United States built 104 aircraft carriers during the war, while the Japanese only built 14. Both a very fine introduction and also a book for the serious well-read student of World War II. Excellent at mixing diplomatic activity, strategy, tactics, and geopolitics. Some useful observations:

1. Most of the major countries had a considerable variety of options as to whether and how to make war.
2. The Italians really were disastrous in terms of dragging Germany into North Africa, Greece, Yugoslavia, and so on, all of which were unnecessary for the sine qua non defeat of the Soviets.
3. The Nazi's "racial social Darwinism" was a huge impediment to their victory in the east, tying up vast resources unnecessarily.
4. Although the Japanese wanted to wait until 1946, Hitler promised Japan he would declare war on the United States if Japan attacked. He wanted their "big navy" for use against the United States, although Japanese naval authorities had already decided on a war with the United States *before* the oil embargo.
5. German fears of a repeat of the home-front collapse during World War I greatly impeded their mobilization efforts.
6. Far from being "totalitarian" in character, the Nazi governmental structure was "authoritarian anarchy."
7. The disastrous Dieppe raid convinced the Allies that they should land on beaches and bring their own harbors with them, while it gave the Germans great hope they could defeat any eventual Allied effort to seize a true port.
8. Japan built 26,000 planes in 1944 and had 5,000 held back to attack any landings on the home islands.

If you are only going to read one book about the intertwined military, diplomatic, and geopolitical dimensions of World War II, then this is the one.

Note: For an engaging counterfactual look at World War II on its global scale, see David Downing, *The Moscow Option: An Alternative Second World War* (London: Greenhill Books, 1979). In this account, the key reverse points are (1) the Germans ignore everything else in 1941 to capture Moscow, (2) Rommel wins in North Africa by avoiding Tobruk, (3) the Japanese have a new code at Midway and defeat the Americans both there and off the Panama Canal, (4) the Germans take Stalingrad from the north, and (5) the Japanese lunge into Siberia. Intriguing, no?

Other engaging works on World War II in the European and American theaters include:

Rick Atkinson, Liberation Trilogy (*An Army at Dawn*, *The Day of Battle*, and *The Guns at Last Light*) (New York: Henry Holt, 2013). A marvelous set of reads based on a wide-ranging assessment of the literature and taking the formation of America's World War II army though North Africa, Sicily, and Italy and then France, Germany, Holland, Belgium, and Luxembourg. Atkinson makes a number of important points. Here are but a few for flavor:

1. The fighting in Normandy was fierce and reminiscent of World War I's static struggle for a long time. A German general would call it a "monstrous blood-mill, the likes of which I have not seen in eleven years of war," in part due to poor planning and field generalship by Montgomery, Eisenhower, and Bradley; the lunge for Brest and Brittany, for example, was useless, and no supplies flowed through it until after the war
2. The Americans suffered 1,334,000 casualties; British Canadians and Poles, 91,000.
3. The American excuse for surprise at the Battle of the Bulge and the difference among different American and German units to its demands: "We wouldn't have attacked under these conditions."
4. US soldiers engaged in black-market activities, rapes, and looting far and wide, with only an occasional reprimand.
5. The Ruhr Pocket yielded 324,000 prisoners, more than Stalingrad or Tunis.
6. Fighting alongside the French in the Vosges area produced difficulties for Patch's Seventh Army, particularly General Jacob Devers, especially battles for Metz, Stuttgart, Strasburg, and Colmar Pocket (Alsace), as French and de Gaulle's political demands often overrode military considerations; "[p]erpetual frustration."
7. The American Communications Zone (COMZ) contained 500,000 troops, and SHAEF headquarters, 16,000, which the author likens to a sea serpent: "Some had seen its head, some its middle, some its tail. No one had seen all of the sprawling mass."
8. The post-Arnhem drops in Operation Varsity are still debated as is the value of paratroops: often huge losses for marginal gains (for parallel, fast-forward to 1950 Korean War drops at Sukchow and Sunchou).

A sober and less-than-joyful finish to the trinity I must say.

Patrick Beesly, *Very Special Intelligence: The Story of the Admiralty's Operational Intelligence Centre 1939–1945* (Garden City, NY: Doubleday, 1978).

The maritime war against Germany is painstakingly covered until each and every one of the German capital ships and all 1,170 U-boats built since 1935 were accounted for, including two that made it to Argentina.

Antony Beevor, *D-Day: The Battle for Normandy* (New York: Viking, 2009). A good follow-on to previous books about the Normandy campaign, focusing a good deal on what happened after the "Longest Day." Importantly points out that Allied bombing and war-making killed 100,000 French civilians in the process.

———, *The Fall of Berlin: 1945* (New York: Viking, 2002). The Russians suffered 100,000 casualties taking the German capital, but they put an end to Hitler and wreaked havoc on the Germans in their path.

———, *Stalingrad* (New York: Viking, 1998). One of the many books about this famous battle by an author who writes well and easily and conveys much of the feeling of the fighting from both the German and the Russian perspectives.

Antony Beevor and Luba Vinogradova (eds.), *A Writer at War: Vasily Grossman with the Red Army 1941–1945* (New York: Pantheon Books, 2005). Divided into five easy-to-digest sections from the beginning of the war to the destruction of Nazi Germans, these eyewitness accounts are searing and penetratingly honest and cover the epic Battles of Stalingrad and Kursk and Operation Bagration in Belorussia in 1944.

Marc Bloch, *Strange Defeat* (New York: W. W. Norton, 1968). Written during World War II by a veteran of both world wars and the French Resistance, this analysis in medias res by a French nonpracticing Jewish scholar is poignant and powerful and very perceptive. He blames the French defeat on a shortage of equipment, the lack of a fighting spirit especially among the French generals, and a debilitating overlay from the slaughter of World War I. Unfortunately, he was captured by the Germans and executed before the end of the war.

Ernle Bradford, *Siege: Malta 1940–1943* (New York: William Morrow, 1986). Another failure of the Italians draws in the Germans, and it is a close-run thing, but according to the author, the arrival of British Spitfighters and the convoy of Operation Pedestal (which was a tactical disaster, losing three cruisers and an aircraft carrier) actually was a strategic turning point for the 100-square-mile island south of Tunis. The German Fliegerkorps X came close but couldn't close the deal.

Rodric Braithwaite, *Moscow 1941: A City and Its People at War* (New York: Knopf, 2006). A very important reappraisal of the war, with the revisionist notion that the gigantic (involving an area as large as Western Europe) battle for Moscow in 1941 was actually the turning point in the war, not Stalingrad (1942) or Kursk (1943).

Paul Brickhill, *The Dam Busters* (New York: Ballantine Books, 1955). A rousing adventure story covering the exploits of the Royal Air Force's Squadron 617, which became famous for destroying the Moehne and Eder Ruhr dams, sinking the battleship *Tirpitz*, and bombing the SS barracks at Hitler's Berchtesgaden. Behind the snappy dialogue, however, is a vivid discussion of the melding of science, technology, military innovation, and sheer courage.

Arthur Bryant (ed.), *The Turn of the Tide 1939–1943: A Study Based on the Diaries and Autobiographical Notes of Field Marshal the Viscount Alan Brooke* (New York: Collins, 1957). Based on the war diaries and commentaries of General Alan Brooke, who in 1941 became chief of the imperial general staff (CGS) at a "time of dreadful disasters," this book provides significant insights into the conduct of World War II, with perspicacious descriptions of Roosevelt and Stalin, whom Brooke saw up close. Previously he had served as head of British forces in France during the summer of 1940. A marvelous account from the man who stood up to Churchill many, many times to cut off at least some of his harebrained schemes. One he could not stop was invading Germany through Norway. Brooke also presided over the "Second Dunkirk," saving more than 140,000 British and almost 70,000 other troops as France was about to surrender by evacuating them from the French northwest ports of Brest, Cherbourg, and St. Nazaire. After all of his fine service, Churchill offered him the position of the head of Allied armies for the Normandy invasion, only to back off instantly when Roosevelt insisted on Ike. Brooke naturally found it a "crushing blow." Despite his rough handling by the prime minister, his diary reflects great admiration for (and great exasperation with) Churchill.

Hal Buell (ed.), *World War II Album: The Complete Chronicle of the World's Greatest Conflict* (New York: Black Dog and Leventhal, 2002). Lavishly illustrated with highly readable maps, this massive compendium is 740 pages in length and contains both a useful chronology of the war as well as numerous photographs from virtually all fronts and all theaters of war. Relatively few of the photographs have been widely published elsewhere, and numerous usually underrepresented dimensions are helpfully fleshed out. This is the ultimate coffee-table book for World War II buffs and a good visual companion to many single-volume accounts of the war.

Michael Burleigh, *Moral Combat: Good and Evil in World War II* (New York: Harper, 2011). A penetrating look at the moral dimensions of World War II. Plenty of blame to go around, including the Hungarians and Romanians, whose "savage depravity embarrassed" even the German Einsatzgruppen (p. 240). There were no provisions made by the Germans for feeding Russian POWs, let alone providing them with housing and clothing. As a result, more than 5 million died. German troops outside of Leningrad used Russian civilians as human shields. Marvels at the expansion of US armed forces: The Army went from 190,000 in 1939 to 8 million in 1945; 2,470 airmen to 80,000; Navy, 1,000 ships to 67,000 ships; and 75,000 aircraft. The US "tail to tooth" ratio was 18 to 1 in the Pacific. There were 2.9 million Jews "killed by men standing a few feet away from them," belying Hannah Arendt's "banality of evil." Four out of five Germans who died did so on the Eastern Front, and the debt of blood owed their victims by the Germans, Japanese, and Soviets was never paid.

J. R. Butler (ed.), *History of the Second World War*, United Kingdom Military Series (London: Her Majesty's Stationary Office, 1956–1972). A very comprehensive, in-depth look at the strategic and tactical effort of Great Britain during World War II, as well as the thinking behind many of the operations. Check out the debacles in Norway, Greece, Crete, and Dieppe, as well as the more famous British successes, such as the Battle of Britain, to get a balanced assessment of that effort. Also, rejoice that many of Churchill's proposed strategic lunges (such as to Rhodes or wanting to substitute a landing at Trieste instead of southern France in August 1944) were thwarted.

Winston Churchill, *The Gathering Storm* (Boston: Houghton Mifflin, 1948). An extremely candid and incisive look at the causes of World War II, especially on the "Locust Years." He argues, cogently in fact, that the British were more than a little responsible for the war because "their unwisdom, carelessness and good nature allowed the wicked to rearm." He also blames a cluster of French politicians for failing to heed the wisdom of Marshal Foch, who, after World War I ended, said, "This is not Peace. It is an armistice for twenty years." It was only that.

Winston Churchill and the editors of *Life*, *The Second World War*, 2 vols. (New York: Time, 1959). Some outstanding photos, although some are from the 1950 *Life's History of World War II* (reviewed in this chapter), but the primary attraction here is Churchill's inside information, consistent biases, and splendid prose.

Robert M. Citino, *The Wermacht Retreats: Fighting a Lost War in 1943* (Lawrence: University Press of Kansas, 2012). This is an engaging look at how the Germans, with a long history of accenting *Bewegungkrieg* ("War of Movement"), of which blitzkrieg was just the latest incarnation, and trying to avoid *Stellungkrieg* ("Static War of Position"), ended up fighting so well and for so long while on the defensive in World War II. The work looks extensively at operations in North Africa, Italy, and the Eastern Front. In the process, the author is intrigued but hard-pressed to explain how, in their Operation Axis in 1943 after the government of Italy surrendered to the Allies, the Germans were somehow able to kill 7,000 to 12,000 Italian soldiers and intern 600,000 others, while 1.5 million Italian soldiers simply melted away. At the time, the Germans had approximately 600,000 troops in Italy and the Greek Islands, while the Italians had almost 3.5 million men under arms.

I also found of particular note the author's efforts to put the Battle of Kursk in a more balanced perspective, but I wish he had dealt more fully with the question of the actual size of the battle because it is often claimed to be the largest tank battle in history. He is skeptical but doesn't give any examples of significant larger ones. By contrast, Valeriy Zamulin, in his *Demolishing the Myth: The Tank Battle at Prokhokovka Kursk, July 1943: An Operational Narrative* (Solihull, UK: Helion, 2011), makes a convincing case that Kursk was not, giving instead considerable documentation for what he believes was *truly* the largest tank battle in history: the July 1941 clash between the Soviet Southwestern Front and the First Panzer army and elements of the Sixth Army in the Brody-Berestechko-Dubno theater. In that epic confrontation, Zamulin claims, the Russians counterattacked with 5,000 tanks and self-propelled guns (of which they lost 2,648), while the Germans used 1,000.

Of considerable interest and import is Citino's final emphasis on the German tradition of *Totenritt*, or "death ride" (with its accent on carrying out a senseless order to the best of one's ability), a concept that helps explain German performance from 1943 to 1945. Citino declares, "[T]o ride the *Totenritt*, you have to substitute faith—blind faith—for rational thought." In this regard, he underscores the performance of the Germans during Field Marshal Albert Kesselring's defensive war in Italy, 1943–1944, going so far as to declare that "Smiling Albert" was a "stander" (*Steher*) and "may well be one of the most important German commanders of all times" for his superb use of *Stellungkrieg* during this particular *Totenritt.*

This is a fine work for those interested in the European and North African portions of World War II.

Pierre Clostermann, *The Big Show* (New York: Random House, 1951). The author flew more than 400 missions during World War II as part of the Free

French and Royal Air Forces and was an ace many times over. From the Battle of Britain in 1940 to the Normandy invasion to the final days of the war, Clostermann kept a diary that faithfully records the ups and downs of the war, as well as the difficulties in transitioning from air-to-air to air-to-ground combat. This very engaging book ranks with Robert Scott's *God Is My Co-Pilot*, Adolf Gallant's *The First and the Last*, Saburo Sakai's *Samurai*, and J. E. Johnson's *Wing Leader* as must-reads from accounts of World War II in the air.

Robert Crisp, *Brazen Chariots: An Account of Tank Warfare in the Western Desert, November–December 1941* (New York: Ballantine Books, 1961). An easy-reading introduction to armored warfare in North Africa as the British defeat the Italians and in turn are defeated by the Germans.

———, *The Gods Were Neutral* (New York: W. W. Norton, 1961). An engaging account of the author's adventures and misadventures in North Africa and the British campaign in Greece, which ended so ignominiously. But his quote from Hitler's diary from February 17, 1945, contains a very important insight by the Führer, who, when reflecting on the Italian misadventures in North Africa and the Balkans, declared, "This compelled us, contrary to all our plans, to intervene in the Balkans, and that in turn led to a catastrophic delay in launching our attack on Russia." Hitler goes on to say that he would have preferred using his paratroops and other special units against Gibraltar rather than the Corinth Canal and Crete.

F. W. D. Deakin, *The Embattled Mountain* (London: Oxford University Press, 1971). A firsthand account of the Second World War in Yugoslavia, with the multifaceted resistance(s) and the internal politics of a country that was not really a country after all, with Serbs, Croats, Bosnians, and Albanians all operating for themselves first and the overall cause second. This resulted in what an understated British colonel called "internal discord." Interesting wartime portraits of Josip Broz (Tito) and Draza Mihailovic add zest. The 1943 battle on the Sutjeska River in Montenegro becomes a microcosm of the British efforts to aid the partisans.

Jim DeFelice, *Oman Bradley: General at War* (Washington, DC: Regnery, 2011). Straightforward account of a general either overpraised or overlooked. DeFelice gives him a great deal of credit for Operation Cobra in July 1944, which led to the breakout at St. Lo, but cuts him a lot of slack for failing to close the bag and also asserts that, while he wasn't as aggressive as Paton, he was more aggressive than Montgomery and Eisenhower. For Bradley, military reality trumped everything else, and he simply wasn't interested in geopolitics. Could be petty and jealous at times.

Leon Degrelle, *Campaign in Russia: The Waffen SS on the Eastern Front* (Torrance, CA: Institute of Historical Review, 1985). I have read many, many books about World War II in the East, but none are like *Campaign in Russia*. The author, a staunch Belgian Catholic, joins and leads the SS Wallonien Legion in a "crusade against Bolshevism" *after* Belgium is occupied by Germany! It was, he believed, a chance "to become the comrades and equals of our conquerors." With stunning candor, naïveté, and self-promotion, Degrelle describes the wild ride of his unit through the Ukraine (where the "greatest gift one could confer on them (the peasants) was a portrait of Hitler. Often they would hang it next to their icons.") and on into the Caucasus and back, eventually breaking out of the Cherkassy Pocket, or *Kassel*, then on to Pomerania and Norway. Defeat is in the cards, but "[n]evertheless, we had to shoulder on and wear the yoke of suffering right to the end." Personally given the Knight's Cross and Oak Leaves by Hitler, Degrelle eventually flies from Denmark to Spain as the war ends. Ideology aside, his descriptions are vivid and unromantic. Equally dispassionate, he describes the Romanians' massacres of Russian prisoners and the Russians dousing their Italian prisoners ("who are so busy eating and fornicating that they are easy to catch") with icy water in the 35-degrees-below-0 weather.

Note: See also Otto Carius, *Tigers in the Mud* (Mechanicsburg, PA: Stackpole Books, 2003), for an unrepentant account of the war by a German whose sense of righteousness, while disturbing, helps to explain a lot. His being presented with the Oak Leaves to the Knight's Cross ("delightful evening") from Heinrich Himmler himself is worth the price of the book. Carius's only regret seems to be the Amis didn't team up with the Panzer SS to fight the Russians.

Wilhelm Deist (ed.), *The German Military in the Age of Total War* (Warwickshire, UK: Berg, 1985). A variety of essays, including "The Impact of the War Economy on the Civilian Population of Germany during the First and Second World Wars" and the very intriguing "My Honor Is Loyalty: The SS as a Military Factor in Hitler's Germany," which gets at the complicated and multidimensional activities of the Waffen SS, which consisted of 38 divisions. About 800,000 men served, and 20 to 25 percent were killed.

———, *The Wermacht and German Rearmament* (Toronto: University of Toronto Press, 1981). A German scholar closely examines the role of the massive German military in the buildup prior to World War II after the Nazis came to power in 1933. Excellent section on Admiral Erich Raeder and his view that any war with Great Britain would be long and drawn out, a view that clashed with Hitler's notion it would be short and snappy. Raeder also felt that Hitler's plan to take on Great Britain and France, as well as their empires, and Russia was, at best, grandiose.

Isabel Denny, *The Fall of Hitler's Fortress City: The Battle for Konigsberg 1945* (London: Greenhill Books, 2007). East Prussia and Konigsberg gave Hitler his highest-winning percentage in 1933, largely on his promises for agricultural subsidies and wiping away the "Shameful Peace of 1918." Unfortunately for the city, German atrocities in Russia and the bad luck of being the first German territory reached by the advancing Soviets resulted in it undergoing a siege from January 30 to April 9, 1945, and being razed to the ground. The Soviets eventually deported the entire population, took possession of the city, and retain it even today as Kalingrad (Mikhail Kalinin was one of Stalin's drinking buddies who died just prior to the end of the siege). An absorbing account of an underreported phase of the war. Several million Germans managed to evacuate by boat and over the ice from East Prussia, but still, more than a million lost their lives in a "vast flood of human misery."

Carlo D'Este, *Bitter Victory: The Battle for Sicily, July–August, 1943* (New York: Harper, 1988). Well-documented and carefully researched, this thick volume covers the 500,000 Allied soldiers—American, British, and Canadian—seamen, and airmen who attacked the island. D'Este laments the fact that 67,000 Italians and 54,000 Germans accomplished the "Great Escape" to Italy proper. Very harsh judgment of Allied commander Harold Alexander and his strategic abilities, or lack thereof. The whole idea of an Italian strategy was deeply flawed.

Jonathan Dimbleby, *Destiny in the Desert* (New York: Profile Books, 2012). Seeking to dispel the "myths and the mysteries" surrounding the Battle of al-Alamein, the author wades through an overwhelming amount of material to revisit the battle and to point out how Mussolini got into a mess; Churchill dangerously diverted four divisions into Greece and Crete just as General O'Connell was winning; and Hitler, against his better judgment, opened yet another front for the Germans. Makes it clear that, with the overwhelming Allied superiority of firepower and manpower—to say nothing of intelligence power—they should have/could have won without Monty. General Richard O'Connell was a likely candidate, but he was in an Italian POW camp by this point. The author is quite harsh on Montgomery and gives the benefit of the doubt to both Wavell and "Auk" Auchinleck.

R. M. Douglas, *Orderly and Humane: The Expulsion of the Germans after the Second World War* (New Haven, CT: Yale University Press, 2012). From 1945 to 1957, the "tragic and unnecessary" mass forced migration of 12 million to 14 million Germans (from Czechoslovakia, Hungary, Poland, Yugoslavia, and Romania, whether "wild" or "organized," proved deadly, with 500,000 to 1.5 million dying in the process. It was the "largest forced population transfer in history." Tragically, few noticed or cared.

Paul Eden (ed.), *The Encyclopedia of Aircraft of World War II* (London: Amber Books, 2007). A stupendous collection of photographs, drawings, and text featuring not only the legendary JU 87 "Stukas," B-17s, JU-88s, JU-52 "Tante Ju," B-29s, and Mk 1 Hurricanes but also virtually all of the aircraft used by the major combatants. Excellent analysis of lesser-known (in the West) aircraft, such as the Russian Ilyushin Il-2 "Bark," which was produced, along with the Il-10, in greater numbers than any other military aircraft in history, and the Japanese Nakajima B5N "Kate" and Aichi D3A "Val" torpedo and dive bombers, which were so effective in the early stages of the war. A valuable research item.

Frank Ellis, *The Damned and the Dead: The Eastern Front through the Eyes of Soviet and Russian Novelists* (Lawrence: University Press of Kansas, 2011). "Russians fought and won only to be cruelly betrayed after 1945. This is one of the most painful outcomes of the Great Patriotic War." Russian authors catalogue dissent, mistakes, horror, ideological debunks, the NKVD, SMERSH, traitors, retribution, and many other subjects, and the author concludes, "War literature is a story without end."

Richard Evans, *The Third Reich at War* (New York: Penguin, 2009). Although Evans doesn't spend too much time on military strategy, he captures the German home front pretty well. A must-read for anyone interested in seeing the Germany behind the bombast of Hitler and the sometimes wild assertions about it being a "totalitarian" system when so much of it was not a system at all. Evans also makes one think that the war would have been over much, much sooner if Hitler had not had Albert Speer around to make production improvements.

Jonathan Fenby, *The General: Charles de Gaulle and the France He Saved* (New York: Skyhorse, 2012). A straightforward, generally favorable, but balanced account of the life and times of Charles de Gaulle. Very helpful in understanding his side of World War II and lots of good material on the Fourth and Fifth Republics. The vignette when Churchill complains to de Gaulle after he has vetoed the British application to join the Common Market is worth the price of the book. De Gaulle "replies with an empty envelope addressed in his own hand, writing on the back: 'In case of absence, please forward to Agincourt (Somme) or to Waterloo (Belgium).'" What a grudge holder.

Harvey Ferguson, *The Last Cavalryman: The Life of Lucian K. Truscott, Jr.* (Norman: University of Oklahoma Press, 2015). The long and varied career of this fine and dedicated officer spanned the expedition against Poncho Villa though Anzio and Salerno and the invasion of southern France through the CIA (where he spied on the spies for President Eisenhower), with a brief stop with Clare Booth Luce. "No sonofabitch, no commander."

Will Fey, *Armor Battles of the Waffen-SS 1943–1946* (Winnipeg, Canada: J. J. Fedorowicz, 1990). This amazingly lavish pictorial record of the armor battles of Das Reich, Wiking, Liebstandarte, and Hitlerjugend has been scrubbed clean of any German atrocities (even those Germans charged in the Malmedy massacre, he claims, were only convicted because of "inhuman interrogation techniques"). But the on-scene battle accounts are superb and fall into three main categories: the Eastern Front; panzer actions in Normandy and the Ardennes; and the last battles in Hungary, Vienna, and Berlin—all featuring the Mark IV, Mark V (Panther), and Mark VI (Tiger) panzers. The work concludes with the words "Long live Germany, our fatherland." No contrition here.

Peter Fleming, *Operation Sea Lion* (New York: Ace Books, 1956). So much of what happened—for good or evil—during World War II in the West occurred because of what was in Hitler's mind. This overarching determinant of events is clearly seen in his thinking about invading the British Isles following the fall of France in 1940. In Hitler's strategic thinking, as Fleming notes, he "neither needed, nor wanted to conquer England," but "he wanted to eliminate her as an opponent." Part of the reason Hitler failed to accomplish either purpose was due, of course, to the steadfastness of the British government headed by Winston Churchill and King George VI (Fleming is quite fulsome in his praise of the monarch during this period) and its people.

But an even larger part was due to Hitler's ambivalence and strategic mistakes—lunging into Norway at a cost of 3 cruisers, 10 destroyers, and 8 submarines and allowing the British to evacuate both Norway and Dunkirk. To this end, the author does a very good job of exploring the "twin mirages in Hitler's mind"—that the British would simply "capitulate" rather than fight on and that, in any case, its fighting capacity would "disintegrate." Hitler believed the "British should have sued for peace but they did not," and that seems to have come as a great surprise to the Führer.

German intelligence, Fleming notes, was strewn with "errors and naiveties," but the British matched them with so many "flimsy or obsolete makeshifts" that it was good that the Germans never came ashore and put everything to the test. At the same time, their abortive plans to land nine divisions on British soil during the first wave on S-Tag seems simplistically optimistic, even within the context of British weakness on land.

The "protracted anti-climax" following the aerial Battle of Britain meant that the British had won and Hitler would call off Operation Sea Lion and move on to the even more grandiose invasion of the Soviet Union. There would be no German invasion of the British Isles, but as Hitler feared in his "Inselwahn" (or "island madness"), there would subsequently be an Allied invasion of the continent.

As this book makes clear, the summer of 1940 was truly a hinge of history. I like Fleming's statement that this was a "strange reality" and that "many things, large and small which combined in the summer of 1940 to give life in the United Kingdom a flavor which distinguishes that brief period from all the rest of British history." For further context, see the précis on Yank Levy.

Robert Forczyk, *Tank Warfare on the Eastern Front 1941–1942* (Barnsley, UK: Pen and Sword, 2013). A most useful view of the Eastern Front by a tanker. Places much more emphasis on production and supply/ordinance issues than others. Shows how Germans succeeded when they did and the Soviets when they did. Argues that, by the end of 1942, the Germans had not lost on the Eastern Front, having inflicted 7:1 casualties on the Russians and Stalin's rash interventions were as costly as Hitler's but Stalin, by 1943, began to trust Zhukov, Konev, Vasilevsky, Vatutin, and Rokossovsky to use armor better and more wisely.

Will Fowler, *Eastern Front: The Unpublished Photographs 1941–1945* (St. Paul, MN: MBI, 2001). One of the best collections of primarily Soviet photographs of the war (although it contains some pictures of Germans and Romanians as well). Many shots of the wide-tracked T-34 and other Soviet equipment (even a shot-down American B-17) as well as much-needed emphasis on often overlooked battles, such as the Crimea and Oder Campaigns, as well as women and partisans in battle (in these cases, closely watched by the NKVD). Engaging.

Gerborg Frick, *Red Cage* (Lawrenceville, VA: Brunswick, 1994). The author spent nine years in the Soviet Gulag archipelago, from 1945 to 1954, and dedicates these memoirs to the 1.2 million German POWs who perished while in captivity. After the death of Stalin, he became a "free citizen of Russia" briefly before being released.

Adolf Galland, *The First and the Last* (New York: Ballantine Books, 1954). From his 70 kills in the European theater to becoming the commander of the Fighter Arm to his eventual work for NATO, Galland managed to lead a charmed life, his own skill a big part of his ability to rise with the Nazis but not fall with them. Compelling.

David M. Glantz, *After Stalingrad: The Red Army's Winter Offensive 1942–1943* (Solihull, UK: Helion, 2008). A superb study with great detail on both German and Soviet sides, skillfully told with the big picture in mind as well. Try not to get bogged down in the moment-by-moment accounts.

Heinz Guderian, *Achtung-Panzer: The Development of Tank Warfare* (London: Cassell, 1992). Originally written in 1937, Guderian's analysis of tank warfare would turn out to be seminal. Pointing out the changed nature of warfare with the introduction of poison gas, the airplane, and the submarine during World War I, he accented the importance of tank warfare with the very crucial addition of motorized infantry and declared that, if panzer forces were "full of verve" and "fanatically committed to progress," then they would "restore the offensive power of the army." Guderian was correct, and the resulting blitzkrieg would turn out to be the revolution he had envisioned.

Note: Germany may have had its Guderian and Great Britain its Hart and Fuller, but America had George S. Patton, who, in the *Cavalry Journal* (May 1920), wrote a splendid little provocative piece entitled "Tanks in Future Wars" (pp. 342–46) about the need to focus on tanks as an armored force and not scatter them in among infantry and artillery. He concluded, "The tank corps grafted onto infantry, cavalry, artillery, or engineers will be like the third leg of a duck; worthless for control, and for combat impotent." Patton was immediately told to cease and desist by the powers that were if he wished to continue in the US Army. For his part, Captain D. D. "Ike" Eisenhower, wrote a contemporary piece for the *Infantry Journal* entitled, "A Tank Discussion" (November 1920), pp. 453–58. In it, Eisenhower tries to make the case for keeping tanks as part of the equipment of the infantry units. Taken together, these two articles show what thin gruel the American musings were after World War I about an armor revolution in warfare. For Patton's war recollections, see his posthumously published *War as I Knew It* reviewed later in this chapter.

———, *Panzer Leader* (New York: E. P. Dutton, 1957). One of the best—and most fortunate—of German generals during World War II. Helped to actualize Germany's blitzkrieg form of warfare and led it to great success in Poland and especially in the campaign against France. Nearly captured Moscow in 1941, as well, but was dismissed by Hitler enough during the war, so, although he eventually became chief of the German army general staff, Guderian had the good luck to surrender to the Americans and help them during the Cold War. Not much repentance in this chap, however, and a haughtiness that can irritate if taken in large doses.

Hope Hamilton, *Sacrifice on the Steppe: The Italian Alpine Corps in the Stalingrad Campaign, 1942–1943* (Havertown, PA: Casemate, 2011). The disaster of the Italian army in Russia (ARMIR) is laid bare and made human, poignant, and insightful all at once. Poorly armed, poorly trained, but bonded by their common experiences as Alpini, these brave men went off to fight a war they never wanted with an ally who never valued them, and most of them perished in a battle that overwhelmed them and the rest of the ARMIR. In December 1941, Mussolini's only worry had been that they might not arrive soon enough to be in at the finish of Germany's invasion of Russia. Amazingly, many Italians who survived did so only because of the kindness of the Russian peasants with whom they interacted and married. See the Italian film *Girasoli* (*Sunflower*). Some 227,000 troops went in, and 85,000 never returned. The descriptions of their midwinter disorganized retreat (100,000 men abandoned everything, including their mess kits and rifles) from the Don are truly mindboggling and well-illustrated. Indeed, the pictures of 25 miles of retreating ARMIR troops with their mules has to be seen to be believed. "By December 25, 1942, infantry divisions of the ARMIR ceased to exist," and only the two Alpini divisions continued to maintain *some* unit cohesion. To add insult to injury, the post-Mussolini government did little to receive them, and the Italian Communist Party insisted that the Russians keep their officers until after the 1946 elections.

Nigel Hamilton, *The Full Monty*, vol. 1, *Montgomery of Alamein 1887–1942* (New York: Penguin, 2001). Eight-hundred-plus pages of great detail and insight into the "Tasmanian Savage" who became a soldier on the Northwest Frontier; was grievously wounded and left for dead in World War I; went back to France and then Ireland; and then read an incredible story about war before, despite his personality and lack of social standing, rose to become the hero of Alamein.

Max Hastings, *Winston's War: Churchill 1940–1945* (New York: Alfred Knopf, 2010). A superb and very useful book, not only putting Winston Churchill into a balanced and valuable perspective but also illuminating many unknown and lesser-known aspects of World War II. Particularly good on the limits of British power and persuasive as to the generally poor fighting quality of their infantry and high quality of their RAF and Royal Navy performances, but tasking these two services for a failure to enmesh with the demands of the ground campaigns. Quite searing on both Roosevelt and Stalin, justifiably. This work is a treasure.

Colin D. Heaton, "Belgian Volunteer in the Waffen SS," *Military History* (November 2006), pp. 46–53. A stunning counterfactual interview with a

Belgian Catholic, Leon Degrelle, who became a general in the SS Panzergrendadier Division Wallonien and received the Oak Leaves to the Iron Cross from Hitler personally. See Degrelle's book reviewed earlier in this chapter.

Alistair Horne, *To Lose a Battle: France 1940* (Boston: Little, Brown, 1969). From the victory parade in 1914 to the German occupation of France, Horne tells the story in splendid prose and incisive vignettes. He makes one understand how the French, having suffered such casualties around a single fort at Verdun, could create the folly of the Maginot Line; he is quite clear that it ended up protecting Germany more than France and was, at the beginning of the war, basically unfinished because it did not include the border with Belgium. Horne unsparingly captures the antimilitarism of the French, the blundering of its naïve and often venal politicians, and a military for whom an order was an "excellent basis for discussion." For those entrusted with the safety of France, there was a mad dash from complacency to panic until the newspaper *L'Epoque* could, on May 19, 1940, tellingly declare, "Every day which elapses without a catastrophe must be regarded as a day of victory." The author also provides the clearest account of the actual tactics and strategy of the Battle of France I have read.

Note: Although most accounts of the failures of the French armed forces circulate widely, it should be noted that 100,000 French servicemen died defending their homeland in 1940 and no fewer than 1,000 German airmen were captured during the Battle of France. Despite Churchill's pleas, these airmen were not sent as POWs to Great Britain but were kept in France and were later liberated by the Germans—many of them to fly again during the Battle of Britain.

Alistair Horne, with David Montgomery, *Monty: The Lonely Leader 1944–1945* (New York: HarperCollins, 1994). Retracing the stops Montgomery's mobile headquarters made from Normandy to the end of the war on the Elbe (Montgomery says 27, Horne says 28, stops). Good map of Monty's TAC HQs from the English Channel to the Elbe. Insightful. Montgomery believes that you must have a battle plan; otherwise you gradually become bent to the enemy's plan. Liked tanks but didn't see them as having more potential than to open up the front lines for the infantry. Really did a stupendous job in getting the Normandy invasion ready and in position to make a breakout. Horne claims that Monty was more popular in early 1944 than Wellington or Nelson in their day but asserts Monty was happiest "in battle surrounded by his young men." Glosses over the agonies of Goodwood and Market Garden, blaming Ike's strategy of "bulling ahead on all fronts." Disputes notions of Montgomery's homosexuality. For a further analysis of this dimension, see Nigel Hamilton, *The Full Monty* (reviewed earlier in this chapter), and his assessment of Montgomery's "homosocial" orientation. In any case, Montgomery had a very cruel mother, who beat him.

David Irving, *Hitler's War*, 2 vols. (New York: Viking, 1977). One of the most engaging books on World War II available in English. In-depth, holistic, interesting, and based on a Nazi insider's diaries and war records, this work captures Hitler's thinking about various countries, leaders, and campaigns and, above all, his day-to-day insecurities and mercurial tendencies. Amazing how important the Balkans were to his thinking and eventual defeat. A wonderful work. Really drills down into the daily conversations Hitler had with his military staff and the wild mood swings often based on faulty intelligence—both ways. No one can truly say they understand Germany's decision making in World War II until they have closely read these two volumes.

Julian Jackson, *France: The Dark Years 1940–1944* (London: Oxford University Press, 2001). Taking the story back to World War I, the author does a marvelous, in-depth analysis and lavishly footnoted study, concluding Vichy was "France" and the Resistance was "France" simultaneously. The chapter on "The Resistance" (pp. 385–505) is well worth reading carefully. Reading this work, I was reminded of the term in Alan Furst's *Night Soldiers* for the French who joined the Resistance only after it was clear the Germans were going to be defeated: *naphtalenes*, or "mothballs." There were many of them, although perhaps not as many as there were collaborators and those French who served the Germans as militia or *milice*.

J. E. Johnson, *Wing Leader* (London: Chatto and Windus, 1956). Group captain J. E. "Johnnie" Johnson was the leading Allied ace of World War II, having shot down 38 German planes by the end of the war, mostly Messerschmitt 109s and Fock-Wulf 190s. Unlike many combat memoirs, *Wing Leader* captures not just the horror and waste of war but also the courage and gallantry of some who fought it. Examples of chivalry abound. For example, early on, Johnson's commander makes it clear that he will court-martial any of his pilots who "shot at the parachutes" of bailing-out German pilots, and when his wing commander Douglas Bader is shot down and captured by the Germans, one of his two artificial legs is smashed in the crash. In their turn, the German airmen offer safe conduct for a small aircraft to fly to their base and drop off a replacement. This is a most modest memoir but one that also provides an intricate and multifaceted portrait of a man of action who was very, very good at his craft and lived to write about it.

Michael Jones, *Leningrad: State of Siege* (New York: Basic Books, 2008). Amazing perseverance under intolerable conditions. An epic story of survival and courage. The Nazis brutally deemed that the city and its inhabitants should die of disease, hunger, and shelling.

———, *The Retreat: Hitler's First Defeat* (New York: St. Martin's Press, 2009). Highlighting the fighting in front of Moscow during October to December 1941 and beyond, the author combs diaries and official statements to capture the sense of near-victory and then crushing defeat and subsequent stabilization of the front for the Germans. In his view, the Wermacht came close to replicating Napoleon's disaster in 1813, and he gives Hitler no credit for his "stand fast" order but credits General Walther Model for taking charge of the German Ninth Army and, through maneuver and counterattack, saving the day.

Jonathan W. Jordan, *Brothers, Rivals, Victors: Eisenhower, Patton, Bradley, and the Partnership That Drove the Allied Conquest in Europe* (New York: New American Library, 2011). An absorbing and minutely detailed account of the conflicted and conflicting relationship among these three key players in the American efforts in the European theater during World War II. Riveting.

Victor Kamenir, *The Bloody Triangle: The Defeat of Soviet Armor in the Ukraine, June 1941* (Minneapolis: Zenith Press, 2008). Puts to bed a number of myths concerning Soviet resistance or the lack thereof. Gives the American reader much useful information from the Soviet point of view.

Vladimir Karpov, *Russia at War 1941–1945* (New York: Fandom Press, 1987). A vivid and often moving pictorial account of the Soviet people at war. Although some of the photos seemed staged for propaganda purposes, the gritty determination of the Russians, civilians, and military alike shows through, as does that pride in accomplishing a victory, albeit at a cost of 10 percent of the entire population of the country.

Paul Kennedy, *Engineers of Victory* (New York: Random House, 2012). Looks at the major efforts that turned the early defeats of the Allies into their eventual successes. Interesting chapters on "How to Stop Blitzkrieg" and "Defeating the Tyranny of Distance." This is problem solving on a grand scale.

Ian Kershaw, *The End: The Defiance and Destruction of Hitler's Germany, 1944–1945* (New York: Penguin, 2011). Focusing not on why Nazi Germany fell but on how it held out so long, the author concludes that, after the July 1944 bomb plot against Hitler, four men stepped up to ensure that defeat was not imminent. This quadrumvirate directed Germany's path: Martin Bor-mann basically took over the Nazi Gauleiter network and made Hitler's words into deeds; Josef Goebbels as plenipotentiary for total war did a monumental job of "combing" out society *and* the armed forces to add a million men to the front lines at a time when the Wermacht was losing 350,000 men per month; Albert Speer in a work of genius kept the Reich running, producing and distributing long after it should have collapsed internally; and Heinrich Himmler tightened the security apparatus and took over the internal discipline enforcement of the constantly retreating German armies (Germany killed 20,000 "deserters" during the war, compared to the British, 40; French, 103; and United States, 146; p. 220. Only 18 cases of desertion were punished by death during the entire First World War) and used the 600,000 Volkssturm to ensure National Socialism did not lose control over the population. Fear of Soviet reprisals and the ongoing nature of total war in the East also staved off defeat for a period at a time when the Soviets had an 11–1 advantage in infantry, a 7–1 advantage in tanks, and 20–1 advantage in aircraft and artillery (p. 168). When the Soviets torpedoed the *Wilhelm Gustloff* in January 1945, 8,000 people were lost, 4 times the number lost in the sinking of the *Titanic*. Great maps of the shrinking Reich during this period.

Albrecht Kesselring, *The Memories of Field Marshal Kesselring* (London: William Kimber, 1953). A straightforward account of Kesselring's time in military service and, after the war, in prison. He cites Goering calling up Ribbentrop after war breaks out with Poland and saying, "Now you've got your war. It's all your doing." States that he was against letting the British and French at Dunkirk be finished off by the Luftwaffe. He believes Germany did not lose the Battle of Britain, but when Operation Sea Lion was called off, Hitler lost interest. He further states that Sea Lion was feasible and has high praise for the Russians who fought with the Germans. He also says he urged Hitler over and over to take Malta. Kesselring is not as harsh on the Italians as some but doesn't have a high opinion of their navy and believes that his command held the Americans and British at bay in Italy; otherwise the war would have been over much earlier.

MacGregor Knox, *Mussolini Unleashed 1939–1941: Politics and Strategy in Fascist Italy's Last War* (Cambridge: Cambridge University Press, 1982). The test of war finds Italy lacking. Resurgent Germany gave Italy a chance to elevate in the international order but the "jackal" was not up to the task. Knox makes a strong case the Hitler rightly blamed Mussolini for dragging Germany into the Balkans and North Africa, detracting from its sine qua non, the defeat of the Soviet Union.

Arthur Koestler, *Scum of the Earth* (New York: Macmillan, 1941). Written at white-hot intensity as the war was going on, this work captures the "progression of despair" and the disorganized, disgraceful collapse of France following the Nazi invasion. Koestler had earlier been interned in the infamous French concentration camp at Vernet and later joined the French Foreign Legion before fleeing from Marseilles to Africa and thence to Portugal and England.

Ivan Konev, *Year of Victory* (Moscow: Progress Press, 1969). A rare gem of a memoir, this account of Konev's successful drive on Berlin and Prague at the end of World War II is self-effacing, illuminating, and a pleasure to read. It provides many interesting insights into his relations with Stalin and Zhukov, and Konev is very generous in his praise of various Soviet generals. In particular, his description of the campaign to capture German Silesia (the "Second Ruhr") without destroying it by allowing the Germans to escape the cauldron is most interesting. He gives significant credit to his subordinates for the successful maneuvers and is proud to quote his adversaries, such as General F. Mellenthin, who claimed that the Russian attack between the Vistula and the Oder in 1945 is "beyond description: nothing like it has been seen in Europe since the collapse of the Roman Empire." He also quotes General Omar Bradley as candidly saying that General Eisenhower was really a political general not a battlefield one, but "his esteem for him was more as a diplomat than a soldier." Well worth reading for any serious student of World War II.

Michael Korda, *Ike: An American Hero* (New York: HarperCollins, 2007). A useful overview of, and introduction to, the American general widely regarded as having ably kept the Anglo-American alliance during World War II on an even keel. As a "political" as opposed to a "battle" general, Dwight David Eisenhower doesn't always get the respect that he deserves for his role during World War II in the European theater. He had many crosses to bear and made some very important decisions—such as bringing in de Gaulle to be an important political and military ally.

Eric Larrabee, *Commander in Chief: Franklin Delano Roosevelt, His Lieutenants and Their War* (New York: Simon and Schuster, 1988). Portraits of Marshall, King, Arnold, MacArthur, Nimitz, and LeMay are drawn with skill and dispatch. Good use of enemy sources, such as Albert Speer's assertion that the Schweinfurt raids could have proven devastating if only the Allies had continued them. Especially good on LeMay and the 1,600 B-29s he ordered up.

Benoit Lemay, *Erich von Manstein: Hitler's Master Strategist* (Philadelphia: Casemate, 2010). This book is a must-read for anyone seeking to understand the military thinking behind many of Germany's campaigns. It is also a perfect example of why a liberally educated person must appreciate the axiom "Two things can be true simultaneously." Von Manstein was the finest strategic mind and battlefield commander of World War II (Russian marshal Malinowski called him "our most dangerous enemy"). His genius came up with "Plan Sickle Cut," which was not only the movement of German panzers through the Ardennes but also, and even as important as, the continuing advance after the Meuse River had been reached. Further, his campaigns in the Crimea and after Stalingrad, saving Army Group A and Don Group, and recapturing Kharkov in the spring of 1943 place him in the top echelon of battlefield commanders. Had he been allowed to conduct Operation Citadel at the time and location of his choosing and had he, not Rundstedt, been in charge of the Western Front at the time of the Normandy invasion, major battles might have turned out differently. But, and it remains a big *but*, he was also complicit in planning the various Nazi wars of aggression and the murderous treatment of Jews, Poles, and Russians (including POWs) on the Eastern Front and of being, at best, politically naïve with regard to Hitler. All this and more is illuminated in Lemay's outstanding work.

Note: The death rate among Russian POWs during World War II was 60 percent compared to 3.6 percent among British and American POWs.

Tony Le Tissier, *Charlemagne; The 33rd Waffen-Grenadier Division of the SS* (South Yorkshire, UK: Pen and Sword Military, 2010). Investigates the strange tale of the French volunteers who, along with their counterparts in the Legion des Volontaires Francais (LVF) both for Hitler and against Communism, fought long and hard on the Eastern Front, ending up, along with the SS Nordland Division, fighting to protect Hitler's Berlin bunker in March 1945. Most survivors were shot out of hand by the Free French troops and the Soviets or were later tried and executed by the French government.

Primo Levi, *The Drowned and the Saved* (New York: Random House, 1989). More amazing insights into the netherworld of the concentration camp centering on an SS guard's statement in 1945: "However this war may end, we have won the war against you." Graphic, incisive, deeply moving. Hitler as a "coherent fanatic."

———, *If This Is a Man* (New York: Penguin Books, 1979). Few books have the power of this one to make you rage and weep. Such a monument to human depravity, such a monument to the human spirit. In excruciating and heartrending detail, this Italian Jew conveys the life of a concentration camp victim and survivor, both within the same poignant cluster of human dimensions, disease, hunger, degradation, and beatings but never losing his own humanity. Captured by the Fascist Militia in December 1943 and taken to Auschwitz, where he was initially able-bodied enough to work, he somehow survived to the end of the war and the liberation of Poland by the Russians after a strange hiatus: "We lay in a world of death and phantoms." The second half of this work deals with his incredible return home to Turin via Krakow, Lvov, Zhmerinka, Bobruisk, Ploesti, Budapest, Vienna, Munich, Innsbruck, Verona, and Milan, one of the 5 percent of Italian Jews who ever returned. You will not easily forget this book if you read even a small part of it.

Life's Picture History of World War II (New York: Time, 1950). A glossy, picture-laden single-volume history of the war and included here simply because it was the first book I ever read about World War II, and I have never forgotten many of its images. The narrative is quite limited, however.

Neill Lochery, *Lisbon: War in the Shadows of the City of Light 1939–1945* (New York: Public Affairs, 2011). "Sitting out the war" in the "City of Lights" or, as it was sometimes known, Casablanca II. Antonio de Oliveira Salazar is an economics professor of great intelligence, extremely hard working, and with a vision for the New State. He struggles mightily to keep Portugal neutral (despite the Anglo-Portuguese alliance from 1373) and uninvaded (by Spain, Germany, or both). Salazar also foresaw and presciently feared the Cold War, although he missed entirely the fragility of Portugal's overseas empire. He ended up leasing the Azores to the British and America and having Portugal get hundreds of millions of dollars of Nazi gold for shipments of wolfram (tungsten) for hardening steel. Salazar would end up running Portugal for 23 more years.

Peter Longerich, *Goebbels: A Biography* (New York: Random House, 2015). An amazing work based on the actual diaries of Josef Goebbels, showing him as a very insecure, totally narcissistic person of considerable talents. Forget his propaganda attributes for a moment and concentrate on his "combing out" talents and his overall commitment to the success of the Reich, and he is head and shoulders above so many of the Nazi "almost" inner circle. An inveterate womanizer who committed suicide (and murdered his own children) is verification that, as the author concludes, "his self-delusion (that he was truly a Hitler insider) had won out"

———, *Heinrich Himmler* (London: Oxford University Press, 2012). A massive tome examining the life and rise of young Himmler, who appears as the ultimate bureaucrat (albeit a murderous one) and who time and time again, starting in 1923, expanded his reach and value to the Führer until his power was unmatched within the Reich.

Keith Lowe, *Savage Continent: Europe in the Aftermath of World War II* (New York: St. Martin's Press, 2012). A "world without institutions," as ethnic expulsions on a continental scale take place, some of the largest population movements ever seen. Seventeen million displaced people were in Germany alone. "Germany in 1945 was a huge ant's nest, everything moving." Returning Jews were persecuted and killed in Poland, Ukraine, and Hungary. Eleven million German POWs were treated roughly, especially in the Soviet Union and Yugoslavia. "The Nazi terror was over, the Communist terror was about to begin." Savage fighting in the Ukraine, with more than 400,000 partisans from 1944 to 1950, and Lithuanian partisans held out into the 1950s, despite savage repression.

Joachim Ludewig, *Ruckzug: The German Retreat from France, 1944* (Lawrence: University of Kansas Press, 2012). While there is a lot of coverage for the period from the breakout at St. Lo to Operation Market Garden, most accounts focus on what the Allies failed to do or did poorly. This German scholar and military officer provides the other side the equation: how the German military did so many things so well during that period of retreat. Field Marshal Walter Model is given more credit here than in many other accounts. The irony, of course, was that the German military would end up giving the Nazi regime enough time to continue the Holocaust.

John Lukacs, *The Duel: The Eighty-Day Struggle between Churchill and Hitler* (New Haven, CT: Yale University Press, 1990). A breezy, insightful account of these two men who controlled so much of the action after the fall of France and before the German invasion of the Soviet Union and the Japanese attack on Pearl Harbor. Concludes with the notion that World War II was a "catastrophe for millions of people but whose outcome spared the world an even worse one."

Ben MacIntyre, *Double Cross: The True Story of the D-Day Spies* (New York: Crown, 2012). The successful landing of 150,000 Allied troops at Normandy was made possible (or at least much more probable) by the elaborate deception in which 23 fake divisions under the "control" of General Patton convinced the Germans to keep 22 divisions of the German 15th Army Group in Calais long enough to establish the beachhead. A good read.

Fitzroy Maclean, *Escape to Adventure* (Boston: Little, Brown, 1950). When World War II broke out, Maclean was prohibited from joining the armed forces, so he quit the diplomatic service to go into "politics" and then immediately enlisted as a private, eventually becoming a brigadier during the course of a lengthy, dangerous war. He went on long-range desert patrols in the Western Desert, including some harebrained schemes to blow up things in Benghazi in the heart of German/Italian-occupied North Africa. Maclean was also sent to Iraq and Iran to seize pro-Nazi leaders and others who would undermine the empire. On the direct orders of Winston Churchill, he was then told to parachute into Yugoslavia in order to make contact with Marshal Tito. Subsequently, Maclean spent a lot of time with the Yugoslav partisans, both on the mainland and among a number of islands, getting chased by the Germans all the while. Despite Tito being an avowed Communist, the British threw their logistical support to him largely because of the positive vetting by Maclean. He had a very "good war" indeed and wrote a splendid memoir to capture that war.

Curzio Malaparte, *The Volga Rises in Europe* (London: Alvin Redman, 1957). Malaparte was an Italian correspondent who saw the initial stages of the war from the Axis side, but with his Communist sympathies, he also caught the essence of Soviet resistance, first during 1941 at the crossing of the Dnieper River and the breaching of the Stalin Line in southern Ukraine and later on the Northern Front at the siege of Leningrad as seen from the Finnish frontier. Malaparte thus provides an unusual but perceptive perspective, as later he was to surface as an American agent. Warning: According to literary critic Dan Hofstadter and others, Malaparte was something of a "shape shifter," having been a Fascist, Communist, and American agent, and may well have "fabricated" some of his war reporting.

William Manchester and Paul Reid, *The Last Lion: Winston Spencer Churchill Defender of the Realm, 1940–1965* (New York: Little, Brown, 2012). This is a marvelous book, one of the most enjoyable about World War II I have encountered. While Churchill is at the center, obviously, the insights into Roosevelt and Stalin and many other figures are superb as well. Roosevelt was a cantankerous dude, nowhere near as smart as he thought he was and who got duped by Stalin. Stalin emerges as a master at realpolitik and someone who strangely enough seems to have had the moral drop on both Churchill and Roosevelt until well into 1944. Truly amazing how much Churchill drank—wine and beer for breakfast and champagne and brandy through the day and night.

Erich von Manstein, *Lost Victories* (Chicago: Henry Regnery, 1958). Perhaps the best strategic mind of World War II and the author of some of the most successful campaigns (the defeat of Poland in 1939 and France and Britain in 1940 and the successful counterattack at Kharkov after Stalingrad in 1943) but whose reputation is somewhat sullied by his anti-Semitism and complicity in the outrages on the Eastern Front.

Evan Mawdsley, *December 1941* (New Haven, CT: Yale University Press, 2011). A close study of the December 1–12 period from the Japanese, German, British, Russian, and American perspectives, ending with Hitler's declaration of war on the United States. Well-researched and fast-paced coverage of the decisions and actions that plunged the world into a true world conflagration and the misperceptions and miscalculations inherent in that process.

F. W. von Mellenthin, *Panzer Battles* (Norman: University of Oklahoma Press, 1956). One of the outstanding books of World War II if one is interested in panzer warfare. Von Mellenthin was a master general, himself fighting in Poland and North Africa, on the Russian front, and at the Battle of the Bulge. A must-read for anyone anxious to learn about panzer warfare as practiced by a general with far more experience in it on a grand scale than Erwin Rommel and other celebrated German blitzkrieg warriors.

Catherine Merridale, *Ivan's War* (New York: Henry Holt, 2006). One of the few volumes in English dedicated to the ordinary Russian soldier during World War II. An outstanding collection of memories and insights. The world of today owes the Ivans of yesterday a great, great debt, a debt often unacknowledged by many Americans.

Allan Michie and Walter Graebner (eds.), *Their Finest Hour* (New York: Harcourt Brace and World, 1941). Compiled during the dark days of the Battle of Britain during the beginning of World War II, this volume captures the British airmen in action. See especially "I Bombed the Barges" (by the captain of a Blenheim bomber; pp. 175–87) for a glimpse of a most underreported aspect of that phase of that war.

Joseph Miranda, "The Iron Dream: A Strategic Analysis of the War in the East, 1941–1945," in *Command*, no. 53 (2000), pp. 44ff. A good introduction to this struggle, which decided the course of World War II in Europe.

B. L. Montgomery, *The Memoirs of Field-Marshal the Viscount Montgomery of Alamein* (Cleveland, OH: World, 1958). A refreshing, candid, and heartfelt account of his life and times by the top British general of the war (with apologies to General Slim). Especially illuminating on World War II, I have long argued and am much in agreement with his criticism of Eisenhower and Marshall's plan for "broad front attacks" and "reluctance to concentrate" forces, strategies that got them into much difficulty from time to time during the European campaign and certainly prolonged the war and added to casualties. Note how Montgomery believes his northern thrust at the end "saved" Denmark from being cut off by the Russians. Says Monty was much better than his press accounts. With a tough childhood and the tragedy of losing his beloved wife, he was very much a lone wolf, but he did have a superb eye for battle.

John Mosier, *Deathride: Hitler vs. Stalin: The Eastern Front, 1941–1945* (New York: Simon and Schuster, 2010). Definitely not the first book one should read about World War II on the Eastern Front, as it goes way out of its way to try to convince. Basically argues that Soviet casualties were much higher than other works suggest and that Stalin was even poorer at grand strategy than Hitler. Nevertheless, it is helpful in correcting some mistakes. For example, the Soviets claim they destroyed 700 Tiger tanks during the Battle of Kursk, yet at the time, there were only 262 Tigers in existence, and only 120 were lost during Operation Citadel.

Rolf-Dieter Muller, *Enemy in the East: Hitler's Secret Plans to Invade the Soviet Union* (London: I. B. Tairos, 2015). A very important book that makes a strong case that Hitler would have settled for an alliance with Poland in order to better attack the Soviet Union, the ultimate enemy. Only when Poland asked for British and French help in 1939 was that avenue permanently shut off. Also, the author indicates that the German general staff (especially General Franz Halder) not only wanted but also planned for an attack on the Soviet Union separate from Hitler's designs. Notes also that the T-34 tank had already been deployed in 1939 by the Soviets but was deliberately not shown in either parades or the war against Finland. The T-34, he claims, was the "greatest surprise for the Wehrmacht during the eastern campaign in 1941." Subsequent analysis of the war in the East will require a thorough exploration of the issues raised in the vital work.

Andrew Nagorski, *The Greatest Battle: Stalin, Hitler, and the Desperate Struggle for Moscow That Changed the Course of World War II* (New York: Simon and Schuster, 2007). This sweeping and crucial battle was fought out on a vast scale covering territory the size of Western Europe and was a close-run thing.

Sonke Neitzel and Harold Welzer, *Soldaten: On Fighting, Killing and Dying* (New York: Alfred A. Knopf, 2012). German POWs, secretly recorded, spill their gut on the joys of killing, their faith in Adolph Hitler (as late as 1944), and the toll on their nerves that the killing of civilians took. Once and for all, this nails down the complicity of some regular German army units (not just the Waffen SS), such as the 707th Infantry Division, the 1st Mountain Division, and the 4th Panzer Division, in atrocities against both civilians and POWs. Chilling.

Jonathan North, "Hitler's Forgotten Victims," *World War II* (January/February 2006), pp. 26ff. The Russians suffered 26 million dead—or more—at the hands of the Germans, and this stupendous figure sometimes gets lost behind the horror of the Holocaust.

R. J. Overy, *The Bombers and the Bombed: Allied Air War over Europe 1943–1945* (New York: Viking, 2014). A sprawling account of the war in the air, which he believes killed 600,000 civilians and wounded more than 1 million. Author concludes, "Bombing was a blunt instrument the Allies knew full well, but its bluntness was more evident and more awkward when the bombs fell outside Germany." Thirty percent of bombs fell on France, Belgium, Norway, Denmark, and the Netherlands.

———, *Goering: The Iron Man* (London: Routledge and Kegan Paul, 1984). Known as the "Iron Man" for "his brutality and utter lack of scruples" as well as his "[r]omantic sense of Germaneness," Goering presided over one over-promising fiasco after another—the failure to destroy the British army at Dunkirk, the failure to win the Battle of Britain, the failed air defense of the Reich, and the failed air supply of the Sixth Army at Stalingrad, among others. The author ably chronicles those failures in his chapter "The Examples of Egotism and Incompetence." Upon reflection, Goering's performance as the number-2 man in the Third Reich, especially from 1939 to 1942, may not have been a good thing for Germany, but it was a very good thing for the Allies.

———, *Russia's War: Blood upon the Snow* (New York: TV Books, 1997). Building on the fine research of John Erickson and David Glantz, the author looks at the Great Patriotic War with skill and dispatch. Most useful are his efforts to balance the blame Stalin has received versus the overriding importance of his drive and will to the war effort. Overy underscores the "uneasy symbiosis" among popular patriotism, native endurance, and the roles of both Stalin and the Communist Party apparatus. He is tantalizingly brief, however, on the ongoing Soviet efforts to wipe out anti-Communist and anti-Russian insurgents in Lithuania, Poland, and the Ukraine, efforts that went on until 1950.

George S. Patton Jr., *War as I Knew It* (Boston: Houghton Mifflin, 1947). From the wartime diaries of General Patton (judiciously edited by his wife Beatrice). An engaging mixture of humor, vitriol, and perspicacious insights, as well as a very careful record of the killed, wounded, and captured (and their equipment) of the Third Army and the Germans, battle by battle. Angry at Germans in GI uniforms, he writes, "One sentinel, reinforced, saw seventeen Germans in American uniforms. Fifteen were killed and two died suddenly." Refreshingly candid in his self-evaluation, going to his grave believing "had I been permitted to go all out, the war would have ended sooner and more lives would have been saved." Sounds right to me.

Robert Paxton, *Parades and Politics at Vichy: The French Officer Corps under Marshal Petain* (Princeton, NJ: Princeton University Press, 1969). France was allowed to keep a 100,000-man "Armistice Army" pending the signing of the peace treaty with Germany. Only a minor fraction of the French officer corps defected to the Free French. New blood had a good chance for advancement. Strong Anglophobia, especially within the navy. Many army officers found "liberation in captivity."

———, *Vichy France: Old Generals and New Order 1940–1944* (New York: Alfred Knopf, 1972). A dispassionate but devastating analysis of those Vichy French leaders who actively sought "collaboration" with the victorious Germans, even putting the term *collaboration* in the armistice agreement and pushing for it thereafter through 1942 and 1943. For Paxton, the Third Republic was a complex phenomenon, based as it was not only on defeat but also on an integral Catholic vision of moral order, xenophobia, veneration of peasant values, and moral bankruptcy of Pierre Laval, Francois Darlan, and Philippe Petain. He concludes that cruel times require one to disobey the state in order to save a nation's deepest values, and this virtue he did not find in those men.

John Pimlott, *Wermacht: The Illustrated History of the German Army in WWII* (Osceola, WI: Motorbooks International, 1997). One of the very best pictorial books on the German army in action that I have seen. The photographs are very well chosen and telling, and the text is written by a man very familiar with the tactics, strategies, and equipment of World War II. An excellent analysis that gives the reader a sense of the vastness of the German enterprise and the wildly erroneous strategic lunges into Yugoslavia, Greece, and North Africa that the Germans undertook, much to the advantage of the Allies.

Barrie Pitt (ed.), *The Military History of World War II* (London: Temple Press, 1986). A quite useful and impressive blend of narrative, maps, and truly first-rate diagrams of all manner of weapons used. A fine work for seeing weapons and strategy in action, with special attention to the equipment of the Axis powers seen side by side with that of the Allies.

John Prados, *Normandy Crucible* (New York: NAL, 2011). An interesting attempt to look at the battle for Normandy from both Allied and German sides. High praise for the heavy-caliber rockets of the Nebelwerfer Brigades (copied from the Russians). Blames Patton for holding up production of the M-26 Pershing with its 90-mm gun in favor of Shermans. The M-26s could have arrived in early 1944 instead of 1945. Rommel kept telling Hitler the war was lost. Allies were infected with overconfidence.

Roger R. Reese, *Stalin's Reluctant Soldiers: A Social History of the Red Army, 1925–1941* (Lawrence: University Press of Kansas, 1996). Debunking the assertions that German superiority or Stalin's malfeasance caused the Soviet disasters at the start of World War II, the author believes that it was not just the Great Purge of 1937–1938 (34,000 officers killed, retired, or sent to the Gulag), but that *plus* the concomitant expansion of the armed forces, that caused the horrible showing in 1941. Reese believes that by March 1942, however, it was a whole new army.

Anna Reid, *Leningrad: The Epic Siege of World War II, 1941–1944* (New York: Walker, 2011). Using survivors' testimonies, the author captures the heart and soul of the longest siege of World War II, the flawed German strategy behind it, and the almost incomprehensible suffering of the Russian people, between 650,000 and 800,000 of whom died. During January, February, and March 1942, 100,000 people a month starved to death. "Caught in a mousetrap," Leningraders faced cannibalism, corruption, daily death, and slow starvation. In fact, there is a description on pages 213 and 214 of what it is like to starve in the bitter cold: "Your protruding bones ache, forcing you to keep changing position." And when it was over, of course, Stalin instituted a purge of many during the mysterious "Leningrad Affair." Also, an interesting take on the surrender of General Andrei Vlasov, who went over to the Germans during the siege when trapped in the Myasnoi Bor "cauldron" in May 1942.

Tim Ripley, *Elite Units of the Third Reich* (London: Lewis, 2002). Some important insights into what made the various elite units so effective in combat, no matter who they were fighting against—or what they were fighting for.

Han Rudel, *Stuka Pilot* (New York: Ballantine Books, 1958). Flying more than 2,500 operational flights, mostly on the Eastern Front, Rudel was to destroy something on the order of 500 tanks, receive special medals from Hitler, and end up feted by British pilots after his capture. An intriguing read, up close and personal on the ever-shifting battlefield of the Russian theater of operations. The author is one of the Germans who got their stories out early and managed to avoid being tarred too much with the heinous legacy of the Third Reich.

Guy Sajer, *The Forgotten Soldier* (New York: Harper and Row, 1967). An extremely insightful memoir by a French-German lad who was drafted into the Wermacht at the age of 16 and in 1942 found himself on the Eastern Front, first in a supply capacity and then, after joining the Gross Deutschland Division, in constant action all the way until the end of the war, when he finds himself, against all odds, transported to Denmark, enabling him and some comrades to surrender to the British rather than the Russians. His description of his part in the Battle of Kursk is outstanding. Despite everything, he concludes, "I cannot regret having belonged to a combat unit. We discovered a sense of comradeship which I have never found again, inexplicable and steady, through thick and thin."

Heinz Schaeffer, *U-Boat 977* (New York: W. W. Norton, 1952). According to the author, of the 720 German submarines that went to sea during World War II, 640 were sunk; of the approximately 40,000 sailors who manned them, 30,000 lost their lives. By the end of the war, the life expectancy of the average U-boat was down to only 40 days. Thus the submarine branch of the Kriegsmarine was the most dangerous service in which to serve during the war, yet it was the one that brought Great Britain closer to defeat than any other during the doom-haunted early days of the war. Schaeffer writes skillfully and interestingly about his entire career, from becoming a submariner to rising to commander of U-Boat 977. Indeed, his wartime adventures and the insights he provides make this book well worth reading, but the story at the end of the war that he and his crew made a three-and-a-half-month trip (66 days underwater) to Argentina is superb.

Ronald Seth, *Stalingrad: Point of Return* (New York: Coward-McCann, 1959). When this work first appeared, it was only one of a handful of books dealing with the Russian side of things in the Battle of Stalingrad. The story of Sergeant Pavlov is a telling one.

John Slessor, *The Central Blue* (New York: Frederick Praeger, 1957). The blue is the royal blue of the Royal Air Force, and Slessor takes us from the Royal Flying Corps of 1915 to his vantage point as chief of air staff in 1952 in this lengthy "mixture of history and biography." Best on World War II I found, it ends on a note that the hydrogen bomb brought hopes of peace.

Jean E. Smith, *Eisenhower in the War and Peace* (New York: Random House, 2012). Finally, an account of Eisenhower at war that rings true from beginning to end. Sees Eisenhower as a superb political general, from liberating Paris and making de Gaulle de facto the government of France so he wouldn't have to run its occupation to writing Stalin to finesse Churchill. But she also indicates that Ike's understanding of the battlefield was "abstract and academic." As such, he stayed wedded to the American (from John Pershing and George Marshall) doctrine of all-along-the-front strategy (of the 750,000

Western Allies casualties, four-fifths came "after the autumn slowdown"). Believes Eisenhower should have given Montgomery control over the land battle and sent 40 divisions north of the Ardennes into Germany before the Wermacht could regroup. Von Rundstedt called his decision not to a "god-send." Eisenhower stood tallest and made the right strategic decisions at the Battle of the Bulge, but the subsequent all-along-the-front follow-up cost tens of thousands of unnecessary casualties and prolonged the war. Smith believes that the Battle of the Bulge, with 80,987 casualties, made it the second-most costly campaign in US military history, right behind Grant's 1964 efforts in northern Virginia. Well-balanced, judicious, and fair, this work is destined to become a classic. Also, for the first time, the author tells the whole truth about Eisenhower's affair with Kay Summersby, applauding the supportive role she played in keeping him in the war and accepting that, after the war was over, he sacrificed her and his love for her to his career.

Timothy Snyder, *Bloodlands: Europe between Hitler and Stalin* (New York: Basic Books, 2010). A must-read to appreciate the horrifying breadth of field involved in the slaughter that took place in Poland, Belorussia, the Ukraine, and the Baltic states between 1933 and 1945. Filled with a number of stunning observations, including:

1. "During the years that both Stalin and Hitler were in power, more people—14 million—were killed in Ukraine than anywhere else in the Bloodlands, or in Europe, or in the world"
2. Mass starvation in the Ukraine during 1933 was the greatest man-made starvation in history, a "premeditated murder."
3. Between 1933 and 1938, the main killers in the Bloodlands were the Soviets; between 1939 and 1941, they were equal; and from 1941 until 1945, the Germans were the worst.
4. German POW camps were more deadly than their concentration camps.
5. The Soviets sent 18 million people to the Gulag, with 6.6 million dead.
6. In December 1941, Hitler communicated his desire to kill all the Jews of Europe just as Moscow was *not* being taken.
7. In a supreme irony, in early 1945, surviving Jews in Polish concentration camps were brought back to Germany to do the work the killers would have done if they hadn't gone off East to do their killing.
8. Most of the Holocaust took place in the Bloodlands, taking 5 million Jews and 5 million non-Jews. One million starved during the siege of Leningrad, and more than 680,000 were shot in Stalin's Great Terror of 1937–1938.

David Stahel, *Kiev 1941* (Cambridge: Cambridge University Press, 2012). Declaring Kiev "no ordinary battle," the author makes the case that, even though four Soviet armies were destroyed (some 750,000 men), the battle was a huge sign that the German Ostheer (Eastern Army) had failed to break Soviet resistance and that it was a classic example of a pyrrhic victory. Claims that Stalin began to listen more to his generals and Hitler less to his as a result of this battle. German losses were "irreplaceable," and von Manstein's masterstroke in the spring of 1943 only got the Germans back to where they were in August 1941.

———, *Operation Typhoon: Hitler's March on Moscow, October, 1941* (Cambridge: Cambridge University Press, 2013). Field Marshal Fedor von Bock and 2 million men, armed with 1,500 tanks and 1,000 aircraft, try to take Moscow before winter. They fall behind schedule due to Russian resistance; some conflicting strategies; and, the author maintains most important of all, the enormity of the task. The operation falls way behind that of Napoleon (who left Moscow on October 19, 1812, after having been there for 34 days). Excellent maps and some useful tables (pp. 135–139) showing how German tank production and infantry replacements fell seriously behind the losses suffered in the field. Zhukov said they would not take Moscow, and in this, as in so many other dimensions of World War II, he was right.

George Stevens and Max Hastings, *Victory in Europe: D-Day to VE Day in Full Color* (Boston: Little, Brown, 1985). Many previously unseen color photos of this phase of the war. I was surprised to learn how many Russian soldiers went into battle *with* their medals but *without* their helmets, in contrast with Americans, who went into battle mostly without their medals and with their helmets.

Bert Stiles, *Serenade to the Big Bird* (New York: W. W. Norton, 1952). One of those rare war memoirs that haunts you long after you've finished it. Stiles was a 20-year-old copilot of a B-17 during World War II. The B-17, the "Big Bird," flew very dangerous daylight missions over the heart of the Reich. Stiles finished all 35 of his required missions (with casualty rates as high as 20 percent on some missions, that was an unattainable goal for many). But then, instead of rotating out of combat as he was entitled, Stiles transferred to fighter command and learned how to fly the P-51 Mustang. He was subsequently killed on a mission over Germany after he finished this book, which had closed with what turned out to be a vain hope: "So far the lady of luck has let me come through. There is hope as bright as the sun that it will end soon. I hope it does. I hope the hell it does."

Note: The perceptive USAAF officer on bombing during World War II: "The RAF carried out precision attacks on area targets, while the USAAF carried out arca attacks on precision targets" suggests the difficulties of hitting any targets actually aimed for.

Alexander Swanston and Malcolm Swanston, *The History Atlas of World War II* (Edison, NJ: Chartwell, 2007). Large format, clear maps (many with three-dimensional illustrations), and a reasonable amount of text to carry the novice reader along. Good for both this chapter and the following one on World War II in the Pacific.

Frederick Taylor, *Dresden: Tuesday February 13, 1945* (New York: HarperCollins, 2004). The 1,100 British- and American-bomber raid has been widely criticized, but Taylor puts it in some historical and moral perspective. Twenty to 40,000 people were killed, not the 100,000 plus claimed by Josef Goebbels. Nor was Dresden "overflowing" with refugees. Dresden was a communications and transportation hub *and* a major location of the arms industry, producing radar, torpedoes, aircraft parts, artillery shells, and radios for the military. Taylor concludes it was "overkill," but "[b]y the standards of the time, a legitimate military target." Also, it was only undefended because the Germans had moved its anti-aircraft defenses to the West to defend the Ruhr. There is also considerable dramatic irony in the work, as when a Jewish rabbi watching the destruction of his synagogue in 1938 asserts, quite correctly, "The fire will return."

Telford Taylor, *The March of Conquest: The German Victories in Western Europe, 1940* (New York: Simon and Schuster, 1958). Long regarded as a classic account, depicting in clear and understandable jargon-free prose, of the outbreak of World War II and the wave of German conquests in 1940. Argues that the "terrible nature" of Nazi rule in Poland was not hidden and makes the seldom-seen point that Operation Weseruebun (Wesser River Exercise), the occupation of Denmark and especially Norway, far more than German sideshows in Yugoslavia, Greece, and North Africa, shattered the German navy and greatly detracted from Germany's ability to defeat Great Britain when it had the chance during the summer of 1940, leaving Hitler with the strategic dilemma of "*Was nun*?" ("What now?"). I'm not sure the German lunge into the Balkans to bail out Mussolini and the Italians was not ultimately more important, but certainly any realistic thought of invading the British Isles went by the board in 1940.

Arthur William Tedder, *With Prejudice* (London: Cassell, 1966). War memoirs of a marshal of the Royal Air Force, an account that spares few, and his tart comments about a number of leaders, civilian and military, are a tonic. Asserts, I believe correctly, that the four-year campaign in the Mediterranean was essential for the subsequent Allied victory in the West, especially due to the lessons learned concerning combined operations and the importance of air fields and air power.

Raleigh Trevelyan, *The Fortress: A Diary of Anzio and After* (London: Collins, 1956). An easy-to-read look at a British officer during the Italian campaign. Reading Keats and Tolstoy and getting wounded by the "Jerries," he carries, on making "my first kill" and trying to make sense out of the campaign that seemed bollixed up—and was.

Note: For a more in-depth look at the 1940 campaign for Norway, see the splendidly illustrated books by Douglas Dildy, *Denmark and Norway: Hitler's Boldest Gamble* (Oxford, UK: Osprey, 2007), and Francois Kersaudy, *Norway 1940* (Lincoln: University of Nebraska Press, 1998), a most sardonic but informative account of the British, Polish, French, and Norwegian troops who fought the Germans but were overcome. Prior to the invasion, Norway was hypnotized by its preoccupation with a possible *British* invasion. Not that it would have made any difference anyway, as the Norwegians had no tanks, no anti-aircraft guns, no submachine guns, and so on. Ironically, however, in the process of the battle for Norway, the German Kriegsmarine was crippled for the rest of the war, and Hitler's strategic vision remained clouded thereafter, keeping hundreds of thousands of German troops in Norway all during the war, as he feared a British invasion there.

Note: An interesting variation on the Norway theme is found in Maurice Harvey, *Scandinavia Misadventure: The Campaign in Norway, 1940* (Tunbridge Wells, UK: Spellmount, 19990), which uses the euphemism of the title to describe an operation that was too late, too ambitious, and too helter-skelter, "Mistaken Strategy and Faulty Execution."

Debi and Irwin Unger, *George Marshall* (New York: HarperCollins, 2014). An exhaustive and ultimately exhausting look at this very important figure from World War II. While many previous books have overpraised Marshall, this one seems to underpraise him. They give him high marks for calling for America's preparedness for World War II but point out that the celebrated Marshall Plan was really not his creation and his naiveté in China was palpable. They also think he was misguided on the proposed 1943 invasion of Europe, Operation Husky, and believe he grossly underestimated the casualties of invading the Japanese home islands and had only a mediocre record when picking subordinate generals.

Martin van Creveld, *Fighting Power: German and U.S. Army Performance, 1939–1945* (Westport, CT: Greenwood Press, 1982). Makes a persistent point that German armed forces during World War II, whether on the offense or the defense, consistently showed greater "fighting power" than its US counterparts. Painful conclusion but true.

Adrian Weale, *The SS: A New History* (New York: Little, Brown, 2010). A good and comprehensive look at the German SS. "The central idea of National Socialism—that the German people were involved in a Darwinian struggle with the Jews for world supremacy—gave the SS its mission and transformed it from a repressive police force into an instrument of genocide." Quite useful on the Waffen SS.

Russell F. Weigley, *Eisenhower's Lieutenants: The Campaign of France and Germany 1944–1945* (Bloomington: Indiana University Press, 1981). Detailed and straightforward, the work examines Eisenhower's relations with Patton, Clark, Bradley, Montgomery, Hodges, and many others, concluding, "American military skills were not as formidable as they could have been," and, "In the end, it was the preponderance of material resources that carried its army through to victory in World War II."

World War II Day by Day (London: Dorling Kindersley, 2001). An imaginative account of the entire war day by day, as if one were reading the actual events described in the papers of that day. Takes some getting used to, as some events that are reported on a given day were not known until later, but it turns out to be a surprisingly useful way to "live the moment" during the war. Great photos make the stories come alive as if they were today's news. In fact, if one isn't careful, one can start believing that the information was all really available that day.

Valeriy Zamulin, *Demolishing the Myth: The Tank Battle at Prokhokovka Kursk, July 1943: An Operational Narrative* (West Midlands, UK: Helion, 2010). The author is one of the few scholars able to examine all 60,000 pages of the Soviet-era records as well as the German ones. His conclusions verify the importance of the Battle of Kursk, but in the process, he demolishes the notion that the Germans were ever close to winning it, proving, among other things, that the German II SS Panzer Corps did not even win the battle's opening round (which has been widely asserted). He also disabuses us of the assumption (even more widely held) that Kursk was the largest tank battle in history, with a claimed 2,500 tanks and self-propelled guns in action. Instead, his careful and meticulously documented analysis puts the number at no higher than 1,122 (702 for the Soviets, 420 for the Germans), showing that Soviet gunners claimed they destroyed far more tanks than the Germans even had in the theater. Zamulin, however, does give us, for the first time, exciting documentation for the truly largest tank battle in history: the July 1941 clash between the Soviet Southwestern Front and the First Panzer Army and elements of the Sixth Army in the Brody-Berestechko-Dubno area. In that epic confrontation, the Russians had 5,000 tanks (of which they lost 2,648), and the Germans, 1,000. For two years afterward, the Germans took observers to the battlefield to show them the miles and miles of destroyed armor. This book is a must-read for anyone interested in understanding the titanic struggle on the Eastern Front. There are many good combat photos from the battle not found elsewhere, as well as excellent operational maps.

Marshalls Zhukov, Konev, Malinovsky, Rokossovsky, Rotmistrov, Chuikov, and other commanders, *Battles Hitler Lost and the Soviet Marshalls Who Won Them* (New York: Richardson and Steinman, 1986). From Zhukov to Konev, Yeremenko to Chuikov, the Soviet military leaders tell World War II in their own words about their own battles. With clear and easy-to-understand maps and lavish illustrations—with many German POWs on prominent display—the volume gives the reader a most useful inside look at many understudied Soviet actions, from Operation Bagration to dealing with the Japanese in Manchuria.

Chapter Twenty-Eight

World War II

Warcraft in the Pacific, 1941–1945

This chapter is devoted to the principles of force projection and the mighty clash of two immense navies. It is also about clear and strategic insight or what Admiral Yamamoto knew from Hannibal and the Second Punic War, the Battle of Midway and the international espionage subculture, and the decision to drop the atomic bomb as West War reached a logical and horrible conclusion and time was speeded up by events. Note that the Japanese actions in 1941–1942 were the greatest "over-water" offensive of all time until that point, matched only by the American counterthrust during 1942–1945.

In the process, the mass killing of civilians became an accepted means of warfare for the British, Germans, Russians, Japanese, and Americans, perhaps calling into question the notion that warfare is conditioned by culture and not by Mars.

Akira Triye, *The Origin of the War in the Pacific* (London: Longman, 1987). Looks at the Japanese domestic scene and the search for security on Japan's part. Believe that the United States and Great Britain's policy of deterrence "failed miserably."

Hugh Ambrose, *The Pacific* (New York: NAL, 2010). Unfortunately, I read this book after watching the HBO series of the same name (which was one of the most powerful war series I've ever seen; the attack on Peleliu in particular is depicted most graphically), so the more extended stories of the men and their families who suffered through the campaigns on Guadalcanal, Peleliu, and Iwo Jima seemed far less exciting. This is one time I would advise seeing the film and skipping the book.

Noel Barber, *A Sinister Twilight: The Fall of Singapore 1942* (Boston: Houghton Mifflin, 1968). For those readers who imagined that British blundering about in the Balkans, Norway, Greece, or Crete represented the nadir on how to conduct warfare, please take note: The disaster in Singapore in 1942 eclipses all of them and then some. The British "Maginot Line," Fortress Singapore, had all its heavy guns pointed out to sea and was virtually defenseless on the land side. Even after the Japanese landed in northern Malaya and advanced 650 miles at 9 miles a day through "impenetrable" jungle, the British general in charge, A. E. Percival, refused to allow the military to fortify the land side because he felt it would be "very bad for morale." Not quite as bad for morale, one supposes, as the British abandoning its naval base before the Japanese arrived. Nearly 140,000 British and commonwealth forces were lost in this debacle, the worst in British history.

Walter Borneman, *The Admirals: Nimitz, Halsey, Leahy, and King—The Five-Star Admirals Who Won the War at Sea* (New York: Little, Brown, 2012). A long, loving, and incisive look at the men who won the war at sea for the United States. Nimitz, Halsey, and King were early on interested in submarines and aircraft and their roles in naval strategy. King developed the fast-moving carrier battle groups (which often sailed without slower battleships), and Nimitz oversaw the operational significances (and was always in a bureaucratic struggle with MacArthur). Leahy had the most important behind-the-scenes role as confidant to both Roosevelt and Truman. He opposed using the atomic bomb and the establishment of Israel. This volume contains amazing figures on *net* American ship levels during World War II (458–459), 790 to 6,768 vessels and 383,150 to 3,405,525 personnel. There was also a net of 6 battleships, 21 fleet carriers, 70 escort carriers, 35 cruisers, 206 destroyers, 361 destroyer escorts, 120 submarines, and 3,604 auxiliary craft.

John Boyle, *China and Japan at War 1937–1945: The Politics of Collaboration* (Stanford, CA: Stanford University Press, 1972). A fascinating effort to put in perspective the life of Wang Ching-wei and other Chinese who, along with many other local leaders and warlords, cooperated with the Japanese during World War II. Losing out to Chiang Kai-shek in a struggle to control the Kuomintang after the death of Sun Yat-sen, Wang ended up cooperating with the Japanese, and while not as subservient to them as other true "puppets," such as Emperor Klangte in Manchukuo, he nevertheless assisted them in running the area of China around Nanking and thus deserves our "healthy ambivalence." The author also suggests that the Chinese Communists during this period grew from 50,000 to 500,000.

Joseph Bryan III, *Aircraft Carrier* (New York: Ballantine Books, 1954). Bryant's account is at once personal and yet explanatory in its detail—the harsh heat, the sudden call to battle stations, the Japanese bomb penetrating below decks, and bouts with malaria. There is a large measure of the daily scuttlebutt so prevalent aboard ship, with many concomitant CICs ("Christ I'm confused") moments. There are skunk (surface contacts) alerts and bogey (air contacts) alarms and wry humor: The chaplain explains, "Today is Palm Sunday. Last Sunday was bomb Sunday."

This is really not a work of grand strategy. Instead, it is a charming tale of young men in a serious war, or as Admiral William F. Halsey called it, "This yarn in diary form." Nevertheless, it resonates with power and insight and casts in sharp relief life aboard the attack carrier battle group, which even today remains the enduring paradigm for force projection at sea, then seen in its initial, muscle-flexing stage.

R. V. Burgin, *Islands of the Damned: A Marine at War in the Pacific* (New York: W. W. Norton, 2010). Not as lyric or moving as *With the Old Breed* but a matter-of-fact account of a Marine in the First Division who saw action on Cape Gloucester, Peleliu, and Okinawa, the latter battle taking 82 days and costing US forces 12,000 dead and 38,000 wounded—a casualty rate of almost 30 percent. Proud of being a Marine and glad he met and later married an Australian woman, Burgin gives the reader a Mr. Marine portrait through thick and thin and being wounded.

Robert Buton, *Tojo and the Coming of the War* (Stanford, CA: Stanford University Press, 1961). A judicious account of the career of Hideki Togo, who went from war college to being prime minister and chief instigator of the war against the United States and being hung after a two-year war-crimes trial. In the run-up to the war, the United States was considered a "selfish meddler, interfering unreasonably with Japan's natural aspirations." Ousted from heading the imperial government after the fall of Saipan in July 1944, his career took a downward trajectory.

Raymond Callahan, *The Worst Disaster: The Fall of Singapore* (Newark: University of Delaware Press, 1977). A scholarly but no-less-scathing treatment than Barbers of the worst disaster in British military history, with 140,000 British and commonwealth troops killed, wounded, or captured. Not that the troops were well-led or prepared, lacking as they did antitank guns, tanks, shore-based anti-aircraft guns, and so on. Nor were there *any* plans for evacuation. Churchill may have been distracted before and during the Singapore crisis, but he was not after it: "In all the war I never received a more direct shock." The rest of the Royal Navy wisely and quickly fled to Ceylon until the Americans could once again make the South Pacific safe for them return.

Clayton K. S. Chun, *The Doolittle Raid of 1942* (Oxford, UK: Osprey, 2006). An in-depth, technically oriented account. Good map of the Japanese Areas of Defense Responsibility in April 1942, when the 16 B-25Bs of the Doolittle raid struck a number of Japanese cities, much to the surprise of the Japanese high command. This raid certainly challenged the "Victory Disease" of the Japanese high command, and although all planes were lost, it gave the United States a morale boost. Chun cites the Chinese figure of 250,000 civilians killed by the Japanese in retribution (p. 83). Other sources give 100,000. In any case, a lot.

Geoffrey Evans and Antony Brett-James, *Imphal: A Flower on Lofty Heights* (London: Macmillan, 1965). "The greatest defeat on land ever suffered by the Japanese in the course of their history." This was an epic battled designed by the Japanese in order to prevent the invasion of Burma and ran from March to July 1944. Odd strategic imperative from the Japanese point of view. Malaria was a real scourge for both sides.

James J. Fahey, *Pacific War Diary 1943–1945* (Boston: Houghton Mifflin, 1963). Serving aboard the cruiser *USS Montpelier*, seaman Fahey would log 200,000 miles and see New Caledonia, the Solomons, Marshalls, Marianas, the Philippines, China, Okinawa, and eventually Japan. John Keegan would love this work, as it is war seen from an enlisted man's perspective. His diary entries point out that he saw only two women in one year; believed that 90 percent of war material was going to Europe, not the Pacific; picked up many pieces of Japanese pilots from the deck; suffered the ubiquity of heat rash; lived on cookies and pie whenever possible; and noticed the friendliness of Australians. He also believed in his cause and loved America.

George MacDonald Fraser, *Quartered Safe out Here: A Recollection of the War in Burma* (London: HarperCollins, 1992). A wry, highly recommended account by a man who fought in the 17th Black Cat Division of the British 14th Army under General William Slim and traveled the long road from India to Rangoon. No political correctness here: "It was then I smelt Jap, rank and nasty." And when asked whether the soldiers in the field were expendable rather than dropping the atomic bomb on Hiroshima, he replied, "No, never and that goes for every Indian, American, Australian, African or other soldier whose life was on the line in August 1945. So drop the bomb."

Harry Gaitey, *The Liberation of Guam 21 July–10 August 1944* (Novato, CA: Presidio, 1988). This is the only popular account of the battle for Guam that I've seen. Involving the US Marine 3rd Division and the Army's 77th Division, the action cost 10,000 Japanese lives and 1,400 dead Americans. Yet the author is quick to point out that the taking of Guam was not, like the taking of Tarawa or Peleliu, of questionable strategic value because Guam and the Northern Marianas (Guam along with Saipan and Tinian) quickly became the nerve center for Nimitz's Pacific Command and the home of the 20th Air Force, which was soon to devastate Japan. In fact, the Japanese commander of the home defense headquarters said, "The war was lost when the Marianas were taken away from Japan and when we heard the B-29's were coming out." The Guamanians welcomed the return of the Americans, for during Japan's occupation, their schoolchildren had sung, "Oh Mr. Sam, Sam. My dear Uncle Sam. Won't you please come back to Guam." The northern part of Guam was very rugged, which helps to explain why two Japanese survivors didn't surrender until 1960 and a final poor soul was captured in August 1964.

Robert Goralski and Russell Freeburg, *Oil and the War: How the Deadly Struggle for Fuel in World War II Meant Victory or Defeat* (New York: Morrow, 1987). Presents the case for oil or its lack being the independent variable in determining the outcome of World War II for both Germany and Japan. On the face of it, a bit simplistic but certainly correctly underscoring both the Axis's lack of it and the United States' abundance—the United States produced nearly two-thirds of the world's supply of oil in those days.

Carroll V. Glines, *The Doolittle Raid: America's Daring First Strike against Japan* (New York: Orion, 1988). Based on survivor accounts, this work provides a lot of human interest on the subject and claims that the raid goaded Admiral Yamamoto into ordering the Midway operation, thinking the American planes came not from carriers but from the island itself.

Peter Grose, *An Awkward Truth: The Bombing of Darwin, February, 1942* (London: Allen and Unwin, 2011). A breezy but informative account of the two air raids on Darwin by the Japanese. Argues that the proposed invasion of Australia was actually a close-run thing and that the air raids that killed 300 were the "deadliest event in the history of Australia." Also shows the lack of Japanese strategic sense.

Tsuyoshi Hagegawa, *Racing the Enemy: Stalin, Truman and the Surrender of Japan* (Cambridge, MA: Harvard University Press, 2005). The author sees "no heroes but no real villains, either—just men." The author believes that Stalin and Truman were in a race against each other to get Japan to surrender, and the peace faction in Japan was in a race against time. Stresses the importance of the Soviet Union's entrance into the war and belief in and of itself that it could have ended the war, making the dropping of the atomic bombs unnecessary.

Eric Hotta, *Japan 1941: Countdown to Infamy* (New York: Knopf, 2013). The incredible, and I do mean incredible, account of the Japanese side of the run-up to Pearl Harbor. How a country could be led to war by man who did not truly believe they could win that war brings to mind the account of the Germans and Austrians before World War I. One of the most fascinating books about World War II I have read, even if the military side of things is given short shrift. Amazing insights and aspects of the Japanese leaders fumbling and bumbling into war.

Ronald Lewin, Slim: *The Standard-Bearer* (Hamden, CT: Archon Books, 1976). This "Heaven Born Captain" performed very well in India and Burma despite startling realities (in the first 6 months of the war, Britain lost 40,000 men killed or wounded and 282,000 to sickness and disease). Slim's break of the Irrawaddy Line was a masterstroke that the author contrasts with the pedestrian efforts of Montgomery on the Rhine. Sacked by General Lesse after Burma was liberated, he was promoted to replace Lesse by Mountbatten. Praise and veneration followed with the end of the war.

Oliver Lindsay, *The Lasting Honour: The Fall of Hong Kong 1941* (London: Hamish Hamilton, 1978). A spritely account of the Japanese takeover on Christmas 1941, as the small garrison is told, "Hold fast for King and Country." Unlike Singapore, the Hong Kong affair did not receive major criticism from Churchill, who said, "The Colony had fought a good fight. They had won in reality 'the Last Honour.'" Does show the Maginot-Line mentality, however, as the Japanese were much better prepared and far better trained, as well as tactically and strategically more robust.

W. David McIntyre, *The Rise and Fall of the Singapore Naval Base, 1919–1942* (London: Archon Books, 1979). Calling the Japanese defeat of "Fortress Singapore" "Britain's worst disaster since the loss of the American Colonies," the author makes the case that the Singapore naval base was still a work in progress in 1941 (although it was conceived in 1919) and its fall was due to a true farrago of poor tactics, poor strategy, poor equipment and design, poor leadership, and poor training, plus a great deal of pressure from Australia and New Zealand to reinforce (based on a "kith and kin" argument) the base after its fate was effectively doomed.

Frank McLynn, *The Burma Campaign: Disaster into Triumph 1942–45* (New Haven, CT: Yale University Press, 2011). An amazing saga of ineptitude, duplicity, murderous terrain and climate, ridiculous civilian interference with military affairs, and the most dysfunctional "team" of World War II: Lord Louis "Dickie" Mountbatten, and Generals Joseph "Vinegar Joe" Stilwell, Orde Wingate, and William Slim, with Winston Churchill, FDR, and Chiang "Peanut" Kai-shek leading the ongoing interference, all played out in Burma, India, and China. The author rightly praises Slim as an outstanding commander deserving of more credit and fame than he currently enjoys, but his hyperbole about the Battles of Imphal and Kohima (the "Stalingrad of the East" and an "epic fully the equal of the Somme or Passchendaele") and the veneration of Slim ("Slim's encirclement of the Japanese on the Irrawaddy deserves to rank with the great military achievements of all time") detracts somewhat. Still, a vital read for a true understanding of the Pacific War.

Mitsuo Fuchida and Masatake Okumiya, *The Battle of Midway* (Annapolis, MD: US Naval Institute, 1955). After-action battle assessment by the Japanese, with great access to the battle as well as imperial records. Masatake concludes, "I am firm convinced that the Pacific War was started by men who did not understand the sea, and fought by men who did not understand the air." Good overall assessment. Other key mistakes: (1) putting 50,000 miles on the strike force before the battle; (2) overreaction to the Doolittle Raid; (3) not attacking Australia, as imperial army wanted them for use in Siberia when the Germans defeated the Russians in 1942; (4) Yamamoto not understanding carrier warfare and plans that were too complex and dispersed; and (5) Yamamoto not stopping the battle, when he still had four carriers at his disposal if he had concentrated them.

Rana Mitter, *China's War with Japan 1937–1945: The Struggle for Survival* (New York: Allen Lane, 2013). A provocative and much-needed account of how China almost was dismembered and how it fought longer than any other great power during World War II (1937–1945) and endured huge casualties (14 million dead, 80 million refugees). Only in the 1980s had the present government of China begun to pay attention and some homage to the sacrifices by *all* Chinese during the struggle. Good section on the Japanese Ichisu Advance during late 1944 and early 1945, which devastated the Nationalist positions in Hunan and Henan Provinces and caused 750,000 casualties.

Jonathan Parshall and Anthony Tully, *Shattered Sword: The Untold Story of the Battle of Midway* (Washington, DC: Potomac Books, 2005). The overly complicated Japanese plan to capture Midway and some very bad luck destroyed their sense of invincibility and cost them hundreds of hard-to-replace naval aviators. It was a true culminating point of the war. This is one of the most comprehensive studies of the battle thus far.

E. B. Potter, *Nimitz* (Annapolis, MD: Naval Institute Press, 1976). A very readable account of CINCPAC commander Chester Nimitz, who was thrust into command following Pearl Harbor and who guided the United States through the war in the Pacific. This account reads well and provides a fine overview of that war, as well as Nimitz's skillful handling of his superior CINCFLEET Ernest King.

James M. Scott, *Target Tokyo* (New York: W. W. Norton, 2015). The definitive account of the April 22, 1942, Doolittle B-25 raid on Tokyo, which was of no strategic value and resulted in 250,000 dead Chinese (many by germ warfare and depopulation of whole towns) but that also raised American morale at a time when it was most needed. Thoroughly depressing. I had not realized that the aircrew who made it to the Soviet Union eventually escaped to Iran after Stalin refused to let them go home, supposedly for fear of Japanese retribution.

W. G. Sebald, "A Natural History of Destruction," *New Yorker*, November 4, 2002, pp. 66ff. How causing the bombing of civilians became a way of life for many war leaders.

E. B. Sledge, *With the Old Breed at Peleliu and Okinawa* (Novato, CA: Presidio Press, 1981). This was a classic *before* it was made into the miniseries *The Pacific* for HBO. Its stark realism is as good as it gets when he describes the First Marine Division: "All suffered so much for their country. None came out unscathed. Many gave their lives, many their health, and some their sanity . . . but they suffered and they did their duty so a sheltered homeland can enjoy the peace that was purchased at such a high price."

William Slim, *Defeat into Victory* (New York: David McKay, 1961). A straightforward account by the architect of victory in Burma. Self-effacing yet firm in pointing out which battles turned on which factors. Slim had no sympathy for Japanese soldiers and their officers at the end of the war, countermanding MacArthur's orders to let the latter keep their swords. Points out that this was an Asian war against an Asian enemy fought with mostly Asian troops. Well worth reading.

Edward Stafford, *Little Ship, Big War* (Annapolis, MD: Naval Institute Press, 1984). A thoroughly charming work showing all the logistics behind the mammoth battles in the far Pacific. Interesting interlude at Ulithi atoll, where hundreds of ships anchored so far from home.

Craig Symonds, *The Battle of Midway* (London: Oxford University Press, 2011). The 6-carrier, 400-plane Japanese Kido Butai (Strike Force) hits Pearl

Harbor and runs wild for 6 months and 50,000 miles. With the Japanese Zero the best fighter; Kate, the best torpedo plane in the world in 1942; and the Japanese Type 91 torpedo the best, the Japanese had a technological advantage. The United States had Dauntless dive-bombers, Wildcat fighters, and Douglas Devastators torpedo planes. Stage 2 of the Great East Asia War comes a cropper in June 1942 at Midway for the Japanese, however. Symonds feels Nimitz had better intelligence, a better strategic plan (Japanese sent two carriers to the Aleutians to capture the tiny islands of Attu and Kiska), and three carriers plus Midway versus four carriers, so it was not as big a gamble as some have argued.

Note: The vital role played by naval intelligence is superbly covered in Elliot Carlson, *Joe Rocheford's War: The Odyssey of the Codebreaker Who Outwitted Yamamoto at Midway* (Annapolis, MD: Naval Institute Press, 2011). Proving his worth to Admiral Chester Nimitz at the Battle of Coral Sea, Rocheford relied on hard work, intuition, and a real knack for breaking Japanese coded messages. His assumption that Japan was too "rational" to attack the United States proved to be incorrect, but his sense that Admiral Yamamoto was a "firebrand" who could do something rash was correct. Prior to the strike on Pearl Harbor, Yamamoto cleverly disguised signal traffic, and once the Kido Butai set out for Pearl Harbor, there was radio silence—they knew where they were going, making it difficult to know where they were and where they were heading. His role at Midway was seminal, figuring where and when the Japanese would attack. For his work, he was recommended by Admiral Nimitz for the Distinguished Service Medal (DSM), but that was rejected by his opponents in Washington, and he was eventually transferred to the West Coast. Rocheford was finally recognized by President Reagan in 1986 and given his DSM posthumously.

Barrett Tillman, *Whirlwind: The Air War against Japan 1942–1945* (New York: Simon and Schuster, 2010). Tillman argues that atomic bombs were deserved and actually saved many lives, American and especially Japanese. Captures the mindset of the Japanese military when, after the Battle of Okinawa, Admiral Takejiro Onishi concluded, "I can guarantee absolutely that Japan will not lose. . . . [T]he war is just beginning."

John Toland, *The Rising Sun* (New York: Bantam Books, 1970). Rich, varied, expansive (1,000 pages), and of great detail chronicling the sweep of Japanese expansion and contraction under the hammer blows of the United States. Based on many Japanese sources, it exhibits considerable depth. A classic of the Pacific War.

Ian W. Toll, *Pacific Crucible: War at Sea in the Pacific, 1941–1942* (New York: W. W. Norton, 2012). A stunningly good book, very readable, carefully researched, and judicious in looking at this vital period from the American, British, and Japanese perspectives. Highlights the coming of age of American sea warriors but also sheds very important light on the decisions of the Japanese imperial general headquarters. For example, I had no idea the extent to which Yamamoto pushed for Midway over the objections of the naval general staff (NGS)—who then insisted on the Aleutians fiasco as a "compromise." The imperial general headquarters, correct in my view, preferred a "Southern Strategy" to seize New Caledonia, Fuji, Tonga, and Samoa to cut off the Australia, New Zealand, West Coast, and Panama Canal axis. Also, he gives the Battle of Coral Sea more strategic weight than others previously, calling the failure to take Port Mosby truly a missed opportunity for the Japanese. Interesting production statistics: From 1940 to 1943, Great Britain tripled war production, Germany and the Soviet Union doubled theirs, Japan increased by a factor of 4, but the United States increased that production 25 times. This work is essential reading for understanding the war in the Pacific.

Mark Urban, *The Tank War: The Men, the Machines and the Long Road to Victory* (London: Little, Brown, 2013). A breezy, even jaunty, account of the Royal Tank Regiment that fought in France, North Africa, Italy, and Normandy and into Germany. It is also an account of their shifting armor as the men fight with Cromwells, Stuarts (Honeys), Grants, Shermans, and Crusaders.

Hans J. van de Ven, *War and Nationalism in China 1925–1945* (London: Routledge Curzon, 2003). The author makes the interesting point that Chinese nationalism sprung from a variety of sources, not the least (but most overlooked) the role of the Nationalists and their resistance during World War II. Critiquing the Edgar Snow–Theodore White paradigm that Chang was so deserving of losing and Mao so deserving of winning, van de Ven sees the centrality of the war of resistance as central to today's Chinese pride and that this element is getting more traction in post-Mao China.

Brian A. Victoria, *Zen at War* (London: Weatherhill, 1997). A strange volume about how Buddhism itself was hijacked by the Japanese militarists for their own purposes.

John F. Wukovits, *Devotion to Duty: A Biography of Admiral Clifton A. F. Sprague* (Annapolis, MD: Naval Institute Press, 1995). The ship's log of significant American admirals from World War II includes Ernest J. King, Chester W. Nimitz, William F. Halsey, Marc A. Mitscher, Thomas C. Kinkaid, Frank Jack Fletcher, Raymond A. Spruance, and many others. One admiral often overlooked but surely belonging in that proud company is

Admiral Clifton A. F. "Ziggy" Sprague. He was a true sea warrior whose heroic and cool efforts in the Battle of Leyte Gulf during October 1944 were campaign-saving in nature. As a result, his courageous battle actions became legendary within the US Navy, but his career deserves wider coinage with the general public, as his contributions to the Pacific War are often underestimated when the exploits of the fighting admirals are examined. In fact, this is the first biography of Sprague in 50 years. It is a rewarding one. Although the prose is a bit pedestrian, overall the work provides a pleasant blend of background and insight framed against a useful chronology of the war as well as his career. This work does not overpromise or underdeliver. Sprague was a pioneer, serving in naval aviation from 1914 to 1951, who commanded the seaplane tender *Tangier* at Pearl Harbor and the carrier Wasp in the battle of the Philippine Sea and finally served as unit commander of Task Force Taffy Three off Samar Island during the October Leyte campaign.

This was his true moment of glory. When Admiral Halsey recklessly took his powerful Third Fleet off far to the north after the Japanese diversionary fleet, the American landing troops and ships were left exposed and highly vulnerable, as a second Japanese squadron under Admiral Takeo Kurita attacked with a vengeance. Sprague's 6 lightly armored escort carriers and 6 destroyers and destroyer escorts were no match on paper for the attacking Japanese, who had fielded 4 battleships, 8 cruisers, and 11 destroyers. Outgunned 154 to 29 in 5-inch guns and 91 to 0 in larger-caliber weapons, Sprague skillfully maneuvered his smaller force for several hours, ordering a series of aerial and surface-ship attacks, and his command, although suffering significant losses, beat back the Japanese, saving the American landing.

It was a moment of high drama and excruciating pressure. In the timeless spirit of Horatio Nelson, John Paul Jones, and Daniel Farragut, Ziggy Sprague proved more than a match for the demands of the moment. He would go on to see action at Iwo Jima and Okinawa, but it is for his courage and determination at Leyte that he will be most remembered in the annals of naval engagements.

Note: An excellent book that can serve as an introduction to both the end of World War II and the beginning of the era of the Cold War and decolonization is Ian Buruma, *Year Zero: A History of 1945* (New York: Penguin, 2013), which covers the immediate postwar period, with its mass population movements; governmental switches; and societal efforts to come to grips with previous collaboration and going-forward realities of honor, shame, and economic underpinnings. Good on civil wars in Greece, China, Vietnam, the Philippines, Palestine, and others. Great ironies abound.

Chapter Twenty-Nine

The Cold War and Nuclear Weapons, 1946–1992

From the perspective of Mars, during the Cold War between the United States with its allies and the Soviet Union with its allies, World Wars III through IX were fought in terms of technological innovation and proxy wars—both in terms of nuclear weapons and conventional ones. It is also important to remember that the "inevitable" collapse of the Soviet Union and its empire did not appear to be likely, let alone inevitable, to most observers during the period 1946–1992. In any case, the Cold War would be the longest war in US history, lasting three times longer than any previous war.

Richard Aldorus, *Reagan and Thatcher: A Difficult Relationship* (New York: W. W. Norton, 2012). Obliquely but nevertheless trenchantly, this work covers the end of the Cold War. It seems like a very long time ago, but echoes remain.

Anne Applebaum, *Iron Curtin: The Crushing of Eastern Europe 1944–1956* (New York: Doubleday, 2012). Anyone who ever thinks the Cold War was caused by the United States and Great Britain should read this definitive, scholarly, and very well-researched account of the extent to which Stalin and Beria planned, executed, and exulted in its imposition of totalitarian states in Eastern Europe (the Soviet Union began its efforts for Poland in 1941). It's all here: the political, military, economic, even sociological effort and why, in the end, even totalitarian states fell rapidly despite almost everyone's belief that they would not. Usefully, the author points out that the term *totalitarian* was first used in the context of Italian Fascism as *totalitarismo*, which one of Mussolini's critics defined as "[e]verything within the state, nothing outside the state, nothing against the state." As Walter Ulbricht also put it, "It's quite clear—it's got to look democratic, but we must have everything in our control." The NKVD was everywhere, and tens of thousands of democrats from all the countries under its control were placed in the old Nazi concentration camps and into the Soviet gulag. A must-read.

Raymond Aron, *The Century of Total War* (Garden City, NY: Doubleday, 1954). See especially his look at the dynamics of the Cold War in the chapter "The Cold War—Preparation or Substitute for Total War?" (pp. 226–40). He concludes, "Can a third world war be won without becoming total? No one knew." And, thank goodness, nobody ever found out.

John Baylis and John Garnet (eds.), *Makers of Nuclear Strategy* (New York: St. Martin's Press, 1991). A quick introduction to the nuclear thoughts of Brodie, Wohlstetter, Kahn, Kissinger, Schelling, Buzzard, Blackett, Beaufre, and Sokdovskii.

Robert Bigelow, *The Dawn Warriors: Man's Evolution toward Peace* (Boston: Little, Brown, 1969). Looking over humankind's history, the author sees a progression toward peace despite the weapons evolution from hand axe to H-bomb. Makes the good point that nuclear weapons will not hesitate to obliterate humankind without any remorse or the "slightest qualm." Claims humans will either destroy themselves or get rid of nuclear weapons. Quite an all-or-nothing scenario. Would suggest that, 40 years later, we could be living on borrowed time.

Ray Bonds (ed.), *Russian Military Power* (London: Salamander Books, 1980). A comprehensive, well-illustrated book compiled during the height of Soviet power during the Cold War and captures the drastic overspending on military equipment and personnel that helped to give credence to the pithy statement "Ronald Reagan almost spent the United States into bankruptcy but he did bankrupt the Soviet Union." Amazing how widespread and powerful the Soviet military appeared until the moment it cracked.

Bernard Brodie (ed.), *The Absolute Weapon: Atomic Power and World Order* (New York: Harcourt, Brace, 1946). An early and somewhat superficial read of nuclear weapons that produced a profound change in military affairs, political relationships, and international machinery, echoing J. Robert Oppenheimer: "They are weapons of aggression, and surprise, and of terror."

Michael Burleigh, *Small Wars, Faraway Places* (New York: Viking, 2013). A breezy and highly readable account of the Cold War, with some interesting insights into the total strategy and blundering that went on as both sides operated under strange misperceptions and both blundered about unnecessarily, although all of this is easier seen in hindsight.

Kenneth Conboy and James Morrison, *The CIA's Secret War in Tibet* (Lawrence: University Press of Kansas, 2002). Although most histories of the CIA make mention of this campaign, few give any details. This book is now the standard on the campaign that ran from 1949 until 1972. The Tibetan people (Khampa) have always resisted Mongol, Manchus, or Nationalist Chinese dominance, as well as Mao's version of imperialism, so it was not hard to find people to arm. From Saipan to East Pakistan and India, from Colorado Springs to Okinawa, the CIA used a number of locations and received help from a number of sources, including the Dalai Lama (who blessed the operation), the Indians, the Pakistanis, the Nepalese, and the Nationalists in Taiwan. The Soviets even wanted in on the action as a way to undercut the Chinese. Unfortunately for the people of Tibet, their interests were eventually sacrificed for geopolitics when the United States and others saw China as the best bulwark against the Soviet Union and its expansionism. Quite a read.

Gordon Corera, *The Art of Betrayal* (New York: Pegasus Books, 2012). Although a lot of the first part of the book is rehashed accounts of Soviet, British, and American spy history, it does provide a good glimpse into the activities of both sides and the degree to which all were in the dark about most major events that occurred. A frightening set of revelations given the nearness of nuclear war for such a long time. Quite detailed.

Norman Davies, *Rising '44: The Battle for Warsaw* (New York: Viking, 2003). Often confused with the Warsaw Jewish ghetto rising (April 1943), the broadly based popular uprising in Warsaw in August 1944 was, the author believes, the beginning of the Cold War, as Stalin deliberately held back his close-by forces after encouraging the Poles to rise up against the Germans, who put down the rising with great ferocity, eventually leveling the city after a 63-day battle. Survivors often went from German hands into Soviet gulags.

Michel Dobbs, *One Minute to Midnight* (New York: Knopf, 2008). Argues that the United States and the Soviet Union were much closer to Armageddon during the Cuban Missile Crisis than previously thought; that the Soviet ground commanders had armed tactical nuclear weapons and that their troop strength was 42,000, including combat units (not the 10,000 advisors claimed by the United States); and that Valentin Savitsky, captain of the Soviet submarine B-59, considered firing his 10-kiloton nuclear torpedo at the destroyer *USS Beale* and by the overflight of a U-2 over the Chukot Peninsula, which put the Soviets on high alert. Throw in some tactical nuclear weapons on both sides and cruise missiles (pilotless MIG 15s) pointed at Guantanamo Bay, and we were all lucky to avoid incineration. I like the fact that he fingers the Kennedys for trashing Adlai Stevenson for coming up with the very trade for Jupiter C missiles that actually carried the day.

Ian Frazier, "On the Prison Highway: The Gulag's Silent Remains," *New Yorker*, August 30, 2010, pp. 28ff. The author visits the remnants of Siberia's labor camps and wonders why Joseph Stalin, responsible for the deaths of tens of millions of people (the "fifty-five to sixty million Russians who died of unnatural causes as a result of the Bolshevik Revolution, from its beginning in 1917, to the fall of communism, in 1992"), seems to have escaped the opprobrium heaped on the head of Hitler. This is a good question, well asked.

Aleksandr Fursenko and Timothy Naftali, *Khrushchev's Cold War: The Inside Story of an American Adversary* (New York: Norton, 2006). Khrushchev saw Mao as a menace and the United States as beatable in space, time, and production and was willing to risk war over Berlin and put missiles in Cuba. He was overthrown because of his acting "unilaterally" but not before he brought the world as close to nuclear war as it has ever come.

John Lewis Gaddis, *The Cold War: A New History* (New York: Penguin, 2005). This very readable, even breezy, account breaks no new ground but provides one of the best introductions to the subject I have read, and his conclusion that it could have been much worse should be remembered. Hope did triumph over fear.

———, *George F. Kennan: An American Life* (New York: Penguin, 2011). The definitive biography of the man credited with developing the American "containment" policy toward the expansionist Soviet Union under Stalin.

———, *We Now Know: Rethinking Cold War History* (London: Oxford University Press, 1997). Provides a wide-ranging historical perspective, concluding with one fundamental, even inescapable, truth: "As long as Stalin was running the Soviet Union a Cold War was unavoidable."

Robert Gellately, *Stalin's Curse* (New York: Knopf, 2013). A careful and comprehensive study of Stalin's lifelong commitment to Marxist-Leninism, which "[i]nformed everything in his life." Shows what lengths Stalin went to in order to hide his postwar agenda from the Allies. Roosevelt was fooled, and Truman bullied. Amazing amount of ethnic cleansing as the war wound down and afterward, not just in Poland, the Baltic States, and German territory, but also millions of Tartars, Chechen-Anguish, Kalmyks, Crimeans, and Georgians were treated horribly. Makes no bones about saying "It was overwhelmingly Stalin's actions that led to the Cold War" (p. 306). Gellately concludes ominously, "The struggle between the anti-Stalinists and the Stalinists is still going on in Russia."

Colin Gray, *War, Peace and Victory* (New York: Simon and Schuster, 1991). Trades the strategic intertwining of the United States and the Soviet Union during the Cold War and examines the use and nonuse of nuclear weapons in their respective strategic thinking. His chapter "The Perils and Pleasures of Coalitions" makes interesting reading.

General Sir John Hackett and other top-ranking NATO generals and advisors, *The Third World War, August 1985* (New York: Macmillan, 1979) (F). In this futuristic projection, the West won this war, albeit only after the basic scenario was changed in the rewrite. We now know that Soviet doctrine never allowed for *any* escalation foolishness but always and powerfully insisted on an overwhelming preventive first strike even before war was declared. But the author, a retired NATO general, did predict the end of the Soviet Union and its empire.

Jonathan Haslam, *Russia's Cold War: From the October Revolution to the Fall of the Wall* (New Haven, CT: Yale University Press, 2011). The Cold War polarized academic opinion (and opinion in general). The death of Stalin did not immediately end the Cold War, but there is something called an autonomy of a habit. As of 2013, it would appear those in Russia who feared the West or who wanted greater hegemony are on the rise, suggesting it goes beyond Stalin but not necessarily great power status yearnings.

Seymour Hersh, *The Dark Side of Camelot* (Boston: Little, Brown, 1997). A bruising, caustic, devastating catalogue of John F. Kennedy's presidency in terms of "personal excess and recklessness"; the targeting for assassination of world leaders (Operation Mongoose aimed at Fidel Castro cost more than $100 million); a very dangerous handling of the Cuban Missile crisis; and a secret taping of people in the Oval Office, the Cabinet Room, and even the president's living quarters. Hersh makes some especially telling points, reminding us that the Soviet general in charge of forces in Cuba in 1962, Issa Pliyev, was a veteran of Stalingrad, commanded 42,000 combat and support troops, and could have armed the missiles within hours and that both Bobby and Jack Kennedy deceived the American public with regard to the quid pro quo of removing American Jupiter C missiles from Turkey and Italy. But Hersh's final summation on Vietnam is the most devastating: "Whatever Jack Kennedy's intentions were, Vietnam was his war, even after his death."

Jeremy Isaacs and Taylor Downing, *Cold War: An Illustrated History 1945–1991* (Boston: Little, Brown, 1998). Breezy, glossy, and occasionally glib, this CNN TV series companion is still a good place to get an introduction to the 40-year-old political, economic, and military struggle between the United States and the Soviet Union.

Ian Johnson, *A Mosque in Munich* (Boston: Houghton Mifflin Harcourt, 2010). An oddly fascinating book about the Nazis, the CIA, and the Muslim Brotherhood and their interaction during and after World War II. An incisive look at the Cold War in the rearview mirror.

Herman Kahn, *On Thermonuclear War* (Princeton, NJ: Princeton University Press, 1960). Extraordinary work discussing nuclear weapons as they never had been before. His later work, *Thinking about the Unthinkable* (New York: Touchstone, 1985), I believe, sums things up for that era—and indeed into the present. Yet many are thinking about using nuclear weapons even until this moment. And there are a lot of them. He looks into the heart of the nuclear dragon and doesn't blink. Some say he liked the view too much, but he truly feared a nuclear catastrophe.

L. Douglas Keeney, *15 Minutes: General Curtis LeMay and the Count Down to Nuclear Annihilation* (New York: St. Martin's Press, 2011). Covering much of the nuclear standoff between the United States and the Soviet Union, the work is frightening, episodic, and chilling and gives a year-by-year portrayal of the disasters, near-disasters, and errors surrounding the nuclear arms race. Although Keeney argues, "Deterrence was no doubt the riskiest gamble in world history," he also grudgingly admits that LeMay's insistence on having the Strategic Air Command on 15-minute alert when manned bombers were the principle component of that deterrence resulted in a huge payoff: "LeMay's indistinguishable war machine *was* the perfection of deterrence."

Frederick Kempe, *Berlin 1961: Kennedy, Khrushchev, and the Most Dangerous Place on Earth* (New York: G. P. Putman's Sons, 2011). From 1949 until 1961, 2.8 million East Germans fled the German Democratic Republic, and Khrushchev was pushed by the East German Walter Ulbricht to do something about it. Kempe sees Kennedy's weakness leading to the construction of the Berlin Wall and the subsequent Cuban Missile Crisis. Great Britain and France were of no help, according to the author, and the German Konrad Adenauer was even worse.

Note: For a searing look at West German politics of this era, see the novel by Wolfgang Koeppen, *The Hothouse* (New York: W. W. Norton, 1953). In it, the government in Bonn is corrupt and corrupting, having abandoned its mission to create a true liberal democracy. The disgusted member of Parliament and voice of despair, Keetenheuve, commits suicide in response.

George Kennan and John Lukacs, *George F. Kennan and the Origins of Containment, 1944–1946* (Columbia: University of Missouri Press, 1997). Kennan was one of the very first Americans to see the postwar dangers of the Soviet Union, and he developed the notion of "containment" as the Soviet Union took over Eastern Europe with a vengeance. His February 22, 1946, "Long Telegram" highlights the Kremlin's "neurotic" view of the world and Stalin's being "devoid of the capacity for loyalty."

Albert Legault and George Lindsey, *The Dynamics of the Nuclear Balance* (Ithaca, NY: Cornell University Press, 1974). Provides the reader with a sense of the numbers and force demands of nuclear weapons and how the "triad" and other assumptions about nuclear deterrence became "gospel."

Abbas Milani, *The Shah* (New York: Palgrave McMillan, 2011). One might think that this is an odd spot in a bibliography for a book on Iran, yet the Cold War intrigues by the United States, Great Britain, and the Soviet Union are cast in sharp relief in this penetrating biography of the shah of Iran, who, as a self-styled "King of Kings," ruled for 37 years, stole a ton of money, modernized big sections of the economy, and fell in 1979 to a regime that would become far more despotic and tyrannical and a genuine threat to world peace. The events of the 1950s, 1960s, and 1970s seem like shadows on a distant screen before the shah's fall in 1979 and his end as a true flying Dutchman without a place to lay his head. Ironically, only Egypt was forthcoming and then only briefly, as he soon died of cancer.

Arch Oboler, *Night of the Auk* (New York: Horizon Press, 1958) (F). Frightening play that captures the anxiety and stress of the nuclear standoff as the author imagines a first strike from the moon, which, if you think about it, would be the long way around Robin Hood's barn.

Connor O'Clery, *Moscow, December 25, 1991: The Last Day of the Soviet Union* (New York: Public Affairs, 2011). A truly fascinating account of the last moments of the Soviet Union, the titanic (and petty) struggle between Mikhail Gorbachev and Boris Yeltsin, and what Vladimir Putin would come to call a "great geopolitical catastrophe." I did not know that Russians left the federation first or that the Soviets had 27,000 nuclear weapons, the control of which had to be handed over by Gorbachev to Yeltsin, and what a childish standoff took place while that was happening. The Soviet Union was bankrupt, and the United States flew in hundreds of tons of food and other supplies after the handover. What a Christmas Day and what astonishment there was across of the globe. Colin Powell put it best: "The Soviet Union disappeared without a fight, without a war, without a revolution. It vanished with the stroke of a pen." An amazing read and one that catches Putin in midstride in his efforts to return Russia to many of its USSR dimensions. Gorbachev spent much of his last days negotiating his retirement package (and he got $800,000 for his memoirs).

Serhll Plokhy, *The Last Empire: The Final Days of the Soviet Union* (New York: Basic Books, 2014). A well-researched and very interesting examination of the end of the Soviet Union. Surprisingly, Yeltsin plays the most important role, and the United States is slow to understand what was happening. On Christmas Day 1991, the world is turned upside down with the "peaceful dissolution" of the Soviet Union into 15 independent republics. Quite a fascinating study.

Thomas Schelling, *Arms and Influence* (New Haven, CT: Yale University Press, 1966). Looks at a number of key topics, such as the diplomacy of violence, the manipulation of risk, noncommitment bargaining, and the use of armaments as signals and messages (the "weapons speak"), ending with good advice: "To get an adversary's ear, listen."

———, *The Strategy of Conflict* (Cambridge, MA: Harvard University Press, 1966). Focusing on the "international politics of threat" or deterrence, the author looks at the Cold War and nuclear options through the lens of believing that the capability to retaliate is more useful than the ability to attack. In other words, being able to strike back no matter what is a greater deterrence than the ability to resist an attack. Applies game theory to international relations and the use the nuclear weapons. Argues for the concept of "bargaining," both in terms of the threat of nuclear weapons and in the use of conventional weapons. Uses the notion of a game of chicken to explain how irrationality might actually make things safer; that is, the other side is just crazy enough to use nuclear weapons, so let's not risk it by provoking them. Had a big impact on the notion of "escalation" during wartime. This doctrine proved to be a total failure in Vietnam (although Schelling could argue that the calibration on the part of the United States was what was incorrect, not the concept).

Paul Seabury and Angelo Codevilla, *War: Ends and Means* (New York: Basic Books, 1989). See especially "Military Operations in the Nuclear Age" (pp. 143–59). What a world it was and, to some extent, still is.

Neil Sheehan, *A Fiery Peace in a Cold War: Bernard Schriever and the Ultimate Weapon* (New York: Random House, 2009). Nuclear weapons and warheads and the arms race within the arms race, where World Wars III through IX were fought.

Vasily Sokolovskii, *Soviet Military Strategy* (Englewood Cliffs, NJ: Prentice Hall, 1966). A somewhat chilling roll-out of Soviet military doctrine (a very useful chapter on "Soviet Military Strategy 1917–1945") and nuclear strategy, which assumed that capitalism was out to destroy Communism and any war between the United States and the Soviet Union would be to the death. Some insights: (1) nuclear weapons were to be used as an essential part of war-making ("basic arms in future wars"); (2) massive first-strike use of them was preferable to holding them back (the "effectiveness of the application of force at the very beginning of the war"); (3) "escalation" or "flexible response" was not to be considered, as nuclear war was strictly a "go/no go" paradigm; and (4) the Soviet Union would be the ultimate victor.

Sherry Sontag and Christopher Drew, *Blind Man's Bluff* (New York: Public Affairs, 1998). The United States and the Soviet Union duel under the sea with quieter and quieter nuclear submarines. Frightening to read, even so long after the fact, because you know that the United States and China are playing the same game today.

Sheldon Steen, *The Week the World Stood Still: Inside the Cuban Missile Crisis* (Stanford, CA: Stanford University Press, 1987). Listening to all the declassified verbatim tapes of the meetings of the ExCom, the author points out all the very bad advice President Kennedy got and how he fortunately disregarded most of it, luckily for all of us. Still, here is the proof positive that poor old Adlai Stevenson argued for the exact outcome that saved the day, an argument for which he was later pilloried as an appeaser in the "miasma of self-congratulation" that followed Khrushchev's accepting the trade. Khrushchev himself was strongly criticized by Fidel Castro for lacking cojones.

Norman Stone, *The Atlantic and Its Enemies: The History of the Cold War* (New York: Basic Books, 2010). Credits the West, especially under the leadership of Margaret Thatcher of Great Britain and Ronald Reagan of the United States, with having an inner dynamism that transcended both nationalism and the appeal of the collective state to tip the balance in favor of the West. A searingly honest and insightful work. A must-read to understand the realities of the Cold War and its outcome.

Robert Smith Thompson, *The Missiles of October: The Declassified Story of John F. Kennedy and the Cuban Missile Crisis* (New York: Simon and Schuster, 1992). The author claims that Kennedy knew about the missiles in March 1962 and planned to use them as an excuse to invade Cuba, but then the stakes got too high, and America had to back down. He claims that what was supposed to be a "high point in American power was in fact a harbinger of its decline." What decline you may ask? Just don't ask it of a chap whose knowledge of military matters is so sketchy that he doesn't realize that Khrushchev's "victory" was the removal of liquid-fueled, obsolete Jupiter C missiles in Turkey and Italy, which were already scheduled for demobilization. Deliver us.

Douglas C. Waller, *Big Red* (New York: HarperCollins, 2001). A Trident submarine goes to sea and that captain automatically becomes the fourth-most important nuclear power in the world. A very important book for anyone interested in the most powerful weapon of all time and how it hid—and continues to hide—out of sight and sound of all countermeasures.

Albert Weeks, *Assured Victory: How 'Stalin the Great' Won the War but Lost the Peace* (Santa Barbara, CA: Praeger, 2011). The author believes Stalin should get more blame for starting World War II, more credit for helping win it, and more criticism for the Cold War. The author claims the Cold War involved 100 shooting conflicts and cost each side 4 to 10 *trillion* dollars.

Odd Arne Westad, *The Global Cold War: Third World Intervention and the Making of Our Times* (Cambridge: Cambridge University Press, 2005). The United States and the Soviet Union play far and false across the world, with results the author calls "truly dismal," a "dire example" of a pattern of ongoing, cynical, and dangerous intervention. Believes that bipolarity and ideological blinders forced the two superpowers into truly insane rivalries that did great damage to the countries in which they competed. Underlines the absurdity of either or both actually caring who was the prime minister of Burundi. Does give Ronald Reagan considerable credit for pushing the Soviet Union to near-collapse.

William Appleman Williams, *The Tragedy of American Diplomacy* (New York: Delta Books, 1962). From "Imperial Anticolonialism" to "The Impotence of Nuclear Supremacy," the author rails against the American Empire in all of its manifestations, claiming it failed to live up to its own democratic ideals.

Vladislav Zubok, *A Failed Empire: The Soviet Union in the Cold War from Stalin to Gorbachev* (Chapel Hill: University of North Carolina Press, 2007). Based on declassified Soviet archives, this sparkling work accents the search for security and power that dominated the Soviet Empire until its decline and death.

Chapter Thirty

The Korean War; or, The Forgotten War of Great Irony

This war provides valuable insights into presidential power and military command, setting strategic limits, and the terrible price of foreign policy miscalculations (the United States, the Soviet Union, China, North Korea, and South Korea). Was it really the "wrong war in the wrong place at the wrong time against the wrong enemy" for all participants? It is most ironic that all the major participants got the last thing they wanted and didn't achieve their most important core goals. The Soviet Union got a rearmed Germany, China did not get Formosa/Taiwan, neither South Korea nor North Korea got a united Korea, and the United States got the first of two massive land wars in Asia. Quite a list of failures in this war of misperceptions. And among the nonplayers, Japan and Taiwan especially benefited tremendously through no effort of their own.

Dean Acheson, *Present at the Creation* (New York: Norton, 1969). The former US secretary of state relates how the State Department interacted with Congress and the president. He was truly there. Provides some excellent insights into the mindset of not only the secretary of state but also of the US government as a whole during this formative period.

Bevin Alexander, *Korea: The First War We Lost* (New York: Hippocrene, 1982). Breezy account with some interesting insights. Very good on particular American units and where they operated and often buttressed by some overarching strategic points. A balanced view of MacArthur, who was prone "to inspiration and to error." Weak on Chinese and North Korean moves, motives, and strategies. Points out that the Eighth Army's 120-mile march from the Chongchon River to the Imjin River was the longest in US history, and Chesty Puller was clever to call the Marine retreat "breakout to the east."

Colonel Joseph H. Alexander, *Battle of the Barricades: U.S. Marines in the Recapture of Seoul* (Washington, DC: Marine Corps Historical Center, 2000). Passed over in the history books, this phase of the war deserves far more attention. There was a lot of hard fighting after MacArthur declared the city "liberated."

Louis Allen, *The End of the War in Asia* (New York: Beekman/Esanu, 1979). Pages 15992 shed light on US and USSR moves into Korea in 1945, the United States working closely with the Japanese military to move 176,000 of their troops back to Japan. The United States tells Syngman Rhee that "he is not well known in Korea and his 'provisional government'" was "self-constituted and represented no more than a limited group of exiles." Nevertheless, he was to prevail. For Allen, the "liberation of Korea" had sliced the country in two, and within five years, both halves would be at war. Not a pretty picture.

Roy Appleman, *Disaster in Korea: The Chinese Confront MacArthur* (College Station: Texas A&M Press, 1989). It was a disaster, and it was unnecessary. Well-written and insightful from an author who knows the US Army well.

———, *East of Chosin* (College Station: Texas A&M Press, 1987). Fascinating, especially about the disasters in northwest Korea in November and December 1950. A must-read for those interested in the backwaters of war that coexist with the historically focused "main line action stories." Amazing tales off the front pages.

———, *The United States Army in the Korean War: South to the Naktong, North to the Yalu (June–November 1950)* (Washington, DC: Government Printing Office, 1961). The official history of the war, which includes some useful editorial comments. An excellent source.

J. J. Barker, *Fortune Favours the Brave: The Battle of the Hook Korea 1953* (London: Leo Cooper, 1974). The British Commonwealth Division fought a long and bloody set of battles from October 1952 until May 1953 in the World War I–style trench warfare on the central front. The author concludes, perhaps too grandly, that the stand of the Black Watch and the duke of Wellington regiments ushered in a "new Elizabethan Age." But it was a hell of battle, albeit little known in the United States.

Glenn Barnett, "Perspectives: In 1866 the Koreans Successfully Knocked Out America's First Attempt to Invade the 'Hermit Kingdom,'" *Military History* (December 2006), pp. 19–21. The sinking of the former Confederate blockade runner (renamed the *General Sherman*) in 1866 later resulted in a battle for the *Citadel* (1871), which the Americans won, but then they abandoned their attempts to sail up the Han River and bombarded Seoul in the face of a determined Korean resistance.

Jasper Becker, *Rogue Regime: Kim Jong Il and the Looming Threat of North Korea* (New York: Oxford University Press, 2005). Still looming and still threatening. This work helps to put into perspective all the options the rest of the world doesn't have with regard to North Korea.

Clay Blair, *The Forgotten War* (New York: Times Books, 1987). Good history, with lots of detail. A straightforward, unvarnished account, albeit from a basically American perspective.

———, *A General's Life* (New York: Simon and Schuster, 1983). Good insights into General Omar Bradley's views of the American generals he fought with during the various wars.

James Brady, *The Coldest War: A Memoir of Korea* (New York: Pocket Books, 1990). Looking at the winter of 1951–1952 and how difficult it was to wage war in it. The author has written a number of books about the Korean War; some are fictional, some are factual. All give a good sense of what it was like to be in the Korean War. This one covers his tour of duty on the main line of resistance (MLR) near Hill 880 and the Soyang-gang River. Quoting Sergeant Wooten from World War II: "It ain't much but it's the only war we got." A long, bloody stalemate it was indeed.

———, *The Marines of Autumn* (New York: St. Martin's Press, 2000) (F). Gives the reader a real sense of the "Forgotten War" and the entrance of China into the war.

———, *The Scariest Place in the World: A Marine Returns to North Korea* (New York: St. Martin's Press, 2005). The Demilitarized Zone and why it remains one of the most dangerous places on earth in terms of a possible outbreak of nuclear war.

John R. Bruning Jr., *Crimson Sky: The Air Battle for Korea* (Dulles, VA: Brassey's, 1999). Plowing through masses and masses of unorganized documents, the author comes up with a compelling story of the air war. Interspersed with many personal stories, he captures the ebb and flow of the war, finally solving the "Soviet Riddle" with detailed accounts of the Soviet air war. Interestingly enough, on page 206, he puts paid to the 10:1 ratios claimed by the United States (and the Soviets who claimed in 1951 to have shot down 510 UN aircraft when the actual total was 40). The final MIG Alley totals, he believes, were 379 MIGs and 103 Sabres.

Thomas B. Buell, *Naval Leadership in Korea: The First Six Months* (Washington, DC: Naval Historical Center, 2002). How the US Navy saved the day and provided the logistics for the entire campaign. Of particular importance is the way the navy held back shipping capacity during the drive to the Yalu and thereby had the ships and manpower to evacuate all those who needed to be evacuated when the "hunt blew up."

Ian Buruma, "Kimworld: Inside the North Korean Slave State," *New Yorker*, August 22, 2005, pp. 64ff. Bizarre and tragic for most who live there. Even the liberal-leaning *New Yorker* finds this regime hard to stomach.

Charles M. Bussery, *Firefight at Yechon: Courage and Racism in the Korean War* (New York: Macmillan, 1991). The July 20, 1950, battle at Yechon, northwest of the Pusan Perimeter, is often cited as an example of the inability of black units—in this case the 24th Regimental Combat Team of the 25th Division—to fight. The author, who fought at Yechon and was highly decorated for bravery, criticizes such thinking. He cites, among other evidence, the fact that, during the entire war, the unit never lost a single artillery piece.

Tim Carew, *Korea: The Commonwealth at War* (London: Cassell, 1967). Breezy and conversational, this book puts the "Woolworth Brigade" (lacking tanks, medical supplies, engineers) into perspective and gives the reader a good sense of the actions of the First Argyll and Sunderland Highlanders, the First Middlesex Regiment, the Third Royal Australians, the Gloucestershires, the duke of Wellington, Princess Patricia's Light Infantry, the Eighth Hussars King's Own Scottish Borderers, Royal Ulster Rifles, and Royal Northumberland Fusiliers, especially at the battles of the Imjin River (April 1951) and the Hook (October 1952 to May 1953).

Brian Catchpole, *The Korean War* (London: Carroll and Graf, 2001). The best book on British and commonwealth efforts during the war and for actual descriptions of fighting the Chinese in their third-, fourth-, and fifth-phase offensives.

Jennie Chancey and William Forstchen (eds.), *Hot Shots: An Oral History of the Air Force Combat Pilots of the Korean War* (New York: William Morrow, 2000). A very personal description of the air war in Korea by

Americans who flew missions. Strong examples of unpreparedness on all levels. For example, some pilots had to fly wearing their college football helmets because no others were available.

Jung Chang and Ron Halliday, *Mao: The Untold Story* (New York: Alfred Knopf, 2005). A must-read for those interested in any aspect of Mao's career. It not only provides a revisionist account of his political and personal life, but it also gives very useful insights into the way China became a personal instrument for a single man.

Donald Chung, *The Three Day Promise* (Tallahassee, FL: Father and Son Press, 1989). South Koreans impressed into service as the Korean Augmentation to the United States Army (KATUSA). They were involved in the war for far longer than three days.

James Church, *Bamboo and Blood* (New York: St. Martin's Minotaur, 2009) (F). Inspector O continues in his small and quiet way to open up the North Korean regime to our eyes.

———, *A Corpse in the Koryo* (New York: St. Martin's Minotaur, 2006) (F). An inside look at North Korea through the eyes of a North Korean detective. A chilling view no matter how you read it.

———, *Hidden Moon* (New York: St. Martin's Minotaur, 2007) (F). Good follow-up. It's cold and dark in the "People's Paradise," and every little thing is worth something.

———, *The Man with the Baltic Stare* (New York: St. Martin's Minotaur, 2010) (F). Church's most politically ambitious novel to date, one in which the Chinese, the Russians, the Japanese, and the South Koreans all seem poised to carve out spheres of influence in the "changing" North Korea, thinking it is but a "corpse." But Church's central point (and that of Cummings reviewed later this chapter) is that North Koreans have had 70 years of believing in their regime and in the idea of its North Korea and that the regime has developed survival skills we can only imagine. In any case, the "immaculate omelet, made without breaking a single egg," is impossible, so one should be careful for what one wishes.

J. Lawton Collins, *War in Peacetime* (Boston: Houghton Mifflin, 1969). A general's perspective on fighting a "police action" while the other side is fighting a war. This book is useful to capture the frustration of the military men fighting a war with their civilian authorities, who are not sure what the goals are or who shifts those goals during the war itself.

Eugene Clark, *The Secrets of Inchon* (New York: G. P. Putnam, 2002). How the Inchon landings could have gone very wrong and how much luck and North Korean malfeasance were required in order for the Americans to be successful. American casualties were so light (200) as to be astonishing compared to what they might have been.

Bruce Cummings, *Korea's Place in the Sun* (New York: Norton, 2005). Emergence of two important regional powers. Covers the Korean War's impact and ramifications until the near-present.

———, *North Korea: Another Country* (New York: Dimensions, 2004). An intense look at North Korea by a well-known American professor, who provides a quite different perspective than most, preferring to give the rulers of North Korea every benefit of the doubt: "It is their country, for better or worse—another country." Quite positive in spots and upbeat wherever possible, claiming "only" 500,000 (not 2 million) died in famine and 120,000 in the gulags (not 500,000). However, he seems more realistic than most in asserting that North Korea will not simply collapse, a view echoed by Church (see the review of Church, *The Man with the Baltic Stare*, previously in this chapter).

———, *The Origins of the Korean War: The Roaring of the Cataract, 1947–1950* (New York: Dimensions, 1990). World War II left a strange legacy that could not be resolved short of war, and Cummings asks why. All three of Cummings's books are quite good at presenting the events leading up to the war, its conduct, and its aftermath from the point of view of North Korea.

William Dannenmaier, *We Were Innocents* (Carbondale: University of Illinois Press, 1999). A first-person account of one man's experience in Korea, featuring touching and poignant letters to his sister and parents describing the progress of the war.

Jurgen Domes, *Peng Te-huai: The Man and the Image* (London: C. Hurst, 1984). General Peng's life and career is covered, with some interesting material from his time during the Korean War. Most interesting is his earlier career as a counterinsurgent commander dealing with the insurgencies among the Turkmen, Uigaurs, Kazakhs, Kirgizs, and Uzbeks during the latter portions of 1949. Makes it clear that the fourth- and fifth-phase Chinese offensives did not fail because of poor generalship but because of Mao's overly ambiguous insistence on frontal assaults.

Jim Dorschner, "Douglas MacArthur's Last Triumph," *Military History* (September 2005), pp. 31–36. MacArthur had a great strategic vision and. by force of will and cobbling together various units and capabilities, pulled off a major victory but had no real plan for success thereafter.

Bob Drury and Tom Clavin, *The Last Stand of Fox Company* (New York: Atlantic Monthly Press, 2009). Another aspect of the Marines' historic fight at Chosin Reservoir, with insights from the survivors.

Robert Dvorchak, *Battle for Korea: A History of the Korean Conflict*, 50th anniversary ed. (Conshohocken, PA: Combined, 2000). This oversized, lavishly illustrated volume is easy to read and its images hard to forget. An outstanding introduction to the Korean War and one used by several generations of Bowdoin students, including a number of Korean and Korean-American ones.

Nicholas Eberstadt, *The End of North Korea* (Washington, DC: AEI Press, 1999). Eberstadt was, it would now appear, a tad premature in his judgments because, more than a decade later, North Korea is still functioning.

Paul M. Edwards, *The A to Z of the Korean War* (Lanham, MD: Scarecrow Press, 2005). A lot of useful information here in an easy-to-read format.

Kenneth Estes, *Into the Breach at Pusan: The 1st Provisional Marine Brigade in the Korean War* (Norman: University of Oklahoma Press, 2012). A most useful and engaging study, finally putting to rest some of the biggest myths and exaggerations of the battle for the Pusan Perimeter. Some of the debunking includes:

1. The Marine Provisional Brigade was not the only "fire brigade"; there were four others.
2. The operational strength of the brigade did not drop below 90 percent.
3. Marine aviation supported the Army and ROK units, not simply the Marines.
4. General Walton Walker was a superb commander, and the Eighth Army under his direction saved itself.
5. The Eighth Army achieved manpower parity with the North Korean forces around the perimeter by July 22 and significant numerical superiority by the end of August.

The author also skillfully weaves in an account of the Marine counterattacks at Masan and in the Naktong Bulge and points out that the M-26 Pershing tank developed at the end of World War II was put in service during the beginning of the Marines' deployment, and it proved highly useful. This work also contains some excellent, previously unseen photographs. A worthwhile addition to the extensive literature on the Korean War.

T. R. Fehrenbach, *This Kind of War: The Classic Korean War History* (London: Brassey's, 1963). Written from the American point of view but with lots of local color and insights.

Rosemary Foot, *A Substitute for Victory: The Politics of Peacemaking at the Korean Armistice Talks* (Ithaca, NY: Cornell University Press, 1990). The 2 years of negotiation saw 45 percent of total US casualties during the Korean War and the massive destruction of North Korea (635,000 tons of bombs compared with 503,000 against Japan during World War II).

Joe Foss and Matthew Brennan, *Top Guns* (New York: Star Books, 1991). The air war in Korea with all the bravado of fighter jocks. Students should see alternative sources on the air war from the perspective of China and the Soviet Union.

Lloyd Gardner (ed.), *The Korean War* (New York: Quadrangle Books, 1972). A series of essays probe the concept of limited war. John Foster Dulles, Hanson Baldwin, James Reston, George Kennan, and others weigh in.

D. M. Giangreco, *War in Korea: 1950–1953* (Novato, CA: Presidio Press, 2000). A lavishly illustrated history of the war—would be a fine replacement for Dvorchak, if one were ever necessary.

Donald Goldstein and Harry Maihafer, *The Korean War: The Story and the Photographs* (Washington, DC: Brassey's, 2000). Breezy and liberally illustrated, though many of the photos are smaller than in Giangreco and Dvorchak.

Sergei Goncharov, John W. Lewis, and Xue Litai, *Uncertain Partners: Stalin, Mao and the Korean War* (Stanford, CA: Stanford University Press, 1993). An outstanding portrayal of this complex and often bitter relationship that affected world history.

Allan Goodman (ed.), *Admiral C. Turner Joy, Negotiating while Fighting: The Diary of the Korean Armistice Conference* (Stanford, CA: Hoover Institute Press, 1978). The frustration of dealing with the North Koreans and Chinese by a man who spent years trying at Kaesong and Panmunjom and who found it incredibly aggravating.

Joseph C. Goulden, *Korea: The Untold Story of the War* (New York: Times Books, 1982). A long, rambling account but one with a number of memorable nuggets, such as W. Averill Harriman calling it a "sour little war" and Dean Acheson commenting on MacArthur's wild claims about saving the Eighth Army by sending it into the Chinese trap when he did. The most humorous episode for me was when American POWs, facing Chinese charges of "germ warfare," killed a mouse, attached it to a parachute labeled "US Air Force," and threw it over the prison wire. Chinese guards first fled the mouse but later put it in a bottle and added it to their museum Imperialistic American Bacteriological Warfare. Good on US players, weak on Chinese sources, and overly reliable on Khrushchev's memoirs for Stalin's role.

David Green, *Captured at the Imjin River: The Korean War Memoirs of a Gloster 1950–1953* (South Yorkshire, UK: Leo Cooper, 2003). A little-known memoir by a British soldier who was captured by the Chinese. He describes a much nicer POW experience than many others, especially Americans and ROKs.

W. E. B. Griffin, *Retreat, Hell!* (New York: Putnam, 2004). Gung-ho Marines fight hard, and Griffin tells the world all about it with pride and relish.

Russell Gugeler, *Combat Operations in Korea* (Washington, DC: Center of Military History, US Army, 1987). Detailed and textually granular. Descriptions of small-scale engagements with discussion notes following each section as teaching tools for future warriors. Examples from the Pusan Perimeter, Twin Tunnels, Chipyong-ni, Bunker Hill (Hill 800), and Heartbreak Ridge. Powerful lessons painfully learned.

Chen Guidi and Wu Chuntao, *Will the Boat Sink the Water?* (New York: Public Affairs, 2006). A blistering and very negative account of the plight of the Chinese peasants before, during, and after the Korean War.

Bae Ho Hahn and Chae-Jin Lee, *The Korean Peninsula and the Major Powers* (Seoul: Sejong Institute, 1998). Many useful perspectives from the Korean point of view.

David Halberstam, *The Coldest War* (New York: Random House, 2007). Not much new here but well-written and interestingly presented. I have always been harsh toward MacArthur's misjudgments during the Korean War, but Halberstam's portrait is truly drenched in acid.

———, *The Fifties* (New York: Random House, 1993). Lots of the social, economic, and political flavor from the time of the Korean War in the United States.

Kenneth Hamburger, *Leadership in the Crucible: The Korean War Battles of Twin Tunnels and Chipyong-ni* (College Station: Texas A&M Press, 2010). An in-depth look at the February 1951 battles, which, along with the third battle of Wonju, represented the third Chinese culminating point of the Korean War. Very detailed.

Eric Hammel, *Chosin: Heroic Ordeal of the Korean War* (Novato, CA: Presidio Press, 1981). Lively and well-researched account with many personal stories. Sometimes these writers forget that the Chinese actually won the battle and drove the Marines and US Army from North Korea.

T. X. Hammes, *Forgotten Warriors: The 1st Provisional Marine Brigade, the Corps Ethos, and the Korean War* (Lawrence: University of Kansas Press, 2010). A closely argued work that captures the reasons the First Marines were so good in battle—their training, their ethos, and their close workings with the Marine Air Wing. They were the first American unit during the Korean War to drive the North Koreans back (at Sanchon west of Masan) and crucial to the First and Second Battles of the Naktong Bulge. They were later melded into the First Marine Division that landed at Inchon.

Max Hastings, *The Korean War* (New York: Simon and Schuster, 1987). The British take on the war and one of the best introductions to it. Hastings is a superb writer and allows the reader to feel the intricacies and turbulence of warfare.

Francis Heller (ed.), *The Korean War: A 25 Year Perspective* (Lawrence: Regents Press of Kansas, 1977). Not surprisingly, Truman was conflicted about the nature of the war, as his liberal and conservative sides warred. Certainly his impulsive personality played an important role in the initial US involvement.

Walter Hermes, *Truce Tent and Fighting Front* (Washington, DC: US Government Printing Office, 1966). How the negotiations dragged on and on and on while both sides suffered tremendous casualties, by many accounts more in total than before the talks began.

Michael Hickey, *The Korean War: The West Confronts Communism* (Woodstock, NY: Overlook Press, 2000). A British overview, quite good on commonwealth issues; the author actually fought in the Korean War.

William Holinger, *The Fence-Walker* (Albany: Sate University of New York Press, 1984) (F). Prizewinning (who gives these out?), the first novel of young man, an LT, patrolling the Korean Demilitarized Zone in 1966 during the Vietnam War. Boredom, lots of time with Korean prostitutes (one of whom he leaves with child and no child support), and patrolling—ends up finding a dead North Korean and engages in a firefight in a tunnel. Seems disillusioned at the end of his tour.

Edwin Hoyt, *The Pusan Perimeter: Korea, 1950* (New York: Stein and Day, 1984). Carefully presented and unsparing in its criticism of the folly of early US involvement. Some little-known facts emerge, such as the South Korean navy's PC 701's sinking of a North Korean 1,000-ton steamer with 600 troops aboard headed to Pusan to shut down that port in June 1950. Credits the US Marines with stopping the NK 6th Division near Masan and the NK 4th Division in the Naktong Bulge, while the ROK 6th and 8th Divisions get credit for stopping the NK 8th and 15th in front of Taegu. Well worth a read for appreciating what a close-run thing it really was.

Robert Jackson, *Air War over Korea* (New York: Charles Scribner's Sons, 1973). Quite a good account, not only of the battles in MIG Alley and the efforts against railroads and dams in North Korea, but also surprisingly revealing about the Communists' June 1951–February 1952 bid for air superiority, featuring 525 MIGs led by Soviet and Soviet-bloc "honchos." Points out how the air war eventually led to the development of the U-2, F-104, and MIG 21.

D. Clayton James, *Refighting the Last War: Command and Crisis in Korea 1950–1953* (New York: Free Press, 1993). Focuses on the US commanders and their interactions with the rest of the government.

Charles Jenkins, *The Reluctant Communist: My Desertion, Court-Martial, and Forty-Year Imprisonment in North Korea* (Berkeley: University of California Press, 2008). This soldier, worried about being sent to Vietnam while serving on the Demilitarized Zone in Korea in 1964, decided that going to North Korea would be a better idea. Treated somewhat better than the average North Korean and allowed to teach English and study, he was even given a kidnapped Japanese wife, who eventually was his passport to Japan. Calls North Korea a "giant prison," albeit not while he was there.

Chen Jian, *China's Road to the Korean War* (New York: Columbia University Press, 1994). A classic. Very good about giving the reader a sense of the dynamics behind Chinese decision making. Well-researched and quite an advance over other works dealing with this subject when it first appeared.

Gabriel Jonsson, *Towards Korean Reconciliation* (Burlington, VT: Ashgate, 2006). Some hope for the future, not likely to be realized anytime soon.

C. Turner Joy, *How Communists Negotiate* (New York: Macmillan, 1955). How Communists try to get at the conference table what they couldn't get on the battlefield and vice versa. His long, long negotiating efforts taught him that what the Communists needed was "more steel and less silk."

Burton I. Kaufman, *The Korean War: Challenges in Crisis, Credibility and Command* (Philadelphia: Temple University Press, 1986). Politicians need to have a more sympathetic understanding of the military implications of fighting a war, claims the author.

Richard Knott, *Attack from the Sky: Naval Air Operations in the Korean War* (Washington, DC: Naval Historical Center, 2004). A very important aspect of the war, one that doesn't always get the proper amount of attention it deserves. A stunning number of sorties were flown.

Donald Knox, *The Korean War: Pusan to Chosin* (New York: Harcourt, Brace, Jovanovich, 1988). An oral history, with many good stories of the war. It provides a host of observations for those who were there. Their reactions and recollections are very vivid and illuminating.

Jon Latimer, "Holding the Hook," *Military History* (May 2005), pp. 59ff. Interesting account from the "forgotten phase of the forgotten war," about the British units (among them the Royal Fusiliers, the Duke of Wellington's Regiment, and the Black Watch) holding one of the many hills along the battlefront in 1953 against seemingly relentless Chinese shelling and attacks.

Robert Leckie, *Conflict: The History of the Korean War* (New York: Da Capo Press, 1962). This military historian writes vividly and with spirit. Photographs and stories are intermixed with the narrative.

Xiaobing Li, Allan Millett, and Bin Yu, *Mao's Generals Remember Korea* (Lawrence: University of Kansas Press, 2001). Worthwhile reading that gives a sense of the Chinese perspective, with and without propaganda.

Eric Linklater, *Our Men in Korea* (London: Her Majesty's Stationery Office, 1952). One of the first books I ever read about the Korean War. It covers the British forces in action and is quite candid.

Peter Lowe, *The Origins of the Korean War* (London: Longman, 1997). Sees it as a "war of miscalculations" and blames both the American and the Chinese interventions for being "dangerous to world peace."

Douglas MacArthur, *Reminiscences* (New York: McGraw-Hill, 1964). The old soldier tries to set the record straight. Remarkably free of bombast and invective—quite statesmanlike, perhaps surprisingly.

S. P. Mackenzie, *The Imjin and Kapyong Battles: Korea, 1951* (Bloomington: Indiana University Press, 2013). An exhaustive look at the blunting of the Chinese fifth-phase offensive north of Seoul in April 1951. The Chinese turned out to be a "well trained, clever, cunning opponent." The Glorious Glosters were overrun on the Imjin, but the Australians and Canadians were "luckier" at Kapyong, writes the author.

Harry J. Maihafer, *From the Hudson to the Yalu: West Point '49 in the Korean War* (College Station: Texas A&M Press, 1993). Of 574 graduates of the class of 1949, half would fight in Korea, quite a few as pilots. Maihafer was with the 24th Division from the Pusan Perimeter to the drive north of Pyongyang through the later battle of Matthew Ridgeway. Wounded and the recipient of the Silver Star, the author weaves together memories of West Point with the exploits of his classmates in a variety of situations. A class act.

Ben S. Malcom, *White Tigers: My Secret War in North Korea* (London: Brassey's, 1996). An astonishing account of the US and South Korean intelligence-gathering and sabotage operation in 1952 and 1953, operating from the island of Paengnyong-do, 125 miles northwest of Seoul. A forerunner of the Special Forces missions, this one involved at the time of the armistice more than 22,000 Koreans. Their status became a bureaucratic challenge for both the American and the South Korean militaries. Tragically, all the learning from this operation was ignored when the Vietnam War began to involve Americans more and more.

Note: One would think that the works of Mao Tse-tung, such as his *The Art of War* (El Paso, TX: El Norte Press, 2005) would be mentioned here. It is not, for the simple reason that, during the Korean War, there were virtually no copies of his books available in the United States—the first English translations didn't appear until 1954! It might have been a good idea for the US military to have gotten copies while the war was in progress.

Carter Malkasian, *The Korean War 1950–1953* (Oxford, UK: Osprey, 2001). How a "war of maneuver" became a "war of attrition," in the process "avoiding a Third World War." Useful maps.

S. L. A. Marshall, *Pork Chop Hill* (Nashville, TN: Battery Press, 1954). A must-read. Classic for understanding any war, for the moral imperatives are enduring. Over the years, students have responded very positively to its lessons and imperatives. Unfortunately, there is unnecessary focus on the African American malcontent.

Bradley Martin, *Under the Loving Care of the Fatherly Leader* (New York: Thomas Dunne, 2006). An incisive account of North Korea, underscoring what a strange world it truly is. The Fatherly Leader seems neither loving nor caring.

MHQ: The Quarterly Journal of Military History, "Special Issue: Korean War 50th Anniversary" (September 2000). A valuable set of articles in this retrospective, some of which provide new information.

James A. Michener, *The Bridges at Toko-Ri* (New York: Ballantine Books, 1953) (F). Sad account of naval aviation and some heroics that don't pan out. Made into a depressing movie. Duty, honor, and country do not always end up with happiness.

Allan R. Millett, *The War for Korea, 1945–1950*, vol. 1, *A House Burning* (Lawrence: University of Kansas Press, 2005). An outstanding effort to put the outbreak of the war in its true historical perspective. A classic.

———, *The War for Korea, 1950–1951*, vol. 2, *They Came from the North* (Lawrence: University of Kansas Press, 2010). One of the very best books on the tactical and strategic aspects of the war on the ground and in the air. Good material on General Peng and General Ridgeway.

Malcolm Muir Jr., *Sea Power on Call: Fleet Operations June 1951–July 1953* (Washington, DC: Naval Historical Center, 2005). The US Navy played a huge role in the war and Muir, tells that story in a workmanlike fashion. The service seldom gets the credit it deserves.

Bernard C. Nalty, *Long Passage to Korea: Black Sailors and the Integration of the U.S. Navy* (Washington, DC: Naval Historical Center, 2003). A sad story with a happy ending. The US Navy was the most resistant of any of the services to having African Americans treated equally and fairly.

———, *Outpost War: U.S. Marines from the Nevada Battles to the Armistice* (Washington, DC: US Marine Corps Historical Center, 2002). The forgotten phase of the forgotten war is examined.

———, *Stalemate: U.S. Marines from Bunker Hill to the Hook* (Washington, DC: US Marine Corps Historical Center, 2001). Some background information. Not too inspiring, actually.

Kongdan Oh and Ralph Hassig, *North Korea through the Looking Glass* (Washington, DC: Brookings Institution Press, 2000). It does seem, to most outsiders, a strange and bizarre world.

Joseph R. Owen, *Colder than Hell: A Marine Rifle Company at Chosin Reservoir* (New York: Ivy Books, 1996). Frozen Chosin, up close and personal. Amazing courage shown by the Chinese and the Americans. Few Korean soldiers involved on either side during this campaign. Follows a Marine mortar company under a green LT across Korea to the Chosin Reservoir. Really needed some maps to clarify the group's movements and progress, especially during the run-up to the Chosin campaign.

Paik Sun Yup, *From Pusan to Panmunjom* (New York: Brassey's, 1967). Perhaps the best volume in English by a prominent South Korean general—a must-read. Lots of good detail and insights and quite a balancing act between criticizing American actions and appreciating their being there in the first place.

Peng Dehuai, *Memoir of a Chinese Marshal: The Autobiographical Notes of Peng Dehuai, 1898–1976* (Beijing: Foreign Languages Press, 1984). From peasant to Chinese marshal, Peng rose with hard work, a knack for fighting, and learning about warfare and ideological flexibility. Next to Matthew Ridgeway, he was probably the most accomplished general of the war. Chapter 14, "The War to Resist U.S. Aggression and Aid Korea, October 1950–July 1953," is useful in its candor, as Peng quite rightly blames Chinese failure on a lack of air power, long supply lines, and huge casualties—as well as Mao's overreaching dictums—and singles out the amazing resilience of the Chinese war-fighting doctrine that allowed them to achieve a stalemate in Korea. Horrified to see himself declared a "Rightist" and a member of the "Military Club," he refused to recant and would later lose his life for his honesty.

James Pocock, "Task Force Smith Artillery Commander," *Military History* (July/August 2006), pp. 32–38. A soldier who was in Task Force Smith argues against the oft-repeated calumny that the unit did not fight well and was completely unprepared for action in Korea.

William (Bill) Richardson, *Valleys of Death: A Memoir of the Korean War* (New York: Berkley Caliber, 2010). A sergeant in the First Cavalry fights in the Pusan Perimeter; liberates Pyongyang; gets ambushed at Usan on November 2, 1950; and is taken prisoner. Much on life in Chinese and North Korean prison camps, where he survived for 34 months.

General Matthew Ridgeway, *Soldier: The Memoirs of Matthew B. Ridgeway* (New York: Harper and Brothers, 1956). Straightforward and direct in the tradition of Ulysses S. Grant; even a bit dull in spots.

Patrick Roe, *The Dragon Strikes* (Novato, CA: Presidio Press, 2000). Chinese aims and the processes by which Mao moved the country into war. What Mao hoped to achieve and why he didn't succeed.

Martin Russ, *Breakout: The Chosin Reservoir Campaign, Korea 1950* (New York: Fromm International, 1999). An hour-by-hour account of the First Marine Division and its historic breakout. Conveys a good sense of the struggles of Americans, Chinese, and British fighting in 20-degree weather.

———, *The Last Parallel* (New York: Fromm International, 1957). The battle lines are still drawn in Korea and why that should concern everyone.

Andrew Salmon, *To the Last Round: The Epic British Stand on the Imjin River, Korea April, 1951* (London: Aurum, 2009). The 29th Infantry Brigade [1st battalions of the Royal Northumberland Fusiliers (1647), the Gloster Regiment (1694) ,and the Royal Ulster Rifles (1793)], along with a Belgian battalion, makes an effort to hold position on Line Kansas; but after holding up the Chinese in a valiant struggle, they eventually retire to Line Golden and help to save Seoul from the 300,000-man-strong Chinese fifth-phase offensive. The Gloster battalion was wiped out in the process. Good description of prison conditions in the chapter "In Enemy Hands: The Camps."

Stanley Sandler (ed.), *The Korean War: An Encyclopedia* (New York: Garland, 1995). Extremely useful work, especially detailed on order of battle of various units, such as the Fourth NKA Division, the "Kim Chaek, Seoul Guards Division."

———, *The Korean War: No Victors, No Vanquished* (Lexington: University of Kentucky Press, 1999). But a useful tie? Sometimes stalemate is the consummation devotedly to be wished.

Choe Sang-Hun, "Unearthing War's Horrors after Years in South Korea," *New York Times*, December 3, 2007, p. A9. A savage indictment of US soldiers killing South Korean refugees along with North Korean soldiers, some in uniform, some not.

James Schnabel, *United States Army in the Korean War* (Washington, DC: Center of Military History, 1992). Covers the role of the army with an official cast. Many useful insights from that perspective.

Shu Guang Zhang, *Mao's Military Romanticism: China and the Korean War 1950–1953* (Lawrence: University Presses of Kansas, 1995). A very useful look at the way the Chinese made war during this period and the extent to which Mao's often muddled and romantic notions cost the Chinese 390,000 casualties, including 148,000 dead. His refusal to understand American air power, his forces' logistical requirements, and the rudiments of Clausewitzian strategy cost the Chinese dearly, even though the Korean War effort did vaunt Chinese power and prestige globally. Points out that the Chinese commitment of troops lasted until 1958. Especially good on the Chinese third-, fourth-, and fifth-phase offenses and how they came to naught following the "nothing fails like success" dictum, as Mao was convinced that the United States was going to evacuate the peninsula.

Edwin H. Simmons, *Dog Company Six* (Annapolis, MD: Naval Institute Press, 2000). Grim fighting recorded by eyewitnesses. Captures the feeling and tensions of small-infantry formations in mortal combat.

Bill Sloan, *The Darkest Summer: Pusan and Inchon 1950* (New York: Simon and Schuster, 2009). Captures the spirit of those desperate times, when North Korean troops were asked to charge up hills many times and Americans and South Koreans were asked to charge back up them after the North Koreans had prevailed. Gives the reader a good sense of the fragility of the Eighth Army's position.

John Spanier, *The Truman-MacArthur Controversy and the Korean War* (Cambridge, MA: Harvard University Press, 1959). Spanier concludes, "Clausewitzian realism and American liberalism combine to create a limited war strategy that went against the deeply held American attitude that one must fight a war to total victory." Gives a fair account of both sides in the conflict and asks the question, "Would the US have accepted assurances that a Russian army advancing through Mexico would stop at the Rio Grande?"

Elizabeth A. Stanley, *Paths to Peace: Domestic Coalition Shifts, War Termination and the Korean War* (Stanford, CA: Stanford University Press, 2009). Very valuable insights here about how wars in general end and in particular what factors finally brought the Korean War to a close, albeit not to a peace treaty.

Shelby L. Stanton, *Ten Corps in Korea 1950* (Novato, CA: Presidio Press, 1996). An excellent overview of the controversial X Corps. Gives a fair presentation of the pros and cons of its leader, General Ned Almond, a commander who came under great fire for being a sycophant to MacArthur but one who was nevertheless kept on by his replacement, General Ridgeway. A must-read for the subject.

Richard W. Stewart, *Staff Operations: The X Corps in Korea, December, 1950* (Ft. Leavenworth, KS: Combat Studies Institute, 1991). The X Corps extricates itself from northeast Korea and from the Chinese offensive; a close look at the command decisions.

Norman Stone, *The Atlantic and Its Enemies: A History of the Cold War* (New York: Basic Books, 2010). Good background on the history of the Cold War and the underlying assumptions on the part of the United States and the Soviet Union that affected its course. Not much on the Korean War per se but provides much-needed context for that war in the greater scheme of things.

Stories of the Chinese People's Volunteers (Beijing: People's Literature Publishing House, 1960). A follow-up volume accusing the "American butchers" of "wholesale massacres" designed to "exterminate" the Korean people.

William Stueck, *The Korean War: An International History* (Princeton, NJ: Princeton University Press, 1995). One of the best and most helpful views of the Korean War in its international context.

Jae-Jung Suh, *Power, Interest and Identity in Military Alliances* (New York: Macmillan, 2007). A very detailed and incisive analysis of US–South Korean interactions and their mutual and differing interests. It suggests that South Korea was stronger militarily than many believe.

Colonel Harry G. Summers Jr., "The Korean War: A Fresh Perspective" *Military History* (April 1996), pp. 22–28. Also in this issue are other articles about the Korean War, including the 1871 incident in which US forces were repulsed.

Jay Taylor, *The Generalissimo: Chiang Kai-shek and the Struggle for Modern China* (Cambridge, MA: Harvard University Press, 2009). Of considerable significance for the Korean War are pages 435–69, as the author looks at that war from the point of view of Chiang Kai-shek, for whom the "wonderful news for everyone on the island was that America would prevent the PLA's invasion of Taiwan." True to form, MacArthur flew to Taiwan on July 31 (without US State Department knowledge) and promised to aid the Nationalists. He even sent 6 jet aircraft, the first Chiang had ever seen, but the Defense Department quickly sent them back to Japan and later refused to credit Chiang's plan to land 600,000 troops on the mainland with US logistical and air support.

John Taylor, *An American Soldier: The Wars of General Maxwell Taylor* (Novato, CA: Presidio, 1989). Maxwell Taylor's son gives a good account of his father's life and service to his country. Of interest are the pages (172–93) on his Korean War service from February 1953 until early 1954.

Addison Terry, *The Battle for Pusan* (Novato, CA: Presidio Press, 2000). Much-needed attention is given to this pivotal struggle that enabled the United States to eventually liberate South Korea.

Nigel Thomas and Peter Abbott, *The Korean War 1950–1953* (London: Osprey, 1986). Uniforms and arms of the belligerents are presented, along with some interesting commentary.

John Tolland, *In Mortal Combat: Korea, 1950 to 1953* (New York: William Morrow, 1991). A useful overview, now somewhat dated by new information, with very much a pro-American perspective.

Bernard E. Trainor, "Worst-Case Scenario: Suppose North Korea Starts a War," *Boston Globe*, Saturday, May 24, 1997, p. A11. Frightening. Trainor sees how North Korea could destroy the South before itself was destroyed. Little has changed in that equation in the last 20 years, so that scenario is not really dated.

Harry S. Truman, *Memoirs*, vol. 2, *Years of Trial and Hope* (Garden City, NY: Doubleday, 1956). Truman speaks plainly and well.

US Army, *Korea: 1950* (Washington, DC: US Army Center of Military History, 1989). Another look at Korea by the US military. A straightforward account.

Michael Varhola, *Fire and Ice: The Korean War 1950–1953* (Mason City, IA: Savas, 2000). A mini-encyclopedia covering the war amd containing a number of interesting factoids.

A Volunteer Soldier (Beijing: People's Literature Publishing House, 1956). A marvelous blend of propaganda and insights into how China framed the Korean War.

Ansil L. Walker, "Eighth Army Ordeal in Korea," *Military History* (December 1998), pp. 34–40, 97. Provides a very good sense of the biggest "bug out" in US military history and the skill with which the Chinese brought it about. Unfortunately for them, the Eighth Army could retreat faster in trucks and trains than they could follow on foot.

James Webb, *The Emperor's General* (New York: Bantam Books, 2000) (F). A novel that gives a lot of history with its drama. Shows MacArthur in a very harsh spotlight, especially his record in the Philippines before, during, and after the war.

Stanley Weintraub, *MacArthur's War* (New York: Free Press, 2000). It was, but only in part. Really, it was Truman's war, for without his personal and direct action, it would have been over before it began. Whether that would have been for good or for evil, we leave to posterity, but Mr. Weintraub did not think of that dimension.

Allen S. Whiting, *China Crosses the Yalu: The Decision to Enter the Korean War* (Stanford, CA: Stanford University Press, 1960). Great maps on pages 67, 120, and 121, showing the movement of Chinese armies beginning in July 1950 and culminating in October 1950. Good background on why Mao wanted the Korean War and how well it served his many purposes.

Courtney Whitney, *MacArthur: His Rendezvous with History* (New York: Alfred A. Knopf, 1956). Aide, confidante, and friend to the General, Whitney paints a most sympathetic portrait and a good balance to some of the other, more overly harsh assessments of him. Least persuasive when handling the debacle in northern North Korea during October and November 1950: "In less than a month, MacArthur had reached up, sprung the Red Trap and escaped it." Rightly blames "Washington" (both civilian and military) for many contradictory orders and much wishful thinking, however.

Jim Wilson, *Retreat, Hell!* (New York: Pocket Books, 1988). The "Frozen Chosin" campaign in great detail, some parts do an almost hour-by-hour, day-by-day description, sometimes down to the squad level.

Xiaoming Zhang, *Red Wings over the Yalu* (College Station: Texas A&M University Press, 2002). Looking at the Chinese, Russian, and North Korean air efforts. Many mysteries unraveled here.

Chapter Thirty-One

Vietnam

People's War, Long War

When studying the wars in Vietnam, especially keep in mind the wisdom of T. E. Lawrence in *Seven Pillars of Wisdom*, when he said, "To make war upon rebellion is messy and slow, like eating soup with a knife." Important topics to be explored in this section include: making war in a democracy, making a long war in a democracy, civilian–military relations, warcraft by tactics or strategy, escalation as a self-limiting hypothesis, international civil wars, and above all the dangers of ignoring the dictates of Clausewitz (as did the United States) and the desirability of embracing them (as did North Vietnam).

Note: Ken Lopez, a Massachusetts bookseller, once listed 3,500 titles dealing with Vietnam, so please make sure to realize that the relatively few works cited here (and in chapter 40) are but a very modest beginning to understanding the nuanced and multidimensional and very complex war, a war that unfortunately is often dismissed by many in a sentence or two as a "mistake" or a "well-deserved victory."

Note: For the United States, there were two major epochs in the war. From 1965 until 1968, the doctrine of "search and destroy" held sway when US forces sought battle in order to inflict maximum damage on enemy combatants. Then from 1968 until 1972, the main effort was to "clear and hold" to deny the North Vietnamese Army (NVA) and Viet Cong (VC) access to the population of South Vietnam, while from 1972 until 1975, the burden for that task increasingly fell to the forces of South Vietnam.

For the North Vietnamese (see Broylus reviewed in chapter 38), there were four major stages: 1959–1965, in what they called "special warfare" (i.e., guerrilla warfare). Then came 1965–1968, "local warfare" [i.e., NVA regulars fighting the United States and VC fighting the Army of the Republic of Vietnam (ARVN)], followed by 1968–1973, which was dominated by a strategy of "fight and talk," and ending with 1975's "final offensive."

I believe that one way to begin to understand the complexity of the Vietnam War from a military perspective is to look at the different dimensions of it based on who was doing the fighting and where. For example, there were occasions where US forces battled NVA regulars in such situations as Khy San, the Ia Drang, or A Shau Valleys, with few if any ARVN or VC participants or many Vietnamese civilians present (dimension A).

At the other end of the battle spectrum were the struggles primarily between the South Vietnamese villager-manned Popular Forces, with American advisors, and their opponents, the VC, often with NVA advisors. In many parts of the Mekong Delta, these two fought it out largely without major regular American or North Vietnamese forces. This war on a microlevel took place primarily among Vietnamese civilians (dimension B).

In addition to these typologies of battles, there were many, many situations where regular American forces with regular South Vietnamese forces battled mainline VC forces and North Vietnamese regulars, often with civilians right in the middle (dimension C). Later (from 1972 on), there would be extremely important battles, primarily between NVA regulars and ARVN regulars, sometimes without too many civilians but often with many.

For dimension A, I recommend starting with Harold Moore and Joseph Galloway, *We Were Soldiers Once—and Young* (New York: Random House, 1992), and Robert Pisor, *End of the Line: The Siege of Khe Sanh* (New York: W. W. Norton, 1982), as well as John Del Vecchio, *The 13th Valley* (New York: Bantam Books, 1982); for dimension B, start with Bing West, *The Village* (New York: Harper and Row, 1972); for dimension C, start with James Willbanks, *The Battle of An Loc* (Bloomington: Indiana University Press, 2005) (with American air and artillery support) and George J. Veith, *Black April* (New York: Encounter Books, 2012) (without American support).

Dale Andrade, *America's Last Vietnam Battle: Halting Hanoi's 1972 Easter Offensive* (Lawrence: University of Kansas Press, 2001). Important to read him "backward"; that is, see why the South Vietnamese soundly defeated the NVA in the battles for Hue, Kontum, An Loc (Dien Bien Phu replayed but with South Vietnam the winner), and Binh Long and during their Easter Offensive of 1972, not why they failed to stop the NVA three years later having lost US air support, artillery support, and indeed weapons and ammunition. A superb read and very important book about a battle that tells us so much about winning and losing a war. See also the Fleming article described later this chapter.

Christian Apy, *Patriots: The Vietnam War Remembered from All Sides* (New York: Penguin, 2003). A terrible war from just about everybody's point of view. At least the North Vietnamese could say their agony had not been in vain. But an important notion explored: namely you don't have to be on the winning—or losing—side to have given your all.

Dwight Birdwell and Keith William Nolan, *A Hundred Miles of Bad Road: An Armored Cavalry Man in Vietnam, 1967–1968* (Novato, CA: Presidio Press, 1972). Serving on tanks before the war was either irrevocably lost or won, the author plied the road from Saigon through Chu Chi to Tay Ninh near the Cambodian border. Vivid contrasts between life in the rear and life in combat and between services: "Our mess hall is open twenty-four hours a day," said the air force guy.

Tom Campbell, *The Old Man's Trail* (Annapolis, MD: Naval Institute Press, 1995) (F). Outstanding novel by an American but, strangely enough, from the perspective of the Vietnamese side of the war. Much of the description seems spot-on.

Philip Caputo, *A Rumor of War* (New York: Henry Holt, 1977). A gung-ho Marine goes to Vietnam in 1965 and to Danang. "Ambushes are murder and murder is fun" soon becomes less fun as first the VC and then the NVA inflict continuing casualties and refuse to fight set-piece battles. He gradually turns against the war and declares at the end, "We had survived and that was our only victory." Justly reviewed as powerful and honest but not necessarily accurate on the conduct of the total war, especially the 1968–1972 period, although the author arrives as a newsman just as South Vietnam is falling in 1975, missing at both ends the high point of the war from the South Vietnamese and American points of view.

William Colby, *Lost Victory* (Chicago: Contemporary Books, 1989). One of the principle architects of the Vietnam War on the ground makes his case for the might-have-beens. His defense of Operation Phoenix rings true, albeit unpalatable, for many. This is one of a number of revisionist works on Vietnam that need to be sampled by interested students if only to challenge the conventional narratives of the war.

Note: I found a most useful companion to the Colby book in Randall B. Woods, *Shadow Warrior: William Egan Colby and the CIA* (New York: Basic Books, 2013). Woods provides a needed overview of Colby's interaction with Vietnam and the competing strategic decision makers. However, it is always important to remember that the "enemy gets a vote," and there is little countervailing description of what was going on in Hanoi during the comparable period.

Gregory A. Daddis, *No Sure Victory: Measuring U.S. Army Effectiveness and Progress in the Vietnam War* (London: Oxford University Press, 2011). Carefully examines the process by which the Pentagon tried to measure success or failure in Vietnam, concluding that the metrics used (such as body counts or tons of rice seized) were not as accurate as some not used (as to how many village leaders slept in the villages they claimed were "secure"). The author concludes that the Military Assistance Command, Vietnam (MACV) employed and collected vast amounts of statistics, but improper analysis of those statistics and "intellectual rigidity" hurt the enterprise and observes rather wryly that "there is more to winning than counting."

Philip Davidson, *Secrets of the Vietnam War* (Los Angeles: Presidio Press, 1990). Some secrets and lots of good solid information. A good counterbalance to traditional interpretations. Stimulating even when incorrect.

———, *Vietnam at War, The History: 1945–1975* (Novato, CA: Presidio Press, 1988). A comprehensive look at the conflict, with special attention to diplomatic, political, and military detail, concludes that Vietnam was a "defeat unprecedented in the annals of military history," with the United States winning all the battles and the North Vietnamese the war. But then seemingly goes on to answer this conundrum by asserting that it was a "war nobody wanted." Sees the United States losing a war not on moral grounds but to a "superior strategy which took advantage of U.S. political and psychological vulnerabilities."

Van Tien Deng, *Our Great Spring Victory* (New York: Monthly Review Press, 1977). The author, the chief of staff of the People's Army of Vietnam (PAVN) and commander of the Campaign 275, which took over Saigon and South Vietnam during the spring of 1975, is well-positioned to describe in great detail that effort. Once you get beyond his "quislings," "puppets," "lackeys," and "neocolonialism" jargon, it is a very revealing account. First, it puts the struggle in the context of 4,000 years of Vietnamese history, suggesting the greater importance of nationalism rather than Communism as a motivator. Second, the author gives a most-illuminating account of the buildup to the final offensive, including the building of thousands of miles of roads, telecommunications, and even an oil/gas pipeline to support it. Third, the work astutely points out the importance of pinning down the Army of the Republic of Vietnam (ARVN) strategic reserve (marines at Tru Trien and paratroopers at Thoung Duc) and the flooding of South Vietnam with additional North Vietnamese Army (NVA) divisions. These dispatches in *Nhan Dan*, the newspaper of the Vietnam Worker's Party, are well worth reading.

Frederick Downs, *The Killing Zone: My Life in the Vietnam War* (New York: W. W. Norton, 1978). A very moving and insightful account of Downs's service in the highlands with the Fourth Division during 1968. Wounded four times *before* stepping on a mine his last day in the field, he captures the attritional war of search and destroy, the matter-of-factness of death and wounding ("1 Kool Aid, 3 peanuts for the slick"), the frustration of dealing with the Vietnamese in general and the nature of guerrilla warfare in particular (he prefers the mano a mano struggles in the jungle without civilians), and especially the poor strategy that constantly moves units around, forcing them to acclimatize to new surroundings. He loses an arm in Vietnam and, after a year of rehab, encounters a civilian who says "Did you lose that in Vietnam?" and when he says "Yes," the person says "Serves you right."

Bob Drury and Tom Clavin, *Last Men Out* (New York: Free Press, 2011). An exciting but depressing account of the last Americans out of Vietnam, protected by the Marine Security Guards. North Vietnamese general Van Tien Dung is mystified by the "medicine man" sitting by the side of the dying patient.

William Duiker, *The Communist Road to Power in Vietnam* (Boulder, CO: Westview Press, 1996). The author, a former US State Department foreign service officer, writes a standard treatment of the war. He is on firm ground with chapters on "Dialectics of Escalation" and "War of Attrition" but much weaker on the post-Tet period of 1968–1972. The book ends more strongly on the 1973–1975 conclusion of the war, especially the fall of the "Divine Hammer" on Saigon, concluding, "All great revolutions are the product of multiple causes."

———, *Ho Chi Minh: A Life* (New York: Hyperion, 2000). The Father of His Country and the inspiration behind a long, long war to achieve the unification of that country. This is a long, long book and quite complimentary to Ho, concluding, "It is difficult to image the Vietnamese Revolution without the active participation of Ho Chi Minh." Looking at the parallel histories of Laos and Cambodia, he could be right. At the same time, it is quite amazing to revisit how important the material support of the Soviet Union and Lenin's "useful idiots" was in the US Peace Movement, which Ho played like a violin. Well worth reading.

Daniel Ellsberg, *Secrets: A Memoir of Vietnam and the Pentagon Papers* (New York: Penguin, 2002). From the man who outed the Pentagon Papers, making him either a hero or a villain, depending on one's perspective of the war and view of patriotism.

Bernard Fall, *Hell in a Very Small Place* (Washington, DC: Da Capo Press, 2002). Dien Bien Phu up close and personal. A chilling tale of hubris and mistaken strategy on the part of the French and considerable hubris and correct strategy on the part of the Viet Minh. Both the North Vietnamese and the United States would spend hilarious amounts of blood and treasure trying to replicate the battle, albeit, in the case of the United States, with a different result.

———, *Street without Joy* (Mechanicsburg, PA: Stackpole Press, 1994). The classic on the French debacle in Indochina. A must-read. Too bad those who made American war policy either didn't read it or, in their arrogance, assumed they could ignore its lessons.

France Fitzgerald, *Fire in the Lake: The Vietnamese and the Americans in Vietnam* (Boston: Little, Brown, 1972). Strongly antiwar in theme and in reaction to events that occurred during the war. The author is, for example, stunned that "to many American officials in Saigon, Tet was not an end but a beginning." Believing that "U.S. officials had emired Vietnam. . . . By involving the United States in a fruitless and immoral war, they had also corrupted themselves." He looks for the "narrow flame of revolution to cleanse the lake of Vietnamese society from the corruption and disorder of the American War."

Thomas Fleming, "How Nixon Almost Won the Vietnam War," *MHQ: The Quarterly Journal of Military History* (Winter 2012), pp.70–79. A very important look at the Battle of An Loc in March and April 1972. Regarded as a "Dien Bein Phu in reverse," the battle showed that the Army of the Republic of Vietnam (ARVN) was capable of defeating the North Vietnamese Army (NVA) as long as it enjoyed American artillery and air support. *Paris Match* compared it to Verdun and Stalingrad, while an American paper said, "Perhaps the best than can be said is that the city died bravely." And therein lies another tale.

Note: The two works by Vo Nguyen Giap, *People's War, People's Army* and *How We Won the War* are reviewed chapter 8.

Gordon M. Goldstein, *Lessons in Disaster: McGeorge Bundy and the Path to War in Vietnam* (New York: Holt, 2010). The "best and the brightest" weren't. There was a certain hubris associated with being from Harvard, one finds.

Richard Guidry, *The War in I Corps* (New York: Ivy Books, 1998). An African American marine infantryman surveys the war along the Demilitarized Zone 1967–1968, before General Abrams took over. Angry at the tactics and strategy of the American effort, he and his fellow Marines "greeted [the visits of division commanders] with resounding indifference," as "[w]e were fighting the NVA and our own leadership, and both seemed equally determined to kill us." No antiwar pacifist, Guidry favored the invasion of North Vietnam to take the fight to the enemy. "Have a good life" rings out as soldiers complete their year in 'Nam and rotate back to "The World."

David Halberstam, *The Making of a Quagmire: America and Vietnam during the Kennedy Era* (Lanham, MD: Rowman and Littlefield, 2008). This revised edition by Daniel Singal captures all the power and incisiveness of the original version, as Halberstam goes local and sees problems in counterinsurgency from the very beginning.

Michael Herr, *Dispatches* (New York: Alfred Knopf, 1971). A "New Journalism" book mixing a lot of facts, some seeming fiction, and composite characters, all combined to give the reader a sense of the war in Vietnam. It is episodic, focusing on the individual soldiers and other reporters and projecting the surreal nature of that war as he contrasts the American "search and destroy" ("More a gestalt than a tactic") with the Viet Cong (VC) and North Vietnamese Army (NVA) "find and kill."

Charles Hudson, *The Killing Fields* (London: Michael Joseph, 1984). The moving account of the love between Sydney Schanberg, the *New York Times* bureau chief, and his assistant, Dith Pran, during the 1973 destruction of Cambodia as the Khmer Rouge rebels and their reign of terror killed millions. A ghastly backstory.

Phillip Jennings, *The Politically Incorrect Guide to the Vietnam War* (Washington, DC: Regency, 2010). A revisionist account, a bit glib in spots but with a point of view worth examining for the intellectual stimulation of it.

Ward Just, *A Dangerous Friend* (Boston: Houghton Mifflin, 1999) (F). A well-written and deeply evocative novel about the Vietnam War, one that captures the heavy-handedness of the American footprint and decries the military approach as well as the Americans being there in the first place. A most telling irony is captured in the American assessment that the Vietnamese know so little, even about their own revolution "except that it had to be done and it would succeed, thanks to Ho and General Giap and the example of Dien Bien Phu." And, as one reviewer notes, "That's all they needed to know." For the author, America is a dangerous country, thrashing around the world on various missions, enterprises, and military adventures.

Stanley Karnow, *Vietnam: A History* (New York: Viking Press, 1983). One of the most widely read and favorably reviewed histories of the war. Quite weak on the post-1968 events, however, and overall unsatisfying if one is interested in the war qua war. Rode the PBS series wave to respectability. Greatly out of date as of 2012.

John Keegan, *Victory at Any Cost: The Genius of Vietnam's General Vo Nguyen Giap* (London: Recon Press, 1976). What it took and why he took it and who it affected. Keegan sees in Giap that element of true military greatness, his willingness to sacrifice anything to achieve one's goals.

Henry Kissinger, *Ending the Vietnam War* (New York: Simon and Schuster, 2003). His sense of history and what was happening in a war he helped end and therefore for which he has to take both the praise and the blame. He is, quite naturally, a little more reluctant about taking the latter than the former.

Andrew Krepinevich, *The Army and Vietnam* (Baltimore, MD: John's Hopkins University Press, 1986). This Vietnam-servicing soldier courageously argues that the ultimate blame for what happened in Vietnam lay with the military itself, not just the politicians. Shows how unprepared the armed forces were for fighting a guerrilla war. The American military remade itself after the Vietnam War—with little help from its civilian authorities—and forgot all the counterinsurgency lessons it had so painfully and costly learned. But then he shows what the American military did about it afterward to totally remake itself, without any help from America's political leaders. By the time of the Gulf War, it was an altogether different armed force. Luckily for the United States and Kuwait, as things turned out.

Meredith H. Lair, *Armed with Abundance: Consumerism and Soldering in the Vietnam War* (Chapel Hill: University of North Carolina Press, 2011). An extremely valuable reappraisal of the way the "suffocating luxury" in places like the Long Binh Post, with its 12 swimming pools, 81 basketball courts, and 31 theaters (along with others, such as Tan Son Nhut, Cam Ranh Bay, etc.) produced a great disconnect between the worlds—and narratives—of the battle grunts and the REMF's ("rear echelon mother fuckers"). In 1967, for example, only 70,000 of 464,000 US troops were in combat (p. 25). The cornucopia of food, the war against boredom for the REMFs, the paper clip, and the almost-ubiquitous post-exchange war lead the author to conclude that for most, "[a]t the end of it all, the Nam was a wonderland, and, through unsavory to admit, much of what American soldiers experienced there was wonderful" (p. 221). Quite a revisionist overview, this. Well-written and well-documented by the daughter of a two-tour veteran of the war in Vietnam, this work deserves to be read by anyone truly interested in understanding American success and failure in that conflict.

Michael Lanning and Dan Cragg, *Inside the VC and the NVA* (New York: Fawcett Columbine, 1992). Two Vietnam veterans who only wish they had killed more Viet Cong (VC) or North Vietnamese Army (NVA) during their tours of duty examine 2,600 interviews with POWs and defectors, coming away believing that the war was more likely to be won by big-battalion operations than counterinsurgency but end up realizing that the impendent variable was the will of the NVA ("They hung in there until they won"). They conclude with respect for their opponents, promising to buy them the first round of drinks should they ever encounter them in the future.

William Le Gro, *Vietnam from Cease-Fire to Capitulation* (Washington, DC: US Army Center of Military History, 1981). Written by a senior staff member of US Military Assistance Command Vietnam (MACV) from 1972 to 1975, this is a very important document, for it provides the reader with a very compelling if minority point of view. First, he asserts, "Yet the outcome could have been different," allowing for alternative history speculation. Second, he asserts categorically, "Unit for unit and man for man, the combat forces of South Vietnam repeatedly proved themselves superior to their adversaries." Third, he states, "Missing however, were inspired civil and military leadership at the highest levels and unflagging American moral and material support." Very detailed and highly professional, this account does lack an index, which would have been very helpful.

Fredrik Logevall, *Embers of War: The Fall of an Empire and the Making of America's Vietnam* (New York: Random House, 2012). The best book I have read on the origins of American involvement in the war, chockfull of might-have-beens. Seen in this format, it is overwhelmingly sad to see the slow, seemingly *not* inexorable slide toward the tragedy that was the end of the Vietnam War(s). Amazing to see the ad hoc and incremental way the United States got involved and how American presidents constantly saw themselves as prisoners of their predecessors (and how more bellicose Truman and Eisenhower were *after* they were president than when they were!). A must-read to understand this war. Some highlights:

1. Almost 400,000 Vietnamese starved as a result of the Japanese taking much rice from 1944 to 1945.
2. Ho Chi Minh identified himself with George Washington and, in the beginning period during and after World War II, looked to Washington for support.

3. By 1947, there was fighting throughout Vietnam; in 1949, some in the US State Department rightly saw the "dead-end alley."
4. Senator Mike Mansfield was initially a big supporter of the war: "[I]ts loss would start a chain reaction extending to the Persian Gulf"!
5. When the Korean War ended, China wanted to damp down the Vietnam War; the Soviet Union actually proposed partition.
6. Giap came much closer to failure at Dien Bien Phu "than is generally believed," and the author feels the decision to wait to start attack was crucial.
7. China pushed Viet Minh to accept partition as the line was drawn in 1954. Of Zhou Enlai: "He has double-crossed us."
8. The United States believed the Diem government was "untainted" by colonialism and could be a bastion of democracy!

Tom Mangold and John Penycate, *The Tunnels of Cu Chi* (New York: Random House, 1985). Two BBC journalists give an inside look at the vast tunnel complex northwest of Saigon, a complex that baffled those looking for it from the times of the Viet Minh. Situated in III Corps at the terminus of the Ho Chi Minh Trail, the complex contained supply depots, hospitals, staging areas, wells, a myriad of false tunnels, and many other dimensions. Both the tunnels and the tunnel rats who penetrated them speak to the enduring phenomenon of warriorhood under extreme situations. One of the most illuminating books about the Vietnam War I have read.

Karl Marlantes, *Matterhorn* (New York: Atlantic Monthly Press, 2010) (F). One of the best novels about the Vietnam War from the perspective of a US Marine. For a more extensive review see chapter 40.

Robert Mason, *Chickenhawk* (New York: Viking, 1983). A Huey pilot operating in the Central Highland during 1965–1966, when "search and destroy" passed for strategy. Present at the battles in the A Shau Valley near Pleiku, Mason burns out and has a very rough time back in the United States. A sad story from beginning to end but an important one to understand.

Frank McAdams, *Vietnam Rough Riders* (Lawrence: University Press of Kansas, 2013). In 1968, the Marines were in charge around Danang. McAdams, a fresh lieutenant, spent much more of his time fighting his superiors (with good reason!) than the Viet Cong (VC) or North Vietnamese Army (NVA). His troopers put a $5,000 bounty on his immediate superior's head but said leader refused to go on convoy duties, so they never got a real chance.

H. R. McMaster, *Dereliction of Duty: Lyndon Johnson, Robert McNamara, the Joint Chiefs of Staff and the Lies That Led to Vietnam* (New York: HarperCollins, 1997). A searing indictment of the best and the brightest, starting with President John F. Kennedy ("Vietnam is the place to make America's power credible") and his advisors Robert McNamara and General Maxwell Taylor, through President Johnson, Secretary of State Dean Rusk, and especially the Joint Chiefs of Staff ("Five Silent Men"). McMaster spares no one, concluding that the Vietnam war "slunk in on cat's feet." To nobody's credit.

Robert A. McNamara, *In Retrospect: The Tragedy and Lessons of Vietnam* (New York: *New York Times* Books, 1995). Although self-serving in spots, it describes the hermetically sealed world of the Johnson White House quite well.

Marshall Michell III, *The 11 Days of Christmas* (San Francisco: Encounter Books, 2002). One of the most fascinating books about the war and one that can teach the most lessons for the future of conducting warfare. The rigidity of the US Strategic Air Command (SAC) headquarters during this operation seems almost unbelievable—certainly it is very sobering. Stopping the bombing the day after the North Vietnamese ran out of surface-to-air missiles seems, in hindsight, to have been a rather substantial blunder. This book is a must-read for understanding the Vietnam War.

———, *Clashes: Air Combat over North Vietnam 1965–1972* (Annapolis, MD: Naval Institute Press, 1972). Operations Rolling Thunder (1965–1968) and Line Backer I and II (April 1972 until the end of 1972). Michell concludes that the United States never truly achieved air superiority over North Vietnam, that the fighter pilots were North Vietnamese (although with North Korea, Chinese, and Soviet advisors), that the MIG 21 shot down 5 American planes for every MIG 21 lost, that 60 percent of American planes shot down were surprised, and that American Sidewinder missiles failed 56 percent of the time and only 25 percent struck home—the rest missed. Quite a sobering read.

Lt. General Harold Moore and Joseph Galloway, *We Were Soldiers Once . . . and Young, Ia Drang: The Battle That Changed the War in Vietnam* (New York: HarperCollins, 1992). Search-and-destroy strategy at its worst and, in November 1965, the first major Air Cav operation of the war. It is often said that the US military never lost a battle during the entire course of the Vietnam War, but this Ia Drang struggle comes pretty close. Told in a very powerful way, gripping and haunting. You will remember this book long after you have finished reading it.

Ted Morgan, *Valley of Death* (New York: Random House, 2010). The Battle of Dien Bien Phu and why it represents a stunning and enduring lesson on how not to fight a counterinsurgency. A great deal of in-depth description and quotes from survivors on both sides, carefully presented and militarily accurate. Reading it, I was reminded that William Lansdale (in his book *In the Midst of Wars* reviewed in chapter 8) compared the Dien Bein Phu strategy of the French with their earlier emphasis on Verdun and the Maginot, deciding "that here was the place where the enemy could be lured out onto the killing grounds before the guns of the forts and be slaughtered by the thousands with the result that the Vietminh would give up their war." Was never going to happen. And didn't.

Wilbur Morrison, *The Elephant and the Tiger: The Full Story of the Vietnam War* (New York: Hippocrene Books, 1990). Having digested "ten million words" about the Vietnam War, Morrison blames one and all for a national tragedy in an honest, straightforward, and unsparing account. McNamara, for example, is rightly criticized as "inept," but antiwar protesters are castigated for their failure to speak out against the bloody Communist takeovers of South Vietnam, Laos, and Cambodia. Actually does a pretty fair job of presenting the war from beginning to end and beyond. The title is taken from Ho Chi Minh's notion that the elephant could easily stomp the tiger if the tiger stood still and let him. Not a bad parallel, if the elephant wasn't able to use all of his feet and could only use his trunk every third day.

Mark Moyar, *Phoenix and the Birds of Prey: The CIA's Secret Campaign to Destroy the Viet Cong* (Annapolis, MD: Naval Institute Press, 1997). An excellent reality check to so much drivel written about the war in Vietnam. A vital and powerful challenge to customary wisdom of both the right and the left. Makes a strong case that the "majority of villagers actually favored the Saigon government over the Viet Cong [VC] during the latter years of the war," but despite all that has been written, the center of gravity for the North was the North Vietnamese armed forces. Depicting in great detail the "shadow war" during 1967–1972, when the VC infrastructure was essentially wiped out, with 81,742 rallied to the government, killed, or captured through the efforts of the Phoenix Program [the CIA and the Provincial Reconnaissance Units (PRUs)] and the Civil Operation and Revolutionary Development Support (CORDS)—all pacification efforts minus those of the CIA.

———, *Triumph Forsaken: The Vietnam War 1954–1965* (London: Cambridge University Press, 2006). One of the classic revisionist views of Vietnam, well-told and well-documented. Anyone wishing to know what happened in Vietnam and what might have happened needs to read this book.

John Nagl, *Counterinsurgency Lessons from Malaya and Vietnam* (Westport, CT: Praeger, 2002). Excellent and insightful, many of the most important lessons of counterinsurgency come out of this striking success and abject failure and why (because in the midst of the failure were the true lessons of success).

Lien-Hang T. Nguyen, *Hanoi's War: An International History of the War for Peace in Vietnam* (Chapel Hill: University of North Carolina Press, 2012). Using newly available archives, the author provides illuminating insights on the "Talk Fight Situation" at the end of the war. See especially the chapters "War against Détente" and "War for Peace," in which the Soviets and the Chinese favor détente above a North Vietnamese victory, much to the latter's disgust. Nguyen concludes that North Vietnam's "leaders ensured peace could only be won through the barrel of a gun."

Bao Ninh, *The Sorrow of War* (New York: Pantheon Books, 1993) (F). A North Vietnamese soldier and writer, Ninh survived the war to paint its life and its echoes in sharp relief. An overwhelming sense of doom in the midst of war: "The fighting is endless. No one knows when it will stop." And, in the "jungle of the screaming souls," there was "no remedy" except to "[g]ive them some bronze candy." A very powerful work and also one that, in indirect fashion, traces the evolution of the war and its ebb and flow from the perspective of the North Vietnamese Army (NVA).

Keith W. Nolan, *Into Cambodia: Spring Campaign, Summer Offensive, 1970* (Novato, CA: Presidio, 1990). As Vietnamazation proceeds, Nixon decides to send first the Army of the Republic of Vietnam (ARVN) and then US forces into Cambodia to try to capture the Central Office for South Vietnam (COSVN) headquarters and disrupt the supplies coming into South Vietnam via Cambodia. ARVN units, including airborne and the US 9th, 4th, and 25th Divisions, plus the 1st Air Cav, strike the Fishhook with considerable success. Killing or capturing more than 13,000 North Vietnamese Army (NVA; albeit at a cost of 7,000 Americans and ARVN) and a staggering amount of weapons and ammunition, thereby relieving pressure on Tan Ninh Province for six months. But this tactical military victory was not used strategically as part of anything but winding down the war.

Jon Oplinger, *Quang Tri Cadence: Memoir of a Rifle Platoon Leader in the Mountains of Vietnam* (Jefferson, NC: McFarland, 1993). A sardonic yet poignant account of a First Air Cav LT near the A Shau Valley in 1968, told carefully, even lovingly, but with stark realism. After being wounded, he spent seven months in the hospital and discovered that he had been sent to

Vietnam just as the decision was made to eventually withdraw US troops. Very sad description of the ongoing conveyor-belt of casualties. One of the best of the grunt's-eye views of the war.

Laura Palmer, "The General at Ease: An Interview with Westmoreland," *MHQ: The Quarterly Journal of Military History* 1, no. 1 (Autumn 1988), pp. 30–35. An illuminating look at his views of military history and what lessons he tried to apply in Vietnam. Well worth a read.

Rufus Phillips, *Why Vietnam Matters* (Annapolis, MD: Naval Institute Press, 2008). The author, who was in the CIA at the beginning (1954–1956) and periodically thereafter (1964 on), provides us with a poignant, modest, touching, sad, and illuminating account of what might have been and why that was important. Sees American hubris, Westmoreland's strategy (which marginalized the South Vietnamese armed forces during his search-and-destroy operations), and internal South Vietnamese political disarray and corruption as key factors.

Douglas Pike, *PAVN: People's Army of Vietnam* (Novato, CA: Presidio, 1986). The definitive analysis of this organization whose rise is "probably the most astounding military phenomenon of our lifetime." Citing the almost perfect combination of "Armed Dau Tranh (struggle) and Political Dau Tranh," the author focuses on the Vietnam War, not from the mistakes the United States made but how well the "Prussians of Asia" developed, redeveloped, and executed their strategy, combining "messianic leadership," centuries of martial spirit, and a "Spartan social system." A must-read for anyone wanting to understand war as well as the Vietnam War.

John Pimlott, *Vietnam: The Decisive Battles* (New York: Macmillan, 1990). With more than 3,000 books already published on Vietnam, it is hard to recommend just one. But Pimlott's is as good an introduction as any to appreciate the complexity, longevity, and tragedy of the war, along with stunning photographs and especially computer-generated images of the battles, such as Dien Bien Phu, Ian Drang, Khe Sanh, Hamburger Hill, An Loc, and others. You will learn a lot about the war just by perusing this work.

Robert Pisor, *The End of the Line: The Siege of Khe Sanh* (New York: Ballantine Books, 1982). An excellent account of the classic hedgehog battle of Khe Sanh, which, while an American victory, turned out to be a superb North Vietnamese Maskirovka ploy to divert attention from the stunning Tet Offensive of 1968. Although both Khe Sanh and Tet were smashing military victories for the United States and South Vietnam, both were outstanding political and strategic victories for the North Vietnamese Army (NVA). Khe

Sanh absorbed a huge amount of US resources—all its strategic ground-force reserves and unprecedented amounts of air power. But it certainly was a victory for those who were surrounded and protected only by air power and their own sense of warriorhood.

Norman Podhoretz, *Why We Were in Vietnam* (New York: Simon and Schuster, 1982). A sober and judicious point of view, with idealistic arguments now long forgotten by most who look at the war. Even at the time, it was largely ignored.

John Podlaski, *Cherries: A Vietnam War Novel* (New York: CreateSpace, 2010) (F). This is a somewhat unpolished, didactic ("There are two kinds of enemy soldiers in the Nam, Larry"), and overly descriptive work but one that captures (while educating the Cherries) the essence of the war in 1970 in the Bien Hoa, Cu Chi tunnel complex area. "Welcome to Hell" for 365 days (with luck minus a few weeks in a safe area when you were "short"). At the end of his tour, he is seconded to the A Shau Valley with the 101 Airborne and fights North Vietnamese Army (NVA) regulars. More insights follow.

Colin Powell, "Soldiering," in his *My American Journey* (New York: Random House, 1995), pp. 79–232. A compelling account by the man who would become head of the Joint Chiefs of Staff and American secretary of state and how his years in Vietnam conditioned his perceptions about the Powell Doctrine.

John Prados, *Vietnam: The History of an Unwinnable War, 1945–1975* (Lawrence: University Press of Kansas, 2009). Tells why the author thinks it was "unwinnable" by the United States under any circumstances. Was he right, or was he wrong? To answer this vital question, one has to read this book.

Stephen Randolph, *Powerful and Brutal Weapons: Nixon, Kissinger and the Easter Offensive* (Cambridge, MA: Harvard University Press, 2007). One wonders how the Vietnam War would have turned out if Nixon had been directing the war from 1965 on. A truly stunning set of insights.

James S. Robbins, *This Time We Win: Revisiting the Tet Offensive* (New York: Encounter Books, 2010). A strongly revisionist account of the might-have-beens in Vietnam. Although in the aggregate, the author revisits some obvious mistakes—not making the Demilitarized Zone impervious and cutting the Ho Chi Minh Trail in Laos and not having 16 bombing halts from 1964 to 1968—the task still looks Sisyphean. He is dead right about one thing, however: Johnson was a terrible, terrible war president, and Kennedy and the Wiz Kids around him were responsible for developing a counterinsurgency capability without recognizing its inherent weaknesses.

Al Santoli, *Everything We Had: An Oral History of the Vietnam War by Thirty-Three American Soldiers Who Fought It* (New York: Random House, 1981). This kaleidoscope of impressions taken throughout the war will leave a lasting impression on you as the "raw expressions of a human ordeal" come thick and fast. There is the 9–5 war with servants; the Kennedy idealism turning to cynicism; learning that the First Cav took three months to refit after the vicious 1965 battle in the Ia Drang Valley; and the relief conveyed by "Hello, hello, this is Spooky Zero-Two with forty thousand rounds of happiness." "The village was flattened. . . . It was flattened by the VC." "We got into more trouble for killing water buffalo than we did for killing people." There is a thriving black market run by Americans; a naval club with fine china, valor, and gunfights of great courage; American pilots tortured for years by their North Vietnamese Army (NVA) captors; and, at the end, a young medic who had already done a tour in Vietnam and was sent back in April 1975 to assist in the final evacuations: "I cried my ass off. I don't have any more tears left." This collection of impressions will move you.

William Shawcross, *Sideshow: Kissinger and the Destruction of Cambodia* (New York: Pocket Books, 1979). The bombing of Cambodia, worst-case analysis. War involves a great deal of duplicity, and for some strange reason, this bothers Shawcross a great deal. At least this instance and this unrighteous indignation take away from the impact of his findings. He never seems to grasp the central mistake of the entire enterprise, namely that all those bombs should have fallen on North Vietnam, not Cambodia.

Neil Sheehan, Hendrick Smith, E. W. Kenworthy, and Fox Butterfield, *The Pentagon Papers* (New York: Bantam Books, 1971). The center of a titanic press-and-court battle and fascinatingly revealing documents. Most ominous to those who have never studied war before Vietnam came along. Questions remain today about the publication of military secrets, near-secrets, and not-so-secret items all at once.

Frank Snepp, *Decent Interval: An Insider's Account of Saigon's Indecent End Told by the CIA's Chief Strategy Analysis in Vietnam* (New York: Random House, 1977). Rereading this work 40 years later, I was stunned at how much was revealed about the duplicity and ineptness of the American withdrawal from Vietnam and the extent to which tens of thousands of Vietnamese (including many Communist defectors) were left behind. The author drove President Thieu to the airport on his flight to Taiwan and exile and was one of the last Americans out. Has little good to say about the American ambassador (Martin) the American secretary of state (Kissinger) and the American Congress. Also underscores the fluidity of the war from 1973 to 1975. Equilibrium depended on continuing US aid, which was cut off abruptly and with great duplicity.

Lewis Sorley, *A Better War* (New York: Harcourt Brace, 1999). Makes a strong case that the United States finally adopted a winning strategy from 1968 to 1972, and that winning strategy was on the verge of success when the plug was pulled.

———, *Westmoreland: The General Who Lost Vietnam* (Boston: Houghton Mifflin, 2011). A powerful, searing indictment of the US commander in Vietnam, showing his strategy of "search and destroy" was not only incorrect and ill-advised, but also it marginalized the South Vietnamese armed forces from the struggle they had to win. Westmoreland's "find, fix and destroy the big Main Force units" was not the answer. He also cost many American lives with his one-year rotation policy as casualties during the first six months of the rotations were much higher than the second six months. A vain and shallow man in many ways, the author argues, Westmoreland was, many contemporaries said, at best a divisional commander. Sometimes Sorley goes too far in the blame game, seemingly seeing President Johnson and the Joint Chiefs as having less responsibility for failed policies (although they surely have major responsibility for entrusting the nation's fate to "Westy." That said, on balance, Sorley's judgment seems spot-on: "Westmoreland's strengths eventually propelled him to a level beyond his understanding and abilities. The results were tragic, not just for him and for the Army and the nation he served, and most of all of course for the South Vietnamese, who sacrificed all and lost all."

Ronald Spector, *After Tet* (New York: Vintage Books, 1994). Sees the Greek tragedy to its conclusion. Is not happy with many of the things he sees during the 1965–1975 period. Can't blame him. But this is a most important look at the world after Tet, one that is often largely ignored by so many historians, who write off the South Vietnamese and American effort after it.

Shelby Stanton, *Green Berets at War: U.S. Army Special Forces in Southeast Asia 1956–1975* (Novato, CA: Presidio Press, 1985). A comprehensive, very detailed account of the Special Forces operations in Laos, Cambodia, Vietnam, and Thailand during the period of review. Excellent maps show the troops "out on the Rim." We should remember the Green Berets' motto: "To liberate from oppression."

Robert Stone, *Dog Soldiers* (Boston: Houghton Mifflin, 1973) (F). This highly overpraised novel is less about the war itself than a long, languid, drug-fueled account of Saigon and a lot of bad dealing. The war itself seems quite distant, almost irrelevant. I honestly do not get the appeal of this work.

Harry Summers, *On Strategy: A Critical Analysis of the Vietnam War* (New York: Dell, 1984). Argues persuasively that "clear and hold" strategy would not have succeeded unless a parallel "search and destroy" strategy was also employed at the same time. Not clear that the US military of the 1960s could have ever accomplished both missions simultaneously. A major work that should be read as closely as it was by many in the US military.

Truong Nhu Tang, *A Vietcong Memoir* (New York: Harcourt Brace Jovanovich, 1955). "My life as a revolutionary" has this privileged, French-educated South Vietnamese coming under the sway of Uncle Ho Chi Minh and serving in the National Democratic Front and later the Provisional Revolutionary Government of South Vietnam, only to become very disillusioned as the hard-core, Marxist-driven cadres from the North take over the South at independence. After 300,000 are sent to reeducation camps, he flees by boat and makes it to Indonesia and then France. This is a rare look at an interesting dimension of the Vietnam War as the American antiwar movement took it as an act of faith that the North would not in fact take over the South.

Hoang van Thai, *The Decisive Years: Memoirs of Vietnamese Senior General Hoang Van Thai* (Washington, DC: US Government Printing Office, 1987). The North Vietnamese deputy minister for defense provides many interesting insights into the rapidly developing situation in 1974–1975. Clearly indicates that the North Vietnamese Army (NVA) paid no attention to the Paris Peace Accords but instead did a masterful job of preparing for the endgame. Interestingly enough, Hoang and others feared President Thieu would be able to actualize his "enclave strategy," which would have been very difficult to overcome. Give him the final word for the perspicuity of the "entire nation had gone to war," which was, at the end of the day, the key variable for North Vietnamese success.

Lam Quang Thi, *The Twenty-Five Year Century* (Denton: University of North Texas Press, 2001). A true South Vietnamese patriot tells a moving story about South Vietnamese soldiers, a set of insights often ignored by most who write about the war, especially those who were against the war, period. Thi is a most authentic voice.

Robert Thompson, *Make for the Hills: Memories of the Far Eastern Wars* (London: Leo Cooper, 1989). This work is earlier reviewed in chapter 8, but its chapters on Vietnam (pp. 122–89) should be read by all those seeking an understanding of the author's unique perspective. Going to Vietnam in 1960 after being at the heart of the successful counterinsurgency in Malaya, he discovered how little the Americans had studied either the French failures or the works of Mao and that "[t]here was always in Vietnam a constant tendency to militarize civilian functions." He was against the assassination of Diem, was high on Nixon and Creighton Abrams, and thought the dykes should have been bombed and "Harriman's Highway" [a.k.a., the Ho Chi Minh Trail; the book has photos of North Vietnamese Army (NVA) trucks running three abreast, hundreds in the frame] should have been occupied by American troops. He also believes that the country was lost in 1965 and regained in 1970 and that the American enclave strategy used up far too many resources that could have been used to secure the countryside.

Robert L. Tonsetic, *Days of Valor: An Inside Account of the Bloodiest Six Months of the Vietnam War* (Philadelphia: Casemate, 2007). Based on the firsthand accounts of those in the 199th Light Infantry Brigade, which fought north and northwest of Saigon during the period under review, from December 1967 until May 1968. Straightforward and descriptive. Revealing.

George Veith, *Black April: The Story of the Fall of South Vietnam 1973–1975* (New York: Encounter Books, 2012). A meticulous and scholarly study, indeed the definitive one to date, on Thang Tu Den (Black April). Seeks to answer the question "Why did North Vietnam win in 1975 when it couldn't win in 1972?" The answer, the author asserts, is a combination of North Vietnamese will (they broke the peace accords of 1973 within weeks); the actions of the US Congress to cut off all military aid (the last American troops left in March 1973); Nixon's preoccupation with Watergate; the economic situation in Vietnam; and finally the inept military strategy of President Thieu, who removed the South Vietnamese regular army from the Central Highlands, long thought to be the crucial battleground of the war. By contrast, General Giap changed his strategy of attacking on all fronts (1968 and 1972) and concentrated on the Central Highlands and the road to Saigon. This is a marvelous, in-depth work of great historical importance, with its access to so much documentation from all sides.

F. J. "Bing" West Jr., *The Village* (Madison: University of Wisconsin Press, 1972). A very powerful and insightful account of a Marine unit of 12, who from 1966 to 1970 worked with 5,000 Vietnamese in the I Corps village of Binh Nghia. Day-by-day action with the Popular Forces (PFs) fending off the Viet Cong (VC) and the North Vietnamese Army (NVA) and eventually putting them on the defensive until well into the 1970s. The fact that the village progress was eventually overridden by events elsewhere does not diminish the poignancy and price it cost to assert that progress in 1970: "The village is intact. The village has endured." A must-read for anyone wishing to understand the Vietnam War.

William Westmoreland, *A Soldier Reports* (Garden City, MD: Doubleday, 1976). Oddly enough to American ears, some North Vietnamese generals thought that his plan for cutting the Ho Chi Minh Trail was the proper strategy, or at least the one they believed would have given them the most difficulty in overcoming. Interesting counterbalance to the notion that his strategy of "search and destroy" instead of "clear and hold" doomed the American effort in Vietnam and cost South Vietnam a reasonable chance to become "Korea South." What is so astonishing after the all this time is to realize that the problem was not with the American civilian authorities interfering too much with strategy (the way they mucked up the tactics, particularly the use of American air power) but that they didn't interfere enough, challenging the basic assumptions of Westy's search-and-destroy strategy, which wasted too much time and treasure and cost so much blood. Quite a read nevertheless.

Andrew Wiest, *Vietnam's Forgotten Army: Heroism and Betrayal in the ARVN* (New York: New York University Press, 2007). Imagine fighting 20 years for your country and getting no credit for your sacrifice and your heroism. A very important book for getting a more objective view of what really happened during the Vietnam War.

James H. Willbanks, *The Battle of An Loc* (Bloomington: Indiana University Press, 2005). An excellent description of a battle that showed many people many different things by a man who experienced it during the North Vietnamese spring 1972 "Nguyen Hue" campaign, with attacks in the north, central, and capital areas of South Vietnam. Shows the achievement, promise, and limits of Vietnamazation with US advisors; heavy, heavy air support; and very substantial logistical assistance. The B-52s played a vital role, as did the close air support of the Cobras and 2,500 other air support missions. The North Vietnamese Army (NVA) commitments included 3 whole divisions and 78,000 rocket and artillery rounds. Nevertheless, a resounding victory for the South Vietnamese, their "Stalingrad," whose siege lasted 70 days in the "battle that saved Saigon." By contrast, the siege of Dien Bien Phu lasted 56 days. South Vietnamese casualties were more than 30,000; the NVA, 100,000.

Chapter Thirty-Two

Mars Is a Jealous God, Part II

The Lessons of Afghanistan I and II and Iraq II and III

This chapter covers the return of medieval warfare in Afghanistan and Iraq and includes a discussion of the ongoing question, "Where do the wars in Iraq and Afghanistan fit on the space-time continuum of the Template of Mars?" What is the relationship of the international power configurations (bipolar, power centralization, power diffusion, hegemonic) to interstate activity today?

Nation-building, imperial policing, and urban asymmetrical warfare are all covered in this chapter, as is the continuing transformation of the armed forces, not just of the United States, but of other peoples and groups as well. This set of works also explores the notion that "[n]othing fails like success" and concludes that Afghanistan was a poor prologue for the Second Iraq War. Also, various works ask the reader the real meaning of the concept "winning the hearts and minds" during an insurgency.

AFGHANISTAN I AND II

After the attacks of September 11, 2001, on the American homeland, there seems to have been a depressing lack of contingency plans for dealing with Afghanistan except with some cruise-missile launches. So going to war against the Taliban-led government of Afghanistan and al-Qaeda was very much a spur-of-the-moment, plan-as-you-go lunge into the unknown, with the CIA leading a reluctant Pentagon into what could only be described as medieval warfare and warfare by hire along with modern technology.

There were many, many ironies involved in the First Afghan War: the speed of its success and the extent to which the United States did not finish winning the peace but rapidly focused on Iraq, only to have to return to Afghanistan for a second war, this one a counterinsurgency action against a foe with a major sanctuary in neighboring Pakistan. Anyone truly interested in the serendipity of warfare will find fascinating reading outlined here:

Brigadier Nigel Aylwin-Foster, "Changing the Army for Counterinsurgency Operations," *Military Review* (November–December 2005), pp. 2ff. How military doctrine can change with history and change history and why some are better than others in noticing the difference between the two. Afghanistan was to test these and other theories.

Gary Berntsen, *Jawbreaker: The Attack on Bin Laden and al-Qaeda* (New York: Crown, 2005). One of the great might-have-beens of recent military history. The CIA field operative was in the hunt for Osama bin Laden in the drive from Kabul to Tora Bora. Good insights about that operation's failure in 2001 due to the inability of the US military to block bin Laden's escape into Pakistan and the unwillingness of Afghan forces to do so. Valid questions about the American ability to fight wars in high places with short notice.

Eric Blehm, *The Only Thing Worth Dying For* (New York: HarperCollins, 2010). Eleven Green Berets helped Hamid Karzai get to Kabul and become president, while every member of the team was wounded or killed following 9/11. Some would argue the opposite outcome would have been much better.

Gregory Bolich, "Terrorism in the Ancient Roman World," *Quarterly Journal of Military History* (Spring 2006), pp. 52–59. Afghanistan has seen this movie before.

Carl von Clausewitz, *On War* (New York: Alfred A. Knopf, 1993), book 6, *Defense*, chaps. 1–9, 26. See especially chapter 26, "The People in Arms," a superb and incisive account of what happens when people get involved in warfare on a sustained and personal basis. Read it metaphorically for maximum benefit. Clausewitz understood the power of the Spanish ulcer.

Henry Crompton, *The Art of Intelligence* (New York: Penguin Press, 2012). See especially "Afghanistan Strategy," "Afghanistan Operations," and "Beyond Afghanistan" (pp. 169–273). The author, who was in charge of the CIA/Special Forces effort to overthrow the Taliban following the terror strikes of 9/11, gives an insider's view of what worked and what did not. A fascinating companion piece to *Jawbreaker* reviewed earlier this chapter.

Captures the fighting spirit of those charged with avenging the deaths of the almost 3,000 people who perished in the attacks. Brutally honest about failures of intelligence and operations before, during, and after the attacks. Sees a new era when CIA, FBI, and the Departments of Defense and State must all be retasked and redirected to face the demands of the post-9/11 world and worldwide jihadi activities. Page 154 describes a perfect opportunity to kill Osama bin Laden prior to 9/11 and the reasons for its failure. Sobering. Enjoyed the rapidity and cynicism with which the Obama administration adopted the drone program and greatly expanded it.

Norman Friedman, *Terrorism, Afghanistan and America's New Way of War* (Annapolis, MD: US Naval Institute, 2003). The United States and the war in Afghanistan is put in a larger context. The new way of war was also to demonstrate some other dimensions as old as war itself.

Robert Gates, *Duty: Memoirs of a Secretary at War* (New York: Alfred Knopf, 2013). There are few memoirs of defense secretaries that matter. This one does. Every citizen of the republic should be forced to read and absorb his concluding chapter, "Reflections," if not the entire work. Personally, I would put it with U. S. Grant's memoirs as a document of supreme import. Gates's conclusions—about Congress; the $700 billion Pentagon; the short-sighted political nature of the modern presidency; the wisdom of Eisenhower in avoiding war where subsequent presidents, Democrat and Republican, rushed into unnecessary overreaches of American military power—ring so true and so centrally correct that one can only appreciate his efforts. The shallow cherry-picking of the right and the left in their reviews of his work only proves this point. His ultimate dedication to America's fighting men and women is underscored by his conclusion how poorly they are served by their political and military masters. "Our foreign and natural security policy has become too militarized, the use of force too easy for presidents" is a conclusion that should be remembered by all, "hawk" or "dove," but especially by "concerned citizen."

Marcela Grad, *Massoud: An Intimate Portrait of the Legendary Afghan Fighter* (St. Louis, MO: Webster University Press, 2009). This paean delivered in short, staccato bursts reeks with hero worship of Ahmad Shah Massoud, the Tajik leader of the Northern Alliance who fought the Soviets, their puppet Communist government, and the Taliban for 20 years until his untimely assassination just prior to the attack of 9/11. Had he lived—and been Pashtun—he might have led a free and democratic Afghanistan.

Malcolm MacPherson, *Robert's Ridge* (New York: Dell, 2005). War in Afghanistan has never been easy to conduct, and MacPherson tells us why. The

US operation on Takur Ghar Mountain did not go well, and the author is painstaking in his analysis of this operation of the Seals. See also Naylor's *Not a Good Day to Die*, reviewed later this chapter.

Carter Malkasian, *War Comes to Garmser* (London: Oxford University Press, 2010). The author, a US State Department political officer and Pashtun speaker, gives many insights into tribal life during wartime among the 150,000 inhabitants of Garmser in Helmand Province. He concludes that the United States faces many more of these "murky, messy and gray" wars in its future.

Jane Mayer, "The Predator War," *New Yorker*, October 26, 2009, pp. 36–45. Raises some interesting moral questions as the use of drones grows and grows and grows and virtually every commander wants the use of one.

Sean Naylor, *Not a Good Day to Die: The Untold Story of Operation Anaconda* (New York: Berkley Books, 2005). How the US armed forces got an early lesson in how tough the war in Afghanistan would turn out to be. Startlingly candid.

Stephen Pressfield, *The Afghan Campaign* (New York: Doubleday, 2006) (F). What Alexander accomplished and what he did not, parallels and differences between then and now. So much of what happened then had relevance for the war of now, although marrying into the tribal structure of the region was apparently not among them. Yet the importance of the logistical tail and perseverance are clearly underscored.

Donald Rumsfeld, *Known and Unknown: A Memoir* (New York: Penguin Books, 2011). See especially "Into the Graveyard of Empires" (pp. 362–410). Rumsfeld provides a useful, if self-serving, look at decision making during the period right after 9/11 and during the invasion of Afghanistan. A must-read but only when one has digested some more holistic third-party accounts, such as Woodward's *Bush at War* (reviewed later this chapter).

Steve Russell, *We Got Him! A Memoir of the Hunt and Capture of Saddam Hussein* (New York: Threshold Editions, 2011). A surprisingly flat and prosaic account of the Fourth ID's block-by-block search for and capture of Saddam Hussein in his hometown of Tikrit. Some good vignettes of the beginning of the insurgency (as opposed to the end of Iraq II) and one account of how a reporter of the *Washington Post* is shocked, just shocked, when Colonel Russell's troops give insurgents no warnings when they pull weapons and just shoot them. "Do you think that is fair?" she asks.

By contrast, the people of Kuwait cheer the Americans as they return from their mission.

Gary Schroen, *First In: An Insider's Account of How the CIA Spearheaded the War on Terror in Afghanistan* (New York: Ballantine Books, 2005). One of the most interesting books about war I have ever read. Really insightful and exciting in the best sense of that word. Makes war seem manageable.

Anthony Shaffer, *Operation Dark Heart: Spycraft and Special Ops on the Frontlines of Afghanistan—and the Path to Victory* (New York: St. Martin's Press, 2010). A somewhat disjointed account by a former member of the Defense Intelligence Agency (DIA) and his views as to how America's war in Afghanistan is messed up. He is not against it at all in principle, just that it is being fought, in his view, so nonsensically with competing bureaucracies. He sees General McChrystal as a warrior and General Barno as a bureaucrat in uniform, concluding, "The consequences of our failure in Afghanistan and throughout the region would be massive."

Doug Stanton, *Horse Soldiers: The Extraordinary Story of a Band of U.S. Soldiers Who Rode to Victory in Afghanistan* (New York: Scribner's, 2009). A must-read for anyone trying to make sense of war in the Middle Ages redux.

Rory Steward, *The Network* (New York: Harcourt, 2010) (F). Steward has a deep love of Afghanistan and its people, and this book about a British intelligence operation expresses his frustration about the developments of the last 30 years in that country. Nor is he keen on this helter-skelter operation either.

———, *The Places Within* (New York: Harcourt, 2004). A walk across a medieval place sets a good backdrop for any study of Afghanistan. I do not recommend walking across Afghanistan to duplicate Steward's trek however.

John H. Waller, *Beyond the Khyber: The Road to British Disaster in the First Afghan War* (New York: Random House, 1990). The Great Game in full flower, with Russians, British, Persians, Afghans, Sikhs, and Indians plotting, planning, and acting. A remarkable story full of contemporary lessons. Points out that the British expedition in 1839 with its Bengal Army (five camp followers per soldier) and its Bombay Army (three camp followers per soldier) did not travel easily nor swiftly.

Bing West, "How We'll Win in Afghanistan," *Wall Street Journal*, July 29, 2009, p. A15. From the author who most accurately called the "turn" in the Iraqi insurgency.

———, *The Wrong War: Grit, Strategy, and the Way Out of Afghanistan* (New York: Random House, 2011). Candid and brutal: "The strategy in Afghanistan has been muddled since 2001"; "This war will be decided by grit"; "We have fought the wrong war with the wrong strategy. Our troops are not a Peace Corps; they are fighters. Let them fight and let the Taliban fear"; "When avoiding casualties is the achievement, it is time to leave." Points out how the United States has created a culture of entitlement for civilians, not an incentive for defeating the Taliban, and quotes a US Marine: "We must kill to change this war." Clear, hold, and build. West is an excellent and incisive student of war and, unlike many exegetes, quite willing to change his mind as the situation alters.

Robert Wilkinson-Latham, *North-West Frontier 1837–1947* (Oxford, UK: Osprey, 1977). A long history of fighting, feuding, conquest, and rebirth, covering the three Afghan wars (two of which the British won), ending with the British pullout in 1947, leaving the local inhabitants "puzzled" that the British had not been defeated, yet they gave up the land.

Bob Woodward, *Bush at War* (New York: Simon and Schuster, 2002). Bush, the hero. Woodward would later judge the president much more harshly, but in this work, Bush emerges as calm, cautious, and resolute. Most readers will be as stunned as I that there were no coherent plans for dealing with Afghanistan and that the entire operation was made up as the decision makers went along. It is also vital for understanding what was to happen in the second war in Iraq (2003), to observe how the CIA moved into the vacuum outlined earlier and how successful the initial invasion was. Both elements would lead to President Bush taking George Tenant's statement about the case for weapons of mass destruction being in Iraq as a "slam dunk." He had been very right about Afghanistan. Of course, by Iraq II, the Pentagon had any number of plans for an invasion this time.

Abdul Salam Zaeef, *My Life with the Taliban* (New York: Columbia University Press, 2010). This Afghani argues that the "decision to invade Afghanistan and wage war against its people was a mistake, because it drove America and Afghanistan into a quagmire." Argues diplomacy should have been used instead.

See also chapter 8, "Is the Template Obsolete in the Face of Insurgencies?"

IRAQ WARS II AND III

Note that any rational attempt to study the situation in Iraq from 1991 until the present requires an understanding that, in terms of warfare, there are three different wars. The first, in 1990–1991, involves the liberation of Kuwait; the second runs from the American invasion in 2003 until the fall of Baghdad and is *primarily* a conventional war with some irregular elements. The third runs from the summer of 2003 until the end of 2011 and is more like a typical insurgency, although that insurgency, like most insurgencies, had several wellsprings. On the one hand, there is a local/regional Sunni religious/tribal element and, on the other, an international radical Salafist dimension. This section focuses on the third war (Iraq III), with special attention to the Anbar "Awakening" and the "Surge."

Note: When people talk about the war in Iraq, they usually are referring to the US invasion (Iraq II in this volume), and it is always useful to remind ourselves that the US Congress (both the House and the Senate) voted to go to war in 2002 by a margin of 296 to 133 in the House and 77 to 23 in the Senate. One supposes that many wished they had a do-over later in the conflict, as this phase went badly, but the fact remains this was initially a popular war, one definitely sanctioned by the US Congress.

Benjamin Buley, *The New American War: Military Culture and the Political Unity of Force* (London: Routledge, 2008). American military is the best-educated, best-trained, and most careful about civilian casualties in US history. But does that count for enough? See also Linda Robinson, *Tell Me How This Ends: General David Petraeus and the Search for a Way Out of Iraq* (New York: Public Affairs, 2008), who—along with Bing West—captures the tipping point in the war in Iraq.

George W. Bush, "Iraq" and "Surge," in his *Decision Points* (New York: Crown, 2010), pp. 221–71, 355–94. In "Iraq," the president describes his actions, his motives, and his faith that he did the right thing to invade but also considerable disappointment at the way things did not go as planned. A revealing piece.

In "Surge," the president discusses how, in 2006, few in the Washington establishment (including within his own administration) and the American media saw a victory as possible for the United States against insurgents in Iraq. His former secretary of state Colin Powell was saying that the war was lost and the surge could not succeed. Yet Bush insisted it was possible and made that possibility stay open. Illuminating and a must-read for anyone interested in the war in Iraq, regardless of that person's viewpoint on the war itself.

Dick Cheney, with Liz Cheney, "A Nation at War," "Liberating Iraq," Intelligence and Politics," and "Surge," in his *In My Time: A Personal Memoir* (New York: Threshold Editions, 2011). Cheney vigorously defends himself against all the charges leveled against him by anybody on the political spectrum. Dispassionate and perhaps overly descriptive but a necessary component in the mosaic that is historical truth. For example, he is most trenchant in noting that the Iraq Study Group's report on the ongoing war in Iraq mentioned the word *victory* only in one dimension—in connection with an al-Qaeda victory. As he writes, "This was not a strategy for winning the war."

John Duffield and Peter Dombrowski (eds.), *Balance Sheet: The Iraq War and National Security* (Stanford, CA: Stanford University Press, 2009). Ends up declaring that it was a negative in terms of American national security but not by much.

Dennis Filkins, *The Forever War* (New York: Knopf, 2008). Maybe, if history is any guide, this observation is correct. Read how and why.

———, "Strategy Tragedy?" *New York Times Magazine*, February 19, 2006, pp. 27ff. Turned out not to be but no sign that Filkins got the message as yet.

General Tommy Franks, *American Soldier* (New York: Regan Books, 2004). Knew when to head for home. Very weak on fallback strategy after Turkey took the proposed Fourth Infantry Division drive into the Sunni heartland out of the equation.

Matt Gallagher, *Kaboom: Embracing the Suck in a Savage Little War* (Philadelphia: Da Capo Press, 2010). The view from the lieutenant level in 2008, as the surge and the Anbar Awakening took hold and, despite all odds, the American counterinsurgency effort succeeded. Funny and irreverent, the work contains some important insights:

> In conventional warfare, the order of war dissolved into anarchy as time yielded more and more blood. Unconventional, decentralized warfare was the exact opposite. In fluid theory and historical practice, victorious counterinsurgencies served as a shining inverse to said conventionality because, through anarchy and bloodshed, order would eventually be established. This was the war I trained for, brooded over and studied. Then there was the war I fought.

Anne Garrels, *Naked in Baghdad* (New York: Farrar Straus Giroux, 2004). Although it's a war and clash of civilizations, Anne seems to think it's all about her, hence the title, when Saddam's men came to interview her as she watched the war unfold with few clothes on.

Michael Gordon and Bernard Trainor, *Cobra II* (New York: Pantheon Books, 2006). A straightforward account of the first part of the Second Gulf War, or Iraq II, as it is known in this bibliography. Lots of good detail and enough military description to keep me happy throughout.

———, *Endgame: The Inside Story of the Struggle for Iraq from George W. Bush to Barack Obama* (New York: Pantheon Books, 2012). A thick, intricate, and incredibly detailed account of the Iraq story, from the overthrow of Saddam to the withdrawal of the last American forces and the emergence of Nuri al-Maliki as a potential dictator (who had proved his worth in retaking Basra and Sadr City and defeating the Shia militia but then wanted the Americans out, the Kurds subdued, and the Sunnis back in a box). The military playouts are overwhelmed by diplomatic maneuverings and an endless parade of players and want-to-be players. A difficult book to digest but one that needs to be read by anyone wishing to understand the complexity of nation-building during wartime or insurgency time or anytime.

Lee Harris, *The Suicide of Reason: Radical Islam's Threat to the West* (New York: HarperCollins, 2007). To the Salafists, however, antireason is divine reason.

Lesley Hazleton, *After the Prophet: The Epic Story of the Shia-Sunni Split in Islam* (New York: Doubleday, 2009). A very important work for anyone wanting to understand Islam today.

Christopher Hitchens, "Holiday in Iraq," *Vanity Fair*, April 2007, pp. 130ff. He goes to Kurdistan and finds it a most welcoming place, one with statues of George W. Bush.

Robert Jervis, "The Iraq WMD Intelligence Failure: What Everyone Knows Is Wrong," in his *Why Intelligence Fails* (Ithaca, NY: Cornell University Press, 2010), pp. 123–55. A closely argued and very important analysis, one that must be studied by anyone purporting to be knowledgeable about the Second Iraq War and the intelligence failures leading up to it. Argues persuasively that the postmortems are wrong and that reevaluation of the evidence could only have made the intelligence estimates "less certain," not proven them incorrect. States clearly that being wrong is not the same has having made "avoidable and blameworthy errors." At best, given the available intelligence, one could have said, "There was no *firm* evidence of WMDs," not "There were no WMDs." Despite everyone running to cover after the fact, politicization did not override reality; the intelligence community (including the Russians, the Chinese, and the Arabs) all thought Saddam had them. And no one could produce an accurate account of what he actually had, ironically because the Iraqi government's denial and deceptions ("strange and self-defeating") actually seemed to support the assumption he did have them. Concluding that the "paucity of HUMINT was a majority cause of the Iraq debacle" ("Curveball," the German source, spoke little German and was never vetted by the Defense Intelligence Agency, CIA, or anyone besides his German handlers!), Jervis states that the most reasonable conclusion was that "Saddam had a broad range of WMD programs and at least a small store of weapons." Everyone trying to put these events in perspective should read this article.

David Levering Lewis, *God's Crucible: Islam and the Making of Modern Europe, 570–1215* (New York: Norton, 2008). A provocative and interesting account of Christian–Muslim interaction and how it shaped Europe.

Peter Mansoor, *Baghdad at Sunrise: A Brigade Commander's War* in Iraq (New Haven, CT: Yale University Press, 2008). One of the US architects of victory in Iraq tells how it all happened. Quite an important work that should be read by anyone interested in Iraq III.

Jim Michaels, *A Chance in Hell: The Men Who Triumphed over Iraq's Deadliest City and Turned the Tide of the War* (New York; St. Martin's Press, 2010). The Ramadi portion of the equation and one that should be read in conjunction with Bing West, *The Strongest Tribe*, and Matt Gallagher, *Kaboom: Embracing the Suck in a Savage Little War* (both reviewed in this chapter), about the turnaround in Anbar Province and the Sunni triangle, including the Battle of Fallujah, which eventually gave the United States and Iraqis victory. Colonel McFarland finds out that the Iraqis, far from wanting to know when the Americans are going to leave, want to make sure they are going to stay until al-Qaeda is beaten. As for the Sunni tribesmen, "[t]hey fought us for years, they respect our military prowess, and are now very happy that we're all on the same side."

Oliver North, *War Stories: Operation Iraqi Freedom* (Washington, DC: Regency, 2003). A quick, breezy, and easy read of a former Marine and now FOX correspondent embedded in the advancing American forces. Enjoyable shots taken at mainstream media ("What quagmire?"), including Peter Arnett. Rightly praises US forces as the best-educated, best-trained, best-equipped, and most professional in America's history.

Thomas Ricks, *Fiasco: The American Military Adventure in Iraq* (New York: Penguin Press, 2006). Strongly against the US invasion of Iraq and does a good job giving his point of view, although much of what he wrote about the military endgame was largely overridden by subsequent events.

———, *The Gamble* (New York: Penguin, 2009). A good reporter but one whose basic mindset doomed him to missing the "turn" in Iraq and led him to blithely predict that the war would continue for 15 more years.

Linda Robinson, *Tell Me How This Ends: General David Petraeus and the Search for a Way Out of Iraq* (New York: BBS Public Affairs, 2008). Brilliant analysis of counterinsurgency, well told.

Donald Rumsfeld, *Known and Unknown: A Memoir* (New York: Penguin Books, 2011), See especially "Saddam's Miscalculation," "The Occupation of Iraq," "Wartime Detention," "Pulling on Our Boots," and "The Long, Hard Slog." Rumsfeld was actually right in a number of instances during this period, but his unwillingness to take any blame for any major miscalculations or mistakes is irritating. This is more than a little odd, as he blames Paul Bremer—correctly—for the largest blunders of the war but slides around the problem that Bremer in fact worked for him! Still, no one can truly say they understand the Second and Third Iraq Wars without reading this memoir.

Nir Rosen, "Jordan's Jihadist Insurgents," *New York Times Magazine*, February 19, 2006, pp. 54ff. Angry young men looking for action, many subsequently attracted to ISIS.

Karl Rove, "The Surge," in his *Courage and Consequence* (New York: Threshold Editions, 2010), pp. 471–84. An insider's view of a crucial turning point in the history of US warfare and involvement in Iraq. Doubt his account will be read, let alone believed, by many but more's the pity. A good look at power and misperceptions and correct perceptions in action.

Olivier Roy, *Globalized Islam: The Search for a New Ummah* (New York: Columbia University Press, 2004). Probably more than a few internal tensions are working against the new Ummah. Roy captures some of them, not that ISIS cares.

Keith L. Shimko, *The Iraq Wars and America's Military Revolution* (Cambridge: Cambridge University Press, 2010). Very useful overview of Iraq I, II, and III, as well as Afghanistan I and II, on a continuum, all in the context of the revolution in military affairs.

Richard Shultz, *The Marines Take Anbar: The Four Year Fight against Al Qaeda* (Annapolis, MD: Naval Institute Press, 2013). The single-best summary of the war harshly decrying the mistakes and praising the successes. A must-read if anyone is to understand what happened in Iraq II and III. Also reviewed in chapter 8.

William Stuntz, "Doubling Down in Iraq," *Weekly Standard*, November 20, 2006, pp. 32–33. A controversial strategy that worked.

John Tierney, "Can a Popular Insurgency Be Defeated?" *Military History* (March 2007), pp. 33–45. Yes, says the author. Many have been in the past; some are being so dealt with in the present and will undoubtedly be so handled in the future.

Bing West, *The Strongest Tribe: War, Politics, and the Endgame in Iraq* (New York: Random House, 2008). Called the "turn" in Iraq by listening to junior officers he interviewed in the field. One of the few authors who got the counterinsurgency effort in Iraq correct as it was happening. This is one of the best and most-engaging books about the wars with Iraq, especially wars II and II. His analysis of the situation in Afghanistan also looks to be quite prescient.

Bing West and Major General Ray Smith, *The March Up* (New York: Bantam Books, 2004). Fast-paced account of the Second Iraq War from the US Marine perspective. West has very good eyes for what matters in battles and what does not.

Bob Woodward, *Plan of Attack* (New York: Simon and Schuster, 2004) Bush, the menace. A useful read about inside Washington decision making and the perceived need for decision makers to vent with Woodward before others do.

———, *State of Denial* (New York: Simon and Schuster, 2006). Bush, the dope. Bush refused to admit the war was going badly and would be lost. Many people were convinced not only that it was wrong to go to war but also that the United States would be stuck in another "quagmire."

Note: The use of the term *quagmire*" should be banned to all columnists and pundits because it has been used so many times erroneously since *The Making of a Quagmire* about the war in Vietnam was published 30 years ago.

———, *The War Within* (New York: Simon and Schuster, 2008) Bush, the rock. Bush ended up being one of the few supporters of the surge in Washington and was, sorry to say for many pundits, politicians, and others, correct on this aspect of the war. Woodward is so intent on trashing Bush and his policy toward Iraq that he misses the vital turn in the war that results from implementation of the "surge," when everybody, including his own secretary of state and the Joint Chiefs of Staff—to say nothing of many civilian politicians—were urging a pullout. A startling moment in time captured.

Evan Wright, *Generation Kill* (New York: Henry Holt, 2004). Portrait of today's American armed forces in action. Good vignettes of individual soldiers.

Xenophon, *Anabasis* (Cambridge, MA: Harvard University Press, 1998). Greeks get into great trouble in Persia when their patron dies and they have to fight their way home. Many parallels here and lessons, including seeing things through even when they look their worst. Some authorities say that the Persians actually let the Greeks go, making only a minimum effort to stop their retreat.

David Zucchino, *Thunder Run* (New York: Atlantic Monthly Press, 2004). US forces race to Baghdad and create the world's fastest and longest blitzkrieg, but few apparently care to measure success that way. A good case of a military campaign losing credibility due to political factors that came later.

Note: For more works on the subjects of insurgency and counterinsurgency throughout history, see the recommended readings listed in chapter 8.

Chapter Thirty-Three

Weapons in War

> "Equipment can drive tactics. In a nimble military organization it often does."—C. J. Chivers

In the study of this important warfare subject, we need to avoid hyperskepticism ("The stirrup had no impact on warfare at all.") *and* technological determinism ("The stirrup made Middle Age warfare possible."). Of primary interest is how long weapons and weapons systems take to become part of any society's military, the ways in which advances in weaponry travel (i.e. through trade or warfare), and the resistance to innovation by various military elites. Military and political figures, historians, and others are often very excited about the potential of weapons, but sometimes the gestation period between their invention, even their deployment, and their actual impact on warfare is very long indeed. See for example, the German *Vergeltungwaffe* (retaliation) weapons known as the V-1 and V-2, which appeared during World War II; however, the cruise missile V-1 was only able to play a significant role by the time of the Gulf War of 1991 (although the V-2 would be central to the technological developments during the Cold War).

Reviewing some of these works, I was struck by the tremendous human effort that has gone into the refinement of weapons, perhaps as much as has gone into the building of churches, mosques, and shrines, or even world literature. There is a truly amazing panoply of weapons invented by humans stretching back to the dawn of history. Worshiping Mars has taken many forms, but the development of weapons and weapons systems has surely been one of the most consistent and costly.

There is also the element of the sheer volume of weapons that a people can bring to bear on their opponents. Allan Millett and Peter Maslowski in *For the Common Defense* (New York: Free Press, 1984, p. 412), give the statistics for America during World War II, numbers that truly underscore the meaning of the "Arsenal of Democracy" (see table 33.1).

The following is just a sample of works devoted to the impact of weapons and weapons technology on warfare. Interestingly enough, as one examines the role of weapons in history, one needs to pause to consider an important point made by Professor Jonathan Roth of San Diego State University: Most of the weapons used by warriors today were invented 100 years ago; the rifle, the machine gun, the tank, the airplane, the submarine, and the grenade are all still in use, albeit in modern versions, and none have had more of an impact than the internal-combustion engine, itself long past its centennial.

It should also be noted that combinations of weapons have always been part of the ongoing dialectic between armies and cultures. For example, during the early stages of the Gunpowder Revolution, while many armies on the continent were combining pikemen and harquebuses, the English stayed with the combination of longbows and billmen (who used a hooked chopping blade with several pointed projections mounted on a staff) in concert.

Interesting to remember as well that, regardless of time, situation, or culture, the adoption of new weapons is very often impeded by the conservative nature of most officer corps in most militaries.

Also remember that many "new weapons" have important historical antecedents. For example, we think of biological and chemical weapons as "new," but the Mongols practiced biological warfare by tossing Black Death corpses over the walls into besieged cities; Lord Amherst sent smallpox blankets to the Native Americans; and in January 1944, the Germans reflooded the Italian Pontine Marsh (which had been drained and demosquitoed in the 1930s) and reintroduced malarial mosquitoes.

Table 33.1.

Tanks	86,000
Artillery pieces	112,000
Shoulder weapons	14 million
Trucks and jeeps	2.4 million
Combat vessels	1,200
Landing crafts	82,000
Bombers	96,000
Fighters	88,000
Transports	25,000
"Liberty" ships	2,600

Christon Archer, John Ferris, Holger Herwig, and Timothy Travers, *World History of Warfare* (Lincoln: University of Nebraska Press, 2002). A useful tour de horizon of warfare to start one off on the subject of weapons and warfare and their interaction.

M. J. Armitage and R. A. Mason, *Air Power in the Nuclear Age* (Urbana: University of Illinois Press, 1983). A challenging and glaringly realistic assessment of air power seeking—and sometimes claiming—to be the "dominant factor." Some eye-popping statistics: In the Korean War, the United States lost 3,100 aircraft and helicopters, half of them to combat, while in Vietnam, it lost an astonishing 7,800 aircraft and helicopters to all causes. In the former, the authors declare an "almost total failure of interdiction." Quite an indictment. Snappy reading.

James Phinney Baxter, *The Introduction of the Ironclad Warship* (Cambridge: Cambridge University Press, 1933). The "passing of the wooden fleets" is captured in sharp relief. Surprisingly enough, at first, the revolving turrets created more of a stir than the iron armor itself.

Patrick Beesly, *Very Special Intelligence: The Story of the Admiralty's Operational Intelligence Centre 1939–1945* (London: Hamish Hamilton, 1977). Intelligence as a weapon, always remembering Clausewitz's dictum, "A great part of information obtained in war is contradictory, a still greater part is false, and by far the greater part is of doubtful character." The Operational Intelligence Centre (OIC) accounted for each and every German U-boat built between 1935 and 1945 as the Battle of the North Atlantic was fought and won.

Antony Beevor, *Crete: The Battle and the Resistance* (London: John Murray, 1991). This book can also be read for profit on guerrilla and irregular warfare but is reviewed here primarily because it focuses on a military turning point in the use of paratroops in battle. Although the Germans, the Japanese, the British, and the Americans all used paratroops with a range of successes, the German operation in Crete effectively sounded the death knell for using paratroops in pitched-battle situations, so severe were their casualties (although the Germans would drop paratroops in Yugoslavia in 1944 in their fight with partisans). The Germans also planned to try to seize Malta during 1942 using paratroops and Italian infantry, but they abandoned it before the fact.

Donald Burgett, *Currahee!* (Boston: Houghton Mifflin, 1967). Using his regiment's battle cry as the title, the author details his jump into and subsequent battle for Normandy, securing the approaches to Utah Beach with the

101st Airborne, a unit that would suffer 350 percent**[AQ14: Statistic OK? This reads like the whole unit died 3½ times.]** casualties by war's end. Vivid descriptions of friendly fire (from both naval support and aircraft), White Russians charging on horses, almost scalping a dead German with long blond hair, and getting severely wounded. He would later be wounded several times more and receive the Bronze Star.

Peter Harclerode, *Wings of War: Airborne Warfare 1918–1945* (London: Cassell, 2007). Although Leonardo da Vinci drew up plans for parachutes and gliders in the 16th century, it was not until 1918 that Italians dropped from British planes behind Austrian lines on sabotage and intelligence-gathering missions. The Germans would drop airborne troops in Denmark, Norway, Belgium, Holland, and later Crete, while the British would use them in Italy, North Africa, Sicily, Holland, Greece, Germany, Burma, and Malaya. The United States deployed them in Sicily, France, Italy, Holland, Germany, New Guinea, and the Philippines. The Imperial Japanese Army deployed them against the Dutch East Indies and the Philippines. The Soviet Union dropped troops into Romania and all over the western Soviet Union in conjunction with partisans as well as fighting the Germans. General Zhukov also used paratroops in the battle in front of Moscow during early 1942, but in this case, the paratroops suffered extremely heavy losses in German counterattacks. During World War II, most fighting by airborne units for all armies was done by soldiers already on the ground.

John C. McManus, *September Hope: The American Side of a Bridge Too Far* (New York: Penguin, 2012). Follows the US 101st and 82nd Airborne Divisions into Holland using 1,544 C-47s and 478 gliders. Claiming the Americans "fought their hearts out," the author blames Eisenhower rather than Montgomery for putting a "deeply flawed plan" into action for political purposes.

Note: After the official end of World War II, the United States dropped paratroops and OSS personnel into Manchuria near Mukden to liberate American and other Allied prisoners of war (including General Wainwright) in Operation Cardinal. This irritated the Soviet Union somewhat.

General Lloyd Moses, *Whatever It Takes: The Autobiography of General Lloyd R. Moses* (Vermillion, SD: Dakota Press, 1991). Moses took part in Operation Varsity, the largest airborne assault of World War II, crossing the Rhine in 1945 in the battle for the Ruhr Pocket. Along with 1,348 gliders, 1,696 aircraft were involved.

Other Works on Weapons

Jeremy Black (ed.), *War since 1900* (London: Thames and Hudson, 2010). Lavishly illustrated and containing a series of essays on the major weapons and wars of the 20th century, this volume also makes the unintended point that most of the weapons of today—rifle, machine gun, artillery, plane, tank, and submarine—were invented prior to or during World War I.

Jurgen Brauer and Hubert van Tuyll, *Castles, Battles and Bombs: How Economics Explain Military History* (Chicago: University of Chicago Press, 2008). The authors apply six economic principles, including opportunity costs, expected marginal costs and benefits, and diminish marginal returns, to a number of case studies. Interesting but not altogether convincing in terms of predictive value.

Bernard Brodie and Fawn Brodie, *From Crossbow to H-Bomb* (Bloomington: Indiana University Press, 1973). A good, breezy introduction to the role of weapons in warfare, including the "nuclear revolution" and the strategic changes it brought and did not bring.

Stephen Budiansky, *Air Power: The Men, Machines, and Ideas That Revolutionized War, from Kitty Hawk to Gulf War II* (New York: Viking, 2004). A once-over, light, but interesting account of machines that fly and how they have affected war, from biplanes and zeppelins to jets and helicopters and unmanned aerial vehicles.

C. J. Chivers, *The Gun* (New York: Simon and Schuster, 2010). Traces, sometimes in excruciating but often interesting detail, the history of automatic weapons, starting with the Gatling gun, proceeding to the Maxim gun and various submachine guns, before arriving at the AK-47, named for Sargent Mikhail Kalashnikov, whose prototype was accepted by the Soviet military in 1947. The author says, "The Kalashnikov Era had arrived. We are living in it still." Today there are more than 100 million of these sturdy, lightweight, rapid-fire automatic weapons in use. They have become both the symbol of resistance and warfare as well as revolutionary change. The author is right in also seeing them as powerful weapons of suppression and state authority, especially in the failed states of today. Useful contrast with the introduction of the US M-16 and its teething troubles in Vietnam.

John Clements, "Medieval Armor: Plated Perfection," *Military History* (July 2005), pp. 38–44. Counters the usual assumption that medieval knights were clumsy and slow; instead, it argues that armor was much more flexible and the fighting was much faster. He claims that mud, slippery blood, and opponents' weapons, not inflexible armor, were the causes for such defeats of knights as Courtrai (1302) and Crecy (1346).

Bryan Cooper, *The Battle of Cambrai* (New York: Stein and Day, 1968). November 1917 saw the first use of massed tanks at the Battle of Cambrai. Supported by 1,000 guns, 474 tanks blasted through the German lines, eventually penetrating to a depth of 5 miles (unheard of during this period of trench warfare), but the British did not follow up, the German "tank panic" subsided, and the whole effort ended up as a "cruel disappointment." A new technology and its teething problems are the story here.

Arthur Cotterell, *Chariot: From the Chariot to Tank, the Astounding Rise and Fall of the World's First War Machine* (Woodstock, NY: Overlook Press, 2005). The chariot was the defining weapons system for hundreds and hundreds of years in many, many cultures and societies. Eventually done in by infantry with iron swords and cavalry bowmen, it nevertheless was one of the most important weapons prior to 1200 BCE.

Russell S. Crenshaw Jr., *South Pacific Destroyer: The Battle for the Solomons from Savo Island to Vella Gulf* (Annapolis, MD: Naval Institute Press, 1998). The naval campaign during the Battle for Guadalcanal in 1943 is carefully presented, but the "lean months" section and afterword contain very useful information about the poor performance of the US Navy's torpedoes, including the whys and wherefores.

Wilbur Cross, *Zeppelins of World War I* (New York: Paragon House, 1991). A good case study of a weapon that showed promise (when the rigid airship of Count Ferdinand von Zeppelin was invented in 1900) but, after initiating the German terror bombing of cities in June 1915, came a cropper with British fighters, so by 1937, when the Hindenburg disaster occurred, it was safe to say that the "rigid airship was now nothing more than a ghost from the past."

Diagram Group, *The New Weapons of the World Encyclopedia: An International Encyclopedia from 5000 BC to the 21st Century* (New York: St. Martin's Griffin, 2007). With a multitude of pen-and0ink drawings, this work provides a great deal of useful information and especially the opportunity to view the wide range of weapons created by humans to kill and maim other humans. Truly a dismal aggregation from a humanistic point of view, albeit a clear period-by-period examination of the tools of warfare.

Christopher Duffy, *Fire and Stone: The Science of Fortress Warfare 1660–1860* (London: David and Charles, 1975). A richly illustrated (unfortunately only in black and white) volume dealing with all manner of "barriers to intrusion," including castles, forts, citadels, and towns with ramparts. Sees a constant dialogue in the struggle between forces of destroying those fortifications and the development of new forms. Points out that some people eschewed fortifications, such as the Spartans, who, he claims, thought walls would make its soldiers "effeminate and cowardly."

T. N. Dupuy, *The Evolution of Weapons and Warfare* (Indianapolis: Bobbs-Merrill, 1980). The Age of Muscle, the Age of Gunpowder, and the Age of Technology all get a once-over.

John Ellis, *A Short History of the Machine Gun* (New York: Random House, 1975). Amazingly, a nearly sentimental tale of this marvelously horrible firearm. Wonderfully done, some of the quotes and images will haunt you for a long time. Lots of humor and lots of insights. A superior read.

Note: For a droll account of the life of the inventor of the Maxim gun, Hiram Maxim of Sangerville, Maine, see his autobiography, Hiram S. Maxim, *My Life* (London: Methuen, 1915). Although the United States was slow to see the value of his invention, England, France, Russia, Germany, and Turkey did not miss the opportunity for carnage he had created.

George Forty, *M4 Sherman* (Pooles, UK: Blanford Press, 1987). The "Ronson Lighter" (because it often "lit up" on the first strike) is lovingly remembered in this extensive and well-illustrated praise poem. There is a short history of US tanks, including the precursor M-3 Lee. Judged undergunned with a 76-mm cannon and too lightly armored as well as having a too high silhouette, the Sherman nevertheless was considered to be mechanically reliable and having a good fire-control system. Most importantly, more than 49,000 of them were built, and they served in Europe, North Africa, and the Pacific during World War II and thereafter with a variety of forces, such as those in Korea and Pakistan. The 49,000 figure exceeded the total German production of *all* tracked vehicles from 1934 to 1945. Also, in the Yom Kippur War of 1973, the Israelis used a number of "Super Shermans" upgunned with 105-mm guns and additional armor, and they played a significant role in the crucial point of the war during Operation Gazelle (see chapter 39). Alistair Horne says they were excellent in 1942 (especially as a replacement for the undergunned British Centurions) but outdated by 1944.

Note: General Lucius D. Clay, in charge of US military procurement during World War II, put it best: "We were never able to build a tank as good as a German tank. But we made so many of them that it didn't really matter." Except to the American tank crews, perhaps.

J. F. C. Fuller, *Armored Warfare* (Harrisburg, PA: Military Service, 1943). This is a collection of Fuller's lectures that were delivered in 1932 advocating that mechanized warfare was the warfare of the future. Argues that the British Tank Corps of 1918 was "far more an armored force than any existing today" and laments the failure of the British military to appreciate its potential. Detailed maps and diagrams help flesh out his thinking. Fuller wrote more than 40 books and was one of the first to see the advantages of massing tanks in formation and "punching through" defensive lines. His ideas were adopted more readily by the losing Germans than the victorious British and French following World War I.

Terry Gander, *The German 88: The Most Famous Gun of the Second World War* (Barnsley, UK: Pen and Sword Books, 2009). A paean to the Krupp's 88, although the author argues that the ammunition, not just the gun, was crucial to its enormous success as both an anti-aircraft and antitank gun.

John Geoghegan, *Operation Storm* (New York: Crown, 2013). A little-known tale of the huge aircraft carrying submarines (I-400 class) built by the Japanese for an attack on Washington and New York, then the Panama Canal, and then Ulithi. The biggest submarines in the world (until 1961, when the United States launched the Ethan Allen class in 1961). Four hundred feet long and technologically advanced, the I-400s would serve as a prototype platform for the eventual US Navy missile-carrying submarines (first Regulus and then Polaris). Author also claims *kamikaze* was an American popularization of the Japanese term (*tokko*) for suicide tactics.

Richard Guilliatt and Peter Hohnen, *The Wolf: How One German Raider Terrorized the Allies in the Most Epic Voyage of World War I* (New York: Free Press, 2010). As a weapon, the surface raider made for good stories but was quite inefficient compared to submarines.

Holger H. Herwig, "The Battlefleet Revolution, 1885–1914," in *The Dynamics of Military Revolution 1300–2050*, edited by Williamson Murray and MacGregor Knox (Cambridge: Cambridge University Press, 2001), pp. 114–31. At the heart of the estrangement between Great Britain and Germany—and it wasn't all Germany's fault.

Tracey R. Hill, *The Catapult: A History* (Yardley, PA: Westholme, 2007). A very useful weapon in a siege and one with a 1,000-year history to boot. This is an enjoyable read.

Beekman House, *Tanks and Weapons of World War II* (New York: Crown, 1973). Although this work is about the equipment and weapons of World

War II, it can serve as a good introduction to the roles design creation, logistics, and weapons upgrades can play before and during a war. Many fine illustrations, photos, and charts, as well as an easy-to-understand text, make this a good place to start one's education on this subject.

Robert Jackson, *Submarines of the World* (London: Brown Books, 2000). A brief history illustrated with more than 300 undersea weapons and submarines, a list running from Albacore (United States) to Pioneer (Confederate States of America) to Yankee Class (Soviet Union) and Zoea (Italian).

R. V. Jones, *The Wizard War: British Scientific Intelligence 1939–1945* (New York: Coward, McCann, 1978). First-rate and highly readable account of the British versus the Germans in the technology phase of World War II. The intelligence and code-breaking of the Allies ended up giving them a substantial advantage in both the European and Pacific theaters. Shows intelligence as a true weapon of war. Michael Howard said, from the British point of view after they broke the German Enigma codes and produced what they called Ultra, it was "like playing poker with marked cards, albeit against an opponent with a consistently better hand than you."

Larry Kahaner, *AK-47: The Gun That Changed the Face of War* (New York: Wiley, 2007). And made lawless societies more lawless and failed states incurable. Quite an amazing description of one weapon that has had an impact out of proportion to its individual soldier orientation.

John Keegan, *A History of Warfare* (New York: Vintage Books, 1993). A sweeping view of weapons, strategy, and changes in warfare under such topics as "Stone," "Fire," and "Iron."

———, *The Price of Admiralty: The Evolution of Naval Warfare* (New York: Penguin Books, 1990). Keegan looks at four battles—Trafalgar, Jutland, Midway, and the Atlantic—to give insights into tactics, strategy, and technological transformation of naval operations.

Theodore Knappen, *Wings of War* (New York: G. P. Putnam's Sons, 1920). This often-overlooked book is extremely valuable in documenting the way in which a new weapon system, in this case the airplane, came into production in the United States. In 1914, the Germans had 1,200 military aircraft; the French, 300; Great Britain, 250; and the U. S. "practically none." In 1915, the United States did not produce a single military aircraft, and in 1918, it turned out 11,800. This book tells how all that happened.

Milton Leitenberg and Raymond A. Zilinskas, with Jens H. Kuhn, *The Soviet Biological Weapons Program: A History* (Cambridge, MA: Harvard University Press, 2012). In a stunningly holistic and definitive account (running almost 900 pages), this work catalogues the entire history of 65 years of Soviet and Russian biological warfare research, tracking the various civilian and military *Biopreparat* programs that employed as many as 65,000 people from 1928 to 1992 and whose managers resisted the efforts of such Russian leaders as Gorbachev and Yeltsin to shut them down. There is extensive, in-depth coverage, not only of the various civilian and Ministry of Defense (MOD) efforts, illustrated with useful diagrams, but also of the various doctrines for using weaponized pathogens.

The authors document the extensive Soviet and later Russian Federation violations of the 1972 Biological and Toxin Weapons Convention, ending with a chilling reminder that, today, we do not know the current status of all the programs not verifiably terminated—nor why this is the case. "There is no available information as to why the Russian government is so intent on keeping its vast BW program a secret, or why the veil of secrecy is maintained over RF MOD and anti-plague facilities and their operations" (p. 712). This is a very important, disturbing book.

Norman Longmate, *Hitler's Rockets: The Story of the V-2's* (London: Hutchinson, 1985). An engaging account focusing on the role of the V-2 in the war between Germany and Great Britain. Lots of good inside baseball on who did what and who didn't do what. An interesting note: The August 18, 1943, huge British air assault on Peenemunde cost them 40 bombers, and although there was considerable damage, the Germans didn't repair a lot of the above-ground facilities and thus gained 9 months' respite from further heavy raids.

E. W. Marsden, *Greek and Roman Artillery: Technological Treatises* (Oxford: Oxford University Press, 1971). All anyone would ever want to know about these instruments, including how to build your own.

John C. McManus, *September Hope: The American Side of a Bridge Too Far* (New York: Penguin, 2012). See the review earlier in this chapter.

William H. McNeill, *The Pursuit of Power: Technology, Armed Force and Society since A.D. 1000* (Chicago: University of Chicago Press, 1984). A 1,000-year look at the way technology and warfare have been intertwined. Very good on the eras of Chinese and European predominance and the relationship between those successes and the weapons they used. See especially "The Era of Chinese Predominance" (pp. 24–62) and both "The Business of War in Europe, 1000–1600" (pp. 63–116) and "Advances in Europe's Art of War 1600–1750" (pp. 117–43). Very clever in pointing out that Chinese superiority rested on the twin pillars of the ease with which peasants could be taught to use a crossbow (a "couple of hours") and the "high skill" needed in manufacturing reliable trigger mechanisms on which the crossbow depended.

Michael J. Newfeld, *The Rocket and the Reich: Peenemunde and the Coming of the Ballistic Missile Era* (New York: Free Press, 1995). Said rockets were a huge diversion of scarce resources for literally no military payoff during World War II for Germany. But they revolutionized warfare for others later.

Robert O'Connell, *Soul of the Sword: An Illustrated History of Weapons and Warfare from Prehistory to the Present*, illustrated by John Batchelor (New York: Free Press, 2002). Stating that "war is not simply mass violence," the author gives the reader a once-over, light history of weapons arranged under such snappy titles as "Dawn of the Bow" and "False Pinnacle." World War I becomes "Accidental Armageddon," and World War II, "The Grudge Match." The illustrations are most helpful.

Stephen Phelps, *The Tizard Mission: The Top Secret Operation That Changed the Course of World War II* (Yardley, PA: Westholme, 2010). The cavity magnetron saves the Allies, assisting in research and development of jet propulsion, more effective radar, and the atomic bomb, and its transfer to the United States is a turning point in the technological war between the Allies and Axis.

Edwin Pratt, *The Rise of Rail Power in War and Conquest 1833–1914* (New York: J. J. Lippincott, 1916). A paean of praise for the ease and speed with which soldiers could be brought up to the front.

George Raudzens, "War-Winning Weapons: The Measurement of Technological Determinism in Military History," *Journal of Military History* 54 (October 1990), pp. 403–33. A quick look at weapons and their impact, real and imagined, on history.

Dennis E. Showalter, *Railroads and Rifles: Soldiers, Technology and the Unification of Germany* (Hamden, CT: Argon Books, 1975). See especially "The Genesis of the Needle Gun" (pp. 77–90). The needle gun was a seminal invention, a breech-loading rifle in the era of flintlocks and percussion muskets. Proved decisive in the Battle of Koniggratz, which made a Prussian-dominated Germany a reality. After the battle, every major country in Europe adopted breech-loading rifles.

P. W. Singer, *Wired for War: The Robotics Revolution and Conflict in the 21st Century* (New York: Penguin Press, 2009). A thick, hard-to-read, but necessary analysis of what lies ahead on the war front. Singer is truly enamored with these new weapons.

Stephen A. Skubinna, "Hermit Kingdom's Naval Genius: Korean Admiral Yi's 'Turtle Ships' Were the First Ironclads," *Military History* (April 1988), pp. 10ff. A marvelous example of a military innovation from the mid-15th century in Korea that was hugely successful but then never followed up. Remember that the first ironclads were not developed in the United States until the Civil War.

Richard Snow, *A Measureless Peril: America in the Fight for the Atlantic, the Longest Battle of World War II* (New York: Scribner, 2010). A chilling account of how the Unterseeboote (U-boat) almost won the war in the Atlantic and how, had German followed a different strategic course before and during the early stages of the war, things might have turned out differently. Considerable emphasis on the technology of both sides that aided or harmed their respective military efforts.

Hugh D. H. Soar, *The Crooked Stick: A History of the Longbow* (Yardley, PA: Westholme, 2004). How the longbow found fame and fortune among the Welsh and the English but not many others. Took a lot of practice compared to other bows; crossbows; and, later, muskets.

Edwin Tunis, *Weapons: A Pictorial History* (Baltimore, MD: Johns Hopkins University Press, 1999). A clear, simple, and very well-illustrated—with drawings—account of the major weapons over time. Quite enjoyable and easy to understand.

Martin van Creveld, *The Age of Airpower* (New York: Public Affairs, 2011). The author sets out to disprove the impact of air power on war, and he is right to disparage the grandiose claims of its primacy and refocus attention on nuclear weapons, thus the title is most ironic because he goes overboard in trying to disprove airpower's impact. Certainly he cannot have it both ways, and he avoids some aspects that undercut at least part of his central thesis. For example, the Germans were forced to divert more than 2.1 million men in order to try to halt the Allied air offensive (and repair the damage after its raids), which dropped 2 million tons of bombs. His example of Vietnam actually does not add to his case at all because the airpower was so consistently and fatally *misused* by the United States. Still, there is a lot of useful historical information here, if one can get past the somewhat slanted analysis.

———, *Technology and War: From 2000 B.C. to the Present* (New York: Free Press, 1991). An intriguing, stimulating, and in some ways controversial book. Argues, for example, that the Gunpowder Revolution depended as much on writing paper and organization as on the application of the substance itself. Looks at "roads, vehicles, communications and timekeepers" as having done as much as weapons themselves for changing the nature of war.

Weapons: A Visual History of Arms and Armor (New York: DK Publishing, 2006). A lavishly illustrated history of weapons; a bit heavy on bows, pistols, and rifles but well worth a perusal and follow-up on suggested topics.

Robert Wilcox, *Wings of Fury* (New York: Pocket Books, 1996). An exuberant account of air power and how it grew in importance and skill from Vietnam to the First Gulf War, albeit only partially by the new weapons systems.

Frank H. Winter, *The First Golden Age of Rocketry* (Washington, DC: Smithsonian Institution Press, 1990). Although rockets were used in India and China almost 1,000 years ago, it was the rocket revolution of William Congreve, whose rockets were brought by the British to many parties—Naples in 1805, Copenhagen in 1807, the War of 1812, Waterloo in 1814, the Opium War of 1839–1842, and the Crimean War of 1853–1856—before rockets were eclipsed by rapid-firing artillery.

Steven Zaloga, *V-1 Flying Bomb 1943–1952* (Oxford, UK: Osprey, 2005). The "flying bomb" was something of a strategic bust, but it was the most widely used guided missile of the war. It was the world's first cruise missile; its descendants still sit aboard surface vessels, subs, and aircraft, waiting to be used today, 70 years later.

Dwight Zimmerman, *The Book of War* (New York: Tess Press, 2008). A clear, highly readable, well-illustrated introduction, not just to important weapons throughout history, but also to the battles and commanders of impact through history's long arc.

Chapter Thirty-Four

Logistics in War

"Amateurs talk strategy, professionals talk logistics."—General Oman Bradley

"Logistics is the practical art of moving armies."—Henri Jomini

"My logisticians are a humorless lot. . . . [T]hey know if my campaign fails, they are the first ones I will slay."—Alexander the Great

"The battle is fought and decided by the quartermasters before the shooting begins."—Field Marshal Erwin Rommel

"You will not find it difficult to prove that battles, campaigns, and even wars have been won or lost primarily because of logistics."—Dwight D. Eisenhower

Upon reading *Winning at War*, Frank Mahncke of the Institute for Defense Analyses pointed out that "superiority in supply, resources, transport, replenishment and replacement as tailored to the situation" needs to be considered when looking at a country's ability to wage war in a particular time and place. As he pointed out, during World War II, the C-47, the deuce-and-a-half truck, the LST, and the Jeep were all essential to getting what was needed on the battlefields of Europe and Asia to those battlefields.

The books in this chapter provide the student of war with some prominent examples, both of logistical systems that were independent variables and theories about which dimensions of logistical systems were crucial to war-making throughout time and space. Some are about logistics themselves, while others are about campaigns, such as the Falklands and Gallipoli, where logistics clearly stand as independent variables, and still others look at the interplay between weapons and war more generally.

It is interesting to note, at least in passing, that two of the most important American generals, Dwight Eisenhower and U. S. Grant, both started their military careers in logistics and the supply service, gaining the most valuable experience that they would both use later in their careers at the heads of vast armies.

Leslie Anders, *The Ledo Road* (Norman: University of Oklahoma Press, 1995). A straightforward, unvarnished account of the Lend-Lease Road from India (Ledo) to China (Wanting) through northern Burma in order to link up New Delhi to Chungking. "The Road to Victory" cost $150 million, thanks to disease, monsoons, enemy action, and the sheer ruggedness of the terrain ("flat areas" turned out to be swamps; "hills" turned out to be mountains).

Edmund Castillo, *The Seabees of World War II* (New York: Random House, 1963). From Bora Bora to Normandy, this breezy account with useful pictures tells the story of how the fighting men and women were supported by the sinews of war, from Guadalcanal to Normandy, from Casablanca to Tinian and Iwo Jima, written by the chief of the navy civil engineers. In war and in peace, logistics rule, he says.

John E. Edwards, *Combat Service Support Guide* (Harrisburg, PA: Stackpole Books, 1993). It's all here: logistics data, formulas and procedures, formats for order, plans, and estimates. For anyone, anywhere, who ever thought that wars were uncomplicated and settled only on the battlefield, this book will be a revelation.

Adrian Gitraas, *Engineers at War: The United States Army in Vietnam* (Washington, DC: Center of Military History of the United States Army, 1980). An amazingly lucid account of the logical basis for US involvement in Vietnam, from initial buildup to final chapters. Extraordinary photos and maps. Anyone trying to understand the US presence in Vietnam during that war must read this work.

Max Hastings and Simon Jenkins, *The Battle for the Falklands* (London: Michael Joseph, 1983). What better place to analyze the importance of the logistical dimension than in the 1982 war in the Falklands/Malvinas? The British took 28,000 people 8,000 miles to go to war, using 51 warships, 51 fleet auxiliary craft, and 54 merchant ships. Britain got a considerable boost from the United States, as it flew supplies, ammunition, and weapons (including Stinger missiles) to Ascension Island prior to the final showdown. The motto of the war, like so many others, quickly became "You never know."

Note: The British Commodore Amphibious Warfare (COMAW) Michael Clapp has written (together with Ewen Southby-Tailyour) a very detailed account of the war from a British perspective. See his *Amphibious Landing in the Falklands* (Annapolis, MD: Naval Institute Press, 1996). The Royal Navy shows its old "can-do spirit" and uses it against the Argentines. Justifiable pride accompanied by many photographs.

James D. Hornfishcer, *Neptune's Inferno: The U.S. Navy at Guadalcanal* (New York: Bantam Books, 2010). After going into the ships and battles in minute detail, this long-overdue book highlighting the naval campaigns of the United States and Japan in the Solomon Islands and South Pacific during 1942 concludes, "[I]n their zeal to advance the Japanese had stretched themselves beyond the nourishment of their supply train and exposed themselves against an enemy who was proving to be absolutely implacable in defense."

James Huston, *The Sinews of War: Army Logistics 1775–1953* (Washington, DC: Office of the Chief of Military History United States Army, 1966). The definitive account of how the US Army has moved and been moved and supplied over time. I found most useful the chapter "Some Principles of Logistics," featuring such concepts as "First with the Most."

George P. Kennedy, *Germany's V-2 Rocket* (Atglen, PA: Schiffer Military History, 2006). A companion piece to the next work, with even more pictures and diagrams and somewhat clearer prose. Liked the section on "European Roundup" of scientists, support staff, and equipment.

———, *Vengeance Weapon 2: The V-2 Guided Missile* (Washington, DC: Smithsonian Institution Press, 1983). Featuring the Wehrmacht at Peenemunde and the V-2 at war, with engaging photos and diagrams, this work gives a history of the weapon, which first flew at the A-4 in October 1942; was initially shunned by Hitler (until Himmler extolled its virtues); and saw 5,947 produced (largely underground), of which 1,115 hit England and 1,775 fell on such continental targets as Antwerp. Eventually most of the scientists and technical material fell to the British and the Americans, while the Russians got the production facilities (which they quickly transported east of the Urals). The V-2 would eventually morph into the US Redstone and Jupiter ballistic missiles and finally the Saturn. The author believes it was the "technological base from which we entered the space race."

Kenneth Macksey, *For Want of a Nail: The Impact on War of Logistics and Communications* (London: Brassey's, 1989). A series of essays around the notion of "Teeth versus Tail: An Interminable Contest." Some interesting pieces on the Crimean War, the American Civil War, and the "overburden" of World War I.

Chris McNab, *A History of the World in 100 Weapons* (Oxford, UK: Osprey, 2011). Lavishly illustrated, this work places weapons in their strategic and historical contexts. It covers the waterfront, from chariots, sickle swords, triremes, plug-and-socket bayonets, needle guns, MIG 21s, RPGs, and UAVs. Perhaps a bit overly British-oriented in some instances but a good reference work nevertheless.

Martin Middlebrook, *Operation Corporate: The Falklands War 1982* (New York: Viking, 1985). Although lacking access to Argentinean sources, the author does a credible job of presenting a war fought 8,000 miles from British shores with the long, long logistical tail proving invaluable—and the fleet surprisingly vulnerable to Exocet missiles. It was a shock at the time how easily and readily "modern" ships burned and sank. Not sure the lessons of the Falklands War have been learned by the navies of the world.

Note: For readers more interested in the actual fighting on the ground, please check Julian Thompson, *3 Commando Brigade in the Falklands: No Picnic* (Barnsley, UK: Pen and Sword, 1985). "Yomping" through the boondocks, 3 Commando takes various mountains and Port Stanley despite being cold, tired, hungry, and always on the lookout to take high-priced Argentine boots off the living and the dead.

Gus Pagonis and Jeffrey L. Cruikshank, *Moving Mountains: Lessons in Leadership and Logistics from the Gulf War* (Cambridge, MA: Harvard University Press, 1994). Not only a most useful account of the incredible movement of logistical supplies in *and* out of Saudi Arabia during the First Gulf War but also some excellent analysis of the dimensions of the experience that transcend the theater and the campaign. In all, 122 million meals were moved and served and 1.3 billion gallons of fuel used to transport 7 million tons of supplies. At the end of the Gulf War, Pagonis was charged with removing every last item back out of the theater.

Edgar F. Raines Jr., *The Rucksack War: The U.S. Army Operational Logistics in Grenada, 1983* (Washington, DC: Center of Military History United States Army, 2010). The October 1983 invasion of Grenada, Operation Urgent Freedom, is detailed showing how this small operation was a practice run for many of the more extensive force projections of the 1990s and beyond. The 82nd Airborne provided a swift and successful coup de main.

Damon Schechter with Gordon Sander, *Delivering the Goods* (New York: John Wiley and Sons, 2002). Attempting to extract lessons for corporations from wartime situations, this volume looks at a number of dimensions, such as "war is logistics" and the "soldier as customer." Quite right about Napoleon in Russia, arguing that his loss was due to faulty logistics, and enjoys asserting that, compared to Alexander the Great, when it came to logistics, Napoleon was a rank amateur. Liked the concept of "ghost busters," who, in armies and companies, report on "what is or is not there" in a timely fashion.

Martin van Creveld, *Supplying War: Logistics from Wallenstein to Patton* (Cambridge: Cambridge University Press, 1977). Believing that the "logistic aspect of war is nothing but an endless series of difficulties succeeding each other," the author focuses on campaigns in 1630, 1704, 1812, 1870, 1914, 1941, and others to conclude that "more campaigns have been ruined by poor logistics than enemy battles." He urges more study on the "War of the Accountants."

Gary Wetzel, *A-10 Thunderbolt II Units of Operation Enduring Freedom* (Oxford, UK: Osprey, 2013). Good pictures and text concerning the performance of the Warthog. The US Air Force keeps trying to kill this aircraft, while the army keeps trying to keep it—for good reason, it seems to me.

Thomas Wildenberg, *Gray Steel and Black Oil: Fast Tankers and Replenishment at Sea in the U.S. Navy, 1912–1992* (Annapolis, MD: Naval Institute Press, 1996). An exhaustive and important study of how logistics played a key role in many US efforts and none more important than World War II in the Pacific. "Our ships sailed on water, but they moved on oil, and the demand never ceased."

Christian Wolmar, *Engines of War: How Wars Were Won and Lost on the Railways* (New York: Public Attitudes, 2010). From the Civil War through the Franco-Prussian and Russo-Japanese Wars and World Wars I and II, the author takes much evidence of his hypothesis. Certainly the Germans found out in 1941 in Russia how important a functioning rail net—or lack thereof—could be. The author asserts that a double-track railroad carries the equivalent of 1,600 trucks.

Admiral Sandy Woodward, *One Hundred Days: The Memoirs of the Falklands Battle Group Commander* (Annapolis, MD: Naval Institute Press, 1992). Conducting a war after sailing for three weeks, Woodward ran the war from the 29,000-ton carrier *HMS Hermes* but worried most about logistics and the onset of winter—which struck in full fury only 7 hours after the Argentines surrendered. He believes that, if General Winter had come earlier, "he would have finished us. Ships are just as vulnerable as the marching armies of Napoleon and Hitler were in Russia."

Chapter Thirty-Five

The Daughters of Mars

Women at War

Any study of women at war through the ages must immediately start with the cogent judgment of Russian general Vasily Ivanovich Chuikov, who, after successfully defending Stalingrad during World War II, stated categorically, "Women soldiers proved themselves to be just as heroic in the days of fighting as men." Stalingrad, as Andrew Roberts has written, is "deservedly considered to be the most desperate struggle in human history." Regardless of how limited the participation of women in armies has been throughout the course of human history, there is no way around the assertion that there is no a priori reason they cannot be soldiers and good ones at that, regardless of their sex.

There are, of course, many cultural, physical, and sexual arguments—some of them better than others—used by many different societies to oppose the use of women in combat, but none of these is thereby ipso facto sufficient to overcome the fact that women can and have been both useful and successful in battle throughout much of the temporal and geographic range of war.

At the end of the day, however, if we were looking at women in war strictly in terms of the Template of Mars, then it would be difficult to argue with the notion of Dominique Lozzi that "Mars does not look at war through a male-centered lens, and therefore, neither should we." In the future, therefore, it seems likely that the bibliography listed here will grow, and grow substantially.

Karen Abbott, *Liar, Temptress, Soldier, Spy: Four Women Undercover in the Civil War* (New York: Harper, 2014). An engaging tale of women at war in a variety of roles. Of special interest is Emma Edmonson, the soldier who enlisted as Private Frank Thompson in the Second Michigan and fought at Fredericksburg, in the Shenandoah Valley, and at Second Bull Run before deserting and "becoming" a woman again. Her memoir sold 175,000 copies, and she gave all the money to sick and wounded survivors of the Army of the Potomac.

Volundr Lars Agnarsson (ed.), *The Saga of Erik the Red* (New York: Create Space Independent Publishing Platform, 2012). In the 10th century CE, Fredis Eiriksdottir, a shieldmaiden, sees her menfolk losing a battle to the Skraelings, or Native Americans. Although pregnant, she bares her breasts and appears to be sharpening her sword on them as she attacks the Skraelings. They flee. A cautionary tale about the chance encounter between two cultures, as well as the extent to which women warriors in history have been subordinated to their male counterparts.

Note: See also Margaret Elphinstone, *The Sea Road* (Edinburgh, UK: Canongate, 2000) (F), for the saga of Gudrid of Iceland, who traveled to Vineland, bore a son, and witnessed the battles (and misunderstandings) with the Skraelings and who also seemed to see alternatives to fighting.

Miranda Aldhouse-Green, *Boudica Britannia: Rebel, War-Leader and Queen* (Edinburgh, UK: Pearson Education, 2006). Gives a good background on the causes of Boudicca's uprising and the impact it would later have on Roman Britain.

Peter Aleshire, *Warrior Woman: The Story of Lozen, Apache Warrior and Shaman* (New York: St. Martin's Press, 2001). Lozen spent 40 years fighting, a true warrior even among the likes of Cochise, Geronimo, and Victorio in a "war dominated culture." Finding her a "great exception" to the normal sexual hierarchy, the author marvels at her talent, steadfastness, and vision-skills in seeing the enemy. A good source for her whole life.

Svetlana Alexiyevich, *War's Unwomanly Face* (Moscow: Progress, 1985). Despite its title and propaganda intent, this work manages to become a most powerful paean to women at war. During World War II, 800,000 Russian women answered their country's call, either as members of the regular armed forces or partisan bands or both. Moving, poignant, and insightful, it captures the many faces of women in battle, including front-line soldiers. Ordinary women doing extraordinary things as comrades in arms, declaring, "What do you mean, girls—they're soldiers." They were.

Teena Apeles, *Women Warriors: Adventures from History's Greatest Female Fighters* (Emeryville, CA: Auslow, 2003). A breezy and somewhat superficial account but one that leaves the correct impression that a fair number of women warriors have always existed across a variety of societies, from Vietnam and China to Russia to Mexico and among many Native American tribes, including the Apache, Cherokee, Blackfeet, and Gros Ventre.

Kathryn J. Atwood, *Women Heroes of World War II* (Chicago: Chicago Review Press, 2011). A breezy introduction to women who risked their lives in the war, featuring examples from Poland, Denmark, France, Great Britain, Holland, Belgium, and the United States. Unfortunately, no Russian women are included.

Anni Baker, "Daughters of Mars: Army Officers' Wives and Military Culture on the American Frontier," *Historian* 67, no.1 (Spring 2005), pp. 20–42. Army wives in the 1870s and 1880s in the American West identified with their husbands' army and its cultures and traditions. The author believes these women developed the "Cult of Army Womanhood," "incorporating military and masculine characteristics into their behavior, even explicitly rejecting some values of civilian women."

Lolita Baldor, "Few Army Women Want Combat Jobs," *Portland Press Herald*, February 26, 2014, p. A-5. Only 7.5 percent of women in the US Army at the time of this survey said they were interested in combat positions (including field artillery and combat engineers). But an overwhelming number of that 7.5 percent wanted to be Night Stalkers (elite special operations helicopter pilots, navigators, and gunners). Note the comparison with the "Night Witches" of World War II Soviet women.

Juliet Barker, *Conquest: The English Kingdom of France 1417–1450* (Cambridge, MA: Harvard University Press, 2012). See especially "Jehanne D'Arc" (pp. 93–171). As Barker puts it, the story of Jehanne, the Pucelle (or Maid), is "extraordinary almost beyond belief. Her youth, her sex, her background, all militated against what she became: the companion of princes, inspirational military leader, martyr for faith and country" (p. 102). All of those and then some.

Gretchen Bataille and Laura Lisa (eds.), *Native American Women* (New York: Routledge, 2001). A grand listing of many Native American women, from poets and ballerinas to warriors. Some useful examples of Native American warriors include (1) the Other Magpie, who rode with the Crow Wolves, scouts for the US Army, and counted coup four times against the Lakota in the Battle of the Rosebud (1876) ,crying "See my spit is my armor" (p. 309); (2) Running Eagle of the Blackfeet Nation, who entered the Braves Society and fought the Crows and others and was eventually killed by the

Flatheads (p. 258); and (3) Dahteste, an Apache warrior who fought with Geronimo and Lozen and her two husbands against the US Army and later served as a scout with them (p. 83).

Judith Bellafaire, *Women in the United States Military: An Annotated Bibliography* (New York: Routledge, 2001). Quite a compendium of books, articles, and notes on women in the military, ranging from "Early Patriots" to the "All-Volunteer Force and the War on Terror." Many seldom-cited articles are mentioned here. A very useful resource.

Barbara Benton, "Friendly Persuasion: Women as War Icons 1914–1945," *MHQ: The Quarterly Journal of Military History* 6, no. 1 (Autumn 1993), pp. 80–87. Captures the essence of women as icons in, and of, war when used by men to achieve particular goals, whether buying war bonds or supporting the country's armed forces. An interesting array of the actual posters used by a number of countries is provided.

Carol Berkin and Clara Lovett (eds.), *Women War and Revolution* (New York: Holmes and Meier, 1980). A set of essays covering situations (ranging from France in 1789 to Cuba, Italy, and World War I) in which women played a role seeking peace as well as war. Wide-ranging account, with many interesting dimensions presented.

Jane Blair, *Hesitation Kills: A Female Marine Officer's Combat Experience in Iraq* (Lanham, MD: Rowman and Littlefield, 2011). A strong woman warrior who rises through the ranks in the US Marine Corps, sees action in Iraq in the 26-day war and part of the occupation, and certainly shows that women have been assets in combat if allowed to participate. A very insightful portrait of one female "Devil Dog," who remarks, "Once you kill, you can't take it back" and "Our only certain destiny was killing or being killed."

Tanya Blank, *Undaunted: The Real Story of America's Servicewomen in Today's Military* (New York: NAL Caliber, 2013). Since 9/11, more than 250,000 women have served in Iraq and Afghanistan, and more than140 have been killed in action. Blank looks at a brigadier general in the Marines, a drill instructor in the Marines, an army major, and an army MP. The author also gives a very good set of insights into what life in the military is really like for today's women.

Deanne Blanton and Lauren Cook, *They Fought Like Demons: Women Soldiers in the Civil War* (New York: Vintage Books, 2002). An in-depth look at the broader (many more than the normally quoted 250–400) range of women who fought in the Civil War, beginning with First Bull Run and ending with

Appomattox and including the Peninsula campaign, Second Bull Run, Antietam, Fredericksburg, Chancellorsville, Gettysburg (Pickett's Charge!), Vicksburg, the Wilderness, Cold Harbor, Red River, Shiloh, Petersburg, and Sherman's March to the Sea. Also examines the many motivations for enlisting, including patriotism; love of freedom; desire to be with one's spouse, father, or brother; spirit of adventure; economic advancement; and escape from the confines of womanhood in the mid-19th century.

James Blythe, "Women in the Military: Scholastic Arguments and the Medieval Images of Female Warriors," *History of Political Thought* 22, no. 2 (Summer 2001), pp. 242–69. Analyzes the arguments for or against women in war in the political treatises of Ptolemy of Lucca (1236–1347) and Giles of Rome (1243–1316). Opposes not only their "medieval misogyny" but also their echoes among the arguments of some "difference feminists," who use the same arguments against women in combat today.

Laudomia Bonanni, *The Reprisal* (Chicago: University of Chicago Press, 2013) (F). This powerful novel captures the essence of the "warrior woman" with her "fury-like power" and feminine qualities (her heroine, "La Rossa," the red-haired revolutionary, is pregnant and not killed by the Fascist partisans until after giving birth). Takes place in Italy during the waning months of World War II, when Germans, Fascists, and partisan bands of various political hues roamed the land in a Hobbesian all-against-all struggle.

Frederick Brown, "The Battle for Joan," in his *The Embrace of Unreason: France, 1914–1940* (New York: Alfred Knopf, 2014), pp. 76–91. An incisive look at the continuing fascination with the "Maid." She not only was beatified in 1909, but she also became the patron saint of Vichy France, so that the "archenemy was no longer Germany but England."

R. Brzezinski and M. Mielczarek, *The Sarmatians 600 BC–AD 450* (Oxford, UK: Osprey, 2002). The authors claim that Sarmatian women fought in battle and were either (a) not able to marry before killing an enemy in battle or at least (b) not able to marry unless facing an enemy in battle. In any case, they were doing actual fighting against many enemies. Good pictures of women in action so many eons ago. Many excavated graves show Sarmatian warrior women buried with their weapons.

Kimberly Moore Buchanan, *Apache Women Warriors* (El Paso: University of Texas at El Paso Press, 1986). Believes that women were allowed into male activities with "high prestige" in many Native American societies (among them the Pawnee, Mandan, Gros Ventres, Crow, and Sioux) and gives specific examples, such as Water Sitting Grizzly of the Kutenai, Running Eagle of the Blackfeet, Ehyophsta of the Cheyenne, and Chief Earth Woman of the Ojibwa. See especially chapter 6, "Women Warriors: Forgotten Gladiators." The author gives a number of reasons women went to war: revenge of a loved one (such as Running Eagle of the Blackfeet), love (such as Chief Earth Woman of the Ojibwa), and rescue (such as Ehyophsta of the Cheyenne).

Alex Burghart, "Aethelflaed: Iron Lady of Mercia," *BBC History Magazine* 12, no. 8 (2011), pp. 60–63. Answers the question "Where is a woman's place?" with the answer "A woman's place is on the throne" or "in the saddle leading her troops successfully against the Welch and the Norsemen."

Richard Burton, *A Mission to Gelele, King of Dahomey* (New York: Praeger, 1966). Burton goes many places (including Mecca), but he is quite impressed with the Amazons and describes their practices, weapons, dress, and many other aspects of their military lives. Suspect now, as anthropologists have had difficulty corroborating his findings. At the same time, his descriptions project warrior abilities and capabilities upon their existence, which are very verified.

Pierce Butler, "Jeanne de Montfort," in his *Women of Mediaeval France* (Philadelphia: Ritten House, 1907), pp. 285–305. Marguerite, Countess to Jeanne Montfort of Britany (also known as "La Flamme") took over defense of his realm when he was captured. Led her mounted knights to destroy the camp of the besieging French and burned their tents and wagon train, hence her nickname. She later fought at sea with the English against the Genoese (Jeanne de Montfort had pledged allegiance to Edward III), and later her husband escaped his captivity. Their son was eventually recognized as duke of Brittany. This is a real warrior's warrior.

Kevin Cahillane, "The Women of West Point," *New York Times Magazine*, September 7, 2004, pp. 46–59. A contemporary look at how women are being prepared for combat missions beginning in 1976. From 1802 to 1976 there were no women; since then, almost 4,200 have enrolled. In 2015, they made up 22 percent of the incoming class.

D'Ann Campbell, "Women, Combat, and the Gender Line," *MHQ: The Quarterly Journal of Military History* 6, no. 1 (1993), pp. 88–97. Indicates that a US World War II study proved that female soldiers were ready to serve under fire and examines the reasons the United States did not adopt that strategy.

Rick Campbell, *Empire Rising* (New York: St. Martin's Press, 2015) (F). In the near-future, China invades Taiwan *and* Japan and sinks the US Pacific fleet but eventually loses due to computer malware inserted by the United States via a Seal team in Beijing led by National Security Advisor Christine O'Connor, who is twice as smart as any man and better with a gun. A new superhero emerges. Oddly enough, neither side uses tactical or strategic nuclear weapons in this strange work.

Helen Castor, *Joan of Arc: A History* (New York: Harper, 2015). Perhaps the best of all her biographies in putting "La Pucelle's" story in the true context of the existing power struggles between and among the English, the Burgundians, and the House of Valois. She tells the story from the beginning of the period (with great detail) to the exact story of the Maid. Very good on the military side of Joan. Her religious fervor, native sense, and personal courage eventually brought her triumph, but all of these were eventually subsumed by the medieval intrigue of the day—at least in the short term.

———, "The Real Joan of Arc Is Every Bit as Extraordinary as the Myth," *BBC History Magazine* 15, no.10, pp. 50–51. The title says it all, and its content is well-introduced by the telling phrase "In many ways, then, her story is a life told backwards."

———, *She-Wolves: The Women Who Ruled England before Elizabeth* (New York: Harper, 2015). Vilified for their "usurpation" of men's roles and rules, these women—such as Matilda, Eleanor, Isabela, and Margaret—paved the way for Queen Elizabeth. Smoothly flowing stories intrigue the reader.

E.V. Cernenko, *The Scythians 700–300 BC* (Oxford, UK: Osprey, 1983). These horse archers held sway for 400 years and spawned an immortal legend of women warriors. See page 24D for a drawing of a Scythian noblewoman, fourth century BCE. During their centuries of glory, the Scythians defeated the Babylonians, Medes, Persians, Assyrians, and Urartus. A dominant warrior culture in which women played an important role in battle.

Cynthia Cockburn and Zubravka Zarkove, *The Postwar Moment: Militaries, Masculinities and International Peacekeeping Bosnia and the Netherlands* (London: Lawrence and Wishart, 2002). This collection of essays explores the notion that peace is but a "continuum of conflict" with war. Moreover, because war may threaten masculine-dominated society, "[a]fter war, the traditional militarized gender regime endows men with the power in politics and locates women's importance within the family." The authors are justly disdainful of the "soft yet armed, masculinity" of the Dutch peacekeepers who turned over male Bosnian Muslims to the Bosnian Serbs during the siege of Srebrenica. In fairness to "masculine-dominated societies," few male or female warrior cultures would ever exonerate the Dutch peacekeepers for their performance here, for it was pathetic by any standards of warfare. Various contributors comment on the "gendered nature of war," and one, Cynthia Enloe, declares, "Nationalism typically has sprung from masculinized memory, masculinized humiliation and masculinized hope," while Stefan Dudink argues, "War and the military have always been major forces in the making of modern Western masculinity." The "heightened masculinity" of the radial Salafists would make an interesting comparative study because that particular strain is more than 1,200 years old.

Carol Cohn, "Sex and Death in the Rational World of Defense Intellectuals," *Signs* 12, no. 4 (Summer 1987), pp. 687–718. If one can get by the truly bizarre fascination (of both the author and the subjects of her study) with the male phallus as images of the atomic bombing of Nagasaki and nuclear craters as "feminine," the article ends with some common-sense suggestions concerning nuclear deterrence leading to "creating compelling alternative visions of possible futures."

——— (ed.), *Women and Wars* (Cambridge: Polity Press, 2013). A wide-ranging set of essays that operate through a "gender lens" on such diverse topics as "Sexual Violence and Women's Health in War" and "Women and the Peace Process" and "Women 'after' Wars." A very extensive bibliography is included. Strangely, given the title of the work, there is almost nothing about women warriors or war leaders, especially in combat qua combat throughout history, although Cohn argues that "all wars are deeply gendered."

Vanessa Collingridge, *Boudica: The Life of Britain's Legendary Warrior Queen* (Woodstock, NY: Overlook Press, 2006). A long, lingering (100 pages of pre-Boudica and many post-Boudica), meandering look at the reality and the legend, concluding she was a "Queen for All Seasons."

Helen Collinson (ed.), *Women and Revolution in Nicaragua* (London: Zed Books, 1990). Contains a telling map showing the Food and Drug Administration incursions that begin the book, and there are many essays dealing with the women, especially "Fighting for Peace." Quite a worthwhile look through the prism of revolutionary fervor. One of the authors admits to being "stunned" by the "Rightwing" National Opposition Union (UNO) victory at the polls in 1992, as a union of 14 parties defeated the Sandinista Front for National Liberation, but she does not believe that victory will turn back the clock on the revolution.

Dan Connell, *Against All Odds: A Chronicle of the Eritrean Revolution* (Trenton, NJ: Red Sea Press, 1993). A must-read, as it highlights all the ways women served in the revolution against Ethiopia and the extent to which this Red Sea country leveled the playing field for women in combat. However, many women justly claim that, after victory, they were not given the credit and equality they felt they deserved.

Sarah Corbett, "The Women's War," *New York Times Magazine*, March 18, 2007, pp. 41ff. Some insightful glimpses into women in today's armed forces. Shows how close to actual combat many women have come and how some have been in combat, no matter how rigorously that is defined by those not wanting to admit it.

Bernard Cornwell, *The Empty Throne* (New York: Harper, 2014) (F). Set in the early 10th century in what is now England, this is the fictionalized account of how Aethelflaed, daughter of Alfred of Wessex and widow of Aethelred, lord of Mercia, ruled Mercia. A very capable woman who "proved herself to be more of a warrior than her brother." A good read and one that also provides valuable insights into the shield war form of warfare practiced by the Vikings and Saxons (among others) and how men would follow a woman in battle if she had the right birthright and warrior skills.

William Dalrymple, *The Last Mughal: The Fall of a Dynasty: Delhi, 1857* (New York: Alfred Knopf, 2006). A highly sympathetic and in-depth analysis of the Great Indian Rebellion of 1857, with an extraordinary richness of characters and background. See especially the author's take on Rani Lakshmibai, who led her Jhansi troops against the British during the Sepoy Rebellion of 1857–1858. She called on her troops to die in battle if necessary and is now regarded as one of the pioneers for Indian independence more than 100 years later.

Francine D'Amico and Laurie Weinstein (eds.), *Gender Camouflage: Women and the U.S. Military* (New York: New York University Press, 1999). Very informative collection of essays divided into "Women in the Military," "Women *with* the Military," and "Outsiders: Women and the Military." Basically argues that the military always puts men first and that great change is necessary to liberate women *and* men from the core gender hierarchy that overpowers all.

Jeannine Davis-Kimball, *Warrior Women: An Archaeologist's Search for History's Hidden Heroines* (New York: Warner, 2002). Breezy and personal account of the author's discoveries among the Saka, Scythian, and Sarmatian graves from Kazakhstan, Russia, and China, focusing on women in various cultures who were warriors.

Edwin Denig, "Warrior Woman," in *Five Indian Tribes of the Upper Missouri*, edited by John Ewens (Norman: University of Oklahoma Press, 1961), pp. 195–201. Calling her the only woman chief among the Sioux, Arickaras, Assiniboines, Crees, and Crows, the author outlines the fascinating history of "Warrior Woman." This Gros Ventre girl, captured at age 10, had a foster father who allowed her to pursue her passions, which included hunting, counting coup, stealing horses, and using weapons. Upon his death, she assumed command of his family and participated in both warfare and tribal decision making and would acquire four wives before being killed ironically by her own people, the Gros Ventres.

Linda Grant DePauw, *Battle Cries and Lullabies; Women in War from Prehistory to the Present* (Norman: University of Oklahoma Press, 1998). A wide-ranging and very comprehensive look at thousands of years of female participation in the military, including the Hundred Years' War, the Napoleonic wars, and the French and Indian War. Good section on African women leaders in combat.

Christine de Pizan, *The Book of Deeds of Arms and of Chivalry* (University Park: Pennsylvania State University Press, 1999). An amazing document for the 15th century written by an early feminist. Imagine how many cultural and religious barriers she had to overcome to even get her book published. Written by this Italian-born but French-court author, the *Book of Deeds* resurrects many classical writings on war (especially Vegetius) but provides very useful contrasts between medieval Europe war practices and those from antiquity, including just war, siege warfare, chivalry, trickery, and subtlety. Many examples from contemporary Europe as well as campaigns of Scipio, Hannibal, and Hanno. A truly amazing work given the time, the place, and the sex of its author. Belies the notion that women did not appreciate war in the Middle Ages but were only interested in court romance and jewelry.

———, *The Book of the City of Ladies* (New York: Penguin 1999). A fine translation of an extraordinary book first written in 1405. Utterly charming, the book uses the device of a city peopled by women to destroy many myths of the day concerning women. Among her warrior choices are pagan Queen Semiramis (who married her son as the only man worthy of her and who conquered Babylon and led an expedition into Ethiopia), the Amazons of Scythia, Queen Thamiris (who defeated and decapitated Cyrus the Great), Queen Penthesilea of Amazon fame, and Queen Fredegunde of France. Although wrong about Queen Artemisia (who de Pizan has fighting Xerxes instead of with him), the author shows a powerful and courageous inclination to fight against the "literary misogyny of the mediaeval period." Honestly, who wouldn't want to live in this marvelous metaphoric city? The answers were, and are, telling. The author was a most unusual woman herself, earning her livelihood by writing in an age when few women tried or succeeded in doing so.

Kelly DeVries, *Joan of Arc: A Military Leader* (London: Sutton, 1999). Easy to read and understand, this is the best book I have read focusing on her military accomplishments, concluding, "She had completed her mission and sealed it with her blood." Gives the "Maid" credit for turning around the French military situation (with her successes in the Loire Valley, Reims, and Patay) and blames the newly crowned Charles II for a failure to keep using her and her tactics both at Paris and beyond. "Joan of Arc was a soldier plain and simple." Not too clear on how this farmer's daughter acquired the skills of warfare, however, leaving us wanting more information. Men in that era literally took years—if not decades—to learn how to fight effectively in medieval warfare.

Kirstin Downey, *Isabella: The Warrior Queen* (New York: Doubleday, 2014). An in-depth look at the "woman who governs the world from her bed" (i.e., dying of cancer), whose will united Spain, defeated Portugal, recaptured the emirate of Granada, launched the Inquisition, bankrolled Columbus, and led an empire that took 1.5 billion of today's dollars in gold and silver out of the Americas. Isabella was a devout Catholic who turned back the Muslim tide, was beloved by her people, and forced her husband Ferdinand to excel in battle and to do her bidding. "Isabella's influence on the New World cannot be underestimated." She truly was a "warrior queen."

Nora Duff, *Matilda of Tuscany* (London: Methuen, 1909). This loving, lush portrait from an earlier era does outline the importance of Matilda and her relevance to the Italy of her day. Calling her the "Warrior Mind of the Holy Church," the author points out that her early weapons training, her two suits of armor, and her military acumen were all special, as was her protection of the papacy from the designs of the Holy Roman emperor. She also believes that she is most worthy of her inclusion in St. Peters and Dante's Canto 28, "Guardian of the Earthly Paradise."

Robert B. Edgerton, *Warrior Women: The Amazons of Dahomey and the Nature of War* (Boulder, CO: Westview, 2000). The most scholarly and inclusive account of the Amazons of Dahomey, accenting their small-unit cohesion, their vigorous training, their in-group bonding, and their elite stat-us as warriors without sex. If one became pregnant, then she was tortured until she revealed the cause, and both were then killed. Covers their participation in the slave trade and as worthy opponents of the French Foreign Legion as late as 1890s.

Susan Edgington and Sarah Lambert (eds.), *Gendering the Crusades* (New York: Columbia University Press, 2002). An exciting reappraisal of the role of women during the Crusades, which finds them far more central to the process than heretofore acknowledged. See especially Karen Caspri-Reisfeld's "Woman Warriors during the Crusades, 1095–1254." Despite church policy, women played extensive roles in sieges, invasions, looting, and other warrior aspects, including as archers. Much evidence from Muslim sources in various battles from Antioch to Acre make this an excellent and provocative source.

Elizabeth A. Eldredge, "Sources of Conflict in Southern Africa, c. 1800–30: The 'Mfecane' Reconsidered," *Journal of African History* 33, no. 1 (March 1992), pp. 1–35. A devastating critique of the previous revisionist arguments concerning the Mfecane. Very germane background for any study of the rise of Mmanthasatsi (sometimes Manthatisi) and her "horde." For a more traditional explanation of the Mfecane and some of the subsequent literature about it, see Leonard Thompson, "The Zulu Kingdom and the Mfecane," in his *A History of South Africa* (New Haven, CT: Yale University Press, 2001), pp. 80–87. Note: In Sesotho, *Mfecane* is translated *Difaqane*. This study also provides a broader examination of Mmanthasatsi, whose Tolokwa clan of the Sotho people formed the basis of her horde.

Note: Thompson makes the point that there were three "hordes" operating at the same time for some of this period. See Monica Wilson and Leonard Thompson (eds.), *The Oxford History of South Africa*, vol. 1 (London: Oxford University Press, 1964), pp. 391–405.

Jean Elshtain, *Women and War* (New York: Basic Books, 1987). Focusing on the "seduction" of war, she sees women as the "ferocious few and the non-combatant many" and men as the "militant many and the pacific few." Argues that women don't belong in combat (or men either, if she had her druthers).

Cynthia Enloe, *Does Khaki Become You? The Militarization of Women's Lives* (London: South End Press, 1983). Although much of this work is quite dated in terms of specifics and, one could argue, shows a complete misreading of the reasons for the original Amazon legends (the Greeks rightly feared their style of warfare as well as their gender), nevertheless some of the chapters, such as "The Military Needs Camp Followers" and "Some of the Best Soldiers Wear Lipstick," raise important issues. Still, one wonders whether one of her core statements, "Women are being used by militaries to solve their nagging problems of manpower availability, quality, health, morale and readiness," might be made even more relevant by changing or adding "and men" to the quote. What do you think? Unfortunately, there is almost nothing here on women warriors qua warriors.

———, *Globalization and Militarism: Feminists Make the Link* (Lanham, MD: Rowman and Littlefield, 2007). Based on her Tokyo lectures on the eve of the US invasion of Iraq, the author uses the device of a "feminist curiosity" to declare a link between globalization and militarization. She also argues that the woman soldier should not simply be a globalized version of the "modern woman" but should instead push for a "less militarized military (which) would be less imbued with an institutional culture of masculinized violence."

———, *Maneuvers: The International Politics of Militarizing Women's Lives* (Berkeley: University of California Press, 2000). Noting that "[w]omen *in* the military has never been an easy topic," the author argues it shouldn't be. "Sexism, patriotism, violence and the state—it is a heady brew." The author looks the many impacts of war on many women—prostitutes, rape victims, mothers, wives, nurses, and feminist activists. Concerned about masculinity and militarism, she studies how and when women are recruited in her chapter 7, "Filling the Ranks."

———, "Paying Close Attention to Women inside Militaries," in her *Globalization and Militarism: Feminists Make the Link* (Lanham, MD: Rowman and Littlefield, 2007), pp. 63–92. This work serves as a counterbalance to the assumptions of both Bowdoin graduates Santangelo and Petronio (cited in chapter 35) that service in the military is ipso facto liberating if women are treated as equals within it. In fact, Enloe argues that a "less militarized military would be one less imbued with an institutional culture of masculinized violence" and instead would (and should) place more emphasis on rescuing civilians from natural disasters worldwide.

John Ewers, "Deadlier than the Male," *American Heritage* 16, no. 4 (1965), pp. 10–13. Despite its hyperbolic and inaccurate title, this short, pithy article introduces Elk Hollering in the Water, the Blackfeet woman warrior; the Other Magpie of the Crow; Woman Chief of the Gros Ventre (but raised as a Crow); and Running Eagle of the Blackfeet. Suggests that female participation in war and raids was more widespread than previously understood or acknowledged. Young brides, for example, often went with their new husbands on raids. Note also how Running Eagle was thought to have been killed because she violated her oath of celibacy, a widespread projection of male values upon even women warriors, as obviously no similar prohibition existed for male warriors.

Ilene R. Feinman, *Citizen Rites: Feminist Soldiers and Feminist Antimilitarists* (New York: New York University Press, 2000). According to the author, feminist antimilitarism relies on the fundamental connection between patriarchy and war, while feminist egalitarian militarism accents the notion that women can be as good as men in the "be all you can be" ethos of the military. Declaring that "[w]ar is no longer a dick thing," she argues that "women have fought long and hard for inclusiveness and respect in the realm of martial citizenship" and also that liberal feminists hope to democratize the military.

Lorry Fenner and Marie deYoung, *Women in Combat: Civic Duty or Military Liability* (Washington, DC: Georgetown University Press, 2001). Two articulate authors give the pros and cons for American women being used in combat. They both have extensive military experience: Fenner is an air force intelligence officer and favors opening up all aspects of military service to women, while deYoung, an army chaplain, opposes this on many grounds, arguing the recent surge for their use is based purely on political correctness, not a dispassionate examination of all of its dimensions.

Nic Fields, *Boudicca's Rebellion AD 60–61* (Oxford, UK: Osprey, 2011). Fine details and maps of the rebellion and some good commentary. Ends by asking the question, "Why do men fear women warriors so much?" and puts Boudicca's rebellion in the long line of such sentiments going back to the original "Amazonian" concerns of the ancient Greeks.

Will Fowler, *Eastern Front: The Unpublished Photographs 1941–1945* (St. Paul, MN: MBI, 2001). One of the best collections of primarily Soviet photographs of the war (although it contains some pictures of Germans and Romanians as well). Many shots of the wide-tracked T-34 and other Soviet equipment (even a shot-down American B-17), as well as much-needed emphasis on such often-overlooked battles as the Crimea and Oder campaigns and an exploration of women and partisans in battle (in these cases, closely watched by the NKVD). The shots of women in wartime action on behalf of the Soviet Union are most engaging and show their centrality to the war effort. Taken in context, they provide stunning evidence of women warriors in daily combat during World War II.

Linda Bird Francke, *Ground Zero: The Gender Wars in the Military* (New York: Simon and Schuster, 1997). This book looks at the ways in which an element of society and the military tried to restrict the role of women in the military but how over time the stereotypes of males as protectors and women as caregivers lost some of their power.

Antonia Fraser, *The Warrior Queens* (New York: Alfred Knopf, 1990). Well-written, erudite, and entertaining, with a kaleidoscope of historical, psychological, and sociological references. Queen Boudicca, Elizabeth I, Catherine the Great, and the Rani of Jhansi all are colorfully described. Who says women can't be as ruthless as men, in battle as well as elsewhere? Not this author. Contains interesting subthemes about the role of sexuality in the depiction of warrior queens as either voracious breakers of norms or virginal upholders of them, depending on one's perspective.

Deborah Gera, *Warrior Women: The Anonymous Tractatus de Mulieribus* (Leiden, Netherlands: E. J. Brill, 1997). Based on 6 pages of an anonymous tract depicting 14 outstanding women in the 5th–4th centuries BCE, the author expands on these "Women Intelligent and Courageous in Warfare," including Zarinara (Parthian), Rhodegyna (Persian), Pheretine (Cyrene), Tomyris (Massagetae), and Artemisia (Halicarnassian).

Daniela Gioseffi (ed.), *Women on War* (New York: Feminist Press at the City University of New York, 2003). A wide-ranging collection of authors celebrating not female warriorhood but female Cassandrahood. Many women from various cultures decry war and exploitation in all its forms.

Nancy Loring Goldman (ed.), *Female Soldiers: Combatants or Non Combatants* (Westport, CT: Greenwood Press, 1982). Gives the reader portraits of women in war (Germany, Russia, Yugoslavia, Vietnam, and Israel) and as noncombatants (Greece, Japan, Denmark, and Sweden), ending with an argument for women in combat (Mandy Seal) and against (Jeff Tuten). These last two essays are somewhat overridden by events but well worth perusing for historical context and values.

J. S. Goldstein, *War and Gender: How Gender Shapes the War System and Vice Versa* (Cambridge: Cambridge University Press, 2001). It is estimated that 8 percent of Soviet armed forces in World War II were women and that there were three squadrons of night bombers flown by women. Also, approximately 40 percent of doctors in the Soviet armed forces were women, and many of these served in combat. According to Goldstein's estimates, approximately 1 percent of all warriors throughout history have been women.

Mary Gordon, *Joan of Arc* (New York: Viking Penguin, 2002). Although she made none of the major decisions of the various campaigns in which she fought, and although she was eventually convicted and burned alive on a charge of being a transvestite, "[i]s it possible to say that she fought like a knight but otherwise didn't behave like one?" The author answers this question in the affirmative. Still, it is hard to beat this story of a woman at war, at least for me. Joan of Arc was a frightening and overwhelming role model for me as a 10-year-old boy, for I feared I could not live up to her fiery expectations, God or not.

Julie Gottlieb, *Feminine Fascism* (London: I. B. Tauris, 2000). This interesting study is determined to give "women their due as autonomous historical agents" by looking at a "polyphony of women." Puts a spotlight on some understudied women. Useful appendix.

Susan R. Grayzel, "The Role of Women in the War," in *First World War*, edited by Hew Strachan (London: Oxford University Press, 2011), pp. 149–63. Looking at both individuals and groups, the author highlights the important work done by them in a variety of venues, including industrial production (40 percent of Russian, 33 percent of French) and especially the impact of the war in giving women the vote in Austria, Belgium, Britain, Latvia, Hungary, Lithuania, Germany, Denmark, Netherlands, the United States, and Czechoslovakia, among others.

Andrew and Nicola Hallam, *Lady under Fire: The Wartime Letters of Lady Dorothie Fielding MM 1914–1917* (South Yorkshire, UK: Pen and Sword, 2010). Dorothie Fielding was the first English woman to be awarded the Military Medal for bravery in the field. Working as an intrepid ambulance driver and "dodging shells and misogynist officers," she showed great compassion and courage under fire and wrote with dry wit: "It's topping being up near things and so jolly and interesting." Quite an adventurous role model.

Eleanor Hancock, "'Women as Killers and Killing Women': The Implications of 'Gender-Neutral' Armed Forces," in *The Human Face of Warfare: Killing, Fear and Chaos in Battle*, edited by Michael Evans and Alan Ryan (St. Leonards, Australia: Allen and Unwin, 2002), pp. 159–76. Puts in sharp focus a number of the conundrums and differing moral and ethical aspects of "gender neutral" when it comes to warfare.

Kathryn Harrison, *Joan of Arc: A Life Transfigured* (New York: Doubleday, 2014). An extraordinary work examining in great detail many sources, including the thousands of pages from her two trials (one long after her death to exonerate her). Harrison goes into excellent and telling detail about her military activities and weaves into the narrative a parallel construction of the life of Jesus. Anyone doing a project on Joan of Arc should avail themselves of this source.

David Hay, *The Military Leadership of Matilda of Canossa 1046–1115* (Manchester, UK: Manchester University Press, 2008). "The most powerful woman of her time," Matilda of Canossa led in war and battle for 40 years, put popes on the thrown and kept them there, and defeated the Holy Roman Emperor Henry IV in battle at Sorbara (1084) and for six years afterward. Well-documented scholarship enables the author to declare her "[o]ne of the most successful military commanders of the age."

Melissa Herbert, *Camouflage Is Not Only for Combat* (New York: New York University Press, 1998). The author argues that the military is a "gendered institution," and as such, women are often judged to be either "too masculine" or "too feminine" by that culture, and the military's masculine ideology will continue to limit women's participation in the military as long as the military remains a male domain for the achievement of manhood.

Margaret Randolph Higonnet, Jane Jenson, Sonya Michel, and Margaret Collins Weitz (eds.), *Behind the Lines: Gender in the Two World Wars* (New Haven, CT: Yale University Press, 1987). A wide-ranging set of essays looking at various war-related impacts on both women and men. The editors declare that it is a myth "that men are naturally fierce and warlike while women are mothers and have an affinity for peace." Central to the work is Margaret Randolph Higonnet and Patrice L. R. Higonnet's "The Double Helix" (pp. 31–47). The authors assert that men and women are two strands of a linked "double helix," in which, regardless of the level of either, the female strand is made subordinate. Therefore, it asks and answers the question, "When is change not change?" The social and economic roles of many women undergo rapid and radical transformation, both at the onset of war and, in a symmetrically opposed direction, at its conclusion. This chapter underscores the "illusory nature of wartime change in World War I and II" and declares a "feminist revision of the *time* in war can make the history of war more sensitive to the full range of experience of both men and women."

Richard Hingley and Christina Unwin, *Boudica: Iron Age Warrior Queen* (London: Hambledon and London, 2005). Calling Boudicca a "woman of many faces," the authors point out how she is at once an actual ruler, an imperial icon, and an impediment to Rome's civilizing mission. They urge us to look to archeology for the foundation of facts.

Jean Holm, *Women in the Military: An Unfinished Revolution* (Novato, CA: Presidio, 1982). This air force major general provides an extensive history of women and the military from World War I through the First Gulf War, arguing that women deserve the right to be in combat.

Kirsten Holmstedt, *Band of Sisters: American Women at War in Iraq* (New York: Stackpole, 2007). The author looks carefully at the experience of women Marines and finds that most felt they were just doing their job, but with more than 500 women killed or wounded in the Iraq Wars, more of them were clearly in combat than many in American society understood at the time or even now.

Maureen Honey, *Creating Rosie the Riveter: Class, Gender and Propaganda during World War II* (Amherst: University of Massachusetts Press, 1984). Shows how World War II got women into the workforce and ended much prejudice about women performing certain jobs through propaganda. But it also shows how, after the war, the return to the nuclear family overrode some of the gains, albeit not all.

Georgina Howell, *Gertrude Bell: Queen of the Desert, Shaper of Nations* (New York: Farrar, Straus and Giroux, 2007). Billed as the "driving force behind the creation of Iraq," this remarkable woman was the first female officer in the history of modern British intelligence and served with distinction during World War I. King Faisal said she had a "genius for war," and she did. See especially the chapters "War Work" and "Cairo, Delhi, Basra." Gives a good glimpse into the sacrifices made by the peoples of the British Empire to the World War I effort—192,000 were killed, captured, or "went missing" and 300,000 were recruited from India alone.

Stephen Hunter, *Sniper's Honor* (New York: Simon and Schuster, 1914) (F). Women warriors continue to fascinate. Witness this novel centered around a Soviet sniper from World War II nicknamed the "White Witch," Ludmilla "Mili" Petrova.

Duong Thu Huong, *Novel without a Name* (New York: William Morrow, 1995) (F). An absorbing and deeply depressing Vietnamese novel about war in the jungle, lost youth, and the immense problem of keeping one's humanity in the middle of a struggle without end: "In times of war, the future belongs to the combatants." She concludes, "Chants of the months, of the years spent in the Truong Son Mountains. Soldier, the dawn is icy. You fall under the bullets. On the white of the parachute cloth, I see your blood spreading." Powerful and evocative, a fitting counter to the "glories of war," whether in a revolutionary context or not.

Nancy Huston, "The Matrix of War: Mothers and Heroes," in *The Female Body in Western Culture: Contemporary Perspectives*, edited by in Susan Rubin Suleiman (Cambridge, MA: Harvard University Press, 1986), pp.

119–38. The author cautions, “Eliminating women from reproduction and integrating them into destruction tends to obfuscate the psycho-historical matrix of war, and may prevent us from ever understanding it sufficiently to bring it to an end.”

Sherrie Inness, *Tough Girls: Women Warriors and Wonder Women in Popular Culture* (Philadelphia: University of Pennsylvania Press, 1998). Our society seems to expect a great deal of women. And it is not always easy to fulfill those expectations, says the author.

Tami Amanda Jacoby, *Women in Zones of Conflict: Power and Resistance in Israel* (Montreal: McGill-Queen’s University Press, 2005). “The Campaign for Women in Combat in Israel” portion of the work effectively documents the way the entry of women into combat positions overturned much bias against them in the “last bastion of male privilege in the modern world.”

Steven Johnson, “The Longest War: Vietnam’s War of Independence from 111 BC–938 AD,” in his *Unknown Wars of Asia, Africa and the Americas That Changed History* (New York: Atlas, 2013). This work contains an exciting story of the Trung sisters and their fight against the Han Chinese. They led an uprising that was one-third women, and their initial successes in 39 AD were spectacular, although the Hans eventually returned and crushed the revolt.

David Jones, *Women Warriors: A History* (Washington, DC: Brassey’s, 1997). Female warrior heritage is celebrated and judged to be deeper and richer than previously thought. The author concludes, “Women can share equally with men the title of warrior.” In “The Female Martial Heritage” across time and space, the author concludes, “Women’s martial history is much richer and deeper by far than is commonly understood in the West.” For many women, the “sword is my child.” Good section on World War I and Russia’s all-female “Battalions of Death,” which fought for the czar and the provisional government.

Note: See Anne Eliot Griesse and Richard Stites, “Russia: Revolution and War,” in *Female Soldiers: Combatants or Non Combatants*, edited by Nancy Loring Goldman (reviewed earlier this chapter), which provides a most interesting vignette. On October 25, 1917, a detachment of a Petrograd women’s battalion helped defeat the provisional government and (including women in the Red Guards) turn back a Bolshevik attempt to storm the Winter Palace.

Gwyn Jones (trans.), *The Norse Atlantic Saga* (Oxford: Oxford University Press, 1986). Note especially the account of Freydis Eiriksdottir, the Viking woman warrior who drove off the Skraelings by sharpening a sword on her bare breast while pregnant and berating the Viking men, who were being bested by the Skraelings. The Skraelings fled, and the Viking men were presumably embarrassed and ashamed. As well they should have been.

Joyce Kaufman and Kristen William, *Women at War, Women Building Peace: Challenging Gender Norms* (Boulder, CO: Kumarian Press, 2013). Argues that women now expect to play major roles in war as well as peace. Looks at women as agents of resistance and political violence as well as peace-making. Case studies from Northern Ireland, the Israeli–Palestinian conflict, and Sri Lanka.

Thomas Keneally, *Daughters of Mars* (New York: Atria Books, 2013) (F). Two sister nurses go to Gallipoli and the Western Front and confront a deluge of casualties, the horrors of war, and self-exploration. This novel captures the ongoing impact of war across time and space.

Kathleen Khuhnast, Chantal de Jorge Oudraat, and Helga Hernes, *Woman and War* (Washington, DC: United States Institute of Peace Press, 2011). This wide-ranging volume helps us to understand how gender sculpts conflict, causing women to suffer a great deal more than normal through increased rapes, sexual violence, lost economic opportunities, and the like.

Heinrich Kleist, *Penthesilea* (New York: HarperCollins, 1998) (F). A new translation of the 1808 play. An army of one-breasted women (the better to shoot their bows) arrives in the middle of the Trojan War. Penthesilea, daughter of Ares and queen of the Amazons, shows carnal lust for Achilles ("her beloved enemy") and convinces him that she wishes to marry him. While one seldom feels sorry for Achilles, his final words, as she hunts him down and, together with her dogs, tears him apart and then devours portions of him, are "Penthesilea! My Bride! What are you doing? Is this the rosy feast you promised me?" Surely he deserves our consideration and sympathy. Penthesilea then kills herself. This play is not for the faint of heart, but it underscores the ongoing fascination with a true daughter of Mars and how she seems to upset the balance of the "natural" order in the minds of many.

Dr. Heidi Kraft, *Rule Number Two: Lessons I Learned in a Combat Hospital* (New York: Little, Brown, 2007). I have read hundreds of books about war, and occasionally I've gotten choked up or even cried a bit, but reading this book I cried a lot and deeply, too. It's an extraordinary work. The author takes her cue from the Korean War TV show and film *M*A*S*H*, which stated, "There are two rules of war. Rule number one is that young men die. Rule number two is that doctors can't change rule number one." Without melodrama or forced pathos, Kraft describes young Marines dying in her arms and warriors, men and women, young and old, breaking down under the strain of combat and the loss of comrades, feeling ashamed for feeling fear or for surviving while buddies did not. And how she, a devoted mother of two small children, copes halfway around the world from them. A warrior's warrior, she seems to me. And a powerful voice against war.

Krishna Kumar (ed.), *Women and Civil War* (Boulder, CO: Lynne Rienner, 2000). A potpourri of essays covering a "vast territory," including Rwanda, Cambodia, Georgia, Guatemala, El Salvador, and Bosnia. They fight and, in many cases, seem to play a significant role. Why are these women not more widely celebrated in the polyphony of womanhood?

Richard Lapchick and Stephanie Urbang, *Oppression and Resistance: The Struggle of Women in Southern Africa* (Westport, CT: Greenwood Press, 1982). A look at women in the struggle to liberate southern Africa, with introductory observations at their efforts in South Africa, Zimbabwe, and Namibia. Strangely silent on their role in *Umkhonto we Sizwe*, the "Spear of the Nation," however.

Gayle Tzemach Lemmon, *Ashley's War: The Untold Story of a Team of Women Soldiers on the Special Ops Battlefield* (New York: Harper, 2015). Poignant account of Ashley White, a member of the "Cultural Support Teams" in Afghanistan, who was killed in battle. Her training, integration into a traditionally male-only Special Forces unit, and untimely death are all sketched here.

Isaac Levine, *My Life as Peasant, Officer and Exile: The Life of Maria Bochkareva* (New York: Frederick Stokes, 1919). A truly amazing tale of a young woman who survived extreme poverty, abusive men, and incarceration to find liberation of a kind in the Russian armed forces during World War I after petitioning the czar to let her go into the army. Wounded a number of times, she eventually was selected to lead the all-female "Battalion of Death" during 1917. After the revolution broke out, she tried to fight the Bolsheviks, meeting Kerensky, Kornilov, Lenin, and Trotsky along the way. Although she eventually made it to American via Vladivostok, she ultimately returned to Russia, only to be captured and killed by the Bolsheviks.

Drew Lindsay, "Why Not Send Women to War?" *MHQ: The Quarterly Journal of Military History* 25, no. 3 (2013), pp. 50–61. A wide-ranging study from ancient times through World War II and the wars of decolonization, concluding, "[S]ome people will never accept women in battle—at least, that is, until women are needed."

Kathryn M. Linduff and Karen Rubinson (eds.), *Are All Warriors Male?* (Lanham, MD: Altamira Press, 2008). Scholarly anthropological studies of the Eurasian steppe's Iron Age graves from the Black Sea to Afghanistan, Northern Kazakhstan, western Siberia, and Xiongnu says, emphatically, "No." This work uses archaeological evidence from prehistory to look at the woman warrior in fact and fiction, coming up with plenty of evidence that Xena the Warrior Princess was not sui generis and had many real antecedents. Gravesites and grave goods are "living" proof. There is a reason contemporary culture carries the image of strong, powerful women righting wrong and fighting evil.

Pasi Loman, "No Woman No War: Women's Participation in Ancient Greek Warfare," *Greece and Rome* 51, no. 1 (April 2004), pp. 34–54. The author argues that, far from being victims or mere spectators, women in ancient Greece were not pacifists by nature and often showed strong support for a variety of wars.

Lois Ann Lorentzen and Jennifer Turpin (eds.), *The Women and War Reader* (New York: New York University Press, 1998). A variety of essays concerning war's impact on women (and, in some, men as well), taken from Bosnia, Korea, Israel, Nicaragua, El Salvador, Mozambique, Afghanistan, India, and Pakistan, as well as the United States. Unfortunately, there is neither an introduction nor a conclusion, so the various essays float unattached and episodic.

John Lynn, *Women, Armies and Warfare in Early Modern Europe* (Cambridge: Cambridge University Press, 2008). See especially chapter 4, "Warrior Women: Cultural Phenomena, Intrepid Soldiers and Stalwart Defenders." Good, in-depth accounts of women participating in warfare at a variety of levels and in a variety of ways. Many specific cases are illustrated.

———, "Women in War," *Military History* (October 2001), pp. 60–66. A short, stimulating look at female camp followers in the 16th century, describing how the armies of the day often had a nearly 1:1 ratio of men to women and children and "great crowds of women and children were not unusual; they were the rule." They carried food and clothing, treated the sick and wounded, participated in the looting, and protected its rewards, for "[p]illage was also the business of the army."

Marianne Mackinnon, *The Naked Years: Growing up in Nazi Germany* (London: Chatto and Windus, 1987). An engaging tale of a young woman drawn into war, serving in the Hitler Youth, drafted into the Organisation Todt doing farm and construction work, and narrowly escaping death in the Dresden firebombing raid. She ended up in the British sector after the war, ultimately arriving in Oxford.

Emily St. John Mandel, *Station Eleven* (New York: Alfred Knopf, 2014) (F). A marvelous, inventive dystopian novel, with women leaders ("The Symphony") and women warriors of note (Kirsten). One of the best of this genre I have read. Note: Women do not have to be macho to be true warriors.

Arthur Marwick, *Women at War 1914–1918* (London: Fontana, 1977). Using the "Great Spotlight of War" to show the myriad contributions made by British women during World War I, the author shows the wide range of activities, from munitions work to the Women's Forage Department to the Voluntary Aid Detachments (VAD) and the Women's Army Auxiliary Corps (WAAC), and their huge overall contribution to the war effort. Argues that these contributions greatly enhanced the progress of the suffragette movement. Finely illustrated.

Robert Massie, *Catherine the Great* (New York: Random House, 2012). Fascinating woman, fascinating account of her life, especially her actions promoting the Russian Empire. The expansion of Russia into Poland and to the Black Sea set the stage for further expansion in the 19th century, all the way to the Pacific. Interesting dynamic of the backward portions of Russia and their spectrum flow into new lands. Also good on the political machinations of Frederick the Great, France, England, and Austria in the Seven Years' War and beyond. This was a woman who reigned, ruled, and drove her country onward. See especially Catherine as military strategist and leader in "The First Partition of Poland and the First Turkish War" and "The Second Turkish War and the Death of Potemkin."

Gustave Masson, *The Story of Mediæval France from the Reign of Hugues Capet to the Beginning of the Eighteenth Century* (London: T. Fisher Unwin, 1893). See chapter 13, "Charles VII—End of the Hundred Years' War," for a 19th-century take on La Pucelle and Charles VII.

Jerry Matney, with D. A. Gordon, *Woman War Chief: The Story of a Crow Warrior* (Bloomington, IN: First Books, 2002) (F)**[AQ17: It's a novel, correct?]**. A long, thinly disguised praise poem about Woman Chief, the Gros Ventres woman who was captured by the Crows at age 10 and grew to womanhood with them. She worked hard to master the arts of soldiering and became a first-class warrior, especially against the Blackfeet. She was invited to join the Big Dog Society, normally an honor reserved for males; became a rare woman chief; and took that name. Ironically, she was eventually ambushed and killed by a Gros Ventres raiding party.

Adrienne Mayor, *The Amazons: Lives and Legends of Warrior Women across the Ancient World* (Princeton, NJ: Princeton University Press, 2014). Holistic, scholarly, well-illustrated, overarching, and multidisciplinary, this is the work on Amazons that must be consulted by those doing any research on this fascinating subject. Both the realities and the myths are breathtaking. Most illuminating is her close examination of the 1,000 years of history across 4,000 miles of geography, from the Don Basin to China, concluding, "Between Greece and China stretched the vast homeland of nomadic horsewomen archers, the equals of men, whose heroic lives and deeds inspired awe, fear, respect, and desire in all who knew them."

Madeline Miller, *The Song of Achilles* (New York: HarperCollins, 2012) (F). Putting Achilles in a homoerotic frame due to his love of Patroclus, the author accounts for his success with speed (shades of John Boyd, great American strategist of the 20th century) and the "best warrior of his generation," who goes on to immortality in story and song. Achilles is not frightened of war, for it is "what I was born for." This account of Achilles can certainly help us understand how important the legend of Penthesilea was (and remains) from a feminist perspective because Achilles alone among the Greek warriors could kill her in mortal combat. So many exegetes have missed the symbolic vitality of her killer; it is worth pondering why.

Molly Moore, *A Woman at War* (New York: Scribner's, 1993). The author, a reporter for the *Washington Post*, had access to General Walter Boomer, commander of the Marine expeditionary force, and thus was on the inside of many of the key field decisions of the First Gulf War. Written at a time when it was still unusual for American women to be at war and writing about it up close and personal. Clearly shows women in harm's way and worthy of equal opportunity to remain so, *if* they so choose.

Robert Mugabe, *Women's Liberation in the Zimbabwean Revolution* (San Francisco: John Brown Book Club, 1979). See especially pages 14–16, "The Armed Struggle Stage," for the role of the Special Women's Detachment of the Zimbabwe National Liberation Army.

Kate Muir, *Arms and Women* (London: Sinclair-Stevenson, 1992). An overview of the usual suspects but with some useful historical background.

Clare Mulley, *The Spy Who Loved: The Secrets and Lives of Christine Granville* (New York: St. Martin's Press, 2013). The absorbing story of a Polish woman (born Christine Skarbek) who served various spy organizations before, during, and after World War II and undertook some extremely dangerous missions into Nazi-occupied Europe, including Poland, Hungary, Bulgaria, Yugoslavia, and France. Makes a strong case for the importance of the resistance in Poland and the fact that Poland provided more support for Great Britain than previously credited. Christine was ultimately killed by a man who "loved" her but felt that killing her was the ultimate way of controlling this woman who lived her life to the fullest without boundaries and without remorse. In this sense, her tale is a sad reminder of the problems many men have with independent women and why control issues lie at the heart of so many domestic violence issues.

Rosaria Munson, *Artemisia in Herodotus* (Berkeley: University of California Press, 1988). Properly puts Artemisia in historical perspective and emphasizes the symbolic role of Artemisia, which goes far beyond a "woman in battle." It would seem that her story should be part of feminist canon but for some reason is not.

Bruce Myles, *Night Witches: The Untold Story of Soviet Women in Combat* (Chicago: Academy Press, 1997). A fascinating story of the Soviet women who flew against the Germans during World War II. Features women who dropped bombs and provided close air support and assistance to partisans and ground forces and especially those in the "free hunter" units, who targeted German aircraft. Many personal stories of women in action are presented. Anyone interested in the subject of women in combat should read this work, for it is a soaring paean to women as heroes.

Amy Nathan, *Yankee Doodle Gals: The Women Pilots of World War II* (Washington, DC: National Geographic Society, 2001). Lavishly illustrated and breezily written, this salute to the WASPs (Women Airforce Service Pilots) reports that 25,000 applied, 1,102 served, and 38 died. They towed targets for male pilots to shoot at and also delivered 12,000 aircraft of 77 kinds (including the "tough to fly biggies," the B-25, B-17, and B-29) to the major theaters of war. These "gals" delivered the goods and deserve high praise.

Mary Lou Colbert Neale, "Women of War," *Military History* (December 1993), pp. 35ff. Astonishingly enough, 800,000 Soviet women served at the front during World War II. This short article gives a good overview of what they did, including most of the roles normally reserved for males in the American military of the period. Contains a very interesting interview with "Tamar Pamyatnikh, Soviet Heroine." There is not much to honor about the Soviet system, but their need for warriors gives us a powerful example of how successful women can be when given a chance that women are denied in other societies.

Tim Newark, *Women Warriors* (London: Blandford, 1989). From Amazons of legend and fact to Jeanne, countess of Montfort, and Christine de Pizan, female military leaders are examined. A useful introduction for those not used to seeing women in battle. See especially the chapters "The True Amazons," "Amazons of the Jungle," "Braver than Her Husband," "Celtic Queens," "Women of Christ," and "Hundred Years War Women."

Sharon Newman, *Defending the City of God* (New York: Palgrave, 2014). A useful look at Melisende, the first hereditary queen of Jerusalem (who was Melisende Regina Sola from 1143 to 1149). Offers insights into what it took to be a Christian female ruler in a sea of Muslims and predatory male Christians. "A true ruler."

David Nicolle, *Yarmuk AD 636: The Muslim Conquest of Syria* (Oxford, UK: Osprey, 1994), pp. 70–72. Contains an engaging portrait of Hind Bint 'Utba, mother of the future Caliph Mu'awiyah, who, at the battle of Yarmuk, leads other women in rallying the Muslims when the Byzantines reach the Muslim camp. Contains a great admonishment from the camp followers:

> We are the daughters of the Night;
> We move among the cushions,
> With the grace of gentle kittens
> Our bracelets on our elbows.
> If you attack we shall embrace you;
> And if you retreat we will forsake you
> With a loveless separation. (p. 72)

Anne Noggle, *A Dance with Death: Soviet Airwomen in World War II* (College Station: Texas A&M University Press, 1995). A big, bold, and exciting story of the Soviet pilots who were female. A gripping tale, especially the sections on the female night fighters. They were true warriors by any account.

Wale Ogunyemi, *Queen Amina of Zazzau* (Ibadan, Nigeria: University Press PLC, 1999) (F). Another male author obsessed with the legend of Amina's taking a young man to bed whenever her army captured a city and then killing him in the morning. This play does celebrate her warcraft as well, however, calling Amina a "strategist for all times." It would be interesting if her tale were told in high school to boys and girls and their teachers, men and women.

Organization of Angolan Women, *Angolan Women Building the Future: From National Liberation to Women's Emancipation* (London: Zed Books, 1984). This somewhat polemical work suggests the validity of the "double helix" dynamics of women's suppression offered by Margaret Higonnet and Patricia Higonnet (see reference earlier in this chapter).

Note: For a more scholarly and holistic treatment of the decolonization struggle and liberation of Angola, see John Marcum, *The Angolan Revolution*, vol. 1, *The Anatomy of an Explosion (1950–1962)* (Cambridge, MA: MIT Press, 1969), and especially his *The Angolan Revolution*, vol. 2, *Exile Politics and Guerrilla Warfare (1962–1976)* (Cambridge, MA: MIT Press, 1978).

Iris Origo, *War in Val D'Orcia 1943–1944* (Boston: David R. Godine, 1947). A touching, illuminating, and very satisfying diary account of an English woman married to an Italian and living on a Tuscan property with dozens of farms and a castle. Amazing chaos in Italy after the king turns sides (July 1943) but doesn't join the Allies until the Germans have taken a measure of control, especially over the Italian armed forces. An almost unbelievable situation with refugees, escaped Allied prisoners, partisans, Fascists, Fascist militias, monarchists, Communists, and deserters (both German-Austrian and Italian, as well as false German-Italian ones) roaming around and all coming to her household for food, clothing, shelter, and often directions (a Moroccan escaped POW is headed north instead of south). Lots of ineffectual Allied bombing is recorded as well. This is a fine read.

Reina Pennington, "Reaching for the Sky: Hanna Reitsch and Melitta Schiller," *Military History Quarterly* (Autumn 2009), pp. 33–43. Two of Hitler's favorite pilots, one of whom was Jewish, were female. An amazing story of virtue rewarded and unrewarded. A cautionary tale on many levels.

———, *Wings, Women and War* (Topeka: University Press of Kansas, 2007). Women aviators deserve more publicity than they have received. This book is a start. There are quite a few superlative, if relatively unknown, Russian and German female pilots.

Captain Katie Petronio, "Get over It! We're Not All Created Equal," *Marine Corps Gazette* (July 2012), pp. 29–32. A recent Bowdoin graduate, after serving tours of duty in Iraq and Afghanistan, concludes that women should *not* be placed in *all* combat situations. Her personal experience is cogent.

Gerhard Pollauer, *The Lost History of the Amazons* (New York: Didactic Press, 2014). This work ransacks antiquity and the Middle Ages to find stories of and evidence for Amazonian warfare.

Christian P. Potholm, *Winning at War* (Lanham, MD: Rowman and Littlefield, 2010). Analyzes the seven ingredients necessary for success in warfare, regardless of time or place. These include: superior discipline, superior technology, sustained but controlled ruthlessness, protection of capital from people and rulers, superior will, receptivity to innovation, and the belief there will always be another war. Students should consult this work when doing their research papers to help explain the success or failure of particular women warriors and warrior queens by putting them in broader contexts. The nature of warfare changes throughout history, even though the ingredients for success do not.

James Powell, "The Role of Women in Fifth Crusade," in *The Horns of Hattin*, edited by B. Kedar (ed.) (Jerusalem: Israel Exploration Society, 1992). Shows how women served in many capacities, including that of guards, and fought Muslims at the siege of Darnietta.

Grace Pratt, "Female War Chiefs of the Blackfeet," *Frontier Times* (1971), pp. 22–23, 46. Despite the claims of others, the Blackfeet did have women warriors, including Running Eagle, who worked her way up into the Braves' Society by counting coup and fighting the Crows. Eventually killed in battle by the Flatheads.

Richard Pressfield, *The Last of the Amazons* (New York: Doubleday, 2002) (F). This tale portrays the Amazons as beset by such "bad" Greeks as Heracles and Theseus, who envy and fear their freewheeling lifestyles and the prominence of their women leaders. Most useful in providing the Greek perspective on Amazons; that is, their exploits end badly versus the much more positive perspectives of the Central Asian peoples, who portray Amazons as often triumphing in warfare no matter how much myth and cosmology the Greeks bring to bear.

Sayyid Ahmad-Ullah Qadri, *Memoirs of Chand Bibi: The Princess of Ahmednagar* (Hyderabad, India: Osmania University Press, 1939). A long praise poem to the Sultana who lived from 1550 to 1599, fought off the Moguls, and put down various rebellions, ultimately taking her own life by filling a well with acid and then jumping into it as the Moguls closed in. Other accounts have her killed by her own troops for negotiating with the same Moguls.

Quintus of Smyrna, *The Trojan Epic Posthomerica*, translated by Alan James (Baltimore, MD: Johns Hopkins University Press, 2004). Written 1,000 years after the *Iliad* but based on a long oral tradition, it was compiled by Quintus of Smyrna (on the west coast of Asia Minor) and covers what happens after the death of Hector and the *Odyssey*. Book 1 accents the warrior and daughter of Mars, Penthesileia, and her death at the hands of Achilles.

Dark Rain and James Thom, *Warrior Woman* (New York: Ballantine, 2003) (F). Despite the title, this is actually about the life of Nohelna, the Shawnee woman chief who worked hard for peace and ended up aiding the American Long Knives. Odd title, then.

Countess Ranfurly, *To War with Whitaker: The Wartime Diaries of the Countess of Ranfurly 1939–1945* (London: William Heinemann, 1994). Bound and determined to follow her husband to war, the countess keeps a most engaging diary as she helped to make war in the Mediterranean theater, meeting Wingate, Wavell, Eden, Churchill, Auchinleck, and Patton, as well as the kings of Greece, England, Egypt, and many others, while her poor husband languished in Italy as a prisoner of war, having been captured in the Western Desert early in the war. He later escaped and joined her. This was one resourceful woman. Some of her revelations are quite arresting and even amusing, as when she reports that General Maitland "Jumbo" Wilson went duck hunting in the Nile Delta and he and his party shot 2,300 ducks in a single day! Jumbo later became supreme Allied commander, Mediterranean Theater, and subsequently ambassador to the United States.

Roger Reese, "Soviet Women at War," *Military History* (May 2011), pp. 44–53. Soviet women, motivated by patriotism and revenge, proved themselves in battle, not just as medics and battlefield surgeons, but as machine gunners, snipers, and pilots as well. Women in combat were all volunteers, and training was often more rigorous than for men.

Stephen W. Richey, *Joan of Arc: The Warrior Saint* (Westport, CT: Praeger, 2003). See especially his extensive study of her military accomplishments (both strategic and tactical) in chapter 6, "Joan's Achievements as a Military Leader" (pp. 45–88). He sees her as an "essential factor" in the rise of the Valois dynasty and the ultimate French triumph in the Hundred Years' War. Believes her breaking of the siege of Orleans in 1429 was proof positive of her military leadership.

Sherry Robinson, "Lozen," in her *Apache Voices* (Albuquerque: University of New Mexico Press, 2000), pp. 3–15. This work shows how the myth is real and that "Woman Warrior" Lozen, who "could ride, shoot and fight like a man," did so effectively with both the Chiricahua Apaches and the Warm Springs Apaches. Even among legendary Apache warriors, such as Cochise, Germonimo, Tuh, Nana, and Victorio, she stood out.

Lucia St. Clair Robson, *Ghost Warrior* (New York: Forge Books, 2012) (F). Celebrating the life and times of Lozen, a warrior's warrior, this novel presents her 30-year struggle on behalf of the Dineh. After campaigns in the New Mexican and Arizona Territories and the Mexican states of Sonora and Chihuahua, she and her tiny band of 17 are pursued and finally cornered by 9,000 American and Mexican troops and others. "The odds made Lozen proud." A true and brave warrior well worthy of further study.

Guy Rothery, *The Amazons* (Charleston, SC: Bibliobazaar, 2014). Seeks to buttress the legends of the Amazons with examination of artwork, especially sculpture. Explores Amazons in Central Asia, the Caucasus, and Kurdish country, as well as Africa and America.

Michael Rustad, *Women in Khaki* (New York: Praeger, 1982). Using the microcosm of a single American base in West Germany, the author looks at the emergence of the role of enlisted women in the US Army. Notes the rise of women in the military from 1 percent in 1972 to 8 percent in 1979. See especially "Her Army" (pp. 138–80).

Jessica Amanda Salmonson, *The Encyclopedia of Amazons: Women Warriors from Antiquity to the Modern Era* (New York: Paragon House, 1991). A most thought-provoking collection of historical and mythological references from Aba, the first-century-BCE warrior daughter of Xenophanes, to Zoulvisia of Armenian legend—they are all here. A treasure trove to stimulate your exploration of women warriors throughout the ages.

Thomas Sankara, *Women's Liberation and the African Freedom Struggle* (New York: Pathfinder, 2007). The former leader of Burkina Faso (1983–1987) who was assassinated while in power in 1987, using the "Marxist understanding of human society," declares how women are an important part of the struggle for liberation. This work also contains an interesting photo of women anti-aircraft gunners from Angola.

Note: This pattern of military men seizing power and claiming to be inspired by Marxist-Leninism was a widespread phenomenon in Africa during the 1970s and 1980s as they sought to provide instant albeit often superficial legitimacy to their regimes. For a more in-depth look at this pattern, see C. P. Potholm, "Marxist Modernizers," in his *The Theory and Practice of African Politics* (Englewood Cliffs, NJ: Prentice Hall, 1979), pp. 212–46.

Shelley Saywell, *Women in War* (Baltimore, MD: Penguin Books, 1986). Wide-ranging account of women who have fought and led in battle. A lot of interesting vignettes here.

Stacy Schiff, *Cleopatra: A Life* (New York: Little, Brown, 2010). A superb analysis of one of the most interesting and intriguing women in history, whose true strategic and tactical skill overwhelms her historical image as a courtesan. Schiff uses Cleopatra to give vivid and lasting insights into the world of Rome when it teetered between a republic and an empire and the forces for the latter won out.

Note: Interestingly, while such classical-era writers as Plutarch blame Cleopatra (and citing her "female and Egyptian" attributes) for Mark Antony's disastrous defeat at Actium (Greece) in 31 BCE, modern military scholars give her more credit for the strategic decision to escape from that defeat together, with the Ptolemaic treasury and her naval squadron (held in reserve) leading the way. See David Califf, *Battle of Actium* (Philadelphia: Chelsea House, 2004), and especially Si Sheppard, *Actium 31 BC* (Oxford, UK: Osprey, 2009).

James Schultz, *Running Eagle: The Warrior Girl* (Boston: Houghton Mifflin, 1917) (F). A charming novel making much of the Blackfeet woman in what is now Montana who counted coup and led in battle, ultimately becoming a leading warrior. Eventually she was killed by the Flatheads but not before she etched herself into Native American legend.

Simone and Andre Schwarz-Bart, *In Praise of Black Women I: Ancient African Queens* (Madison: University of Wisconsin Press, 2001). An amazingly, lavishly—and lovingly—illustrated volume of many African warrior-queens, including Tiye, queen of Egypt; the queen of Sheba; Yennenga of the Mossi; Heleni of Ethiopia; and Beatriz Kimpa Vita, the Joan of Arc of Kongo.

Paul Sealey, *The Boudican Revolt against Rome* (Buckinghamshire, UK: Shire, 1997). A good account of the sacking of Colchester, London, and Verulamium (outside present-day St. Albans) by "Boudican hordes" in 60 CE before those hordes were in turn crushed by Roman legions in the Midlands in 61 CE. Much archeological evidence is promulgated herein.

Anita Shreve, *Stella Bain* (New York: Little, Brown, 2013) (F). An American woman serving as an ambulance driver in France is wounded and suffers severe shellshock but eventually recovers. Along the way, this novel gives a good sense of the life at the front from a woman's perspective, filling in what was traditionally a man's job during wartime in many cultures. Illuminating.

Rosemaire Skaine, *Women at War: Gender Issues of Americans in Combat* (Durham, NC: McFarland, 1999). Wide-ranging work dealing with how military service exposes the fault lines of contemporary society, feminism, and the demands of the military.

Kenneth Slepyan, *Stalin's Guerrillas: Soviet Partisans in World War II* (Lawrence: University Press of Kansas, 2006). Insurgent and partisan in Soviet Union–occupied territory, some under the control of Moscow, others not. Useful insights into the day-to-day struggle against the Germans and their allies and the tensions within and among partisan banks. See the excellent chapter on "The Crisis of Partisan Identity, 1943," with superb maps on pages 189 and 192. Partisans went from 100,000 to 181,000 during that year. Good section on women partisans can be found in this chapter.

Helen Solterer, "Figures of Female Militancy in Medieval France," *Signs: Journal of Women in Culture and Society* 16, no. 31 (1991), pp. 522–49. Argues that *Li Tournoiement as Dames* represents the "disturbing multivalence of the figure of the woman warrior," and thus it "provides a scenario for realizing a female martial ambition."

Pat Southern, *Empress Zenobia: Palmyra's Rebel Queen* (London: Continuum, 2008). See especially the chapters "Septimia Zenobia Augusta" and "Aurelian and the Roman Recovery." Palmyrene expansion into Arabia, Egypt, and briefly Asia Minor occurs while the Persians were back on their heels and the Roman struggle with the invasions of the Goths came a cropper when Aurelian came calling in 272 CE. The Romans won the Battles of Immae and Daphne and captured Zenobia as she fled on camel. Provides lots of alternative sourcing for the sketchy record of Zenobia's time on earth.

Note: For additional background on Rome and its adversaries in the region, see also David Nicolle, *Rome's Enemies (5): The Desert Frontier* (Oxford, UK: Osprey, 1991).

Matthew Stibbe, *Women in the Third Reich* (London: Arnold, 2003). See especially chapter 7, "From Total War to Defeat and Military Occupation" to examine the calumny that the "German soldier fought for six years and the German women for only five months."

Judith Stiehm (ed.), *It's Our Military Too!* (Philadelphia: Temple University Press, 1996). Provides a set of interesting perspectives, including "The Enemy Doesn't Care If You're Female"; "Duty, Honor, Country: If You're Straight"; and "Gender and Weapons."

——— (ed.), *Women and Men's Wars* (New York: Pergamon Press, 1983). Arguing that it is "men who plan, prepare for, conduct, conclude, describe and define war," the editor provides a wide set of essays from women in pacifism, women in the military, and women in national liberation struggles.

Laurie Stoff, *They Fought for the Motherland: Russian's Women Soldiers in World War I and the Revolution* (Lawrence: University Press of Kansas, 2014). Although women fought in World War I in Russia under the czar, the huge impetus for women in combat came under the provisional government from February to October 1917, when that government created separate all-female military units. See especially the chapter "Russia's First All-Female Combat Unit." As many as 6,000 women were involved. Most famous was Maria Bochkareva, who was in the First Russian Women's Battalion of Death. Called the "Russian Joan of Arc," she fought for the provisional government, was wounded numerous times, eventually captured by the Bolsheviks, and made it to the United States by way of Siberia, only to return to fight for the Whites. This time, she was captured and shot. All told, 80,000 women fought for Russia in World War I and the revolution. The Bolsheviks kept women in the army but broke up the sexually segregated units and integrated them into their male units. Note also, the author mentions Flora Sandes, the only British woman to have served officially in the military, and she went to Serbia as a nurse and then joined the Serbian army, eventually rising to the rank of lieutenant.

Elizabeth Stone (ed.), *Women and the Cuban Revolution* (New York: Pathfinder Press, 1981). Fidel Castro speaks and makes the case for the participation of women in war in his essay "The Revolution within the Revolution," finding them "[d]oubly exploited, doubly humiliated" and hence highly motivated for change.

Amy Goodpaster Streve, *Flying for Her Country: The American and Soviet Women Military Pilots of World War II* (Washington, DC: Potomac Books, 2009). A well-deserved paean to the women who flew during World War II. In that war, 400,000 women were in the military, including many in the Women Airforce Service Pilots (WASP), the Women's Flying Training Detachment (WFTD), Women Army Corps (WAC), and Women's Army Auxiliary Corps (WAAC). Particularly useful is the chapter on "Gender Issues."

David Sweetman, *Women Leaders in African History* (London: Heinemann, 1984). Breezy and sometimes lacking in definitive sources but a good introduction to this understudied subject. Some prominent African women leaders, such as Amina of Hausaland and Nzinga of Angola, stand out in both warfare and diplomacy. Written 40 years ago, this work needs to be checked against more current sources but should stimulate further research.

Craig Taylor (ed. and trans.), *Joan of Arc: La Pucelle* (Manchester, UK: Manchester University Press, 2006). A rich mosaic of 105 primary documents covering the life and times of the "Maid." Heavy focus on her various trials and contemporary letters about her. Not much on her military activities per se, however. The author concludes that her impact on the eventual outcome of the Hundred Years' War was due more to the breakup of the Burgundian-English alliance than to the activities of La Pucelle, although she is given credit for offering national hope after the English were stopped at the Loire and that they could eventually be ejected from French soil.

Steris Tepper, *The Gate to Women's Country* (New York: Doubleday, 1988) (F). This science-fiction novel describes an ecotopia, where women are in the process of breeding out the warrior strain in men. In the process, many ironic dimensions and developments emerge.

Vicky Thomas, *The Naga Queen: Ursula Graham Bower and Her Jungle Warriors, 1939–1945* (Stroud, UK: History Press, 2012). Carrying a stern gun and a .38, Ursula Bower became the senior woman of V Force in India, leading her Naga fighters against the Japanese. She appealed to General William Slim for equipment before the Battle of Imphal, when the Imphal plain was "[c]rawling with Japanese," and became a legend.

Helen Thorpe, *Soldier Girls* (New York: Scribner's, 2014). This account follows three young women who joined the Indiana National Guard and served in both Iraq and Afghanistan. Their deployments are of note, but so, too, are their reentries into American society. I was struck at how their reasons for entering the military paralleled so closely that of men.

Sharon W. Tiffany and Kathleen J. Adams, *The Wild Woman: An Inquiry into the Anthropology of an Idea* (Cambridge, MA: Schenkman, 1985). Why can't women be warriors without being considered "wild"? It is a somewhat challenging book to read and understand but a rewarding one. Let it provoke your thinking.

Stephen Turnbull, *Samurai Women 1184–1877* (Oxford, UK: Osprey, 2010). A highly interesting set of revelations about the extent to which women warriors were an integral part of the samurai tradition until 1877. Well-illustrated and often excitingly presented.

Karen Gottschang Turner and Phan Thanh Hao, *Even the Women Must Fight: Memories of War from North Vietnam* (New York: John Wiley and Sons, 1998). The Vietnam Wars against the French and the Americans and South Vietnamese through the eyes of the women who served in the North Vietnamese armed forces. Ironically, many women would end up suffering just as their male counterparts have throughout history. "The Return of the Warrior" is never truly easy, whether the warrior is female or male.

Stephanie Urdang, *Fighting Two Colonialism: Women in Guinea-Bissau* (London: Monthly Review Press, 1979). Believes that women are an "explicit and integral part of the overall revolution" there, but in addition to fighting the Portuguese colonialism, they also need to struggle against traditional male-dominated society, economy, sexuality, and politics. Believes attitudes would change faster if more women carried guns and were more deeply entrenched in the cadre class.

US Marine Corps, "Women in Combat? Insights Worth Repeating," *Marine Corps Gazette* (November 1997), p. 73. Are the demands of the Marine Corps different from other services? Strong arguments against women in combat, at least for the type the Marines practice. Very challenging are the words and views of (now) Bowdoin's own government professor, Jean Yarbrough!

Christine B. Verzar, "Picturing Matilda of Canossa: Medieval Strategies of Representation," in *Representing History, 900–1300: Art, Music, History*, edited by Robert A. Maxwell (University Park: Pennsylvania State University Press, 2010), pp. 73–92. Well-illustrated, this chapter provides a look at the many dimensions to this extraordinary woman, only one of two ever to have a sepulchral monument for her in St. Peter's. Features her centrality to the Investiture Crisis and shows her receiving the Book of Prayers and Meditations from Bishop Anselm of Canterbury on his way back from Rome.

Jeanmore Vickers, *Women and War* (London: Zed, 1993). More interested in seeing women out of war than in doing well in war, the author does provide some accents on "Women in Action," "The Impact of War on Women," and "Moving toward a Non-Violent World."

Jocelyn Viterna, *Women in War* (London: Oxford University Press, 2013). Based on more than 200 interviews with El Salvadorian women, the author looks at "Gender, Violence and the Micro-Processes of Mobilization" to conclude that it's not easy being a female in a wartime situation. This work is very interested in the gender truths that 12 years of war threw up.

William Vollmann, *The Ice-Shire* (New York: Viking, 1990). A curious blend of the sagas and travelogue to real places and dreamscapes along the Viking trail to Greenland and Vineland. See especially "Freydis Eiriksdottir; or, How the Frost Came to Vineland the Good" (pp. 129–340). As the pregnant Freydis confronts the Skraelings, she shouts, "Oh, you'd like me to take my shirt off, would you? I'll show you, you savage thralls, you Hell meat." "Sharpening" her sword on her breast, she attacks, and the Skraelings flee, but the Vikings decide the future is not on their side.

Martin van Creveld, "The Great Illusion: Women in the Military, *Journal of International Studies* 29, no. 2 (2000), pp. 429–42. Von Creveld is not a big fan of women in combat or even in the military, and he minces no words about his views. Using the example of the Israeli Defense Force and other combat formations, he makes a strong case against the use of women in war. Agree or disagree, he is one of the military historians who commands a broad audience on this highly freighted subject, so his arguments need to be considered. And he has observed women in action in a number of important contexts.

Marguerite Waller and Jennifer Rycenga (eds.), *Frontline Feminism: Women, War and Resistance* (New York: Garland, 2000). A wide-ranging collection covering ethnic and gender violence, militarism and sexuality, feminist resistance to war, soldier and state considerations, and the exploration of women in and by war.

William Ware, *Zenobia, Queen of Palmyra* (Boston: Estes and Lauriat, 1838) (F). A historical romance celebrating the life and times of Zenobia, queen of Palmyra, and her eventual defeat in the third century CE at the hands of the Emperor Aurelian, who laid siege to Palmyra and eventually captured her. Very praiseful of her, "Julius Caesar himself, Piso, never displayed a better genius than this woman."

Graham Webster, *Boudica: The British Revolt against Rome AD 60* (Lanham, MD: Rowman and Littlefield, 1978). Puts the revolt of the Iceni on a par with the Jewish revolt of 70 AD as being sparked by both religion and terror. Admits that the bravery of Boudicca and the Iceni was undercut by their lack of discipline and their continuing reliance on outmoded chariot warfare and poor weapons. Good map on page 92 of Roman forts of the era—there were quite a few. Not much about Boudicca or her military prowess per se, but the work is heavily grounded in archeological finds of coins, pottery, and other artifacts.

Batya Weinbawn, *Islands of Women and Amazons: Representation and Reality* (Austin: University of Texas Press, 1999). Sees the Amazonian legends everywhere from Homeric tales though the medieval period, as well as in Chinese, Indian, Native American, and Pacific Islanders, and popular culture, up through and including "leisure primitivism."

Judith Weingarten, *The Chronicle of Zenobia: The Rebel Queen* (Cambridge, MA: Vanguard Press, 2006) (F). From Tadmor-Palmyra in what is now Syria, Zenobia led a rebellion against the Romans from 269 to 272 CE. When her husband, the ruler, died, she led her armies to victory, eventually taking over Arabia and Egypt, until she, like Boudicca, was defeated by the Romans, who decided that allowing a rebellion to succeed was not an option. This long, somewhat fanciful novel gives the reader something of a flavor for her life and times.

Lyn Webster Wilde, *On the Trail of Women Warriors: The Amazons in Myth and History* (New York: St. Martin's Press, 1999). Amid some flights of fancy (their "bright, burning vital power"), the author does pin down the grave of the earliest woman warrior, found in Georgia dating from 1200 BCE, and explores the shores of the Black Sea for Scythian and Hittite women warriors.

Bel Irvin Wiley, *The Life of Billy Yank: The Common Soldiers of the Union* (New York: Bobbs Merrill, 1951). While the author declares, "Not all who wore blue were men," he really doesn't make much of an effort to find out who or what they were. He also seems genuinely perplexed about the whole concept, relying mostly on somewhat lurid press accounts of the time. Seems happy that he didn't unearth any examples from the Confederacy.

———, *The Life of Johnny Reb: The Common Soldiers of the Confederacy* (Baton Rouge: Louisiana State University Press, 1978). An updated version of the 1943 edition, which was way ahead of its time in its accent on "common folk" and "little people." Pages 334–335, "Women in the Ranks," shows, however, that historians previously often did not look too hard for signs of real women in real combat.

June Willenz, *Women Veterans: America Forgotten Heroines* (New York: Continuum, 1983). Profiles women in the service during World War II and looks at what happened to them afterward.

Amrit Wilson, *The Challenge Road: Women and the Eritrean Revolution* (Trenton, NJ: Red Sea Press, 1991). "Remember the women who have been martyred fighting" begins this work, which contains a number of life stories of women in the struggle. See especially the chapter "Fighters" (pp. 87–110).

Rex Winsbury, *Zenobia of Palmyra: History, Myth and the Neo-Classical Imagination* (London: Duckworth, 2010). The third-century-CE warrior queen of Syria was defeated by Emperor Aurelius, originally from Serbia and a superb warrior in his own right ("The Achilles of His Era"?). See especially the chapter "Arms and the Woman: Zenobia Goes to War." Zenobia plays in the broad seam between the Persian and Roman Empires, both under duress when she staged her rebellion. Gibbon, Chaucer, and a host of others made of Zenobia much, and much of that was very romanticized. In reality, Zenobia was a tough woman in a world where toughness mattered, but she and Palmyra obviously had certain limits when taking on the most powerful military actors in the Western world.

James E. Wise and Scott Baron, *Women: Iraq, Afghanistan, and Other Conflicts* (Annapolis, MD: Naval Institute Press, 2006). A very good collection of first-person female accounts of their service at war. Examples from the Coast Guard, Marines, US Army, and US Paratroops. A very poignant appendix listing all the women wounded in these wars. The work clearly suggests that there is now a very blurred distinction between "combat" and "noncombat," with many women in transportation, supply, and other areas already in "combat" due to IEDs and ambushes and urban violence against "safe areas." "The voices of these women deserve to be heard." Agreed.

Note: About 7,400 US women served in Vietnam, mostly as nurses, compared with 195,600 women who have served in Afghanistan and Iraq (as of February 2008), with 25,000 currently in war zones. One hundred five American women were killed in Afghanistan and Iraq.

Jane Yolen, *Sea Queens: Women Pirates around the World* (Watertown, MA: Charlesbridge, 2008). Who knew "[s]ome of the greatest pirates ever known were women"? The author gives us Alfhild, Grania O'Malley, Anne Bonney, Mary Read, Rachel Wall, Mary Anne Talbot, and Ching Isao from Holland, England, America, and China.

David Zabecki, "Artemisia at Salamis," *MHQ: The Quarterly Journal of Military History* (Summer 2014), pp. 38–43. A short account of the Battle of Salamis and the role of the first recorded female admiral in it. The very fact that she was trusted with a portion of the entire fleet suggest great competence, and her insights prior to the battle proved to be prophetic.

Yasmine Zahran, *Zenobia: Between Reality and Legend* (London: Stacey International, 2010). This work portrays Zenobia as a Hellenized Arab, one who ruled toward the end of Palmyra (which ran from the third century BCE to the third century CE) and led her conquering armies in Arabia, Egypt (which at the time produced one-third of Rome's grain), and Mesopotamia. The author argues that existing Arab sources have her committing suicide rather than enduring the shame of being paraded at Aurelian's Roman victory parade. Another version has her dying en route to Rome. Either Arab version spares her the parade.

Chapter Thirty-Six

Homosexuality and Warfare

This chapter examines sexual orientation and warfighting. Note the importance of what Nigel Hamilton calls the "homosocial" dimension to studying men at war. There are various recorded historical examples of homosexuals at war, such as the famous Theban band of homosexual lovers commemorated by Alexander the Great. There are also a number of prominent military figures—Frederick the Great, Alexander the Great, Baron von Steuben, and Charles Gordon—who have been thought to be homosexual or bisexual by some observers. There is also the practice of "manly love" lardered through accounts of Greek warriors, samurai, and Persian forces. Many of the books in this chapter hint at, or more openly explore, these phenomena.

Allan Bérubé, *Coming Out under Fire: The History of Gay Men and Women in World War Two* (New York: Free Press, 1990). A wide-ranging, thought-provoking account of those who "marched to a different drummer" during World War II, risking exposure, section 8 discharges, and violence, as gays fought two wars, one against America's opponents and one against the society of the time, both in and out of the armed forces. The author concludes, "The massive mobilization for World War II propelled gay men and lesbians into the mainstream of American life."

Demetrius Boulger, *The Life of Gordon* (London: T. Fisher Unwin, 1896). Not a whiff of homosexuality here, seeing him simply as "a "slave to duty" and a Christian Victorian hero of the first rank. Quite a read from the point of view of heroic portraiture. Still, there was that slave boy.

John Boyne, *The Absolutist* (New York: Other Press, 2011) (F). A carefully crafted tale of homosexuality in the trenches of World War I. Hard to imagine the tensions of life in the trenches without the added anxiety of ambiguity and fear of being found out; then add the strains of discovering you don't believe in war. This is a powerful book, not one for the faint of heart.

A. R. Burns, *Alexander the Great and the Hellenistic Empire* (New York: Macmillan Company, 1948). See especially the Theban Golden Band and the Battle of Chaeronea in 338 BCE.

Byron Farwell, *Eminent Victorian Soldiers: Seekers of Glory* (New York: Routledge, 1992). See especially the chapters on Herbert Kitchener, Hector Macdonald, and Charles "Chinese" Gordon for some possible homosexual generals and military figures.

Homer, *The Iliad* (see review in chapter 11).

Madeline Miller, *The Song of Achilles* (New York: HarperCollins, 2012) (F). Puts Achilles in a homoerotic frame of mind due to his love of Patroclus. The author accounts for his success with speed (shades of John Boyd!), and the "best warrior of his generation" goes on to immortality in story and song. Achilles is not frightened of war, for it was "what I was born for." Of course later, in Hades, he did have second thoughts.

Mary Renault, *The Persian Boy* (New York: Pantheon, 1972) (F). Alexander's alleged bisexuality explored in this novel from the perspective of one of his imagined lovers. This forms the literary device that enables the author to chronicle his life and times.

M. C. Scott, *Rome: The Eagle of the Twelfth* (New York: Bantam Press, 2012) (F). A convoluted account of the Roman legion that was defeated by the Parthians but later was reconstituted and began the suppression of the Jewish revolt of 67 CE. Seen through the eyes of a gay Roman legionnaire, Demalion the Macedonian, and his lover, Tears. The main focus is on the training, discipline, and equipment of the Romans and their enemies, and there are some fine descriptions of the Parthian cataphracts in action.

Dan Vyleta, *The Crooked Maid* (New York: Bloomsbury Press, 2013). These returning warriors—and the women who wait for them—are an Austrian doctor and a Czech soldier who have both served on the Eastern Front and served time in Soviet POW camps. Both men receive a less-than-heroic welcome in postwar Vienna. Central to the plot is a wartime homosexual affair between the Austrian doctor and his Russian camp commander, with the latter paying a huge price for his "indiscretion." Also reviewed chapter 38.

José Zuriga, "My Life in the Military Closet," *New York Times Magazine*, July 11, 1993. Why "Don't Ask, Don't Tell" should be repealed, claims the author. And it was!

Chapter Thirty-Seven

Warriorhood

"Our Heaven is the Great War on Earth. We live in battle, our eternal life."—Waffen SS (*Schutzstaffel*) Division Liebstandarte Adolf Hitler

"When I knew war I passionately loved it. . . . I shall never cease to love it, for all the splendor in which it has clad the most humble."—Captain La Tour Du Pin, *Le Creuset*

"There is no need for a warrior to accept an honorable defeat in shame."—Admiral Heihachiro Togo

"He would sometimes recall . . . the compulsion to press close to death, to feel the tight purity that followed."—James Salter, *The Hunters*

"That joy. That madness. The gods must feel this way every moment of every day. It is as if the world slows. You see the attacker, you see him shouting, though you hear nothing, and you know what he will do, and all his movements are so slow and yours are so quick, and in that moment you can do no wrong and you will live forever and your name will be blazoned across the heavens in glory of white fire because you are the god of battle."—Bernie Cornwell, *Death of Kings*

"That stern joy which warriors feel in foemen worthy of their steel."—Walter Scott, *The Lady of the Lake*

"To be a warrior is much more than a philosophy of life, it is often something akin to a religious calling, and it involves killing, and killing others with skill."—Christopher Coker, *Men at War*

What makes some humans fight their fellows? What makes some humans *have* to fight their fellows? What makes some humans want to fight their fellows? What makes some humans happy only when they are preparing to or actually doing so or reflecting on having done so? What is the essence of the drive to war, and what forms the bonds among those who do fall into these categories? What is the joy that comes to many warriors in battles, a joy that transcends all others?

In so many societies across so many miles of space and eons of time, there is repeated and multiple adherence to what Edward Luttwak has termed the warriors' creed: "We are all mortal anyway so let us fight valiantly." From the *Iliad* to *Beowulf* to *The Centurions*, warriorhood has been highlighted and celebrated. In this regard, the Salafist warriors and their adversaries exist in a realm far beyond that of you and I perhaps but one grounded in the nature of humankind.

There is, of course, another subtheme to warriorhood, and that is the price, sometimes great, sometimes small, but always present, that a warrior must pay for leaving behind nonwarriorhood. As Homer wrote in the *Iliad*, when Agamemnon, king of Mycenae, offers his daughter Iphigenia in sacrifice to the gods in exchange for fair winds to take him to Troy, sometimes these losses are self-inflicted, but they are almost always present.

In this chapter are listed some of the books that directly or indirectly help us to understand the answers to these questions as well as the questions themselves and why they are relevant to some observers of war and not to others.

David Adams, *Foobit* (New York: Black Cat Press, 2012) (F). Seeking to provide this generation with a *Catch-22* antic description of the counterinsurgency war in Iraq, the author focuses on those who are not front-line warriors but "foobits," who at their core are marshmallows ("where their hearts should be beating with a warrior's courage and selfless regard, you'd find a pale gooey center"). In other wars, these support troops might be referred to as REMF ("Rear Echelon Mother Fuckers"), but now the forward-operating bases declare there is no real rear echelon, except perhaps in Kuwait. For most foobits, "[g]oing to war could be a real pain in the ass," despite all the comforts of home at the forward-operating base (FOB). But their real aim is to avoid combat at any cost. Antiheroes worthy of the name.

Antonio Lobo Antunes, *Fado Alexandrino* (New York: Grove Weidenfeld, 1990) (F). An antiwar, antiwarriorhood novel that leads the reader in no doubt that there is no glory in war or in revolution, for that matter. Portuguese soldiers return from Mozambique (Antunes served in Angola) disillusioned, depressed, and impotent. Belongs here as an anecdote to many of the more vibrant and uplifting tales of warriorhood in this chapter.

Aristophanes, *Lysistrata* (New York: Dover Editions, 1990) (F). This fifth-century-BCE marvelous play catches warriorhood in full motion, as Lysistrata gets the women of Greece to withhold sex from men. After various trials, the men make peace. Sexual deprivation has, at least momentarily, frustrated Mars.

Erwin Bartman, *Fur Volk and Fuhrer* (Solihull, UK: Helion, 2013). In diary form, this soldier of the Liebstandarte SS Adolf Hitler writes of *Kameraden* and "obedience unto death" and means it. An elite unit in constant action. This apprentice baker was personally recruited by Heinrich Himmler. Fascinating insights abound.

Timothy Bax, *Three Sips of Gin* (Solihull, UK: Helion, 2013). What makes a warrior? Bax slips into warriorhood but cannot leave it behind, even when he is grievously wounded with the Rhodesian Selous Scouts (a multiracial unit) in Mozambique (where FRELIMO and the Portuguese soldiers share a bordello). Later he went to South Africa (southwest Africa and the Transkei). A hard-drinking, rollicking account of life and war.

Chris Bellamy, *The Gurkhas: Special Force* (London: John Murray, 2011). A somewhat dry and highly descriptive piece that ultimately fails to capture the true warrior spirit of these fine bringers of battle. Still, it is possible to marvel at the places they have fought during the last 200 years (winning no fewer than 26 Victoria Crosses): Afghanistan in 1838 and again in 2005; Burma, 1824 and again in 1942; Palestine, Gallipoli, and Palestine during World War I; North Africa and Italy during World War II; and Borneo, Greece, Malaya, the Falklands (1982), and, as members of the Indian army, in all three wars against Pakistan. As for their service today, the author argues, "It works. Don't fix it." These are true warriors for the ages.

Christian Cameron, *Killer of Men* (London: Orion Books, 2012) (F). A rousing, fine tale of great interest to those seeking the nature of warriorhood. Portrays the life and times of Arimnestos of Plataea, who became a warrior, a slave, then a warrior again, and a fine and successful one at that, during the time of the Ionian revolt against Darius and the Persians. "I was a killer of men. I still am. Once you have been there, you can never go back." Captures the essence of the love for battle true warriors have and their occasional regrets of those they have slain. Well worth a read.

Orson Scott Card, *Ender's Game* (New York: TOR, 1977) (F). A rough and tumble science-fiction tale but one that recognizes the specialness of the true warrior whose battle qualities separate him or her from others.

Brian Castner, *The Long Walk: A Story of War and the Life That Follows* (New York: Doubleday, 2012). An explosives expert who detonates or disarms IEDs in Iraq comes to grips with the peace as he did with the war, despite being "crazy" and suffering PTSD and then not suffering PTSD. The brotherhood welcomes him home, and he works out his anguish and craziness. Also, there is the very telling line in the middle of the book: "Everything about Iraq sucked. I loved it."

Howell D. Chickering Jr. (trans.) *Beowulf: A Dual Language Edition* (Garden City, NY: Anchor Books, 1977). Although there are disputes about the age and authorship of *Beowulf*, the can be no doubt that this oldest poem in Old English (most likely written in the seventh century, although the earliest written version was from West Saxony in 1000 AD) is about Beowulf, the warrior and a prototypical hero in warfare: "There is no shame for a hero to seek glory"; indeed the "aim was glory—'the fame of a good name' after death." This is not by chance, says the *Iliad* and many subsequent works.

Robert Citino, "Meet the *Freikorps*," *Military History* (November 2012), pp. 60ff. "What if a man cannot stop being a soldier, even after wading through blood for four long years?" While the *Freikorps* are rightly primarily seen as the forerunners of the Nazis, nevertheless the author makes a number of telling points about the difficulties some men face, longing as they do for the comradeship felt so intensely in battle.

Christopher Coker, *The Warrior Ethos* (London: Routledge, 2007). Examines such topics as "Achilles and the Warrior Soul" and "Emerson and Self-Trust," as well as such concepts as the US Marine motto "Death before dishonor," and argues that "death before dishonor" is the essence of true warriorhood. Concludes, "Whether or not we think that warriors like war too much, we need them more than ever."

Bernard Cornwell, any of his "The Saxon Tales," such as *Sword Song* (New York: Harper, 2007), *The Burning Land* (New York: Harper, 2010), *The Last Kingdom* (New York: Harper, 2006), *The Lords of the North* (New York: Harper, 2008), and *Death of Kings* (New York: Harper, 2011) (F). Can be read and enjoyed for their appreciation of the zest of battle and the extent to which some, even many, men are driven by a joy in warfare, a joy not matched in other dimensions of their lives. As the narrator puts it in *Death of Kings*, "Men like a battle-tale, which is why we employ poets to entertain us at night with tales of warriors and swords and shields and axes."

Antoine de Saint-Exupery, *Flight to Arras* (New York: Reynal and Hitchdale, 1942). "Why in the midst of defeat do I and my kind continue to take the risk of death," the warrior asks. "It came to this. I was working at my trade." The doomed struggle of French airmen in May 1940 is captured, as is the essence of warriorhood in a somewhat overly philosophical analysis. Still, the battles in the sky ring very true.

Carlo D'Este, *Patton: A Genius for War* (New York: Harper, 1995). If ever there was a man to symbolize a love of battle and zest for war, one who understood modern war and was as deeply flawed as any warrior, it was George Patton Jr., who said of war, "By God, I love it so."

David Donovan, *Once a Warrior King: Memories of an Officer in Vietnam* (New York: McGraw Hill, 1986). There are thousands of books about Vietnam. All capture aspects and dimensions about the war, and many can fit like pieces in a mosaic, providing an overall holistic portrait. But, just as the pieces of a mosaic are part of the whole, they are but one portion of that whole. Donovan's story (set in the IV Corps military region of the Mekong Delta in 1969) provides a vital, powerful perspective on one small piece—what that war was like at the local level, with the Popular Forces doing their fighting with the Vietcong. He and his comrades serve in a five-man Mobile Advisory Team (MAT) embedded in a village. There are no main force American units in action nor North Vietnamese regulars in this war. It is a war, simple, local, direct, and very personal. The insights of the author tumble out one after the other, illuminating our understanding of both that portion of the war and the nature of war as war. His conclusion? "If there was immorality in the war in Vietnam, it was that a democratic nation called her citizens to war, had them killed by the tens of thousands, and then, like a faithless lover, turned and scorned the survivors. Oh, perfidious nation." A sampling of the many insights the author:

1. The warrior's insight after his first firefight: "I asked myself what it was I found so valuable that I was willing to risk death for it, what it was I found so valuable I was willing to kill for it."
2. "We held ourselves to be better men" than those who spent their war in the rear echelons with hot showers, clean sheets and mess halls. "In a sense, we looked upon them as an under caste, pussies, men without the blood of warriors."
3. Fighting with a huge blood price on his head, he nevertheless earns and receives the trust of the villagers he protects and for whom he is the protective warrior-king.
4. Back in the States, he always crushes his beer cans (which in Vietnam were made into hand-grenade casings by the insurgents) because "[i]f I don't, I get this funny feeling in my stomach and a cold breath runs down my spine."

Arthur Conan Doyle, *The Complete Brigadier Gerard* (London: Canongate, 1995) (F). Includes *The Exploits of Brigadier Gerard* (1896) and *Adventures of Gerard* (1903), the rollicking, well-written short stories of a French brigadier in the time of Napoleon.

Shannon E. French, *The Code of the Warrior: Exploring Warrior Values Past and Present* (Lanham, MD: Rowman and Littlefield, 2005). What it takes to be a warrior in various societies and the warrior self-image in them. French outlines the need for a code of honor, one that holds the "warrior to a higher ethical standard than that required for an ordinary citizen within the general population." "The code is not imposed from outside" but is internalized within the warrior culture. The author argues that this is essential to shield the warriors' humanity: "The code restrains the warrior. It sets boundaries on his behavior. It distinguishes honorable acts from shameful acts."

S. C. Gwynne, *Rebel Yell: The Violence, Passion, and Redemption of Stonewall Jackson* (New York: Scribner, 2014). Ah, there was a true warrior who believed totally in God's plan and saw himself as merely fulfilling that plan. Perhaps only matched by Sherman on the Union side as wanting to destroy the enemy totally. Jackson urged the darkest of retribution if the North invaded and did his best to deliver it.

Victor Hanson, *The Savior Generals* (New York: Bloomsbury Press, 2013). Useful and occasionally compelling chapters on those generals who proved decisive when things were going badly: Themistocles at Salamis, Flavius Belisarius in Byzantium, William Tecumseh Sherman in Georgia, Mathew Ridgway in Korea, and David Petraeus in Iraq. What it takes to turns things around when you are a leader of warriors when war itself is unchanging.

Stephen Harding, *The Last Battle: When U.S. and German Soldiers Joined Forces in the Waning Hours of World War II in Europe* (Boston: Da Capo Press, 2013). Ten Americans, 1 Waffen SS man, and 14 Wehrmacht soldiers hold off a determined attack by SS formations at an Austrian castle housing high-value French detainees, including Paul Reynaud, Maxime Weygand, and Michel Clemenceau—the day after World War II "ended."

Mike Hoare, *Congo Mercenary* (London: St. Edmundsbury Press, 1967), *The Seychelles Affair* (New York: Bantam, 1986), *The Road to Kalamata* (Lanham, MD: Lexington Books, 1989), and *Congo Warriors* (Boulder, CO: Paladin Press, 1991). All depict the motivation and dimensions of mercenary action. Set in the Congo, three of the memoirs capture the Conrad's true "heart of darkness" on both sides as well: "This was Central Africa. There was no Geneva Convention here." Call them "Les Affreaux" (the "frightful ones") or the "Wild Geese," depending on your perspective. The Seychelles work, its comic opera dimensions notwithstanding, show mercenary motivation and leadership in crisis.

Homer, *The Iliad*, translated by Robert Fagles (New York: Penguin, 1990). The *Iliad* is, of course, many things to many people and cultures, but its portrayal of Achilles, the indomitable warrior, remains at the heart of the work. Achilles has probably been written about more than any other warrior in history (both positive and negatively). Deeply flawed and self-centered, he is still the warrior against whom all are measured.

Richard Homes, *Soldiers: Army Lives and Loyal Ties from Redcoats to Dusty Warriors* (London: Harper Press, 2011). A massive compendium that takes a look at the "social lives of soldiers," from the British armies of kings to the national army to women soldiers. The section on "Tribes and Totems" dealing with the British regimental system is quite good, comparing as it does the barracks to the public schools in terms of the small-group inculcation of values. "I have loved Tommy Atkins . . . since first I met him." And it shows.

Samuel Hynes, *The Soldiers' Tale: Bearing Witness to Modern War* (New York: Penguin, 1997). Explores the nature of warriorhood in modern war by dispensing with what he terms the myths of the "Good War," the "Bad War," and the "Necessary War." Samples and discusses such writers as Philip Caputo, Michihiko Hachiya, Eric Lomax, Farley Mowat, Siegfried Sassoon, and E. B. Sledge.

Conn Iggulden, any of his novels about Genghis Khan, such as *Genghis: Birth of an Empire* (New York: Delacorte Press, 2007) or *Genghis: Lords of the Bow* (New York: Bantam Books, 2010), Iggulden captures the essence of what warriorhood is all about and, in the process, casts considerable doubt about the oft-cited dichotomy between "warrior" and "soldiers" because the Mongol warriors are truly soldiers and vice versa. Page after page of battles and the excitement they generate for the warriors in question are on full display.

Jeff Jonada, *Saga: A Novel of Medieval Iceland* (Chicago: Academy Chicago, 2005) (F). While not actually a novel about war as war, this tale set in 10th-century Iceland describes a culture and way of life that is so strife-ridden and claustrophobic and filled with never-ending blood feuds and marginal existence that one can only view war—and Viking war especially—as a blessed, blessed relief. Explains much of the Viking outthrusts of conquest and raiding.

John Keegan, *The Face of Battle* (New York: Viking, 1975). By focusing on the nature of battle from the soldiers' points of view, Keegan looks at war in a different way than many other historians, and his portraits of the battles of Agincourt, Waterloo, and the Somme capture the essence of war.

John Keegan and Richard Holmes, *Soldiers: A History of Men in Battle* (New York: Viking, 1986). Providing us with insights into the "timeless fascination" of men with war and warriorhood, this work captures the very ambivalence and ambiguity of that experience, with the horror and desolation penetrating the adrenaline rush, concluding, "But war usually retains its puzzling fascination, even for those who knew it best."

Alex Kershaw, *The Liberator* (New York: Crown, 2012). Felix Sparks, who would rise from second lieutenant to the rank of lieutenant colonel while serving in Europe with the 45th "Thunderbird" Division, was a natural leader of men, a caring, heroic, highly decorated man who spent 500 days in combat in Sicily, Anzio, the Apennines, Rome, southern France, and Reipertswiller; crossed the Siegfried line and the Rhine; and liberated Dachau, where he stopped his men from shooting all of the SS guards. Called by George Patton a "damned fine soldier." He was a true warrior.

Ian Knight, *Brave Men's Blood: The Epic of the Zulu War 1879* (London: Greenhill Books, 1990). The Zulu War of 1879, stretching as it does through the battle of Isandlwana, Rorke's Drift, and Ulundi, is an extremely well-illustrated (270 photos and sketches) account of the true warriors who fought to the death and often took no prisoners on the high veldt of southeast Africa, leaving King Cetshwayo to lament, "All what is mine in Zululand has been overturned and spoiled."

Dr. Heidi Kraft, *Rule Number Two: Lessons I Learned in a Combat Hospital* (New York: Little, Brown, 2007). See the review in chapter 35.

Jean Larteguy, *The Centurions* (New York: Dutton, 1961) (F). A powerful and moving account of the state of warriorhood as seen in French paratroopers and mercenaries. Soldiers and POWs on various sides show the stresses, strains, and betrayals of wartime. Set in Vietnam after the Battle of Dien Bien Phu and later in war-torn Algeria, this work was a worldwide sensation when it was first released. Less praised now.

———, *The Hounds of Hell* (New York: Dutton, 1966) (F). A thoroughly Gallic novel of unrelenting cynicism and comradeship, as the mercenaries sent by France to support the independence of Katanga Province encounter the realities on the ground. A startlingly acid portrait of the first of many declines in the former Belgian Congo, concluding with this epitaph for all mercenaries: "And the mercenaries, Monsieur le President, what is to become of them?" "The mercenaries? A mere detail! That's all they ever were."

Note: For a more prosaic account of mercenaries in action in the Congo during the following period, see Mike Hoare, *Congo Mercenary* (London: Robert Hall, 1967). A firsthand account by the head of Commando 5 on a "rabbit shoot" to liberate the eastern Congo (including Stanleyville). The actual campaign turns out to have been much more difficult and problematic but shows that, in insurgent situations, speed is of the essence in tactical terms but is most appropriate when coupled with strategic purpose. Accents discipline and weeding out "alcoholics, drunks, booze artists, bums and layabouts," who seem drawn to mercenary activity in that (and any other) era. Shows how "[c]olumn warfare from an infantryman's point of view is most unsatisfactory." Rough justice dealt out, not just to the "Simbas" but also to mercenaries who act up (one who rapes a Congolese woman and then kills her has both of his big toes shot off by Hoare). Stanleyville is indeed the "'Inner Station' of Conrad's Heart of Darkness." The book ends with the military coup of Joseph Mobutu, who ousts President Kasavubu and Prime Minister Moise Tshombe. Mobutu would rule the Congo for decades, loot it of billions of dollars, and never pay his mercenaries or his army enough to keep them from behaving lawlessly. Bad scene, that!

———, *The Praetorians* (New York: Dutton, 1963) (F). Some of the same characters carry on here in Algeria as the counterinsurgency in Algiers succeeds but the overall war is lost, and by December 11, 1960, the FLN flag flies over Algiers. Searing portraits of many real participants, both in the military and the civilian sector, as the warriors shoulder on despite interference and muddling through by others. "In this war you do the opposite of what the rules lay down: in fact you only learn the rules in order to do the opposite." Decay throughout the entire French system and government is shown in sharp relief.

Karl Marlantes, *Matterhorn* (New York: Atlantic Monthly Press, 2010) (F). A long and very detailed novel about Marines in northwest Vietnam who are fighting with North Vietnamese regulars in 1969 in the hellish jungle wilderness south of the demilitarized zone. A young lieutenant gets a baptism of fire and starkly portrays the life and death of Bravo Company in and around the mountain and firebase known as Matterhorn. Bravo is torn by racial divisions, dislike, and lack of respect for their alcoholic battalion commander and the down drag of rotations as men come and go, many getting killed and wounded in the process. There is a timeless quality to the work: You could change the names and the locations, but the infantry, the tip of the military spear, would be familiar to those who have fought throughout time and space, from the Hittites to the Greeks to the Mongols. Ultimately, it is about the comradeship that binds men (and women) to a common purpose when death walks beside them. As the narrator puts it, as the central character Mellas joins an outgoing patrol when he doesn't have to

> Running to rejoin them, he felt overwhelming joy. It was as if he were coming home from a lashing winter storm to the warmth of his living room. The sky seemed brilliantly blue and clear, although he knew it was overcast. If he didn't move his legs faster, his heart would outpace his feet and burst. His heart, his whole body, was overflowing with an emotion that he could only describe as love.

———, "The Truth about Being a Hero," *Wall Street Journal*, August 20, 2011, p. C1ff. Heroism main-lined and very readable. Will challenge many of your assumptions about warriorhood.

Lauro Martines, *Furies: War in Europe 1450–1700* (New York: Bloomsbury Press, 2013). His "war mosaic" depicts a world at war, with mercenaries from all over Europe fighting all over Europe for everybody who is anybody in Europe. This results in a "massive bleeding of money," as sieges and sacks of cities as well as the plunder and killing of rural areas, results in continent-wide slaughter.

Charles McDonald, *Company Commander* (Short Hills, NJ: Burford Books, 1947). An almost moment-by-moment account of an American captain and his unit in northeast Europe during 1944 and 1945, crossing the Siegfried Line, enduring the Battle of the Bulge, and ending up in Czechoslovakia, when the "news came by radio that the war was over. There was no defining our joy." The author ends by paying tribute to the fallen dead, "those GI Joes who that others might live themselves became the fallen fruit of a harvest for the devil." Excellent description of battle action and ongoing motivation for warriors.

John C. McManus, *The Dead and Those About to Die* (New York: NAL Caliber, 2014). Omaha Beach, June 6, 1944, the First Division, "the Big Red One," was there, and its warriors fought in a narrow killing ground with great distinction.

Dakota Meyer and Bing West, *Into the Fire: A Firsthand Account of the Most Extraordinary Battle in the Afghan War* (Ganjugal) (New York: Random House, 2012). An extraordinary look into the life and combat of a contemporary warrior who received the Congressional Medal of Honor. The reader can experience his highs and lows and zest for battle: "The more fucked up things got (in the firefight), the more Rod and I started laughing."

Interesting Afghan views of the Taliban as "dushmen," bandits, and murderers, not true Muslims. His tour of duty over, Meyer refuses to give into depression: "I would always regret not saving my guys, but, as for the people I killed, that is what gave me what little peace of mind I had. They were the enemy. That's what you do to the enemy. That part was not going to upset me."

Allan R. Millett, *In Many a Strife: General C. Thomas and the U.S. Marine Corps 1917–1956* (Annapolis, MD: Naval Institute Press, 1993). How warriorhood gets enshrined in tradition despite the efforts and actions of bureaucrats or how Washington, DC, requires constant monitoring by true warriors.

Douglas Porch, *The French Foreign Legion: A Complete History of the Legendary Fighting Force* (New York: Harper, 1991). Many recruits are running from their past, many running toward their future, but in any case, warriorhood is prized and praised and often on display. Unit loyalty remains the key, but "morale in a mercenary force is often very fragile." Interesting accounts of the legion's actions in Spain, Mexico, Indochina, Dahomey, Madagascar, World Wars I and II, Indochina again, Algeria, and finally Desert Storm in 1991.

E. B. Potter, *Bull Halsey* (Annapolis, MD: Naval Institute Press, 1985). Captures the essence of a true naval warrior in the tradition of Admiral Horatio Nelson and John Paul Jones: "As long as I have one plane and one pilot I will stay on the offensive." Moved his carrier task force with skill and dispatch, except at Leyte Gulf and during a typhoon that followed, but was on the *USS Missouri* when the Japanese surrendered. Thereafter, a fish out of water. Very interesting background on the early attacks of the aircraft carriers *Enterprise* on the Marshalls (Kwajalein) and *Yorktown* on the Gilberts (Makin) right after Pearl Harbor, and a fine map shows the seldom-seen movement of the US Third Fleet from July 1 to August 1945 coming into Japan to receive the surrender.

Steven Pressfield, *The Profession* (New York: Crown, 2011) (F). Although this is less a novel than a comic book, it captures the essence of what it means to be a true warrior by describing the key elements of warriorhood: (1) acceptance of a field of constant conflict, (2) acceptance of comrades and brothers in arms, and (3) acceptance of a leader worthy of him- or herself and those comrades. Lessons of honor and glory abound, even though the story line would embarrass many more proficient science-fiction writers.

Thomas Ricks, *The Generals* (New York: Penguin, 2012). Although there is little new information here, the author does put his finger on one of necessities of warfare, namely you have to fire, discharge, and replace generals who don't make it as warriors. You would think this is a simple prerequisite for success in warfare, and Mars certainly seeks to make it so, but as the American examples provide, not firing incompetent generals can have disastrous consequences. Ricks does make the useful point that the jump from colonel to brigadier general is "one of the largest psychological leaps an officer can take." Not all make it well. True warrior generals do.

Ross Ritchell, *The Knife* (New York: Blue Rider Press, 2014) (F). A thinly disguised novel covering a Special Forces team operating in Afghanistan. Strangely flat narrative despite the subject. Ends with the thoughts of Shaw: "Shaw knows there will always be Stag1's to chase until they killed them all, and they never will. Never could. So he can go on chasing them forever and they him until his ghosts have all left him. And they never will. He'd rather charge among them than flee only to be overrun in the end."

James Salter, *The Hunters* (New York: Harper Brothers, 1956) (F). A fighter pilot during the Korean War captures the essence of warriorhood flying F-86s against MIG-15s, concluding, "In this war he was more certain than ever, we would find his peace, as men found love in loneliness and acclaim in anonymity." He dies fighting high in the skies.

E. D. Smith, *Valour: A History of the Gurkhas* (Woodstock, NY: Overlook Press, 1998). Written by a decorated Gurkha, this work captures the essence of warriorhood by a people who, having fought so bravely against the British in the war of 1814–1816, were then actively recruited for the next 200 years, even existing in today's truncated version, always believing, "It is better to die than be a coward." More than one British officer concluded, "I never saw more steadiness or bravery exhibited in my life."

Caroline Varin, *Mercenaries, Hybrid Armies and National Security: Private Soldiers and the State in the 21st Century* (London: Routledge, 2014). Examines the various roles and accomplishments of mercenaries, contract professionals, and other nonstate warriors in the modern world of war, such as the French Foreign Legion, American contractors in Iraq, and a variety of actors in various countries in war-torn Africa.

Bing West, *One Million Steps* (New York: Random House, 2014). So it's the wrong war. Most wars are. So the warriors are badly served by their leaders. Most have been. So the mission is confused and confusing. Such is the nature of warfare. Battalion Three-Fifths of the Fifth Marine Regiment suffered the highest number of casualties in the war in Afghanistan, operating in the most godforsaken portion of that country. Want to learn about true warrior spirit?

Read this book and conclude, with Aristotle, "We become brave by doing brave acts"; "This war's stupid. Well, so what? Our country's in it"; and "You kill one of us, we kill twenty of you."

Owen West, *The Snake Eaters* (New York: Free Press, 2012). Somewhat lacking in the style of his father, Bing, the author does convey an interesting variation on the "born" warrior as he describes the evolution of the cops and CPAs and store managers in the National Guard who not only become warriors themselves (with little help from anybody) but also turn frightened, ill-trained Iraqi *jundis* (enlisted men) into warriors themselves. A total vindication of *Auftragstaktik* in the process.

Andrew Wiest, *The Boys of '67* (London: Osprey, 2012). Follows 45 men of Charlie Company, drafted in 1966 and sent to Vietnam with the 9th Infantry Division in the Mekong Delta. The war is still popular in the United States, and although the Vietcong holds 25 percent of the delta, they set about working their craft of war. By the time they have been through the "steady drumbeat" of war and are rotated home on the "Freedom Bird," only 6 of the 45 are unscathed. Two are dead, and the rest are wounded. Some find a welcome, some do not, and all become "lost faces of a distant war." Quite a poignant account of warriors who served the polis well, even if the polis put them there erroneously.

James Wilde, *Hereward* (London: Bantam Press, 2011) (F). "It was the beginning of End-Times," and there was a three-cornered struggle to see who would be king of England. In 1066, William the Bastard would triumph, but the true warrior would continue to resist, for Hereward is only truly happy for much of this novel when he is fighting. "He felt the thing inside him rise up, the other Hereward, born of rage and bloodlust, unconstrained by human values, and he welcomed it. The Pain of his wounds vanished. As strength flooded into his weary limbs, he reacted with a speed that made the mercenaries seem lead-footed in comparison." He is a warrior, not a soldier, at that moment.

Cecil Woodham-Smith, *The Reason Why* (London: Constable, 1953). The curious life and times of Lords Lucan, Cardigan, and Raglan show the charge of the Light Brigade in startling relief as 700 rode into the valley of death with fire coming from 3 sides, and 195 made it out, most wounded. Lord Cardigan was back on his yacht that night, with a good bath and bottle of champagne and, pleading illness, made it back to England and a hero's welcome (although later, the destruction of his abandoned troops caught up with him). The description of the actions of men facing death and staying true to their code is both macabre and stirring, although on balance, the actions of the lords is enough to shake the faith of Mars, for sure, but the warriorhood portrayed by the charge will last as long as people study war.

Any discussion of warriorhood and glory in battle and war must, of course, grapple also with the nuanced view of them as portrayed by Homer in book 11 of the *Odyssey*, when Odysseus encounters Achilles in Hades and hears his agony:

> But you, Achilles,
> there's not a man in the world more blest than you—
> there never has been, never will be one.
> Time was, when you were alive, we Argives
> honored you as a god, and now down here, I see,
> you lord it over the dead in all your power.
> So grieve no more at dying, great Achilles."
>
> I reassured the ghost, but he broke out, protesting,
> "No winning words about death to me, shining Odysseus!
> By god, I'd rather slave on earth for another man—
> some dirt-poor tenant farmer who scrapes to keep alive—
> than rule down here over all the breathless dead.

Chapter Thirty-Eight

The Return of Warriors

> "After we had sacked the steep city of Priam, and had departed in our ships . . . even then, Zeus was fashioning for us a ruinous doom."—Homer, *The Odyssey*

> "Muse, tell me why, for what attaint of her deity, or in what a vexation did the Queen of Heaven drive one so excellent in goodness to circle through so many afflictions, to face so many toils. Is anger so fierce in celestial sprits?"—Virgil, *The Aeneid*

> "There is a deeper side to coming home. The returning warrior needs to heal more than his mind and body. He needs to heal his soul."—Karl Marlantes, *What It Is Like to Go to War*

After the battle, after the war, exiting the realm of Mars, the returning warriors encounter a most-wide-ranging set of circumstances and reactions. In *The Odyssey*, for example, there is the contrast between Odysseus, who returns to a faithful wife but has a struggle to deal with her suitors. Agamemnon, on the other hand, has an unfaithful wife, who, together with her lover, kills him. The popular culture has always played on these themes of returning warriors from the stories of Odysseus and Agamemnon to the return of Vance in the Elvis Presley film *Love Me Tender*. Many returning warriors have fared badly on their return, whether it is unrequited love, alienation, or sheer menace.

In addition to various rewards and punishments greeting the returning warriors, there is the stark contrast between what they have been and done and what they must now be and do. For some, despite its horrors and trials, life in battle has a rush that cannot be duplicated in civilian life and must be reattained through going off to war again, compelled to take war in all its glory and misery in order to feel truly alive, as the central character shows in the film *The Hurt Locker*.

Of course, many, perhaps even most, who go to war return and end up leading normal lives, reintegrated back into society. But happy stories with happy or at least prosaic endings do not necessarily attract the attention of publishers or the reading public. In any event, the return home is tied up with warriorhood itself. What makes for a successful warrior reintegration throughout time and space? What links those from one era and one culture to another? What is the relationship between the civilian and the warrior, and how do you turn one into the other and then back again? Note that, in these questions, there is always considerable tension (reflected in this chapter and the one that follows) between a functional, wartime approach (i.e., whatever works in combat should be used, what conduct of the warrior that leads to victory is acceptable) and the perceived need by others for a moral code of conduct that encompasses but also transcends the battlefield. Thus for this approach, "to do or die" becomes only the first step of true warriorhood because the surviving warrior must be reintegrated into the society from which he or she sprung, both for the warrior and the society.

From the harsh bleakness of the French Foreign Legion motto "Hardship, friendship, death" across cultures to the need for a moral code to shield the soldier's humanity as suggested by Shannon French in order to distinguish "honorable acts from shameful acts," the tensions between performance and an ongoing sense of selfhood remain constant across cultures and ages.

Posttraumatic stress disorder is not new; therefore, it was always there, but it now has a medical description and is being recognized for what it is. Thus, beyond the chapter on "Warriorhood," we have "The Return of Warri ors," where all the moral and ethical ambiguities of war come home to rest—for good or for evil, for strengthening and peace to weakening and turbulence.

The following are some of the works, fiction and nonfiction, that capture the spectrum of reactions among the returning warriors:

Aeschylus, "Agamemnon," in his *Oresteia*, translated by Robert Fagles (New York: Penguin, 1966), pp. 99–172. Agamemnon does not get the homecoming he'd hoped for, being slain by his unfaithful wife and her lover. This enduring tragedy for one and all has been replayed throughout history in many, many cultures, as the homecoming of the warrior unfolds.

Robert Bittlestone, *Odysseus Unbound: The Search for Homer's Ithaca* (Cambridge: Cambridge University Press, 2005). Looking for the "true location" of Odysseus's Bronze Age palace is quite a chore, and its location—the island of Cephalonia in the Ionian Sea—is an island that split off during a volcanic eruption from the present-day island of Ithaca off the Peloponnese.

Shows how far Odysseus had to travel from Troy. No wonder it took him so long. Not an easy sail in those days. Read and ponder.

William Broylus Jr., *Brothers in Arms: A Journey from War to Peace* (New York: Alfred A. Knopf, 1986). Sometimes the returning warrior needs to reengage the war at its source and location. In this Vietnam memoir, the warrior does just that, finding that "[w]e had tried to kill each other but we were brothers now." And speaking of those who suffered: "They deserved better but then again, so did we all." My enemy equals myself.

Brian Castner, *The Long Walk: A Story of War and the Life That Follows* (New York: Doubleday, 2012). A poignant, even heart-wrenching, tale of a warrior in the Second Iraq War whose memory is impaired by the myriad explosions and carnage he experienced while disarming bombs there. "I died in Iraq. The old me left for Iraq and never came home. . . . I liked the old me." Very sobering.

Edgar Chen and Emily Van Buskirk, "The Czech Legion's Long Journey Home," *MHQ: The Quarterly Journal of Military History* 13, no. 2 (Winter 2001), pp. 42–53. "Like Odysseus before them, the simple task of returning home was complicated by forces beyond their comprehension. They became strangers in many strange lands. But after six years of wandering, they finally made it home." For them, World I blended into the Russian Revolution and a long, long way home.

David Finkel, *Thank You for Your Service* (New York: Farrar, Straus and Giroux, 2013). The soldiers from Iraq returned and faced enormous struggles with the legacy of war, societal indifference, and political posturing, for as the author says, the "after-war continues, as eternally as war itself." Poignant descriptions of real-life soldiers who return to stresses comparable to those they thought they left behind in the war.

Ben Fountain, *Billy Lynn's Long Halftime Walk* (New York: Ecco, 2012) (F). Throughout history, the warrior returns to society only to find that, not only does society not understand his or her experience, but also there is always someone—often many someones—seeking to use the hard-earned warriorhood for their own purposes. The Dallas Cowboys lose to the Chicago Bears, and the men of Bravo Squad lose their chance to be in the movies, despite their genuine heroics in Iraq. At the end of the novel, Crack puts it all in perspective: "Take us someplace safe. Take us back to the war."

Cara Hoffman, *Be Safe, I Love You* (New York: Simon and Schuster, 2014) (F). Laura Clay returns home from Iraq to prove "[h]ome is not always the safest place for a returning warrior." The book concludes with her saying, "I did terrible things," and she is then told, "Of course you did. Don't let anyone tell you otherwise." Depressing.

Homeland (Showtime, 2012) (F). The ripples of battle, captivity, and the various elements of healing and destruction that accompany the return of the warrior are all sharply delineated.

Homer, *The Odyssey*, translated by Robert Fagles (New York: Penguin Books, 1996) (F). In addition to the many vignettes of warriors and warriorhood in the entire work, there is also the theme of the returned warriors and what may befall them after having survived battle, as in book 1, where Agamemnon returns home only to be killed by his wife Clytemnestra and her lover Aegisthus (who in turn is killed by Agamemnon's son Orestes). Or in the case of Odysseus, where the returning warrior finds a still-faithful wife and family. The "return of the warrior," of course, is a leitmotif of many of the novels reviewed in this chapter. See also Aeschylus, *Oresteia*, reviewed earlier in this chapter, for a fuller and more powerful account of Agamemnon's return and perfidious welcome.

James Jones, *Whistle* (New York: 1958) (F). The third volume of Jones's war trilogy (the first two are *From Here to Eternity* and *The Thin Red Line*) looks at the returning warriors who have cheated death, escaped from combat, and now find themselves alive after having written themselves off as dead. But escaping alive is not the same as escaping undamaged, as these wounded warriors verify as they play out their lives in the hospital of Luxor (Memphis, Tennessee). As dreary and melancholic as the real lives of many returning wounded veterans, the book ends with the suicide of Johnny Strange in the cold North Atlantic. The irony of the title is overwhelming ("If you want anything just whistle for it").

Philip Keith, *Blackhorse Riders* (New York: St. Martin's Press, 2012). How the 300 men of Alpha Troop, 11th Armored Cavalry Regiment, rode to the sound of guns to save their comrades, not because they were ordered to but because these "ordinary men" rose to the occasion. Busting through the jungle on March 26, 1970, in Tay Ninh Province (III Corps northwest of Saigon). Led by a few Sheridan tanks (overgunned with 152-mm cannon and underarmored with aluminum) and armored cavalry assault vehicles (ACAVs).

William Manchester, *Goodbye Darkness* (Boston: Little, Brown, 1979). Revisiting the islands on which he found himself during World War II—Guadalcanal, Tarawa, Saipan, Guam, Peleliu, and Okinawa—Manchester draws the reader into his own ambivalence toward the war, the Marine Corps, and the battles that were fought. Poignant accounting of the men he knew who died and what they wanted to be before the war struck them down. A haunting work.

Karl Marlantes, *What It Is Like to Go to War* (New York: Atlantic Monthly Press, 2011). While this book could easily fit under the topic "Warriorhood," it also fits the theme of this chapter most powerfully. Outlining the warrior's inner stew of pride, honor, anger, rage, elation, relief, loyalty, guilt, grief, numbness, fatalism, confusion, and contradictions, Marlantes shows how the dark pit of "kill or be killed" leaves a lasting impression that must be dealt with because the "violence of combat assaults psyches, confuses ethics and tests souls."

Toni Morrison, *Home* (New York: Alfred A. Knopf, 2012) (F). The warrior—wounded, battered, broken—who comes home to a country riddled with racism and despair, and yet he finds a purpose and something of a salvation in his mission to save his sister. Regrettably, I have never enjoyed any of Morrison's previous works, and this is no exception, but it does suit the topic.

Scott Phillips, *The Adjustment* (Berkeley, CA: Counterpoint, 2011) (F). A returning veteran, ironically wounded by a pimp in Rome during World War II, not battle, still finds civilian life too tame, too ordinary, and too confining, so he leaves his beautiful wife and unborn child to rejoin the army and go to occupied Japan, concluding, "A year wasted on civilian life had brought me nothing but boredom relived occasionally by frenzy and mayhem as I clung tight to the miserable delusion that life as a husband, father, and citizen was my inevitable fate. Now the scales had fallen from my eyes to reveal a destiny worthy of my talents."

Kevin Powers, *Yellow Birds* (Boston: Little, Brown, 2012) (F). Captures the sense of disconnection between combat, "where everything matters," and returning home and the sense of isolation and powerlessness that accompanies the transition.

Lewis B. Puller, *Fortunate Son* (New York: Grove Weidenfeld, 1991). The Marine son of fabled Marine Corps general "Chesty" Puller wins the Silver Star, loses his legs in Vietnam, and finds reentry into American society most difficult. He loses his race for Congress but wins his battle with alcoholism, but he only finds peace with "Welcome home, brother." Haunting, gripping, and heart-wrenching. Odysseys, even Agamemnon, were more fortunate.

Thomas Ricks, *Making the Corps* (New York: Scribner, 1997). Although Ricks can be quite arrogant and often wrong (he wrote confidently in 2005 that the United States would be in Iraq for 15 more years), this work, which follows 63 Marine recruits to see how the Marines make them into warriors and how the warrior ethos is central to everything that makes a Marine, is worth reading. For them, the brotherhood of combat requires brotherhood throughout everything else, as well. Ricks notes that half the Marine Corps recruits enlist under a "waiver for crimes, drug use, and medical and psychological problems."

Roxana Robinson, *Sparta: A Novel* (New York: Farrar Straus and Giroux, 2013) (F). A classic tale of the returning warrior, whose reentry into civilian life does not go well when he returns from Iraq. Conrad knows "[h]e was safe, alive. But instead of being a relief, this was faintly sickening. It led somewhere he didn't want to go. There was some unnamed weight attached to his being here. He was here. Olivera was not, and never would be and how had that happened." How much of this angst is due to the failure of the society to adequately reward and cherish his duty (as opposed to the adulation in real-life Sparta) is left unclear and unsaid. The title and the labored analogy remains off-putting and detracts from the author's intent.

Mike Scotti, *The Blue Cascade: A Memoir of Life after War* (New York: Grand Central, 2012). A Marine who served two tours in Iraq had a helmet cam and captured many scenes in battle (see his subsequent film *Severe Clear*). Suffering from PTSD, he struggles to integrate "what he was been into what he is now" against the backdrop of 6,000 US veteran suicides a year. However well-meaning, the people back home cannot appreciate what he has endured. His purge is cathartic, even for nonveterans.

Jeff Shaara, *A Blaze of Glory: A Novel of the Battle of Shiloh* (New York: Ballantine, 2012) (F). Wordy and didactic in places, but still, this novel gives a sense of camp and battle life, especially the diseases, dysentery, and accidents that took more lives than the actual fighting (at least by some accounts). Sherman and Grant are just beginning to jell, and General Albert Sidney Johnson seems to doom the Confederacy in the West, even though it is only April 1862.

Anita Shreve, *Stella Bain* (New York: Little, Brown, 2013) (F). See the review in chapter 35.

Virgil, *The Aeneid*, translated by J. W. Mackail (London: Kessinger, 2010). Aeneas, a defeated warrior during the Trojan War, shows that there is often life after defeat, and he has many more adventures as he leads survivors from

Troy to Carthage and eventually Italy, where he becomes the legendary founder of Rome, not without much more fighting. His lover in Carthage, Queen Dido, kills herself after his departure, symbolizing the pain and price paid by the myriad of those left behind when duty and new wars call. The work does offer the hope of redemption through new adventure and accomplishments, however.

See also many of the novels reviewed in chapter 40 for similar themes of the warrior after battle and war.

Chapter Thirty-Nine

Other Wars

This chapter is for other wars that do not easily or conveniently fall into any of the previous typologies but that seem worthy of additional highlighting and study.

Simon Dunstan, *The Yom Kippur War 1973: The Golan Heights* (Oxford, UK: Osprey, 2005). Well-illustrated with excellent maps, this account depicts the Israeli holdout for 48 hours against the Syrian Soviet-inspired broad-front attack until their mobilization could take place. The Syrians fought very well, however, despite losing 1,150 tanks, and the Iraqis, 200. Both Syria and Israel required a truly massive resupply of ammunition and missiles from the Soviet Union and the United States to complete the battle.

———, *The Yom Kippur War 1973: The Sinai* (Oxford, UK: Osprey, 2005). A massive Egyptian logistical and tactical success as they break through the Bar Lev line, but once their tanks got outside their SAM umbrella and the Israelis brought up armor reinforcements and counterattacked with Operation Gazelle, the tide of battle changed, and the Israelis were able to push to the Suez Canal and beyond, surrounding several Egyptian armies. But the massive casualties and military squabbling during the war hurt the Israelis, as well. Egypt lost 1,000 tanks and many of their 2,000 artillery pieces; the Israelis, 400.

David Eshel, *Mid-East Wars: The Yom Kippur War* (Hod Hasharon, Israel: Eshel-Dramit, 1978). Lavishly illustrated with excellent diagrams, maps, and photos (albeit mostly Israeli). Good insights into the Egyptian crossing of the Suez Canal and ends with the question, “The Yom Kippur War was the biggest of all the Mid-East wars; will it be the last?”

Chaim Herzog, *The War of Atonement October, 1973* (Boston: Little, Brown, 1973). An Israeli major general gives us a most in-depth and substantive account of the war. Does a good job of taking the reader inside the "onslaughts" that occurred in the Golan Heights and the Sinai, concluding that, as so often happens, the victors in the 1967 war (Israelis) misread its lessons, while those defeated (Egyptians and Syrians) learned the most and made the best adjustments. Still, he concludes that, within the broad sweep of 4,000 years of Jewish history, the success of the Israeli Defense Force (IDF) in the Yom Kippur War of 1973 was noteworthy. Great maps. Wish we had a corresponding look in English at the war from the Egyptian or Syrian sides.

Abraham Rabinovich, *The Yom Kippur War: The Epic Encounter That Transformed the Middle East* (New York: Schocken Books, 2004). A detailed and incisive look, primarily with in-depth Israeli sources and photographs and mindsets. "Israel emerged from the war more chastened than triumphant," as their military people lost the haughtiness they had developed since the Six Day War of 1967 and their assumed derogatory mindset: "We are fighting Arabs, not Germans." The author describes the initially successful Egyptian attacks as a "feat of arms not seen since Saladin's victory over the Crusader Army near the Sea of Galilee in the 12th Century." Gives good details about arguments within Israeli high command and provides an excellent map showing the end of war with the Egyptian Second Army in the Sinai north of Ismailia and the Israelis on the west side of the Suez Canal to the south, including the city of Suez and the Egyptian Third Army trapped to their east. Notes that, unlike the Six Day War, in this one, the Israeli Air Force's impact was "marginal" and credits Kissinger with coming up with a solution that allowed the Egyptians to gain "pride" and the Israelis "proportion," leading to the eventual peace treaty. Of course, for his peace achievements, Anwar Sadat was killed.

Chapter Forty

War Novels, Short Stories, Memoirs, and Poetry

> "War literature is a story without end."—Frank Ellis, *The Damned and the Dead: The Eastern Front through the Eyes of Soviet and Russian Novelists*

There are many, many novels about war and the human beings caught in its embrace. Some are antiwar, some are prowar, some are realistic, and some are fantastic. I believe that, among those, the following works should be read by those seeking a true liberal arts education in understanding war.

Richard Aldington (British), *Roads to Glory* (London: Chatto and Windus, 1930) (F). A series of short stories by a survivor of the trenches in World War I make up one of the most powerful antiwar books one could ever read. The aching sadness of the survivor casts a haunting set of reverberations that the reader will not soon forget. A classic. Devotion to duty in the face of virtually certain death is hard to understand as a singular moment in one's life, but to confront such moments over and over again is beyond belief to those of us who have never experience it.

Antonio Lobo Antunes (Portuguese), *The Land at the End of the World* (New York: W. W. Norton, 1979) (F). The New State of Antonio de Oliveira Salazar (who ruled it from 1933 to 1968) faces revolutionary outbreaks in Angola, Mozambique, and Guinea Bissau from 1962 on, culminating in an army coup in 1974 that led to decolonization in 1975. Antunes writes of an army unit in the eastern region of Angola and captures the languid, disintegrating, surrealist war, one with no purpose from the perspective of his unit, but he still soldiers on for 27 months in the "asshole of the world." For me, there are strong echoes of Buzzati's *The Tartar Steppe* in this work.

Isaac Babel (Russian), *Red Cavalry* (New York: W. W. Norton, 2002) (F). Reprint of his short story and diary accounts of the abortive Communist drive into Poland in 1920, a drive that was supposed to carry the revolution beyond the borders of the Soviet Union. Poignant, sad vignettes of war's incalculable harm on innocents. "Comrades. This makes me want to hang myself." Stark portraits of the Cossacks, Jews, Poles, Ukrainians, and Russians, all caught up in a dance of death.

J. G. Ballard (British), *Empire of the Sun* (London: Gollanz, 1984) (F). Young Jim is trapped in a new world when the Japanese take over Singapore during World War II and his parents disappear. Prisoner-of-war camps and the atomic bomb punctuate this engrossing "coming of age in the midst of war" tale. Shows the need for stability even within the constantly changing and disequilibrating landscape of ongoing war.

Henri Barbusse (French), *Under Fire: The Story of a Squad* (New York: E. P. Dutton, 1917) (F). Perhaps the finest account of small-unit trench warfare of the French poilu in World War I. It's all here: sleep deprivation, mud, water, vermin, helplessness, chronic anxiety, rage, noise, boredom, numbing barrages, and stark terror repeated over and over; the relief of wine or coffee rations; the stultifying overhang of death; and the blessed rotation that only prolongs the agony. Perhaps the most haunting image is that of the French Alsatian poilu who is shown his home village by the Germans, and he sees his wife smiling among his children and German officers before being escorted back to his line. "War is frightful and unnatural weariness, water up to the belly, mud daring and infamous filth."

Pat Barker (British), *The Ghost Road* (London: Penguin, 1995) (F). The South Sea islanders' "culture of death" is transported to 1918 France, and a wounded lad insists on returning to combat. Head hunting is banned in the islands but apparently not permanently. Leaves the reader with a question as to whether death in war at the end is worse than death in war at the beginning or in the middle.

Renzo Biasion (Italian), *The Army of Love* (New York: Alfred Knopf, 1954) (F). This is not really a novel but a somewhat interconnected set of short stories, primarily about the Italians in Greece during the World War II. It is a marvelous collection. One feels the heat and the boredom, the sense of being off stage while great events are happening elsewhere as the Italian troops obsess—and I do not use that term lightly or without feeling—about sex. Women and getting inside them is what seems to keep the Italians going from one day or week or month to the next. The Greek women and girls, for many of whom prostitution is an accepted profession or at least an accepted means of survival during wartime, flow through the stories regardless of age, pulchritude, or size. "'She really did make love,' they would say, and adored her for it." The stories capture the hatred the Italians felt for the Germans and their own leaders and, albeit tangentially, the chagrin and frustration the Germans felt at having such allies. For, after all, why are the Germans in Libya and Crete and the Greek Islands and Yugoslavia and all the other places they have been forced to occupy due to the military incompetence of those allies? For the Italians, the "lack of women was as great a torment as a lack of water." They are, in this novel at least, lovers, not fighters.

Edmund Blunden, *Undertones of War* (London: R. Cobden-Sanderson, 1928). Set in 1917 France near Ypres, this lyrical, poetic education in and of the trenches contains many moving vignettes, such as "So the attack on Boer's Head closed, and so closed the admirable youth or maturity of many a Sussex worthy." He felt that 1918 did not look more promising. Luckily, for him at least, this was not correct.

Laudomia Bonanni (Italian), *The Reprisal* (Chicago: University of Chicago Press, 2013) (F). This powerful novel captures the essence of the "warrior woman" with her "[f]ury-like power" and feminine qualities (her heroine "La Rossa," the red-haired revolutionary, is pregnant and not killed by the Fascist partisans until after giving birth). Takes place in Italy during the waning months of World War II, when Germans, Fascists, and partisan bands of various political hues roamed the land. A searing portrait of the "chain of vengeance and murder for its own sake," the human condition.

William Y. Boyd, *A Fight for Love and Glory* (Miami: Warner Brothers, 2003) (F). Quite a melodramatic love story with some pretty stock characters and scenes, but the battle scenes against the Germans in Sicily and France during World War II ring true and give the reader a good sense of the chaos and randomness of death and violence in small-scale encounters.

———, *The Gentle Infantryman* (New York: St. Martin's Press, 1985) (F). A much-praised coming-of-age-in-war book of an American replacement soldier pressed into service during the 1944 drive through Alsace, Nuremburg, and Munich. Will witnesses a huge number of casualties, eventually becoming a sergeant by the time the war ends. Somewhat talky and a bit formulaic.

———, *An Ice-Cream War* (New York: William Morrow, 1983) (F). The ironic title notwithstanding, the World War I campaign in Tanganyika, Kenya, Mozambique, and Rhodesia was a horrible series of blunders; sharp, short engagements; and inspired leadership of the German colonel Von Lettow-Vorbeck. This work captures the death, disease, hunger, malnutrition, and utter human waste, as well as the amazing polyglot force (British, Indian, Nigerian, King's East African Rifles, Rhodesians, South Africans, and Portuguese) that pursued the Germans and their faithful askaris until even after the armistice was signed.

Vera Brittain (British), *Testament of Youth* (New York: Seaview Books, 1980). Autobiographical story of a young, sensitive women who loses her brother, her best male friends, and her fiancé during World War I. Extremely poignant and heart-wrenching about the loss of an entire generation. No wonder she became a pacifist. More powerfully antiwar than *Catch-22* and other more acclaimed works.

Mikhail Bulgakov (Russian), *The White Guard* (London: Collins and Harvill, 1971) (F). This 1925 work is set in the midst of the Russian Revolution, where some make it to Kiev, where switching allegiances led to rumors and facts of war; the czar is dead or in Denmark with the kaiser, and there is ebb and flow, flow and ebb of fortunes: "But in days of blood as in days of peace the years fly like an arrow." Kiev and the Ukraine are the heart of the study, where Germans, Bolsheviks, Poles, Ukrainian nationalists, and Whites struggle and the Family Turbin tries to figure out what is going on.

John Horne Burns, *The Gallery* (New York: Harper and Brothers, 1947) (F). A powerful, evocative novel that captures—better than so many others—the interplay between civilians and the military, not as conquering heroes but as consumers of all manner of products of the defeated. Also, deals with gay soldiers by integrating them totally into the fabric of the life and times portrayed. The setting is the Galleria Umberto in Naples, but flashbacks and digressions give the reader a strong sense of Casablanca and Algiers. One regret is that the author did not see combat and therefore did not write about it.

Dino Buzzati (Italian), *The Tartar Steppe* (New York: Noonday Press, 1965) (F). Reminds me of the existential mindscape of Albert Camus's *L'Etranger*. The narrator, Giovanni Drogo, spends two years in a fort on the edge of the steppe but dies of disease just as war finally comes. There is only the heroism of doing what one is told and seeing it through regardless. Hauntingly captures the feeling of alienation and helplessness the Russian steppes produce in so many armies—or not—as in the case of the Mongols, who came at them from the other direction.

Javier Cercas (Spanish), *The Soldiers of Salamis* (New York: Bloomsbury, 2001) (F). By focusing on the near-death of the Falangist ideologue Sanchez Mazas and the man who could have killed him after he escapes a firing squad but didn't, Cercas explores the many dimensions of civil war in general and the Spanish Civil War, in particular. An enjoyable read as well as a lesson in multifaceted and overlapping stories during wartime.

Christophe Coker, *Men at War: What Fiction Tells Us about Conflict, from the* Iliad *to* Catch-22 (London: Oxford University Press, 2014). A broad-ranging and stimulating look at fiction about war and its heroes, villains, and victims.

Stephen Crane (American), *The Red Badge of Courage* (New York: Pocket Books, 1942) (F). Long-lasting (it first appeared in 1895) and still in print, this classic tale by a man who had not been to war when he wrote the novel tells the story of a young Union recruit, Henry Fleming, who fights and runs but is then deeply ashamed of his cowardice. He envies the wounded he encounters and their "red badge of courage," and the next time his unit is in action, he fights "like a lion" and finds peace of mind: "He had rid himself of the red sickness of battle. The sultry nightmare was in the past. He had been an animal blistered and sweating in the heat and pain of war. He turned now with a lover's thirst to images of tranquil skies, fresh meadow, cool brooks—an existence of soft and eternal peace."

E. E. Cummings (American), *The Enormous Room* (New York: Boni and Liveright, 1922) (F). Not so much a book about war as one about war in the background. Although the narrator served in the ambulance corps, he ended up in prison, the "enormous room," on suspicion of spreading tales of the French mutinies during World War I. "La guerre" is always around, causing strange permutations, and as it turns out, pacifism is a hard-held belief in wartime, even someone else's.

Selma Dabbagh, *Out of It* (London: Bloomsbury, 2011). Gaza and the Palestinians are trapped by history, religion, and a fitful but seemingly never-ending (67 years and counting!) war. The Intifada, she sees, is "war as a cage fight."

John Del Vecchio, *The 13th Valley* (New York: Bantam, 1982) (F). Though sometimes prolix when describing the soldiers' inner monologues and debates over race, democracy, and life in general, this remains one of the most powerful and, according to many veterans, one of the most accurate descriptions of the fighting between NVA regulars and the 101st Airborne. The first time I read it, I cried at the end. The second time, even knowing what was coming, was even harder. Powerful.

Eilis Dillon (Irish), *The Bitter Glass* (London: Faber and Faber, 1958) (F). Set in the "time of troubles" during Ireland's civil war from 1922 to 1923, this novel takes place in the West Country of Ireland and shows the civilians caught between the Republicans and Free Staters, underscoring how freedom in the midst of civil war is most problematic.

John Dos Passos (American), *Three Soldiers* (New York: George Doran, 1921) (F). This is supposed to be a powerful antiwar novel about World War I, but it is such a wretched, wordy bore with so little actual "war" in it that one cannot really be sure why anybody refers to it positively at all. Is it possible to imagine that Hemingway was, if anything, too kind in many of his comments about Mr. Dos Passos?

Slavenka Drakulic (Croatian), *S.: A Novel about the Balkans* (New York: Praeger, 1991) (F). In other countries, this novel went by the title *As If I Am Not There*, which is far more illustrative and emblematic of the story depicted. A teacher in Bosnia, S, is gang-raped multiple times during the break-up of Yugoslavia. An overwhelming, sad story of the murder of civilians, "armies" without discipline or mercy, and a savage warning at how thin the veneer of civilization really is. Yet S's bonding with her newborn child, a product of mass rape, ultimately restores our faith in humanity. Gripping and haunting.

Alfred Duggan (British), *Conscience of the King* (New York: Coward-McCann, 1951) (F). This novel gives a marvelous feeling of the slow disintegration of Roman Britain and the emergence of Saxon Wessex in the fifth and sixth centuries, while providing a most incisive look at the difference between the Roman way of war and the Germanic. The difficulties in creating and maintaining a comitatus are writ large.

Jean Echenoz (French), *1914* (New York: New Review, 2012) (F). A lyrical, light, and yet most poignant account of five Frenchmen caught up in the war from "On to Berlin" until the last gun is fired. Whimsically and profoundly sad.

J. G. Farrell (British), *The Singapore Grip* (London: Weisenfeld and Nicolson, 1978) (F). This highly praised, even overpraised, novel captures the sense of the life and times as the British Empire came to an end in Singapore with the Japanese invasion of Malaya and subsequent capture of the city that was really only defended to the seaward—the British equivalent of the Maginot Line. The British residents and military ignore the "distant thunderstorms of the Sino-Japanese war" until it becomes a "dreadful acceleration."

Timothy Findley (Canadian), *The Wars* (New York: Delacorte Press, 1977) (F). Alternately antic and depressing, it follows the World War I experiences of a Canadian at the battle of Ypres, including being shelled, gassed (and saved by the ammonia in a urine-drenched cloth, which blocks the chlorine), almost buried alive, and nearly killed in a horse stampede, as well as becoming a victim of a homosexual gang-rape and going mad. Not for the faint of heart.

Ken Follett, *Fall of Giants* (New York: Dutton, 2010) (F). Sweeping from one end of Europe to another, this vast novel depicts the changing of the age and the collapse of the Russian, Austrian, and German Empires as World War I undoes what centuries of history have created. Quite readable and with a popular, easy style as the author, who is Welsh, brings an interesting familial dimension to this "war to end all wars," when Russians, Welsh, Germans, English, and Americans find time speeded up by events and overwhelming revolutionary change.

———, *Winter of the World* (New York: Dutton, 2012) (F). A big, sprawling novel that follows various Russian, American, Welsh, and English families during World War II. Although the story is often contrived, it manages to give the reader a fair view of the war from a variety of viewpoints.

Ford Maddox Ford (American), *Parade's End* (New York: Knopf, 1986) (F). Four related novels published between 1924 and 1928. Traces the life of Christopher Tietjens from England into the trenches of France as his inner and outer lives overshadow the events of the war. A powerful warning of what modern warfare had become by the time of World War II:

> There will be no more Hope, no more Glory.
> Not for the Nation, not for the World. I daresay
> There will be no more parades.

C. S. Forester (British), *Death to the French* (London: Michael Joseph, 1933) (F). With Wellington in Portugal before the Spanish ulcer really began to suppurate, Matthew Dodd, rifleman, captures the "offensive spirit" of the British as they help the Portuguese to bring "*morran os Franceses*." This is war at the squad and even individual level. It gives the reader a sense of participation, whether it's the starving Dodd eating raw horseflesh while on the run or reveling in having salt with his food when he finally rejoins his unit after being trapped behind enemy lines.

———, *The General* (Annapolis, MD: Nautical and Aviation, 1947) (F). This American edition followed the earlier 1936 version, in which Forester accurately foresaw the coming of World War II after the tragedy of the First World War. Although he was to write *The Good Shepherd* and *The African Queen*, as well as the extremely popular Horatio Hornblower series, this is my favorite of all his novels. With deft irony and yet some sympathy, Forester traces the career of Herbert Curzon, who, by dint of class and a favorable marriage to the daughter of a duke, rises among the incompetence of the British officer corps during World War I. But there remains something quite endearing about his earnest efforts to do well in battle and do well by his men. Of course, one retains far more sympathy for the tens of thousands of men who died in the seemingly futile battles of the Somme and Ypres.

———, *The Gun* (London: Michael Joseph, 1933) (F). Although Forester is perhaps best known now for his "Hornblower" novels about a British sea captain during the Napoleonic wars, *The Gun* is a true gem, capturing in a series of glances the true nature of guerrilla warfare as the Spanish rise up—with the help of the British and priests of the Catholic Church—to smite the French. Especially telling is the need to turn irregulars ("Long live King Ferdinand and let's go robbing") into disciplined soldiers who fight and flow and fight again but in order.

Julia Franck (German), *The Blindness of the Heart* (New York: Grove Press, 2007) (F). As disturbing as war itself, this novel revolves around a woman who abandons her child on a railway platform and everything that preceded and followed that act. World War I and World War II and the Cold War are all in the background, bleeding into the lives of the young and the civilian, as the history of Germany unfolds with all its manifestations.

Alan Furst (American) (F). The novels listed here do a superb job of capturing the essence of individuals caught up in the grand sweep of events generated by warfare. Furst concentrates on the coming, duration, and ending of World War II in Europe, especially the usually underrepresented people and portions of *Mittleuropa*, all by Random House:

The Blood of Victory (2003)
Dark Star (2002)
Dark Voyage (2005)
Kingdom of Shadows (2001)
Midnight in Europe (2014)
Midnight in Paris (2013)
Night Soldiers (2002)
The Polish Officer (2001)

Red Gold (2002)
The Spies of the Balkans (2008)
The Spies of Warsaw (2011)
The World at Night (2002)

Graham Greene (British), *The Quiet American* (London: Heinemann, 1955) (F). In this novel, the quiet and naïve American turns out to be the CIA station chief and, according to the author, responsible for setting in motion a "Third Force," neither French nor Communist, in Vietnam and willing to go to any length, including blowing up civilians in the middle of Saigon, to achieve his purposes. The model for the central character was Colonel Edward G. Lansdale, who had been highly successful in the Philippines, developing the strategy that defeated the Communist Hukbalahaps.

Henry Green, *Caught* (London: Hogarth Press, 1950). One man's war in the London Blitz of 1940 in the Auxiliary Fire Service showing the dull, the incidental, and the seemingly inconsequential aspects of war as human lives simply move along.

Andrew Greig (British), *That Summer* (London: Faber and Faber, 2000) (F). A poignant, evocative, bittersweet portrait of love between a Hurricane pilot and a member of the Women's Auxiliary Air Force (WAAF) in the middle of the Battle of Britain during the summer of 1941. The staining power of assumption prevails: Planes will burn but not this plane. Men will be killed but not this man. Love will sour but not this love. "The dead ends are dense all about us," said the pilot and soon he was one of them.

W. E. B. Griffin (William Butterworth III) (American) (F). Griffin has written 38 books mostly dealing with war and warriors (although some deal with police and other law enforcement agents). Has a huge following among military personnel, although I have always found his books less than gripping. Nevertheless, their focus on duty, honor, and country is consistent and obviously relevant to the study of those who wage war because so many who do identify with his characters and their lives. His Brotherhood of War series (9 books) features the US Army; the Corps series (10 books), the US Marine Corps; and the Men at War series (6 books), the Office of Strategic Services (OSS).

Vasily Grossman (Russian), *Life and Fate* (New York: Harper and Row, 1980) (F). Epic novel on the order of *War and Peace* following families all touched by the war and especially the Battle of Stalingrad. The gigantic upheaval of war, with its multitude of reverberations, is vividly portrayed. Stalinist terror is juxtaposed with Nazi terror as the purges, trials, and horrible slaughters play out like out-of-control brushfires, and ordinary men and women seek comfort and survival wherever they can find it. The author died of cancer in 1964, having been told by the Soviet authorities that his novel would never be published for hundreds of years! Fortunately, they were wrong. Considered by some to be the best Russian novel of the 20th century.

Aaron Gwyn, *Wynne's War* (Boston: Houghton Mifflin Harcourt, 2014) (F). If only we could prevent those who would like to write about war from seeing *Apocalypse Now* or from reading *The Heart of Darkness*. Here we go again. Rogue American captain leads horse soldiers into Pakistan for gold and treasure. Gets nearly everybody killed and can't get treasure out, so blows it all up. Bummer.

Emily Hahn (British), *China to Me* (Garden City, NY: Doubleday, 1944) (F). How war came slowly but inexorably to Shanghai, the Chinese, the Europeans, and the Japanese. Good portrait of the China that was before World War II, which comes slowly and quietly as if on little cats' feet in the fog, but come it does.

Jaroslave Hasek (Czech), *The Good Soldier Svejk* (New York: Thomas Crowell, 1974) (F). The author had as varied and checkered career as a Svejk, being a Bohemian anarchist, soldier in the Czech Legion in Russia, and later a Bolshevik, before writing his antic, rollicking, complex account of World War I on the Austrian/Russian front. His personal Anabasis is at once irreverent, comedic, and touching, ending with the ironic assertion, "Patriotism, fidelity to duty, victory over oneself, these are the weapons that matter in warfare."

Willi Heinrich (German), *Cross of Iron* (Newport Beach, CA: Noontide Press, 1988) (F). Quite simply one of the best war novels of all time: deeply felt, brutally honest, and capturing the essence of German *Auftragstaktik* on display on the Eastern front during World War II. A true classic.

———, *The Savage Mountain* (London: Corgi, 1958) (F). Savage, brutal, partisan fighting in Czechoslovakia, as German forces are forced back by massive Russian advances in 1945. Depressing to have been a Czech, a Russian, or a German caught up in the end-of-war bitterness. A small masterpiece at capturing the endgame of war.

Mark Helprin, *A Soldier of the Great War* (New York: Harcourt Brace Jovanovich, 1991) (F). A long, loving portrait of an intellectual but deeply feeling Italian soldier reminiscing about his World War I experiences in the Alps. While it does capture the enormous impact war can have on those who experience and survive it, I found the novel much overpraised and not one I would strongly recommend. But be advised: It has gotten a lot of critical acclaim, so I may be missing something.

Ernest Hemingway (American), *Across the River and into the Trees* (New York: Charles Scribner's Sons, 1950) (F). A poignant, end-of-life portrait of war reflected through the lens of a dying man in Venice. Enduring image of the Battle of Huertgen Forest described as "Passchendaele with tree bursts."

———, *A Farewell to Arms* (New York: Charles Scribner's Sons, 1929) (F). The Italian/Austrian front during World War I is the setting for this novel about the corruption, ineptitude, and folly of war as shown through the eyes of a man who abandons war for Switzerland.

———, *The Fifth Column* (New York: Charles Scribner's Sons, 1969) (F). Short, spare, and insightful, especially "Night before Battle" and "The Butterfly and the Tank." One can feel the tension of looming battle and beyond.

———, *For Whom the Bell Tolls* (New York: Charles Scribner's Sons, 1946) (F). A poignant love story set within the Spanish Civil War, which Hemingway covered as a reporter. Good at catching the human dynamics in small partisan units as well as the general feelings of being at war. This, Hemingway's longest novel, was called the "greatest political novel in American literature" by critic Jeffrey Meyers. More importantly for our purposes, perhaps, Fidel Castro indicated (see chapter 18) that this work was the war manual for his guerrilla band when they were in the Sierras and they learned much from it and referred to it often, trying, one supposes, to live up to the ethical standards of the characters. The American captain Jordan, his central character, dies a heroic death in a war where that was possible, if Hemingway is to be believed.

——— (ed.), *Men at War* (Boston: Houghton Mifflin, 1942). Not a novel but has a number of chapters of works from novels, such as Tolstoy's *War and Peace*, and many short pieces of nonfiction capturing many dimensions of warfare. An uneven and Eurocentric collection but of and for its time. Illuminating. Can be perused with profit for those seeking insights into the innumerable aspects of combat and its impact on individuals and groups.

A. P. Herbert (British), *The Secret Battle* (New York: Alfred A. Knopf, 1920) (F). Field Marshal Bernard Law "Monty" Montgomery once called this work the best novel he'd ever read for capturing the life of a soldier in the trenches and, I believe, the moral ambiguities of battle. *The Secret Battle* is outstanding, in a class with Henri Barbusse, *Under Fire* (North Charleston,

SC: CreateSpace, 2011), and Bao Ninh, *The Sorrow of War* (New York: Pantheon, 1995).

On the face of it, it is a simple story of a young man who serves his country faithfully and well in the trenches of Gallipoli and France during World War I, experiencing the nightmare of boredom, terror, exhilaration, overarching anxiety, mood swings, confusion, disease (especially dysentery), digging, and more digging, while witnessing death upon seemingly random death.

Harry is loyal, truthful, and brave and serves his country well, even returning to battle after being seriously wounded and being offered a staff position safe in England. But he goes out again to be with his unit, and this time, the accumulated horrors and nightly nightmares make him stop in the middle of a battle and return to a dugout, where he is met by another officer, ironically one he earlier upbraided for cowardliness in the field when they were in Gallipoli together. Harry is court-martialed, and although mercy is recommended (he is still only 23 and has an unblemished record), none is shown, and he is shot. The novel ends with the powerful indictment of the entire affair: "That is the gist of it: that my friend Harry was shot for cowardice—and he was one of the bravest men I ever knew." It is a truly fine work, worthy of a much wider contemporary readership.

Mischa Hiller (British), *Sabra Zoo* (London: Telegram, 2010) (F). It is 1982, and Beirut is being torn apart by the PLO, other Palestinian factions, the Lebanese army, the Phalangists, and various other factions. The US Marines are withdrawing, and the Israeli Defense Force is arriving. The Sabra and Shatila refugee camps are the scene of great man-made suffering. The futility and indiscriminate nature of warfare is summarized in the statement "I think if they showed the real effects of war we wouldn't have it any more."

Michael Horbach (German), *The Betrayed* (New York: William Morrow, 1960) (F). Jagged and jangled, episodic and jumbled with flashbacks and fragments, this novel vividly describes the end of war, in this case, World War II on the Eastern Front. The victorious Russians are advancing along a 1,000-mile front. Small groups of German soldiers fight an unending series of rearguard skirmishes, always losing, always retreating, sometimes staying ahead of the Soviets, sometimes not, as attrition thins their ranks and moment-by-moment decisions of others decide their fate as men are wounded or killed, often at random. Civilians, too, seek to flee East Prussia but many fail, suffering the same vicissitudes of fate. There is utter relentlessness as the war grinds on to its destructive finish irrespective of individual wishes or efforts.

Ironically, the title refers, not to the victims of Nazi aggression or racial imperatives and destruction, but to those Germans who were led into a war they did not win. Included are those who supported Hitler or didn't, those who knew of the horrors of the concentration camps and persecutions of the Jews and others or didn't, those who wanted a strong but not conquering Germany or who wanted conquest to restore the balance of power lost after World War I or wanted neither. Now they are all trapped in the all-encompassing embrace of war. There is anger here, a deep, diffusive anger that permeates virtually every scene in the book, which is filled with nervous energy from start to finish.

Christopher Hudson (British), *Where the Rainbow Ends* (New York: Atheneum, 1987) (F). Also published as *Colombo Heat* (London: Pan Books, 1987). Although this is a war romance about the British ex-pat community, it captures the panic, indecision, and confusion of Ceylon as the Japanese approach during 1941 and 1942. One also gets a sense also of the rising nationalism and anticolonialism among the Ceylonese (and Indians), as the Japanese show the weaknesses of the British, French, and Dutch. *Rainbow* also does an excellent job at capturing the grandiosity of the Japanese imperial high command, with their simultaneous plans to seize Madagascar as well as Ceylon, moving on to link up with the Germans while at the same time invading Midway and the Hawaiian Islands.

Duong Thu Huong (Vietnamese), *Novel without a Name* (New York: William Morrow, 1995) (F). An absorbing and deeply depressing novel about war in the jungle, lost youth, and the immense problem of keeping one's humanity in the middle of a struggle without end: "In times of war, the future belongs to the combatants." She concludes, "Chants of the months, of the years spent in the Truong Son Mountains. Soldier, the dawn is icy. You fall under the bullets. On the white of the parachute cloth, I see your blood spreading."

David Ignatius (American), *Bloodmoney* (New York: W. W. Norton, 2011) (F). Despite its romantic core idolizing the Pashtun revenge ethos, this work captures the multilevel and interconnected nature of the various wars in Pakistan. Easy to read yet profound in its insights, summed up with the fact that, in Pashto, the word for *cousin* is the same as for *enemy*. "Wars begin with *badal*, an assault on a man's honor and self-respect."

Sebastien Japrisot (French), *A Very Long Engagement* (New York: Farrar Straus Giroux, 1993) (F). Five French soldiers are followed through much of World War I in this stark, fast-moving novel ending with the this piercing vignette: "Here lie five French soldiers, who died with their shoes on, chasing the wind." They were joined by many others from many nations.

James Jones (American*),* *The Thin Red Line* (New York: Charles Scribner's Sons, 1962) (F). As with his previous *From Here to Eternity* and his subsequent *Whistle*, Jones focuses on the impact of war on individuals and groups. Based on his Pacific experiences with the 25th Infantry Division in the South Pacific in the battle for Guadalcanal, the novel captures the intense feelings experienced in war; the depths of depression, longing, and isolation; and the highs of triumph and survival. In the film version, seemingly more Japanese surrender in this one campaign than in the whole Pacific War.

Ernst Junger (German), *Copse 125* (New York: Howard Fertig, 1988) (F). This reprint of the 1930 work gives stunning insight in the trench warfare of 1918 and shows the testing of values "in fox's earth with a foot of soil over my head." Fighting over a tiny piece of land and enduring gas and shelling, a true test of manhood that produces survivors, "this new and iron race tempered in every fiber."

Ward Just (American), *A Dangerous Friend* (Boston: Houghton Mifflin, 1999) (F). While not often regarded as a "war" novel, this work does a masterful job in showing how basic, fundamental, troubling strategic errors by civilians doomed the American effort in Vietnam. But by extension, it can be read with profit about all civilian authorities who put soldiers in harm's way and, in doing so, often hand those soldiers impossible assignments because of the grandiosity and hubris of those who do not have to fight the wars themselves.

Nikos Kazantzakis (Greek), *The Fratricides* (New York: Simon and Schuster, 1964) (F). Kazantzakis is perhaps better known for his *Zorba the Greek* and *The Last Temptation of Christ*, but *The Fratricides* could be his finest work, capturing as it does, the essence of civil wars, often the most vicious and heartrending of all wars. His poem "God Speaks," from Sidna Ali, sets the tone:

> Whoever seeks me finds me,
> Whoever finds me knows me,
> Whoever knows me loves me,
> Whoever loves me, I love,
> Whomsoever I love, I kill.

Thomas Keneally (American), *Confederates* (New York: Harper and Row, 1979) (F). I almost put this novel in the section on the American Civil War (where it fits perfectly), but on reflection, I elevated it because of its perceptive dealing with war in general, human suffering, and the role of civilians as war washes over them and pulls away their loved ones, returning only some. Ostensibly, this is a tale of the men of the Shenandoah Brigade who follow

Tom Jackson through Second Manassas and Antietam during 1863, but it contains a deeper story. There is much good dialogue and emphasis on the Irish who fought for the Confederacy and droll humor: When confounded by Johnny Pope's failure to discern Stonewall Jackson's corps astride his flank at Second Manassas, President Lincoln remarks, "That's just about the same as going to a circus and failing to notice the damn elephant."

Philip Kerry (American), *Field Gray* (New York: G. P. Putnam's Sons, 2011) (F). One in a series of Bernie Gunther police novels but worth including here for its description of various prisoner-of-war, concentration, and returning-POW camps (German, Soviet, American, and French), primarily to give the interested reader a sense of what life was like in them before, during, and after World War II. Noteworthy for its claim that the men in the French camps at Gurs and Le Vernet were in worse conditions than those in Dachau. This is not a charge seen often or made lightly.

Yasmina Khadra (Algerian), *The Attack* (London: William Heinemann, 2006) (F). The wife of an Israeli-Arab doctor turned out to be a suicide bomber in the world where "[r]eason has a mouth full of broken teeth," and another round of savage cruelty begins (or continues). "War among the people" seems even more dreadful than war among the armies.

Jerzy Kosinski (Polish), *The Painted Bird* (New York: Houghton Mifflin, 1965) (F). A powerful, surreal, yet realistically harrowing story of a young child wandering about ravaged Europe during and after World War II, a world in darkness and misery, where evil rules.

A. Anatoli Kuznetsov (Russian), *Babi Yar* (New York: Pocket Books, 1971) (F). A truly amazing and awe-inspiring book centered on the ravine outside Kiev, Babi Yar, where the Germans killed tens of thousands of Jews, Ukrainians, and others during their occupation. Told through the vehicle of a small boy's eyes, the book registers the initially positive welcome the Germans received, their heinous brutality that soon alienated most, the forced migrations to Germany, the rise of Ukrainian partisans, the eventual "liberation" of Kiev by the Russian army, and then the ironic and deadly imprisonment of the Ukrainian partisans by the Soviets. The author is unsparing toward the NKVD as well as the Gestapo and evenhandedly presents the good side of Vlassov's army—whose members were sent to the mines of Siberia after the war with the proviso that they never again be allowed to see the sun.

Margaret Leroy (British), *The Soldier's Wife* (New York: Hyperion, 2011) (F). A frothy romance set on the island of Guernsey during the German occupation of the British Channel Islands during World War II. Not much to recommend it, save the description of panic on the island after the British troops leave and before the German troops arrive. That little piece of writing is a treasure.

Vaino Linna (Finnish), *The Unknown Soldier* (New York: G. P. Putnam's Sons, 1957) (F). The Finnish soldiers in this outstanding work want revenge against the Russians for their previous valiant defeat in the Winter War of 1939–1940 and the concomitant loss of territory, but they instinctively feel that, by joining with Germany in 1941 in the "Continuation War," they are ultimately doomed. This sense of inevitability casts its pall over the entire work and permeates the day-to-day actions of the individuals and the unit giving their actions great poignancy. One of those rare war novels that captures its essence for both the individual and the small unit of which he is a part.

Johnathan Littell (French), *The Kindly Ones* (New York: Harper, 2009) (F). Massive, overblown, but deeply haunting view of World War II through the eyes of a half-French, half-German homosexual *Standartenfuhrer* (colonel) Aue of the SD, who is involved in all aspects of the Final Solution, starting with service in an *Einstzgruppe* on the Eastern Front and ending up assisting Albert Speer in maximizing the use of concentration-camp victims before their deaths. At Sobibor, he is told, "Little men and little women, it's all the same. It's like stepping on a cockroach." Acid and ironic portraits of the Nazi hierarchy exemplify the banality of evil: "In a state like ours, everybody has an assigned role: you the victim and you the executioner, and no one had a choice."

Le Luu (Vietnamese), *A Time Far Past* (Amherst: University of Massachusetts Press, 1997) (F). Written in 1984 by a soldier who was a veteran of the Ho Chi Minh Trail and who fought a long, long war (10 years) for the fatherland, this work captures the dedication and suffering of the NVA, as well as the painful contradictions in postwar Vietnam.

Norman Mailer (American), *The Naked and the Dead* (New York: Holt, Rinehart and Winston, 1948) (F). Maybe because this was one of the first war novels after World War II or maybe because Mailer caught the fancy of critics, this overpraised novel is often mentioned as among the best of books about war. For my part, I have always found it verbose, prissy ("fug" indeed), self-conscious, and even self-indulgent, as well as being more about men at war than war itself. Still, he captures both the impersonality and complexity of modern war and the continuation of small-group solidarity as the principle wellspring of the fighting unit. Still, certainly not on par with

Cross of Iron or *The Betrayed* or the *Unknown Soldier* or even *The Thin Red Line*.

Allan Mallinson, *A Close Run Thing* (New York: Bantam Books, 1999) (F). A jolly good war, what? The Peninsula and Belgium later after a stint in Ireland, practicing. Beating Boney and having the duke say, "Plans? I have no plans sir; I shall be guided by circumstance." All's well that ends well, what? War as described by Jane Austin.

Andre Malraux (French), *Man's Fate* (New York: Simon and Schuster, 1934) (F). An existential novel about the failed Communist uprising in Shanghai in 1927, when the Mao and the Chinese Communists attempted to follow the Leninist notion of urban proletariat-based revolution, only to have it crushed. After this abject failure, Mao transformed Communist doctrine, making the rural peasants the primary revolutionary cadre. Interesting use of suicide bombings foreshadows our world of today.

Frederic Manning (British), *Her Privates We* (New York: G. P. Putnam's Sons, 1930). An unforgettable tale of World War II as a private, later lance corporal, named Bourne endures all there is to in the front lines with humor, good grace, and a refusal to leave his mates in the trenches. The reader gets a good sense of the importance of rum and tea to morale and the constant need to be shifted from the front line to the rear and back again. The mundane, the ordinary, the day-to-day simply underscore the doom-haunted overhang of what will be, of death in the trenches postponed but not evaded. It is "hell, bloody hell." Period.

Olivia Manning (British), *The Fortunes of War* (New York: Penguin Books, 1960–) (F). This work actually consists of two trilogies, the Balkan Trilogy and the Levant Trilogy. *Fortunes of War* is a huge, sweeping, magnificent achievement, capturing World War II from its beginning until its end in Eastern Europe and the Middle East. Ironic, tart, and insightful, the work enthralls and appalls as British diplomatic, military, and intelligence figures blunder about as war sweeps over them: "It's my belief, if those goofs had kept out, there'd have been no attack." Portraits etched in acid and hence fresh, even today.

Bobbie Ann Mason (American), *The Girl in the Blue Beret* (New York: Random House, 2011) (F). Informative and poignant, this is the story of a downed pilot of a B-17 who was saved by the French resistance, retires, and seeks to retrace his escape route across France and the Pyrenees. In the process, he reconnects with the girl who had guided him on part of the journey, and she shares the horrors and human courage of surviving German concentration camps at Ravensbruck and Konigsberg, as well as the radiating penalties so many French suffered for helping the escaping aviators.

John Masters (British), *Now, God Be Thanked* (New York: McGraw Hill, 1979) (F). A multifamily, sprawling novel that captures some of the battlefield during the beginning of World War I and a lot of the home front, both at the poacher's hut level and the highest corridors of power. Highlights the many changes in English society as the war progresses. Like his books set in India, *Now, God Be Thanked* ultimately rests on his powers of description, and there are some gems: "Boy" Rowlands leading his men during Second Ypres and admonishing them to walk, not run, toward the Germans, and the description of the trenches on Christmas Day 1915, when the Germans heavily shelled the British, thinking they were planning an attack. There follows then the delicious irony when the British, being heavily bombarded, assume the Germans are about to attack, and they fire huge barrages to prevent that from happening. In the trenches, one German waves a white flag and wishes one and all a "Merry Christmas," only to be shot twice, "one bullet from in front and one from behind."

Pvlos Matesis (Greek), *The Daughter* (London: Arcadia Books, 1990) (F). A marvelous portrait of rural Greece under German and Italian occupation during World War II as seen through the eyes of a young woman whose father went off to fight in Albania and never returned. There are a number of harshly haunting scenes, foremost among which is one where the Germans smile as they smash over and over again the hand of her little brother when he picks up a potato that has fallen from one of their trucks in the village square. The constant struggle to get enough food during wartime is movingly presented.

Simon Mawer, *Trapeze* (New York: Other Press, 2012) (F). Chick-lit does World War II in occupied France. British agent is blown, the French collaborate or fight bravely, but who can tell in the "cancer" that is Paris? But our heroine, "Squirrel," refuses to fly out with her target/love and is captured by the Germans and their French minions. Luckily, the book ends before her interrogations begin. You might think that a British agent who had just killed two German intelligence officers and had a plane taxiing to fly her to safety in England might, just might, get on that plane. But no. This is a silly piece of business.

Veijo Meri (Finnish), *The Manila Rope* (New York: Alfred A. Knopf, 1967) (F). A rollicking, bitter-sweet series of vignettes depicting the absurdity and pathos of war. Finnish soldiers tell their stories in short episodes. Several reverberate powerfully—especially the battlefield whose corpses are devoured by a massive number of pigs who eat the bloated bodies in a grotesque but telling way, as if they were war itself. In turn, they are finally killed and eaten by the troops (although some soldiers were quite squeamish considering what the swine had recently dined on). Set during the Finnish Winter War and the Continuation War, during which Finland fought alongside the Germans to recover land lost to the Soviets during the Winter War:

> Anemone and Edelweiss
> Flowers of Finland and Germany
> Your blue shades beckon us to battle
> Comrades all, under a single banner.

Anton Meyer (American), *Once an Eagle* (New York: HarperCollins, 1968) (F). A long, sprawling account of American warfare in the 20th century, with great and lasting emphasis on duty, honor, and country. Written during the Vietnam War, this novel became a reading staple at West Point and the Naval War College and among the Marines. Much acclaimed by professional military people as both accurate and inspiring.

James Michener (American), *The Bridges at Toko-Ri* (New York: Ballantine Books, 1953) (F). Close air-support missions from a carrier and "greater love hath no man but that he lay down his life for his fellow." A hauntingly sad book based on a true story.

Mark Mills (British), *The Information Officer* (New York: Random House, 2009) (F). Set in Malta during World War II, this mystery novel somehow captures with great depth and feeling the intertwined nature of military and civilians at war while giving the reader an inside view of an important strategic battle.

Nicholas Monsarrat (British), *The Cruel Sea* (New York: Alfred A. Knopf, 1951) (F). A long, carefully crafted story of a corvette captain and his first lieutenant, who learn to patrol the North Atlantic by escorting convoys across the Atlantic and to Gibraltar and sinking U-boats during the entire course of World War II. Great detail and ongoing sadness make you thankful the war is finally over. As Captain Ericson said after five years of war, "I must say I'm damned tired." Churchill wrote at the time, "It is a war of groping and drowning, of ambuscade and stratagem, of science and seamanship."

For a broader overview of the entire Atlantic naval campaign, see Richard Snow, *A Measureless Peril: America in the Fight for the Atlantic, the Longest Battle of World War II* (New York: Scribner, 2010). A number of important vignettes strung together give the reader a true sense of the scope of that long, long campaign, indeed the longest of World War II in Europe.

Kirk Munroe, *"Forward, March": A Tale of the Spanish American War* (New York: Harper and Brothers, 1899) (F). A sprightly tale of plucky Americans who charge up San Juan Hill and generally thrash the Spanish, although observers compare it to the Charge of the Light Brigade: "It is not war—it is suicide" cries the foreign attaché. Black regiments, coming of age, and heroism are also celebrated.

Iris Murdock (British), *The Red and the Green* (New York: Viking, 1965) (F). The April 1916 Rising in Dublin is captured in the tension between "word of honor" and "betraying King and Country." Murdock underscores the irony of civil war as Andrew Chase-White survives the Rising only to be killed at Passchendaele in 1917. She was later to soften her praise of the rebellion's leaders, who were shot or hung: "They died for glorious things, for justice, for freedom, for Ireland," but the essence of that remains.

Bao Ninh (Vietnamese), *The Sorrow of War* (New York: Pantheon Books, 1993) (F). Soldier and writer, Ninh survived the war to paint its life and its echoes in sharp relief. An overwhelming sense of doom in the midst of war: "The fighting is endless. No one knows when it will stop." And, in the "jungle of the screaming souls," no remedy except to "Give them some bronze candy." A very powerful work and also one that, in indirect fashion, traces the evolution of the war and its ebb and flow from the perspective of the NVA.

Tim O'Brien (American), *Going after Caggiato* (New York: Delacorte Press, 1973) (F). Wildly and widely acclaimed, this surrealist and phantasmagoric novel has a patrol following a deserter to Paris. Antic and humorous despite being overlaid with many powerful insights into the futility of war in general and the Vietnam War, in particular. Reminded me of some of the scenes in the film *Apocalypse Now*; that is, not historically accurate but amusing and telling, nevertheless.

———, *The Things They Carried* (New York: Franklin Library, 1990) (F). A novel comprised of a number of interconnected short stories the author wrote for such diverse publications as *Playboy*, *Esquire*, and the *Massachusetts Review*. Episodic and powerful, it asserts that all soldiers "carried ghosts" and that the "Lives of the Dead" were an intrinsic part of the living for soldiers and veterans alike. Some profound insights, such as "The war wasn't all terror and violence. Sometimes things could almost get sweet" and that Vietnam was the "Garden of Evil. Over here, man, every sin's real fresh and original." In the end, though, it is the memory of the death of his 9-year-old girlfriend when he was a child that is the most powerful thing he carried.

Cothburn O'Neal (American), *Master of the World* (New York: Avon Books, 1952) (F). A lusty fictionalized account of the war, making career of Timor the Lame, Tamerlane, and his conquests, which rivaled those of Genghis Khan. Useful in underlying the extent to which the desire for loot and captives motivated so many of the various armies at play in Eurasia around 1400. Timor's cynical use of the contemporary religions and his ability to absorb various technologies (but seemingly not gunpowder) and methods, including Indian war elephants, are detailed in very bright colors. His philosophy that "war is the way to salvation" seems prescient indeed.

Julie Orringer (American), *The Invisible Bridge* (New York: Alfred Knopf, 2010) (F). A powerful, very moving story about a Hungarian Jewish boy and man who survives the war and the Holocaust while working on labor battalions. Interesting inside view of the logistics involved on the Eastern Front and the chaos that accompanied the German retreat from Russia through Hungary.

Ralph Peters (American), *Cain at Gettysburg* (New York: Forge Books, 2012) (F). Not as good as *The Killer Angels* but better than *Of Gods and Generals*, this novel captures the grit, grime, and gore—as well as the petty animosities and simmering anger—among troops and their leaders. Exceptionally graphic descriptions of men in battle. Quite mean to O. O. Howard, Richard Ewell, Win Hancock ("Hancock the Superb" and "Hancock the Bull Headed"), and Bobby Lee. The author does give General George Meade his much-deserved due and brings into focus Henry Hunt, head of Union artillery, giving him much-needed ink to counter the extensive good press of Porter Alexander of the Confederate guns at Gettysburg. Points out that the 26th North Carolina regiment lost 85 percent of its soldiers, compared to the much more widely quoted 1st Minnesotans, who suffered 80 percent casualties.

Kevin Powers, *The Yellow Birds* (New York: Little, Brown, 2012) (F). This highly praised—and I must say, I fear, overpraised—novel of Iraq III does capture some of the true essence of war, although there are a lot of stock elements present: the revenge for a dead comrade, the blurring of the rules of war under stress, the macho sergeant who keeps the men in the hunt but is screwed up himself, and so on. Some images prevail, however, such as the trooper on guard who rubs Tabasco sauce in his eyes so he won't fall asleep and the sarge filling in the words for the politically correct LT: "You were saying, sir, that hopefully they'll be too scared to shoot before. . . . Before we fucking kill the hajji fucks."

Stephen Pressfield (American), *The Afghan Campaign* (New York: Doubleday, 2006). A most marvelous description of a long military campaign and a paean to the will and genius of Alexander the Great. Gives a strong sense of logistics and perpetual battle in a campaign seemingly without end. Perhaps a metaphor for American involvement in the contemporary Mideast and South Asia?

———, *The Tides of War* (New York: Doubleday, 2000) (F): As seen through the eyes of his dying captain Polemides, Alcibiades (451–404 BCE) and his service for and betrayal of Athens, Sparta, and Persia are recounted, ending in the ominous but often unheeded dictum, "Tyranny is a splendid roost but there is no step down from it."

Simon Raven, *Sound the Retreat* (New York: Beaufort Books, 1971) (F). A poignant, bittersweet, yet antic work covering the last days of the Raj as seen from the point of view of a British group of cadets. Marvelous scenes from the mess and the local house of pleasure against a backdrop of rising communal violence and the withdrawal of British military power on the subcontinent.

Liciu Rebreanu (Romanian), *Forest of the Hanged* (New York: Duffield, 1930) (F). What a splendid piece of work. Although occasionally filled with somewhat melodramatic dialogue and artificial scenes, it nevertheless powerfully and emotionally captures the many conflicted loyalties that arose in the multinational conflict involving the polyglot—and soon-to-be-rivened—Austro-Hungarian Empire. It also deals effectively with the multitude of demands any state places on individuals under its control during wartime. Apostol Bologa is a Romanian sublieutenant in the Austro-Hungarian army of World War I. He fights bravely and loyally for the Austrian Empire when they are fighting the Russians, even voting in a court-martial to hang a fellow Romanian who is caught trying to defect. But in 1916, when Romania unwisely joins the Russians (and the Allies) after their one successful offensive, the Austro-Hungarian army then turns its attention to the Romanians and, with German help, rapidly captures much of the country. In the process, Bologa's unit is ordered to the Romanian front. Eventually, Bologa tries to defect, is captured, and also is hung, his sad tale starkly underscoring the irony and tragedy of war. In reality, the brother of the author also suffered a similar fate, refusing to fight against fellow Romanians despite the powerful dictum of the establishment: "A soldier's duty in wartime is to do any job that comes to hand. That's why war is war."

Erich Maria Remarque (German), *All Quiet on the Western Front* (Berlin: Propylaen, 1929) (F). The daily slaughter on the Western Front during World

War I is the backdrop for this telling novel in the voice of young German, Paul Baumer, through whose eyes we see the squalor and devastation and tedium of life in the trenches. He is finally killed by a sniper, and his death is ironically portrayed as taking place on a day that was "all quiet." To put this in perspective, on the day the armistice finally went into effect, November 18, 1918, 10,000 men lost their lives on the Western Front. That was thought to be a "quiet" day, as well. While this deserves to be one of the books one should read about war, too many critics have not read enough of the other works of fiction about war and too often overpraise this one.

John Reneham, *The Valley* (New York: Dutton, 2014) (F). Take a bit of *Apocalypse Now*, add some *Platoon*, throw in a mini-*My Lai*, remember what you can of *Heart of Darkness*, set it in Afghanistan, throw in an REMF, and coat with FUBAR from World War II, and you've got a highly praised novel that held my attention but seemed, ultimately, phantasmagoric. Really needed a good editing job, which Dutton did not provide. Intriguing but ultimately disappointing.

Derek Robinson, *A Splendid Little War* (London: Maclehose Press, 2013) (F). An engaging account of British (and Irish) aviators involved in support of the Whites under Generals Wrangle and Denikin in southern Russia during that country's civil war. Based on actual events, especially the Battle of Tsaritsyn, much of the day-to-day ups and downs of war ring true. Also, the dogfights between the British Camels and the Red Air Force are well-presented, although the entire novel has a bit too much of the "I say old chap, spot of bother that" irony and insouciance.

Jules Romins, *Verdun: The Prelude and the Battle* (New York: Knopf, 1939) (F). A huge, sprawling Proustian novel with multicharacter involvement, providing a variety of views of this amazingly destructive battle in a "cemetery of stinking clay which only waited for the dead to fall before closing over them." A million fell, although the clay did not get them all.

Siegfried Sassoon (British), *Memoirs of a Fox-Hunting Man* (New York: Coward-McCann, n.d.). An insightful, self-effacing work of great sensitivity that sets the stage for the author's subsequent military career. When he joins the army as an enlisted man, really enjoying the run-up to combat, he says ironically, "And to one who had never heard the hiss of machine gun bullets, there was nothing imaginatively astounding in the notion."

———, *Memoirs of an Infantry Office* (London: Faber and Faber, 1931). A stunningly revelatory but thinly fictionalized account of his heroic fighting in the trenches of World War I at the Battle of the Somme. Sassoon was awarded the Distinguished Cross but turned against war and published "Finished with the War: A Soldier's Declaration," which was printed in the *London Times* and which reads in part, "I am not protesting against the conduct of the war, but against the political errors and insincerities for which fighting men are being sacrificed." The army wanted to court-martial him but finally sent him to a hospital for shellshock. Sassoon eventually became England's poet laureate.

Mike Scotti, *The Blue Cascade: A Memoir of Life after War* (New York: Grand Central, 2012). See the review in chapter 38.

Igor Sentjurc (German), *The Torrents of War* (New York: David McKay, 1962) (F). Reading this account of small-group action on the Russian front during the 1943–1945 phase of the war makes one feel frustrated, horrible, jangled, irritated, overwhelmed, and dissatisfied. The way war makes one feel, as well. Not a pleasure read, but the ironies of who gets killed, who gets wounded, and who gets to survive make the effort worthwhile.

Jeff Shaara (American), *A Blaze of Glory: A Novel of the Battle of Shiloh* (New York: Ballantine, 2012) (F). See the review in chapter 38.

———, *No Less Than Victory* (New York: Ballantine, 2009) (F). Quick read from the American soldier's point of view, a bit wooden in places, but conveys a sweep of the history and a relatively painless way for the uninitiated to learn about the war in Europe from 1944 to 1945 on the Western Front. Not great literature, but an engaging way to pick up a little history.

Irwin Shaw (American), *The Young Lions* (New York: Dell Books, 1948) (F). One of those war novels that impressed early (high school) and faded later in life. Still, it presents a "spirit of the times" story of Americans and Germans, both in action and in life and love. "The roar of the young lions will live forever"—or so they imagined.

Mikhail Sholokhov (Russian), *And Quiet Flows the Don* (New York: Alfred Knopf, 1934) (F). This 1920 novel powerfully and emotively captures the ebb and flow of life, military and otherwise, among the Cossacks of the Don basin. Great soaring events occur as World War I comes, and in the middle of that upheaval, the Russian Revolution spreads and engulfs the region of the Don basin and its people, when "[t]ime entangled the days as the wind a horse's mane."

Konstantine Simonov (Russian), *Days and Nights* (New York: Ballantine Books, 1945) (F). Although perhaps lacking the soaring historical arcs of

Mikhail Sholokhov's *And Quiet Flows the Don*, Simonov describes real Russians in a real war (which ended up killing 20 million of them) with skill and feeling and surprisingly, through the eyes of Captain Saburov and Nurse Anya, provides a moving portrait of the Battle of Stalingrad up to the time of the victorious Soviet offensive, which surrounded the German Sixth Army and changed the course of the war.

Aleksandr Solzhenitsyn (Russian), *August, 1914* (New York: Farrar, Straus and Giroux, 1989) (F). A huge, sprawling, sometimes verbose, mixture of fiction and history covering the Russian invasion of East Prussia during the early stages of World War I, this work captures the gross unpreparedness of the czarist regime for modern warfare and the rising tide of revolution as many Russians decided, "Only the ax can deliver us, and nothing other than the axe."

Elizabeth Speller (British), *The Return of Captain John Emmett* (New York: Houghton Mifflin Harcourt, 2011) (F). A long, deep, and well-researched novel ostensibly about a court-martial and execution of a British officer during World War I and what happened to the men who participated in it. Only 3 British officers were ever killed in this fashion during the whole war (although 300 other ranks were). But beyond that, it is a powerful and moving testimony to the ongoing—and seemingly unending—reverberations that impact those who go to war and see combat. One very perceptive insight: "It was only if you weren't there that you could think in terms of battles. The Battle of the Somme, the Third Battle of Ypres and so on. It wasn't really like that—there were all-out attacks and unexpected skirmishes, and they all led one into another." The stream of individual, unit, and national consciousness flowing forward but always tied to what came before.

Stendhal (French), *The Charter House of Parma* (Baltimore: Penguin Classics, 2007) (F). Written in 1838 by a French author who fought in the Napoleonic wars and even survived the march on Moscow, this novel captures the confusion and chaos of battle as a young Italian nobleman from Lake Como finds himself in the Battle of Waterloo. The battle scenes are often overlooked by reviewers but remain insightful and gripping.

Mario Rigoni Stern (Italian), *The Sergeant in the Snow* (London: MacGibbon and Key, 1954). This marvelous work captures the almost incredible tale of the Army of Italy in Russia (AMIR) caught in the big battles for the Don bend and especially the catastrophic retreat across much of Russia in the dead of winter during 1942–1943 in a desperate attempt to get out of the "Bag," into which the Soviets were encircling the entire expeditionary force. Of the 230,000 men trapped, 90,000 ended up missing and dead and 45,000 frost-bitten and wounded, as the men walked, fought, walked, and fought on and on and on across a bleak and hostile landscape. The terror and helplessness and overpowering sense of loss overcomes the reader, and one forgets the geopolitical considerations and only feels for the human beings on both sides. The wind was "blowing, immense and free, off the limitless steppes. In the cold darkness it found us, poor creatures lost in the war, knocked us about, made us stagger." The relief one feels when Rigoni finally escapes the "Bag" is palpable. This is a most insightful book on war.

Harriet Beecher Stowe (American), *Dred: A Tale of the Great Dismal Swamp* (Boston: Phillips, Sampson, 1856) (F). A tangled tale of a sister (Nina) and brother (Tom) who are left on a plantation and slaves, including Harry, the blue-eyed half-brother of both. An antislavery novel of note but included here for its portrait of Dred, an escaped slave who does not become a protorevolutionary but simply a dead, runaway slave. Illustrates the extent to which Stowe had moved beyond "gradual emancipation," "back to Africa," or even internal revolt. The liberation of the slaves would lie outside their power or their southern location.

William Styron, *The Long March* (New York: Random House, 1952). Marine reservists called up for the Korean War are out of shape, but Marines are supposed to be Marines whether reservists or not. A 36-mile hike brings out the best and the worst in the men.

Mario Tobino (Italian), *The Deserts of Libya* (New York: Alfred Knopf, 1955) (F). Dedicated to "all who did not malinger," this novel is set in Libya during World War II at an aid station. The Italians were pushed back by the British with heavy casualties and massive surrenders. But for Tobino, "[f]aced with death, they rejected hatred, and quietly accepted their fate. Noble soldiers without a flag—nothing could be sadder. But there was no flag they could fight for. And yet, there were heroes." Poorly equipped, poorly trained, and even more poorly led, these Italian soldiers suffered much and for a long time, leaving their dead by the tens of thousands in East Africa, North Africa, Russia, Greece, and Albania. A sad tale all around.

Leo Tolstoy (Russian), *Hadji Murad* (New York: Dodd Mead, 1912) (F). Set in the Caucuses in 1851, this romantic tale tells the story of a devoted Muslim Chechnyan warrior, Hadji Murad, who comes to the Russians when the head of the resistance fighters, Shamil, quarrels with him and takes his family hostage. Hadji Murad offers to help the Russians pacify the area if they will exchange prisoners with Shamil. But due to bureaucratic failure, jealousy in the capital, a Great Russian feeling of cultural superiority, and general inertia, his offer is not accepted. Hadji Murad then tries to flee and is cut down by Cossacks, his bloody head brought back to the Russian garrison. Vivid accounts of search-and-destroy missions, the failure to show generosity in counterinsurgency operations, and a lack of appreciation of the nature of medieval warfare all make this an outstanding set of insights into irregular warfare and the thin line between banditry and guerrilla campaigns. His death is depicted in truly powerful and long-lasting detail and imagery.

———, *War and Peace* (New York: Vintage, 2008) (F). Some have called this the greatest of all the war novels, but for what it is worth, some have always found it hard to get beyond the speeches and long-winded soliloquies. But in terms of describing the motivation for the people for war and resistance, it is quite telling, even marvelous. Also, the descriptions of the Battle of Borodino are very vivid and realistic. In its own way, it represents a most biting and even savage criticism of Clausewitz and his veneration of the climactic battle that decides a war.

Dalton Trumbo (American), *Johnny Got His Gun* (New York: Lippincott, 1938) (F). This truly depressing antiwar novel features Joe, a young American soldier wounded on the last day of World War I. A quadruple amputee, he lies in a hospital bed, his life doomed beyond repair. This work makes *Slaughter House Five* (reviewed later this chapter) read like a nursery rhyme.

Brian Turner (American), *Here, Bullet* (Farmington, ME: Alice James Books, 2005). A slender volume of poems reflecting on the nature of warfare, with numerous observations from the Qur'an. Some powerful imagery throughout as "Observation Post #798" reports, "The Civil affairs office, Lt. Jackson, stares at his missing hands, which make no sense to him, no sense at all, to wave these absurd stumps held in the air where just a moment before he'd blow bubbles out the Humvee window."

Leon Uris (American), *Battle Cry* (New York: G. P. Putnam's Sons, 1953) (F). This novel about Marines during World War II strongly reflects the author's experiences on Guadalcanal, Tarawa, and Saipan and captures both the feel of battle and the small-group solidarity associated with the US Marines. The scenes prior to departure and on leave in New Zealand as well as the descriptions of men with malaria will stay with the reader for a long time.

Kurt Vonnegut (American), *Slaughterhouse-Five; or, The Children's Crusade: A Duty-Dance with Death* (New York: Delacorte Press, 1969) (F). A phantasmagorical and antic antiwar novel about the fire-bombing of Dresden, mixing science fiction, memoir, fact, and artifact to make a hyperbolic statement that war is very harmful to human beings—and perhaps inevitable. One horror after another, "[s]o it goes."

Evelyn Waugh (British), *Men at Arms* (Boston: Little, Brown, 1952) (F). Guy, a 35-year-old English aristocrat, wanders around wartime England seeking to play a role in World War II and, in an antic display, goes through what passed for training in the war and ends up with a loony brigadier landing in Dakar and getting cashiered. However, he does avoid getting trapped with the British forces during the fall of France. As Evelyn Waugh puts it, "The discipline of the square, the tradition of the mess, would work their magic and the spirit de corps would fall liked blessed unction from above." Also made into an enjoyable movie, *Sword of Honor*.

———, *Officers and Gentlemen* (Boston: Little, Brown, 1955) (F). The further adventures and misadventures of Guy Crouchback from Scotland to Cape Town to Egypt and Greece as he tries to do his bit in the war, which seems always one step ahead of the English class system.

James Webb (American), *Fields of Fire* (New York: Bantam, 2001) (F). Written by a Marine veteran who became US secretary of the Navy and a US senator. Shows good action scenes as seen from the inside of small units, in this case a Marine platoon in 1968. Also displays a healthy and considerable antagonism for antiwar groups, which, from the point of view of Mars, seems understandable. Honestly ambivalent about the experience of war, "[h]e sensed that beyond the terror that was today, there was a fullness that no other thing in the remainder of his life would ever equal." This book is regarded as an enduring classic by almost all US Marines.

Rebecca West (British), *The Return of the Soldier* (London: Nisbet, 1918) (F). Written while World War I was still in progress and devouring a whole generation of young men, *Return* is about an officer who has severe shell-shock and can't remember 12 years of his earlier life. The ironic tragedy depicted is that he can be cured, but when he is well, he will be sent back to "No Man's Land where bullets fall like rain on the rotting faces of the dead."

William Woodruff, *Vessel of Sadness* (Carbondale: Southern Illinois Press, 1969) (F). A poignant novel based on this British author's experiences during the North African and Italian campaigns (Anzio and Casino) in 1942–1943. Sees war as killing, random, senseless, and capricious: "Look, stop getting excited and sentimental about a life, lives are lost all the time." The work ends with the sentence "Well, be satisfied; you can't have a victory without a sacrifice. There's always got to be a sacrifice."

Herman Wouk (American), *The Caine Mutiny* (Garden City, NY: Doubleday, 1951) (F). Explores the microcosm that is shipboard life and the aspects of leadership (as well as the role of those who would be leaders). Although much of the spotlight is on the *Caine*'s captain, Lieutenant Commander Philip Queeg, it is the actions and rise of Willie Keith that ground us in the tale. Set in the Pacific during World War II, the novel captures the essence of the tensions and atmosphere of a "small ship," both in action and in typhoons.

———, *The Winds of War* (Boston: Little, Brown, 1971) (F). Although Wouk also wrote *The Caine Mutiny*, *The Winds of War* is a novel much larger in scope, covering the run-up to the war, as well. Combining real people and a fictionalized central character named Victory "Pug" Henry, the work captures the ongoing decision-making tension at the highest levels and the enormous strains war takes on human relationships. It became a highly acclaimed TV miniseries in 1983, which, when viewed today, remains fresh and vivid.

Arnold Zweig (German), *Education before Verdun* (New York: Viking Press, 1936) (F). Powerful antiwar novel focusing on fault lines within the German army, foretelling the eventual persecution of the Jews. Vivid descriptions ("The smell of ruined masonry, human excrement, stale explosions and dried blood") combined with soul-searing pessimism ("They won no more ground than they could cover with their corpses . . . where the French and British had staged a similar fraud").

Chapter Forty-One

War and Antiwar Films, Including Revolutionary Warfare

One useful starting point in terms of war cinema is the *Military History* magazine's *100 Greatest War Movies* (Leesburg: Weider History Group, 2005). Many of these are included in the following suggestions. War is not the most favored topic among filmmakers, but there are quite a number of films dealing with its many dimensions. Students often ask, "Which film is the most realistic in showing war?" Perhaps a more important set of questions revolve around asking which dimension of war is the author, director, or producer trying to show? What perspective does the film provide about which dimension of war in general or this war in particular? All war films have points of view, and some are more valid—or interesting—than others to one student or another. But taken together, the films listed in this chapter round out a fairly representative composite of the various experiences connected to war, in fact or fantasy or the altered but supposedly mystic cords of memory.

FILMS: MY PERSONAL RATINGS (0 * TO 5 *)

9th Company (Russian) (2005)
'71 (British)
300 (2006)
633 Squadron (British) (1964)
1612 (Russian) (2007)
The Alamo (1960, 2004)
Alexander (2004)
Alexander Nevsky (Russian) (1938)

All Quiet on the Western Front (1930)
All the King's Men (1949, 2006)
Ambush (Finnish) (1999) **
The Americanization of Emily (1964)
Andersonville (1996)
Anzio (1968)
Apocalypse Now (1979) ***
Apocalypto (2006) **
Army of Crime (French) (2009)
Arn (Swedish) (2007)
Attila (2001)
Avatar (2009)
Back Door to Hell (1964)
Ballad of A Soldier (Russian) (1959) ***
Band of Brothers (series) (2001) *****
Battle Cry (1955)
Battle Ground (1949)
The Battle of Algiers (French) (1966) ****
Battle of Britain (British) (1969)
Battle of the Bulge (1965)
Battle Royale (Japanese) (2000)
Battleship Potemkin (Russian) (1925)
Beaufort (Israeli) (2007) **
Beau Geste (1939)
Before the Fall (2004)
Behind Enemy Lines (2001)
Ben Hur (1959)
The Best Years of Our Lives (1946)
Big Red One (1980)
Birdsong (2012)
Black Book (Dutch) (2006)
Black Hawk Down (2001)
Blessed by Fire (Argentina) (2005)
Body of Lies (2008)
Born on the Fourth of July (1984)
Braveheart (1995)
Breaker Morant (Australian) (1980) ****
Bridge at Remagen (1969)
Bridges at Toyo-Ri (1954)
Bridge over the River Kwai (1957) *****
A Bridge Too Far (1997) ****
A Bright Shining Lie (1998) ***
The Brylcreem Boys (British) (1998)

The Caine Mutiny (1954)
California Dreaming (Romanian) (2007)
A Canterbury Tale (1944)
Captain Corelli's Mandolin (2001)
Cast a Giant Shadow (1966)
Castle Keep (1969)
Catch - 22 (1970)
Centurion (2010)
The Charge of the Light Brigade (British) (1968)
Charlie Mopic (1989)
Charlotte Grey (2001)
Che (2008)
City of Life and Death (Chinese) (2009) ****
Closely Watched Trains (Czech) (1966)
Colloden (British) (1964)
Come and See (Russian) (1985) ***
Coming Home (1978)
Command Decision (1948)
Commissar (Russian) (1987)
Company of Heroes (2012)
Courage u nder Fire (1996)
The Cranes A re Flying (Russian) (1957) ***
Crimson Tide (1995)
Cromwell (British) (1970)
Cross of Iron (1977) *****
Cup Final (Israeli) (1991)
Curse of the Golden Flower (2006)
Dark Blue World (Czech) (2001)
Das Boot (German) (1981) *****
Dawn Patrol (1938)
The Day After (1983)
D-Day: 1944 (1956)
Decision at Dawn (1951)
The Deer Hunter (1978)
The Desert Fox (1951)
The Desert Rats (British) (1946)
Desert Victory (1943)
Dieppe (British) (1993)
The Dirty Dozen (1967)
Dirty War (British) (2004)
Downfall (German) (2004) *****
Dresden (2006)
Dr. Strangelove (1964)

Dr. Zhivago (1965)
El Alamein (Italian) (2002)
Empire of the Sun (1987)
The Empire Strikes Back (1980)
Enemy At the Gates (2001)
Enigma (British) (2001)
Escape from Sobibor (1987) **
Europa (Danish) (1991)
Europa, Europa (German) (1990)
Everyman's War (2009)
Excalibur (1975)
The Fallen (at a cost of only $600,000!) (2004) **
Field of Lost Shoes (2014) **
Five Days of War (2012)
Flags of Our Fathers (2006) **
Flame and Citron (2008)
Flowers of War (Chinese) (2011) ***
Flyboys (2006)
Force Ten from Navarone (1978)
Foreign Correspondent (1940)
Forrest Gump (1994)
For Whom the Bell Tolls (1943)
Four Feathers (four versions! 1921, 1929, 1939, 2002)
Foyle's War (British) (2003) **
From Here to Eternity (1953)
From the Halls of Montezuma (1950)
Full Metal Jacket (1987) **
Fury (2014)
The Gallant Hours (1960) **
Gallipoli (Australian) (1981) **
Generation Kill (series) (2008)
Geronimo (1993)
Gettysburg (1993) *****
Girasoli (*Sunflower*) (Italian) (1970)
Gladiator (2000)
Glory (1989)
Gods and Generals (2003)
The Good Earth (1937)
The Good German (2006)
Good Morning, Vietnam (1987)
Go Tell the Spartans (1978)
The Grand Illusion (French) (1937)
The Great Escape (1963)

The Great Raid (2005)
The Green Berets (1968)
The Green Zone (2010)
Guadalcanal Diary (1943)
Gunga Din (1939)
Gunner Palace (2004)
Guns at Batasi (British) (1964)
Guns of August (1964)
Guns of Navarone (1961)
Hamburger Hill (1987)
Hamsun (Norwegian) (1996)
Hanoi Hilton (1987)
Hanover Street (1987)
Hart's War (2002)
Heaven and Earth (1993)
Heaven Knows, Mr. Allison (1967)
Hell in the Pacific (1968)
Hell's Angels (1930)
Henry V (British) (1989) **
Hiroshima Mon Amour (French) (1959)
H.M.S. Defiant (1962)
Hostile Waters (1977)
The Hunters (2011)
The Hunt for Red October (1990)
The Hurt Locker (2008) ****
Ike: Countdown to D-Day (2004)
Il Generale Della Rovere (Italian) (1959)
I Live in Grosvenor Square (British) (1945)
The Immortal Sargent (British) (1942)
In Darkness (Polish) (2009)
Inglourious Basterds (2009)
In Harm's Way (1965)
Into the White (Norwegian) (2012)
In Tranzit (Czech) (2008)
In Which We Serve (British) (1942) ***
Ironclad (2011)
Island at War (British) (2004)
Ivanhoe (British) (1952, 1982)
Jarhead (2005)
Jodhaa Akbar (Indian) (2008)
Johnny Got His Gun (1971)
John Rabe (German) (2009)
Joyeux Noel (*Merry Christmas*) (French) (2005)

Kagemusha (Japanese) (1980) ***
Katyn (Polish) (2007)
Khartoum (1966)
The Killing Fields (British) (1984)
King Arthur (2005)
The Kingdom of Heaven (2005)
Kippur (Israeli) (2000)
Land of Blood and Honey (2011)
Language of the Enemy (Israeli) (2008)
The Last Drop (British) (2006)
The Last of the Mohicans (1992)
The Last Samurai (2003)
Lawrence of Arabia (1962)
Lebanon (Israeli) (2009) ***
The Leopard (Italian) (1963)
Letters from Iwo Jima (2006) **
The Liberator (Spanish/Venezuelan) (2014)
The Life and Death of Colonel Blimp (British) (1943)
Lines of Wellington (French) (2014)
The Longest Day (1962) ***
Lost Command (French) (1966)
MacArthur (1986)
Major Dundee (1965)
Maluda (South African) (2001)
The Man Who Would Be King (1975)
Mary Silliman's War (1994)
*M*A*S*H* (1972)
Massacre in Rome (Italian) (1973) ***
Master and Commander (2003)
A Matter of Life and Death (British) (1946)
Max Manus (Norwegian) (2005)
Memorial Day (2012)
Memphis Belle (1990)
Men i n War (1957)
Merry Christmas Mr. Lawrence (1983)
The Messenger (2004)
Michael Collins (1996)
Midway (1976) ***
Mi Mejor Enemigo (*My Best Enemy*) (Chilean) (2005) *
Miral (French) (2010)
The Missiles of October (1973)
Mongol (2007)
Monuments Men (2014)

Mosquito Squadron (British) (1969)
The Naked and the Dead (1958)
Napoleon and Love (British) (1974)
No Man Is an Island (1962)
No Man's Land (Serbo-Croatian) (2001) ***
North West Frontier (British) (1959)
Occupation: Dreamland (2005)
Overlord (British) (1975)
Over There (2005)
Pacific (series) (2010) ****
Paths of Glory (1957) **
The Patriot (2000)
Patton (1970) ***
Pearl Harbor (2001)
Pillars of the Earth (British) (2010)
Platoon (1986) ***
Prisoners of the Mountains (Russian) (1996) ****
Prisoners of the Sun (Australian) (1996) ***
The Purple Plane (1955)
The Quiet American (Michael Caine) (2002)
The Quiet American (Audrey Murphy) (1958)
Quiet Flows the Don (Russian) (1931, 1957)
The Raid (1954)
Raiders of Leyte Gulf (1963)
Ran (Japanese) (1985) ***
The Red and the White (Russian) (1968) ***
Red Badge of Courage (1951)
Red Cliff (Chinese) (2008)
Red Tails (2012)
Rescue Dawn (2007)
Restrepo (2010)
Retreat, Hell (1952)
Return of the Soldier (British) (1983)
Return to Bataan (1945)
Revolution (1985)
Robin Hood (2010)
Rob Roy (1995)
Rome (miniseries) (2005)
Rough Riders (1997) (longest battle scenes ever; higher casualties than the Somme!)
Run Silent, Run Deep (1958)
Saints and Soldiers (2003)
Saladin (1941)

Sands of Iwo Jima (1949)
Saving Private Ryan (1998) *****
Sayonara (1957)
Schindler's List (1993)
The Searchers (1956)
Season of the Witch (2011)
Shadow of the Sun (1965)
She Wore a Yellow Ribbon (1944)
Slaughterhouse Five (1972)
Spartacus (1960)
Spartacus: Blood and Sand (miniseries) (2010)
Spring 1941 (Polish) (2008)
Stalag 17 (1953)
Stalingrad (German) (1993) *****
Star Wars (1977)
Steel Tigers (1995)
Story of G.I. Joe (1945)
Sword of Honour (British) (2001)
Taegukgi: The Brotherhood of Taking Chance (2009)
Taras Bulba (1962)
Tears of the Sun (2003)
They Were Expendable (1945)
The Thin Red Line (1998) **
Thirteen Days (2000)
Thirty Seconds o ver Tokyo (1944)
The Three Kingdoms (Chinese) (2010)
Three Kings (1999)
The Tin Drum (1979)
Tinker, Tailor, Soldier, Spy (British) (1980) **
To Hell and Back (1955)
Tora, Tora, Tora (1970) ****
The Train (1964)
The Trench (1999)
Triage (2009)
Troy (2004)
Tunes of Glory (British) (1961)
Two Women (Italian) (1960) ****
U-571 (2000)
The Unbearable Lightness of Being (Czech) (1988)
The Unknown Soldier (German) (2007)
Upstairs Downstairs, season 4: "1914–1918" (British) (1974)
Valkyrie (2008)
Victory (British) (1981)

The Vikings (1958)
Vincere (Italian) (2009)
A Walk in the Sun (1943)
Waltz w ith Bashir (2008)
War (South Korean) (2004) ***
War and Peace (1956, 2007)
War Game (British) (1965)
The War Lords (Chinese) (1965)
The War Lover (1962) ***
The War Room (1993)
Wartime in Winter (Dutch) (2010)
War Witch (Canadian) (2012)
The Water Diviner (Australian) (2015) **
The Way to the Stars (British) (1945)
We Were Soldiers (2002) ***
When Trumpets Fade (1998)
Who ' ll Stop the Rain (*Dog S oldiers*) (1978)
The Wild Geese (1978)
Winds of War (1983) **
Windtalkers (2002)
The Wind T hat Shakes the Barley (Irish) (2006)
Wings (1927)
Wings (Russian) (1977)
Winter in Wartime (British) (2008)
The Winter War (Finnish) (1984) ***
A Woman in Berlin (German) (2008) ****
Yanks (British) (1979)
The Young Lions (1958)
Zulu (British) (1964) **
Zulu Dawn (1969)

Chapter Forty-Two

Just War/Moral War

This final chapter of *Understanding War* both concludes this volume and anticipates volume 3, *Preventing War*, which in turn completes the War Trilogy. "Just War" has the connotation—in many societies and cultures—of sanctioning war by setting conditions for its employment and its implementation. We have included but a few of the most commonly used texts here, as the broader range of similar works are housed in the final volume. Nevertheless, the books listed in this chapter represent the heart of the arguments for just war and for the just implementation of war. Readers who have gotten this far in the work will, perhaps, marvel at the human efforts to justify war as much as their desire to curb both its frequency and its harmful impacts.

Thomas Aquinas, *Summa Theologica* (London: Burns Oates and Washburn, 1920). Following Augustine, he came up with three criteria for just war. The war had to be decided by the ruler, it must be fought for a just cause, and the intentions of the ruler or decision maker must also be just and must seek peace. The war should be proportional to the grievance. Seems to have ended up broadening the number of decision makers who could declare war and thus paved the way for the nation-state. Aquinas was not keen on "Holy War," having seen the earlier Crusades in action.

Augustine, *The City of God against the Pagan*, edited and translated by R. W. Dyson (Cambridge: Cambridge University Press, 1998). This fourth-century theologian, with his heavenly "City of God," made of Christianity an empire and gave it a justification for war. This was a huge overall shift from the Christianity of its first 300 years and overrode the dictum of Jesus of Nazareth to "turn the other cheek." The notion of *jus ad bellum*, "just recourse to war," was for many Christians a strikingly new concept (although the Book of Revelations already projected human war up into heaven and then back down to earth).

Hugh Grotius, *The Rights of War and Peace* (Westport, CT: Hyperion Press, 1901) reprint. What rules should govern conduct in war? Grotius wrote in the 17th century and was one of the founders of international law following the Treaty of Westphalia, which ended the religious wars by stating that the religion of the prince should be the religion of the people. The author focused on *jus in bello*, the "just conduct of war": The war should be just, the war should be a last resort, only the proper state authorities may declare war, and the war must be waged proportionally (no all-out destruction for a minor reason). Among the relevant chapters of this book are "On the Unjust Causes of War" and "On Doubtful Causes," the lens through which can be run a variety of war.

Michael Howard, George Andreopoulos, and Mark Shulman (eds.), *The Laws of War: Constraints on Warfare in the Western World* (New Haven, CT: Yale University Press, 1994). A wide-ranging set of essays looking at these constraints, from "Classical Greek Times" through "The Age of Chivalry" and "Nuclear War Planning." Human-imposed limits on Mars seldom work, although we keep trying. The editors believe that "failure is not a reason to stop trying."

Sohail H. Hshmi (eds.), *Just Wars, Holy Wars and Jihads* (London: Oxford University Press, 2012). Twenty authors opine on these topics, looking at the ways the three faiths view wars that are "morally justified as legitimate." Jewish, Christian, and Muslim thinkers have always sought—and continue to seek—divine sanctification for war. Frightening.

Brian Orden, *The Morality of War* (Peterborough, ON: Broadview Press, 2006). Looks at just war in the context of a number of recent conflicts over the past 20 years, weaving in an analysis of *jus bello*, *jus ad bello*, and *jus post bello*. Meant to be a follow-up to the work of Walzer (reviewed next).

Michael Walzer, *Just and Unjust Wars : A Moral Argument with Historical Illustrations* (New York: Basic Books, 1977). From the Melian Dialogue through "Three Accounts of Agincourt," submarine warfare and the Nuremburg trials and many other instances, the author wrestles with the situations involving war, just and unjust, concluding, "[T]he world of war is not a fully comprehensive, let alone a morally satisfying place," but one must work for a universal order where nations and peoples "would not be threatened." The author also looks at the moral realities of war, the types of aggression, force-short-of war, the dilemmas of war, and many other related topics. He analyzes historical writings from Karl Marx, Jean-Paul Sartre, Chairman Mao, and Winston Churchill. The section "Just War for Regime Change" seems quite topical.

End Note

So the study of war, like the practice of war, must always include the caveat "There will always be another war," and although that war will have some differences from other wars, it will always have much more in common with previous wars than not. The study of the past will thus always provide useful insights into the future.

And the study of the wars of the past and the present can, if pursued vigorously and realistically, provide some guidelines for *avoiding* wars in the future. Or such is the hope, anyway.

Index